Global safety of fresh produce

Related titles:

Improving the health-promoting properties of fruit and vegetable products
(ISBN 978-1-84569-184-4)

Fruit and vegetable flavour
(ISBN 978-1-84569-183-7)

Handbook of organic food safety and quality
(ISBN 978-1-84569-010-6)

Details of these books and a complete list of titles from Woodhead Publishing can be obtained by:

- visiting our web site at www.woodheadpublishing.com
- contacting Customer Services (e-mail: sales@woodheadpublishing.com; fax: +44 (0) 1223 832819; tel.: +44 (0) 1223 499140 ext. 130; address: Woodhead Publishing Limited, 80, High Street, Sawston, Cambridge CB22 3HJ, UK)
- in North America, contacting our US office (e-mail: usmarketing@woodheadpublishing.com; tel.: (215) 928 9112; address: Woodhead Publishing, 1518 Walnut Street, Suite 1100, Philadelphia, PA 19102-3406, USA)

If you would like e-versions of our content, please visit our online platform: www.woodheadpublishingonline.com. Please recommend it to your librarian so that everyone in your institution can benefit from the wealth of content on the site.

We are always happy to receive suggestions for new books from potential editors. To enquire about contributing to our Food Science, Technology and Nutrition series, please send your name, contact address and details of the topic/s you are interested in to nell.holden@woodheadpublishing.com. We look forward to hearing from you.

The team responsible for publishing this book:

Commissioning Editor: Nell Holden
Project Editor: Elizabeth Moss
Editorial and Production Manager: Mary Campbell
Production Editor: Mandy Kingsmill
Project Manager: Annette Wiseman, RCL
Copyeditor: Eilidh McGregor
Proofreader: Simon Webber
Cover Designer: Terry Callanan

Woodhead Publishing Series in Food Science, Technology and Nutrition:
Number 260

Global safety of fresh produce

A handbook of best practice, innovative commercial solutions and case studies

Edited by
J. Hoorfar

WOODHEAD PUBLISHING

Oxford Cambridge Philadelphia New Delhi

Published by Woodhead Publishing Limited,
80 High Street, Sawston, Cambridge CB22 3HJ, UK
www.woodheadpublishing.com
www.woodheadpublishingonline.com

Woodhead Publishing, 1518 Walnut Street, Suite 1100, Philadelphia, PA 19102-3406, USA

Woodhead Publishing India Private Limited, 303, Vardaan House, 7/28 Ansari Road, Daryaganj, New Delhi – 110002, India
www.woodheadpublishingindia.com

First published 2014, Woodhead Publishing Limited

British Library Cataloguing in Publication Data
A catalogue record for this book is available from the British Library.

Library of Congress Control Number: 2013950760

ISBN 978-1-78242-018-7 (print)
ISBN 978-1-78242-027-9 (online)
ISSN 2042-8049 Woodhead Publishing Series in Food Science, Technology and Nutrition (print)
ISSN 2042-8057 Woodhead Publishing Series in Food Science, Technology and Nutrition (online)

The publisher's policy is to use permanent paper from mills that operate a sustainable forestry policy, and which has been manufactured from pulp which is processed using acid-free and elemental chlorine-free practices. Furthermore, the publisher ensures that the text paper and cover board used have met acceptable environmental accreditation standards.

Typeset by RefineCatch Limited, Bungay, Suffolk
Printed by Lightning Source

Contents

Contributor contact details

(* = main contact)

Editor and Chapter 27

Jeffrey Hoorfar
Technical University of Denmark
Mørkhøj Bygade 19
2860 Søborg, Denmark

E-mail: jhoo@food.dtu.dk

Chapter 1

Inge Bodil Jochumsen
AgroManagement
Kirketoften 5
Gamtofte
DK-5610 Assens, Denmark

E-mail: ibj@agromanagement.dk

Chapter 2

Annette N. Jensen* and Dorte L. Baggesen
National Food Institute, Technical University of Denmark
Division of Food Microbiology
Mørkhøj Bygade 19
2860 Søborg, Denmark

E-mail: anyj@food.dtu.dk

Chapter 3

Erick A. Snellman
Food and Drug Administration
Center for Food Safety and Applied Nutrition
Produce Safety Staff, HFS-317
5100 Paint Branch Parkway
College Park, MD 20740, USA

E-mail: Erick.Snellman@fda.hhs.gov

Chapter 4

Mieke Uyttendaele*, L. Jacxsens and S. Van Boxstael
Department of Food Safety and Food Quality
Faculty of Bio-Science Engineering
Ghent University
Coupure Links 653, 9000 Ghent
Belgium

E-mail: mieke.uyttendaele@UGent.be; Liesbeth.Jacxsens@UGent.be; Sigrid.VanBoxstael@UGent.be

Chapter 5

Declan J. Bolton
Food Safety Department,
Teagasc – Ashtown Food Research Centre
Ashtown
Dublin 15, Ireland

Tom S. Edrington, David J. Nisbet, and Todd. R. Callaway*
Food and Feed Safety Research Unit
Agricultural Research Service/US Department of Agriculture
Southern Plains Agricultural Research Center,
2881 F&B Rd
College Station, TX 77845, USA

Email: todd.callaway@ars.usda.gov

Chapter 6

Keith Warriner* and Azadeh Namvar
Department of Food Science
University of Guelph
Guelph
Ontario, Canada N1G 2W1

E-mail: kwarrine@gmail.com

Chapter 7

Yaguang Luo*
US Department of Agriculture
Agricultural Research Service
Environmental Microbial and Food Safety Laboratory
Beltsville, MD 20705, USA

E-mail: Yaguang.Luo@ars.usda.gov

David T. Ingram
Food and Drug Administration
Center for Food Safety and Applied Nutrition
Produce Safety Staff, HFS-317
5100 Paint Branch Parkway
College Park, MD 20740, USA

E-mail: David.Ingram@fda.hhs.gov

Karan Khurana
Pulse Instruments and AquaPulse Systems
16117 Covello Street
Van Nuys, CA 91406, USA

Chapter 8

Tommy Licht Cederberg
National Food Institute
Technical University of Denmark
Division of Food Chemistry
Mørkhøj Bygade 19
DK-2860 Søborg, Denmark

E-mail: TLCE@food.dtu.dk

Chapter 9

Cristobal Chaidez*
Centro de Investigacion en Alimentacion y Desarrollo
Inocuidad Alimentaria
CIAD Culiacan
Sinaloa, Mexico 80129

E-mail: chaqui@ciad.edu.mx

Marcela Soto
Department of Soil, Water and Environmental Science
The University of Arizona
Arizona, USA

E-mail: johanas@email.arizona.edu

Maribel Jimenez
Universidad Autonoma de Sinaloa
Facultad de Ciencias Quimico Biologicas
Mexico

E-mail: mjimeneze@uas.edu.mx

Chapter 10

David Pimentel* and Michael Burgess
Departments of Entomology and Ecology and Evolutionary Biology
College of Agriculture and Life Sciences
Cornell University
Ithaca, NY 14853, USA

E-mail: dp18@cornell.edu

Chapter 11

Juergen Zentek*, Fanny Knorr, Anneluise Mader
Institute of Animal Nutrition
Freie Universität Berlin
Department of Veterinary Medicine
Königin-Luise-Str. 49
14195 Berlin, Germany

E-mail: juergen.zentek@fu-berlin.de

Chapter 12

Shige Koseki
Hakkaido University
Kita 9 Nishi 9
Kita-ku, Sapporo 060-8589, Japan

E-mail: Koseki@bpe.agr.hokudai.ac.jp

Chapter 13

Shima Shayanfar
German Institute of Food Technologies (DIL)
49610 Quakenbrück, Germany

E-mail: s.shayanfar@dil-ev.de

Chapter 14

Kieran Jordan* and Aidan Casey
Teagasc Food Research Centre
Moorepark
Fermoy, Co. Cork, Ireland

E-mail: Kieran.Jordan@teagasc.ie

Andreas Hoehl
BOKU – University of Natural Resources and Life Sciences
Department of Food Science and Technology Muthgasse 18
A-1190 Vienna, Austria

Geert Bruggeman
Nutrition Sciences N.V.
Drongen, Belgium

Chapter 15

Alvin Lee
Institute for Food Safety and Health
Illinois Institute of Technology, Moffett Campus
6502 South Archer Road
Bedford Park, IL 60501-1957, USA

E-mail: alee33@iit.edu

Chapter 16

Suresh D. Pillai*
National Center for Electron Beam Research
Texas A&M University
College Station, Texas, USA

E-mail: s-pillai@tamu.edu

Carl Blackburn
Joint FAO/IAEA Programme of Nuclear Techniques in Food and Agriculture
International Atomic Energy Agency
Vienna, Austria

Carlos Bogran
National Center for Electron Beam Research

and

Department of Entomology
Texas A&M University
College Station, Texas, USA

Chapter 17

Luminita Ciolacu and Anca Ioana Nicolau*
Faculty of Food Science and Engineering
Dunarea de Jos University of Galati
Romania

E-mail: Anca.Nicolau@ugal.ro; Luminita.Pricope@ugal.ro

J. Hoorfar
Technical University of Denmark
Mørkhøj Bygade 19
2860 Søborg, Denmark

E-mail: jhoo@food.dtu.dk

Chapter 18

Will Daniels
SVP Operations and Organic Integrity
Earthbound Farm
1721 San Juan Highway
San Juan Bautista, CA 95045, USA

E-mail: WDaniels@ebfarm.com

Chapter 19

Zhihua Ye*, Song Chen, Fang Wang, and Tianjin Chen
Institute of Quality Standards & Testing Technology for Agro-Products
Chinese Academy of Agricultural Sciences
Beijing 100081, China

E-mail: zhihuaye@mail.caas.net.cn; yezhihua@caas.cn; c_t_jin@263.net; wangfang1125@126.com

Liang Xiao
Department for Testing and Inspection
National Certification and Accreditation Administration
Beijing 100088, China

E-mail:xiaol@cnca.gov.cn

Chapter 20

J. Hoorfar
Technical University of Denmark
Mørkhøj Bygade 19
2860 Søborg, Denmark

E-mail: jhoo@food.dtu.dk

Peter Feng*
US FDA, HFS-711
5100 Paint Branch Parkway
College Park, MD 20740, USA

E-mail: Peter.Feng@fda.hhs.gov

Geraldine Duffy
Teagasc Food Research Centre
Ashtown
Dublin 15
Ireland

geraldine.duffy@teagasc.ie

Burkhard Malorny
Federal Institute for Risk Assessment
Berlin
Germany

burkhard.malorny@bfr.bund.de

Rachel Binet
US Food and Drug Administration
College Park
USA

rachel.binet@fda.hms.gov

Chapter 21

Rocío Morales-Rayas* and Mansel W. Griffiths
Canadian Research Institute for Food Safety
University of Guelph
Guelph, ON, Canada

E-mail: moralesr@uoguelph.ca

Anna Charlotte Schultz
Division of Food Microbiology, DTU Food
Technical University of Denmark
National Food Institute
Denmark

Chapter 22

Sarah M. Markland and Kalmia E. Kniel*
Department of Animal and Food Sciences
University of Delaware
044 Townsend Hall
531 South College Avenue
Newark, DE 19716, USA

E-mail: kniel@udel.edu

Doris H. D'Souza
Department of Food Science and Technology
2605 River Drive, Room 102 FSPB
University of Tennessee-Knoxville
Knoxville, TN 37996-4591, USA

E-mail: ddsouza@utk.edu

Chapter 23

Michelle A. Smith
US Food and Drug Administration
Center for Food Safety and Applied Nutrition,
5100 Paint Branch Parkway (HFS-317)
College Park, MD, 20740
USA

E-mail: michelle.smith@fda.hhs.gov

Chapter 24

Mansoor A. Baloch
Environmental Health Services Branch DEEHS/NCEH
US Centers for Disease Control and Prevention
Atlanta, GA 30341, USA

E-mail: mbaloch@cdc.gov

Chapter 25

Petra Luber
Federal Office of Consumer Protection and Food Safety (BVL)
Mauerstr. 39-42, D-10117 Berlin
Germany

E-mail: petra.luber@bvl.bund.de

Chapter 26

Eduardo Henrique Miranda Walter
Embrapa Food Technology
Av. das Américas, 29.501
Guaratiba, 23020-470
Rio de Janeiro, RJ, Brazil

E-mail: eduardo.walter@embrapa.br

Arnaldo Yoshiteru Kuaye
Department of Food Technology
Faculty of Food Engineering
CP 6121
State University of Campinas
UNICAMP, 13083-862
Campinas, SP, Brazil

J. Hoorfar
National Food Institute
Technical University of Denmark
Mørkhøj Bygade 19
DK-2860 Søborg, Denmark

Chapter 28

Zengtao Xing* and Xiaoyan Zhao
Supervision and Testing Center for Edible Fungi Quality (Shanghai)
Ministry of Agriculture of China

and

Shanghai Academy of Agricultural Sciences
1000 Jinqi Road
Shanghai, 201403, China

E-mail: xingzengtao@sina.com

Chapter 29

Anca Ioana Nicolau
Faculty of Food Science and Engineering,
Dunarea de Jos University of Galati
Romania

E-mail: Anca.Nicolau@ugal.ro

Editorial advisors

Woodhead Publishing Series in Food Science, Technology and Nutrition

1 **Chilled foods: A comprehensive guide** *Edited by C. Dennis and M. Stringer*
2 **Yoghurt: Science and technology** *A. Y. Tamime and R. K. Robinson*
3 **Food processing technology: Principles and practice** *P. J. Fellows*
4 **Bender's dictionary of nutrition and food technology Sixth edition** *D. A. Bender*
5 **Determination of veterinary residues in food** *Edited by N. T. Crosby*
6 **Food contaminants: Sources and surveillance** *Edited by C. Creaser and R. Purchase*
7 **Nitrates and nitrites in food and water** *Edited by M. J. Hill*
8 **Pesticide chemistry and bioscience: The food-environment challenge** *Edited by G. T. Brooks and T. Roberts*
9 **Pesticides: Developments, impacts and controls** *Edited by G. A. Best and A. D. Ruthven*
10 **Dietary fibre: Chemical and biological aspects** *Edited by D. A. T. Southgate, K. W. Waldron, I. T. Johnson and G. R. Fenwick*
11 **Vitamins and minerals in health and nutrition** *M. Tolonen*
12 **Technology of biscuits, crackers and cookies Second edition** *D. Manley*
13 **Instrumentation and sensors for the food industry** *Edited by E. Kress-Rogers*
14 **Food and cancer prevention: Chemical and biological aspects** *Edited by K. W. Waldron, I. T. Johnson and G. R. Fenwick*
15 **Food colloids: Proteins, lipids and polysaccharides** *Edited by E. Dickinson and B. Bergenstahl*
16 **Food emulsions and foams** *Edited by E. Dickinson*
17 **Maillard reactions in chemistry, food and health** *Edited by T. P. Labuza, V. Monnier, J. Baynes and J. O'Brien*
18 **The Maillard reaction in foods and medicine** *Edited by J. O'Brien, H. E. Nursten, M. J. Crabbe and J. M. Ames*
19 **Encapsulation and controlled release** *Edited by D. R. Karsa and R. A. Stephenson*
20 **Flavours and fragrances** *Edited by A. D. Swift*
21 **Feta and related cheeses** *Edited by A. Y. Tamime and R. K. Robinson*

56 **Frying: Improving quality** *Edited by J. B. Rossell*
57 **Food chemical safety Volume 1: Contaminants** *Edited by D. Watson*
58 **Making the most of HACCP: Learning from others' experience** *Edited by T. Mayes and S. Mortimore*
59 **Food process modelling** *Edited by L. M. M. Tijskens, M. L. A. T. M. Hertog and B. M. Nicolaï*
60 **EU food law: A practical guide** *Edited by K. Goodburn*
61 **Extrusion cooking: Technologies and applications** *Edited by R. Guy*
62 **Auditing in the food industry: From safety and quality to environmental and other audits** *Edited by M. Dillon and C. Griffith*
63 **Handbook of herbs and spices Volume 1** *Edited by K. V. Peter*
64 **Food product development: Maximising success** *M. Earle, R. Earle and A. Anderson*
65 **Instrumentation and sensors for the food industry Second edition** *Edited by E. Kress-Rogers and C. J. B. Brimelow*
66 **Food chemical safety Volume 2: Additives** *Edited by D. Watson*
67 **Fruit and vegetable biotechnology** *Edited by V. Valpuesta*
68 **Foodborne pathogens: Hazards, risk analysis and control** *Edited by C. de W. Blackburn and P. J. McClure*
69 **Meat refrigeration** *S. J. James and C. James*
70 **Lockhart and Wiseman's crop husbandry Eighth edition** *H. J. S. Finch, A. M. Samuel and G. P. F. Lane*
71 **Safety and quality issues in fish processing** *Edited by H. A. Bremner*
72 **Minimal processing technologies in the food industries** *Edited by T. Ohlsson and N. Bengtsson*
73 **Fruit and vegetable processing: Improving quality** *Edited by W. Jongen*
74 **The nutrition handbook for food processors** *Edited by C. J. K. Henry and C. Chapman*
75 **Colour in food: Improving quality** *Edited by D. MacDougall*
76 **Meat processing: Improving quality** *Edited by J. P. Kerry, J. F. Kerry and D. A. Ledward*
77 **Microbiological risk assessment in food processing** *Edited by M. Brown and M. Stringer*
78 **Performance functional foods** *Edited by D. Watson*
79 **Functional dairy products Volume 1** *Edited by T. Mattila-Sandholm and M. Saarela*
80 **Taints and off-flavours in foods** *Edited by B. Baigrie*
81 **Yeasts in food** *Edited by T. Boekhout and V. Robert*
82 **Phytochemical functional foods** *Edited by I. T. Johnson and G. Williamson*
83 **Novel food packaging techniques** *Edited by R. Ahvenainen*
84 **Detecting pathogens in food** *Edited by T. A. McMeekin*
85 **Natural antimicrobials for the minimal processing of foods** *Edited by S. Roller*
86 **Texture in food Volume 1: Semi-solid foods** *Edited by B. M. McKenna*
87 **Dairy processing: Improving quality** *Edited by G. Smit*
88 **Hygiene in food processing: Principles and practice** *Edited by H. L. M. Lelieveld, M. A. Mostert, B. White and J. Holah*
89 **Rapid and on-line instrumentation for food quality assurance** *Edited by I. Tothill*
90 **Sausage manufacture: Principles and practice** *E. Essien*

91 **Environmentally-friendly food processing** *Edited by B. Mattsson and U. Sonesson*
92 **Bread making: Improving quality** *Edited by S. P. Cauvain*
93 **Food preservation techniques** *Edited by P. Zeuthen and L. Bøgh-Sørensen*
94 **Food authenticity and traceability** *Edited by M. Lees*
95 **Analytical methods for food additives** *R. Wood, L. Foster, A. Damant and P. Key*
96 **Handbook of herbs and spices Volume 2** *Edited by K. V. Peter*
97 **Texture in food Volume 2: Solid foods** *Edited by D. Kilcast*
98 **Proteins in food processing** *Edited by R. Yada*
99 **Detecting foreign bodies in food** *Edited by M. Edwards*
100 **Understanding and measuring the shelf-life of food** *Edited by R. Steele*
101 **Poultry meat processing and quality** *Edited by G. Mead*
102 **Functional foods, ageing and degenerative disease** *Edited by C. Remacle and B. Reusens*
103 **Mycotoxins in food: Detection and control** *Edited by N. Magan and M. Olsen*
104 **Improving the thermal processing of foods** *Edited by P. Richardson*
105 **Pesticide, veterinary and other residues in food** *Edited by D. Watson*
106 **Starch in food: Structure, functions and applications** *Edited by A-C. Eliasson*
107 **Functional foods, cardiovascular disease and diabetes** *Edited by A. Arnoldi*
108 **Brewing: Science and practice** *D. E. Briggs, P. A. Brookes, R. Stevens and C. A. Boulton*
109 **Using cereal science and technology for the benefit of consumers: Proceedings of the 12th International ICC Cereal and Bread Congress, 24–26th May, 2004, Harrogate, UK** *Edited by S. P. Cauvain, L. S. Young and S. Salmon*
110 **Improving the safety of fresh meat** *Edited by J. Sofos*
111 **Understanding pathogen behaviour: Virulence, stress response and resistance** *Edited by M. Griffiths*
112 **The microwave processing of foods** *Edited by H. Schubert and M. Regier*
113 **Food safety control in the poultry industry** *Edited by G. Mead*
114 **Improving the safety of fresh fruit and vegetables** *Edited by W. Jongen*
115 **Food, diet and obesity** *Edited by D. Mela*
116 **Handbook of hygiene control in the food industry** *Edited by H. L. M. Lelieveld, M. A. Mostert and J. Holah*
117 **Detecting allergens in food** *Edited by S. Koppelman and S. Hefle*
118 **Improving the fat content of foods** *Edited by C. Williams and J. Buttriss*
119 **Improving traceability in food processing and distribution** *Edited by I. Smith and A. Furness*
120 **Flavour in food** *Edited by A. Voilley and P. Etievant*
121 **The Chorleywood bread process** *S. P. Cauvain and L. S. Young*
122 **Food spoilage microorganisms** *Edited by C. de W. Blackburn*
123 **Emerging foodborne pathogens** *Edited by Y. Motarjemi and M. Adams*
124 **Bender's dictionary of nutrition and food technology Eighth edition** *D. A. Bender*
125 **Optimising sweet taste in foods** *Edited by W. J. Spillane*
126 **Brewing: New technologies** *Edited by C. Bamforth*
127 **Handbook of herbs and spices Volume 3** *Edited by K. V. Peter*
128 **Lawrie's meat science Seventh edition** *R. A. Lawrie in collaboration with D. A. Ledward*
129 **Modifying lipids for use in food** *Edited by F. Gunstone*

130 **Meat products handbook: Practical science and technology** *G. Feiner*
131 **Food consumption and disease risk: Consumer–pathogen interactions** *Edited by M. Potter*
132 **Acrylamide and other hazardous compounds in heat-treated foods** *Edited by K. Skog and J. Alexander*
133 **Managing allergens in food** *Edited by C. Mills, H. Wichers and K. Hoffman-Sommergruber*
134 **Microbiological analysis of red meat, poultry and eggs** *Edited by G. Mead*
135 **Maximising the value of marine by-products** *Edited by F. Shahidi*
136 **Chemical migration and food contact materials** *Edited by K. Barnes, R. Sinclair and D. Watson*
137 **Understanding consumers of food products** *Edited by L. Frewer and H. van Trijp*
138 **Reducing salt in foods: Practical strategies** *Edited by D. Kilcast and F. Angus*
139 **Modelling microorganisms in food** *Edited by S. Brul, S. Van Gerwen and M. Zwietering*
140 **Tamime and Robinson's Yoghurt: Science and technology Third edition** *A. Y. Tamime and R. K. Robinson*
141 **Handbook of waste management and co-product recovery in food processing Volume 1** *Edited by K. W. Waldron*
142 **Improving the flavour of cheese** *Edited by B. Weimer*
143 **Novel food ingredients for weight control** *Edited by C. J. K. Henry*
144 **Consumer-led food product development** *Edited by H. MacFie*
145 **Functional dairy products Volume 2** *Edited by M. Saarela*
146 **Modifying flavour in food** *Edited by A. J. Taylor and J. Hort*
147 **Cheese problems solved** *Edited by P. L. H. McSweeney*
148 **Handbook of organic food safety and quality** *Edited by J. Cooper, C. Leifert and U. Niggli*
149 **Understanding and controlling the microstructure of complex foods** *Edited by D. J. McClements*
150 **Novel enzyme technology for food applications** *Edited by R. Rastall*
151 **Food preservation by pulsed electric fields: From research to application** *Edited by H. L. M. Lelieveld and S. W. H. de Haan*
152 **Technology of functional cereal products** *Edited by B. R. Hamaker*
153 **Case studies in food product development** *Edited by M. Earle and R. Earle*
154 **Delivery and controlled release of bioactives in foods and nutraceuticals** *Edited by N. Garti*
155 **Fruit and vegetable flavour: Recent advances and future prospects** *Edited by B. Brückner and S. G. Wyllie*
156 **Food fortification and supplementation: Technological, safety and regulatory aspects** *Edited by P. Berry Ottaway*
157 **Improving the health-promoting properties of fruit and vegetable products** *Edited by F. A. Tomás-Barberán and M. I. Gil*
158 **Improving seafood products for the consumer** *Edited by T. Børresen*
159 **In-pack processed foods: Improving quality** *Edited by P. Richardson*
160 **Handbook of water and energy management in food processing** *Edited by J. Klemeš, R. Smith and J-K. Kim*
161 **Environmentally compatible food packaging** *Edited by E. Chiellini*
162 **Improving farmed fish quality and safety** *Edited by Ø. Lie*
163 **Carbohydrate-active enzymes** *Edited by K-H. Park*

164 **Chilled foods: A comprehensive guide Third edition** *Edited by M. Brown*
165 **Food for the ageing population** *Edited by M. M. Raats, C. P. G. M. de Groot and W. A Van Staveren*
166 **Improving the sensory and nutritional quality of fresh meat** *Edited by J. P. Kerry and D. A. Ledward*
167 **Shellfish safety and quality** *Edited by S. E. Shumway and G. E. Rodrick*
168 **Functional and speciality beverage technology** *Edited by P. Paquin*
169 **Functional foods: Principles and technology** *M. Guo*
170 **Endocrine-disrupting chemicals in food** *Edited by I. Shaw*
171 **Meals in science and practice: Interdisciplinary research and business applications** *Edited by H. L. Meiselman*
172 **Food constituents and oral health: Current status and future prospects** *Edited by M. Wilson*
173 **Handbook of hydrocolloids Second edition** *Edited by G. O. Phillips and P. A. Williams*
174 **Food processing technology: Principles and practice Third edition** *P. J. Fellows*
175 **Science and technology of enrobed and filled chocolate, confectionery and bakery products** *Edited by G. Talbot*
176 **Foodborne pathogens: Hazards, risk analysis and control Second edition** *Edited by C. de W. Blackburn and P. J. McClure*
177 **Designing functional foods: Measuring and controlling food structure breakdown and absorption** *Edited by D. J. McClements and E. A. Decker*
178 **New technologies in aquaculture: Improving production efficiency, quality and environmental management** *Edited by G. Burnell and G. Allan*
179 **More baking problems solved** *S. P. Cauvain and L. S. Young*
180 **Soft drink and fruit juice problems solved** *P. Ashurst and R. Hargitt*
181 **Biofilms in the food and beverage industries** *Edited by P. M. Fratamico, B. A. Annous and N. W. Gunther*
182 **Dairy-derived ingredients: Food and neutraceutical uses** *Edited by M. Corredig*
183 **Handbook of waste management and co-product recovery in food processing Volume 2** *Edited by K. W. Waldron*
184 **Innovations in food labelling** *Edited by J. Albert*
185 **Delivering performance in food supply chains** *Edited by C. Mena and G. Stevens*
186 **Chemical deterioration and physical instability of food and beverages** *Edited by L. H. Skibsted, J. Risbo and M. L. Andersen*
187 **Managing wine quality Volume 1: Viticulture and wine quality** *Edited by A. G. Reynolds*
188 **Improving the safety and quality of milk Volume 1: Milk production and processing** *Edited by M. Griffiths*
189 **Improving the safety and quality of milk Volume 2: Improving quality in milk products** *Edited by M. Griffiths*
190 **Cereal grains: Assessing and managing quality** *Edited by C. Wrigley and I. Batey*
191 **Sensory analysis for food and beverage quality control: A practical guide** *Edited by D. Kilcast*
192 **Managing wine quality Volume 2: Oenology and wine quality** *Edited by A. G. Reynolds*
193 **Winemaking problems solved** *Edited by C. E. Butzke*
194 **Environmental assessment and management in the food industry** *Edited by U. Sonesson, J. Berlin and F. Ziegler*

195 **Consumer-driven innovation in food and personal care products** *Edited by S. R. Jaeger and H. MacFie*
196 **Tracing pathogens in the food chain** *Edited by S. Brul, P. M. Fratamico and T. A. McMeekin*
197 **Case studies in novel food processing technologies: Innovations in processing, packaging, and predictive modelling** *Edited by C. J. Doona, K. Kustin and F. E. Feeherry*
198 **Freeze-drying of pharmaceutical and food products** *T-C. Hua, B-L. Liu and H. Zhang*
199 **Oxidation in foods and beverages and antioxidant applications Volume 1: Understanding mechanisms of oxidation and antioxidant activity** *Edited by E. A. Decker, R. J. Elias and D. J. McClements*
200 **Oxidation in foods and beverages and antioxidant applications Volume 2: Management in different industry sectors** *Edited by E. A. Decker, R. J. Elias and D. J. McClements*
201 **Protective cultures, antimicrobial metabolites and bacteriophages for food and beverage biopreservation** *Edited by C. Lacroix*
202 **Separation, extraction and concentration processes in the food, beverage and nutraceutical industries** *Edited by S. S. H. Rizvi*
203 **Determining mycotoxins and mycotoxigenic fungi in food and feed** *Edited by S. De Saeger*
204 **Developing children's food products** *Edited by D. Kilcast and F. Angus*
205 **Functional foods: Concept to product Second edition** *Edited by M. Saarela*
206 **Postharvest biology and technology of tropical and subtropical fruits Volume 1: Fundamental issues** *Edited by E. M. Yahia*
207 **Postharvest biology and technology of tropical and subtropical fruits Volume 2: Açai to citrus** *Edited by E. M. Yahia*
208 **Postharvest biology and technology of tropical and subtropical fruits Volume 3: Cocona to mango** *Edited by E. M. Yahia*
209 **Postharvest biology and technology of tropical and subtropical fruits Volume 4: Mangosteen to white sapote** *Edited by E. M. Yahia*
210 **Food and beverage stability and shelf life** *Edited by D. Kilcast and P. Subramaniam*
211 **Processed Meats: Improving safety, nutrition and quality** *Edited by J. P. Kerry and J. F. Kerry*
212 **Food chain integrity: A holistic approach to food traceability, safety, quality and authenticity** *Edited by J. Hoorfar, K. Jordan, F. Butler and R. Prugger*
213 **Improving the safety and quality of eggs and egg products Volume 1** *Edited by Y. Nys, M. Bain and F. Van Immerseel*
214 **Improving the safety and quality of eggs and egg products Volume 2** *Edited by F. Van Immerseel, Y. Nys and M. Bain*
215 **Animal feed contamination: Effects on livestock and food safety** *Edited by J. Fink-Gremmels*
216 **Hygienic design of food factories** *Edited by J. Holah and H. L. M. Lelieveld*
217 **Manley's technology of biscuits, crackers and cookies Fourth edition** *Edited by D. Manley*
218 **Nanotechnology in the food, beverage and nutraceutical industries** *Edited by Q. Huang*
219 **Rice quality: A guide to rice properties and analysis** *K. R. Bhattacharya*

220 **Advances in meat, poultry and seafood packaging** *Edited by J. P. Kerry*
221 **Reducing saturated fats in foods** *Edited by G. Talbot*
222 **Handbook of food proteins** *Edited by G. O. Phillips and P. A. Williams*
223 **Lifetime nutritional influences on cognition, behaviour and psychiatric illness** *Edited by D. Benton*
224 **Food machinery for the production of cereal foods, snack foods and confectionery** *L-M. Cheng*
225 **Alcoholic beverages: Sensory evaluation and consumer research** *Edited by J. Piggott*
226 **Extrusion problems solved: Food, pet food and feed** *M. N. Riaz and G. J. Rokey*
227 **Handbook of herbs and spices Second edition Volume 1** *Edited by K. V. Peter*
228 **Handbook of herbs and spices Second edition Volume 2** *Edited by K. V. Peter*
229 **Breadmaking: Improving quality Second edition** *Edited by S. P. Cauvain*
230 **Emerging food packaging technologies: Principles and practice** *Edited by K. L. Yam and D. S. Lee*
231 **Infectious disease in aquaculture: Prevention and control** *Edited by B. Austin*
232 **Diet, immunity and inflammation** *Edited by P. C. Calder and P. Yaqoob*
233 **Natural food additives, ingredients and flavourings** *Edited by D. Baines and R. Seal*
234 **Microbial decontamination in the food industry: Novel methods and applications** *Edited by A. Demirci and M.O. Ngadi*
235 **Chemical contaminants and residues in foods** *Edited by D. Schrenk*
236 **Robotics and automation in the food industry: Current and future technologies** *Edited by D. G. Caldwell*
237 **Fibre-rich and wholegrain foods: Improving quality** *Edited by J. A. Delcour and K. Poutanen*
238 **Computer vision technology in the food and beverage industries** *Edited by D-W. Sun*
239 **Encapsulation technologies and delivery systems for food ingredients and nutraceuticals** *Edited by N. Garti and D. J. McClements*
240 **Case studies in food safety and authenticity** *Edited by J. Hoorfar*
241 **Heat treatment for insect control: Developments and applications** *D. Hammond*
242 **Advances in aquaculture hatchery technology** *Edited by G. Allan and G. Burnell*
243 **Open innovation in the food and beverage industry** *Edited by M. Garcia Martinez*
244 **Trends in packaging of food, beverages and other fast-moving consumer goods (FMCG)** *Edited by N. Farmer*
245 **New analytical approaches for verifying the origin of food** *Edited by P. Brereton*
246 **Microbial production of food ingredients, enzymes and nutraceuticals** *Edited by B. McNeil, D. Archer, I. Giavasis and L. Harvey*
247 **Persistent organic pollutants and toxic metals in foods** *Edited by M. Rose and A. Fernandes*
248 **Cereal grains for the food and beverage industries** *E. Arendt and E. Zannini*
249 **Viruses in food and water: Risks, surveillance and control** *Edited by N. Cook*
250 **Improving the safety and quality of nuts** *Edited by L. J. Harris*
251 **Metabolomics in food and nutrition** *Edited by B. C. Weimer and C. Slupsky*

252 **Food enrichment with omega-3 fatty acids** *Edited by C. Jacobsen, N. S. Nielsen, A. F. Horn and A-D. M. Sørensen*
253 **Instrumental assessment of food sensory quality: A practical guide** *Edited by D. Kilcast*
254 **Food microstructures: Microscopy, measurement and modelling** *Edited by V. J. Morris and K. Groves*
255 **Handbook of food powders: Processes and properties** *Edited by B. R. Bhandari, N. Bansal, M. Zhang and P. Schuck*
256 **Functional ingredients from algae for foods and nutraceuticals** *Edited by H. Domínguez*
257 **Satiation, satiety and the control of food intake: Theory and practice** *Edited by J. E. Blundell and F. Bellisle*
258 **Hygiene in food processing: Principles and practice Second edition** *Edited by H. L. M. Lelieveld, J. Holah and D. Napper*
259 **Advances in microbial food safety Volume 1** *Edited by J. Sofos*
260 **Global safety of fresh produce: A handbook of best practice, innovative commercial solutions and case studies** *Edited by J. Hoorfar*
261 **Human milk biochemistry and infant formula manufacturing technology** *Edited by M. Guo*
262 **High throughput screening for food safety assessment: Biosensor technologies, hyperspectral imaging and practical applications** *Edited by A. K. Bhunia, M. S. Kim and C. R. Taitt*
263 **Foods, nutrients and food ingredients with authorised EU health claims** *Edited by M. J. Sadler*
264 **Handbook of food allergen detection and control** *Edited by S. Flanagan*
265 **Advances in fermented foods and beverages: Improving quality, technologies and health benefits** *Edited by W. Holzapfel*
266 **Metabolomics as a tool in nutritional research** *Edited by J-L. Sebedio and L. Brennan*
267 **Dietary supplements: Safety, efficacy and quality** *Edited by K. Berginc and S. Kreft*
268 **Grapevine breeding programs for the wine industry: Traditional and molecular technologies** *Edited by A. G. Reynolds*
269 **Handbook of natural antimicrobials for food safety and quality** *Edited by M. Taylor*
270 **Managing and preventing obesity: Behavioural factors and dietary interventions** *Edited by T. Gill*
271 **Electron beam pasteurization and complementary food processing technologies** *Edited by S. Pillai and S. Shayanfar*
272 **Advances in food and beverage labelling: Information and regulations** *Edited by P. Berryman*
273 **Flavour development, analysis and perception in food and beverages** *Edited by J. K. Parker, S. Elmore and L. Methven*
274 **Rapid sensory profiling techniques and related methods: Applications in new product development and consumer research** *Edited by J. Delarue, B. Lawlor and M. Rogeaux*
275 **Advances in microbial food safety: Volume 2** *Edited by J. Sofos*
276 **Handbook of antioxidants in food preservation** *Edited by F. Shahidi*
277 **Lockhart and Wiseman's crop husbandry including grassland: Ninth edition** *H. J. S. Finch, A. M. Samuel and G. P. F. Lane*

Preface

Healthy but not safe!

Fresh produce is seen as a healthy choice by consumers worldwide; however, health benefits and microbiological safety are independent of one another. Most foods will inevitably encounter potential safety hazards somewhere along the supply chain. Non-processed foods such as fresh produce come with their own safety problems for producers, suppliers and retailers to consider.

The book's main focus is microbiological safety. It was conceived around three key questions related to eating fresh vegetables and fruits:

- What safety problems are we facing?
- Why do such problems occur?
- How can we avoid them?

Although it might at first seem like a simple task to address these questions, a closer look will reveal the complexity of the issue. As well as healthy, safe, fresh consumers also demand organic products, environmentally friendly production systems, locally produced food and minimal waste. Finding a solution to these often contradictory demands will require innovative technologies, sharing of knowledge and dissemination of information about how the problem is being addressed and which strategies are working across the globe.

The book targets university students, food safety specialists, food producers and distributors, quality control managers, food policy regulators, private consultants and auditors, laboratory managers and technicians in control laboratories from all over the world.

In order to comprehensively cover fresh produce production and safety, the 29 chapters are organized in five parts:

- Farm-level production and regulation.
- Environmental issues.

- Commercial solutions.
- Laboratory testing.
- Case studies in real-life situations.

The book begins at farm level, because even a minor effort at this very first step has a major impact further down the production chain. Nevertheless, there is a risk of cross-contamination during the packaging, distribution and retail processes, which is why an entire part of the book has been devoted to the use of innovative technologies to address this issue. Finally, knowledge-sharing through real-life examples is demonstrated in chapters covering outbreaks such as the major sprout outbreak in Germany, the tomato-related outbreak in Denmark and the issue of coconut safety in Brazil. These examples can serve as colourful pictures to facilitate teaching and learning about this complex issue.

The main strength of this book is the combination of theory and practice covered. Across the world, different production systems are used and producers and distributors work under different regulatory structures. We have therefore tried to provide readers with examples of best practice from across the world, including Europe, the USA, South America and China. We have also covered a wide variety of technologies and best practices, including ways to reduce cross-contamination through wash water, application of irradiation technologies, application of edible coatings, implementation of new auditing practices and development of meaningful and rational testing schemes.

In conclusion, this book has brought together highly recognized international experts from centres in different disciplines and different parts of the world, to share a variety of research and ideas on what is needed to avoid fresh produce reaching the consumer that is healthy, but not safe.

Enjoy reading.

J. Hoorfar
Denmark

Foreword

Humans in every civilization and in almost every part of the world have been consuming fresh produce for thousands of years. Undoubtedly, there have been historic cases of foodborne illness associated with the consumption of fresh produce, but it is only in recent years that it has become the focus of worldwide safety concern.

The emergence of fresh produce as a food safety problem was not sudden, but rather a gradual process with many contributing factors. Among these we may include changes in societies and shifts in family structure from single to multiple income households, which placed great time constraints on our daily lives. This in turn led to changes in lifestyles and eating habits, and increased dependency on restaurants, fast foods and delis. It has also stimulated the emergence of 'convenience' food industries, which produce ready-to-eat (RTE) foods that are well suited to consumers' busy lives. A final factor is the healthy living initiatives that have been implemented to reduce obesity and other complications arising from unhealthy diets, which also effectively promote the health benefits of fresh produce and have popularized the consumption of fresh fruits and vegetables.

The combination of these shifts and trends and other factors has increased the demands for fresh produce and probably stimulated the growth of the 'fresh-cut' industry and popularized the consumption of ready-to-eat (RTE) bagged salads and fruit cups. It has been estimated that over 1 billion bags of salad vegetables are sold in the USA each year. This increased consumer demand has also resulted in increased imports and exports of fresh produce. As these products are mass produced, broadly distributed and often marketed worldwide, a single pathogen contamination event can have broadly impacting consequences.

The basic fresh produce production practices have remained essentially unchanged. Most produce is still cultivated in soil in a field and irrigated with available sources of water. The constant exposure of produce plants to the environment makes them susceptible to microbial contamination, which can come from many different sources. As contamination may be sporadic, affected by regional and seasonal variations, always heterogeneous in distribution and not visible, it is nearly impossible to recognize contaminated products. However, even a small number of contaminated plants entering the massive processing chain can end up spreading the contamination to other plants.

This book recognizes that each country's fresh produce production practices and scales of production varies, and thus each setting can have different problems. Comprising 29 chapters divided into five parts, this book is unique in that it examines the problem of produce safety from a global perspective. Part I focuses on the farm, the various problems associated with large-scale production, problems of contamination by animal wastes and existing produce legislation in various parts of the world. Part II examines the impact of environmental factors on produce safety, such as irrigation water quality, effectiveness of washing, the use of sanitizers to minimize the spread of contamination and risk assessment of microbial and chemical contamination of fresh produce. Part III looks at some novel approaches and treatments that may be implemented to improve the safety of fresh produce, such as packaging practices and biocontrol. Also included in this section is a chapter describing the use of e-beam irradiation, which is generated from electricity and may have potential applications in post-production treatment of various types of products, including sprouts. Part IV presents the testing schemes that have been implemented by the produce industry in response to outbreak incidents, the approaches being planned by some foreign governments, novel approaches to testing fresh produce, problems with viral contamination and the testing of specific bacterial pathogens in produce, which all present unique logistical problems. The final part of the book presents several case studies of actual produce outbreak events, such as sprouts outbreaks in the USA and the O104:H4 outbreak in the EU, several non-O157 VTEC outbreaks carried by leafy greens, safety problems with mushroom production and some of the safety practices and control measures implemented by different countries in the world.

This book is written by over 40 scientists worldwide who are experts in the various aspects of fresh produce, from farm practices, production, processing and treatment to legislation, testing and outbreaks. Their experiences, shared through each of the chapters, provide an interesting global perspective on all aspects of fresh produce production and safety.

Peter Feng
USA

Part I

Farm-level production and regulation of fresh produce

1

Best practice in large-scale production of fresh produce

I. B. Jochumsen, AgroManagement, Denmark

DOI: 10.1533/9781782420279.1.3

Abstract: The risk of chemical, biological and physical contamination of fruit and vegetables in large-scale production is outlined. Reducing possible food safety hazards by using a systematic approach such as risk assessment for all steps in the production flow is described. A flow chart for the production of raspberries and the corresponding risk assessment are given as an example. The importance of training and education of the producers is emphasized.

Key words: food safety hazards, risk assessment, microbiological contamination, fruit and vegetable production.

1.1 Introduction

Fruit and vegetable producers are responsible for maintaining hygiene in their products. Food hygiene risks can be minimized, but will always exist as long as produce is grown outdoors in the fields. It is therefore very important that farmers have the necessary tools to help identify where these risks could occur. In order to use these tools successfully, farmers need education and knowledge about how they can minimize risks and improve the level of food hygiene.

One popular method of risk assessment is to draw a flow chart visualizing every step in the production line (Fig. 1.1). In the flow chart, the farmer needs to create a detailed overview of those aspects of the production flow for which he is responsible. It is important to note where in the production chain the farmer's responsibility ends. At each stage, the question of whether there is a chemical, biological or physical risk to the product should be considered.

The flow chart should begin with the land (production location). The farmer must ask himself whether the fields are his own, or new land that he has rented or bought. Next, the farmer should consider the seed or plants that he buys. Where

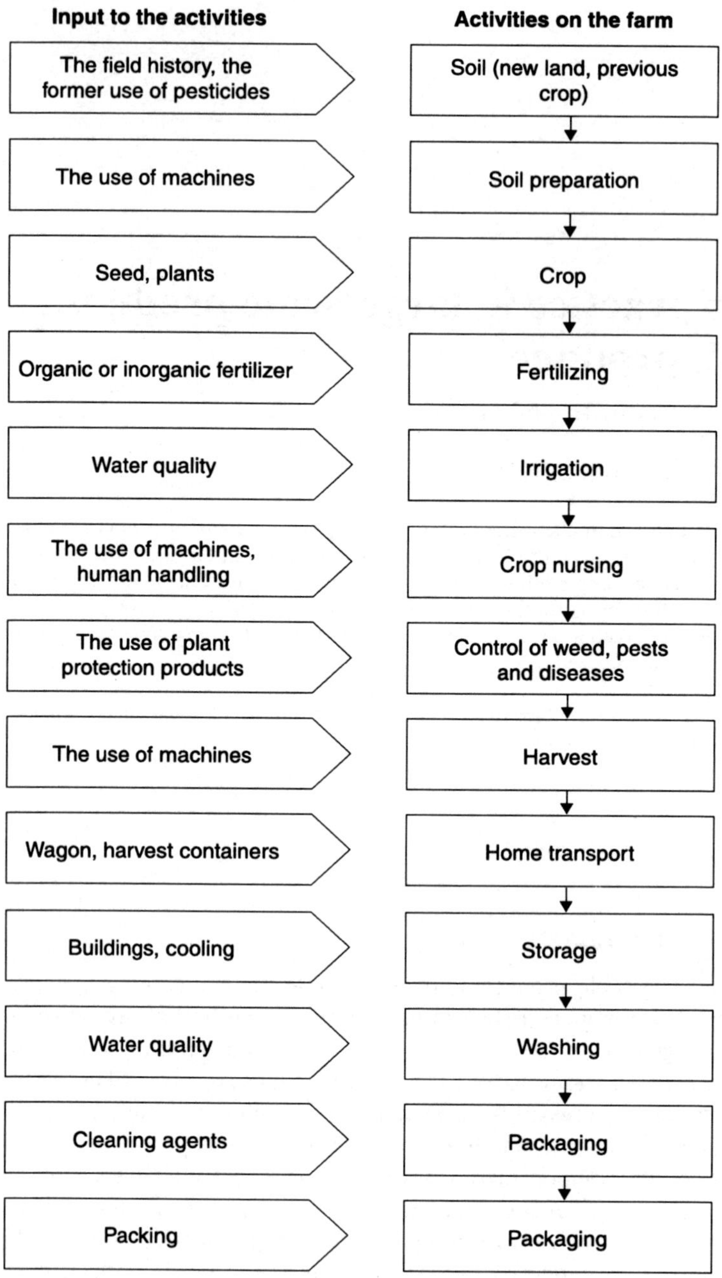

Fig. 1.1 Production flow.

do the planting materials come from? There are also issues that the farmer has to focus on with regard to food hygiene, for example, whether the plant or seed is treated with pesticides and how these are applied if so.

1.2 Risk assessment at the farm level

Every step in a food production chain must be evaluated for any risk of potential chemical, biological or physical contamination. Examples of biological risks include pathogens such as *Escherichia coli*, *Salmonella* and *Listeria*, as well as other microorganisms in the soil. Physical risks could include waste from workers, dust, sand, bullets from hunting activities in the area of the farm, or similar foreign objects. Chemical risks include chemicals in the products with which crops are in contact, chemicals in packaging materials (e.g. boxes or plastic bags) or the pesticides that the farmer or his neighbours use in or near the production field.

When conducting a risk assessment, it can be a good idea to think about what the normal production practices are or should be. The farmer must ensure that new employees receive the correct and relevant information, and that emergency measures are in place to manage any difficulties. For example, as part of the risk assessment the farmer might ask himself what he would do if the irrigation water, that had previously been tested and found to be clean, is suddenly contaminated. Incidents such as this raise crucial questions that have to be answered in a risk management plan focused on a specific production farm. The risk assessment will always be different, depending on the farming methods used and the crops being grown.

The flow chart is designed to answer questions such as those discussed above as part of a risk assessment. If there are critical issues in the production that impose a risk, the farmer must demonstrate in the risk assessment how he will manage the possible risk (Table 1.1).

Depending on the end product, fresh produce farmers have different practices. For example, for potato farmers the risk of biological contamination is not a major issue, because potatoes will always be heat treated before consumption. The potato farmer may therefore have a more relaxed attitude towards managing the biological risks than, for example, a farmer growing leafy greens. In terms of physical and chemical contamination, however, the conditions are equal for all growers of fresh produce.

Postharvest treatments also vary depending on the crop that is produced. For example, in the case of carrots, those destined for deep-freezing will be heat treated, adding a further decontamination step between producer and consumer and minimizing the biological risk. However, where fresh raw carrots are sold directly to the distributor or consumer (i.e. without further treatment), the farmer does not have this intermediate step on which to rely. Even if a bag of carrots is marked 'wash before use', the farmer should not ignore the fact that many consumers do eat carrots directly from the bag, without washing. Another issue to take into account is the packaging and marketing of the product. In recent years,

Table 1.1 Examples for risk assessment for the crop raspberries

Activities	Actual risk	How to manage actual risk
New land	*Chemical:* Use of pesticides in my own crops. If I buy or rent new land, do I ask for information about the previous uses of the land and previous treatments, e.g. with untreated sludge?	Documents that show the former use of plant protection products on the crops from the fields and the use of untreated sludge
	Chemical: Have pesticides previously been used on the land that may persist in the soil?	Pesticides and other products that remain in the soil long after application should not be used
	Chemical: Wind drift from the neighbours	All neighbours are informed of the current use of the land and potential hazards to the current crop from their fields
Soil preparation	*Chemical:* Fields are prepared before planting, and in the autumn. Oil may spill from machines and contaminate the soil	Machines undergo maintenance checks twice a year
Planting	The plants are planted by hand in the autumn before the harvesting year. No risk	
Fertilizing	*Biological:* Organic fertilizers contain different bacteria and viruses	Use of organic fertilizer is stopped 60 days before harvest. Machines are used so that the fertilizer does not come into contact with the crop
	Chemical: The organic fertilizer does not contain any chemicals	The vendor should be instructed as to the requirements for the fertilizer to ensure that it contains no chemicals
	Chemical: Inorganic fertilizer could contain heavy metals	Only government certified inorganic fertilizer is used
Irrigation	*Biological:* Irrigation water could contain *E.coli*, *Salmonella*, etc.	The water is analyzed every year in May
	Chemical: Irrigation water could contain chemicals	Ensure that the water quality is in accordance with WHO advice for irrigation water. Drip irrigation is used, so the fruits are not in contact with the water
	Chemical and biological contamination: Irrigation water could be accidentally contaminated with chemical and/or biological substances	Irrigation is stopped as soon as the problem is discovered. Any products already sold that pose a risk will be recalled, and the water source will not be used again until it is clean
Crop nursing	The old branches are cut down in the fall. There is no risk	

(*Continued*)

Table 1.1 Continued

Activities	Actual risk	How to manage actual risk
Plant protection products	*Chemical:* Only products that are approved by the relevant authorities are used	Preharvest intervals are documented in the journal for the crop. Residue analysis is taken every year in the first week of the harvest
Harvest	*Biological:* During manual harvesting, workers could transfer pathogenic bacteria to the fruit *Physical:* Cigarette ash could get into the product or product storage boxes if smoking is allowed in the field	Instructions for workers regarding handwashing, use of toilets, use of gloves and instruction not to pick up berries that are dropped on the ground. The workers are not allowed to smoke, eat or drink from glass bottles in the field. Pets and other animals should not be touched
Home transport	*Physical:* The harvest receptacles could be contaminated	Instruction is given for cleaning of boxes
	Chemical: The boxes could contain chemicals that are not suitable for food contact	Documentation proving that the boxes are approved for food use must be provided
	Biological: Rainwater from the roads could contaminate the product	Adequate coverings must be provided
Storage	*Biological:* The raspberries are temporarily stored on the farm before transportation	Workers are informed of the necessity to keep berries cold and out of direct sunlight
	Biological: Storage areas are used only for crops. No animals are allowed in the storage area	The storage area must be cleaned if any animal enters it
	Biological: The temperature must not be over 5 °C	An alarm is installed on the cooling system. This is tested every month
Sale	*Physical:* The boxes are covered so no dust can come in contact with the berries	Contract is made with the transport company. The transport company transport other fruit and/or vegetables only

carrot producers have begun to produce small carrots in small bags, which signals to the consumer that this is a ready-to-eat product, although it may still state that the produce should be washed before eating. Factors such as these must be taken into account when conducting a risk assessment.

The highest risk occurs when 'ready-to-eat' products are used without a wash step. Fresh produce such as raspberries, strawberries, apples and even lettuce are eaten without washing by many consumers, or used directly from the freezer in products such as ice-cream or smoothies. To revisit an earlier example, if the irrigation water used by the farmer was contaminated with *E. coli*, *Listeria* or another pathogen, there would be a significant risk to the consumer using the product without a wash step.

1.3 Following the steps in the production chain

It is much more costly and time-consuming to improve the safety of produce further down the production line. Therefore, is it very important that the process of minimizing contamination risk, especially from biological hazards, begins with the farmer. Even then, fresh produce can be damaged during the later steps, for example in warehouses and distribution, where cross-contamination from product to product or unhygienic handling can compromise produce safety.

As an example, in Denmark it is common for supermarkets to exhibit fresh produce in front of their store on the street. It is normal to see potatoes stored directly in the sun, which puts them at risk of solanin production. Cauliflowers are displayed in rain and sun at 20 °C, which is sufficient to multiply *Salmonella*, *E.coli* or *Listeria* in the product.

1.3.1 Consumer habits

The typical consumer of today is less knowledgeable about the correct handling of fruit and vegetables than previous generations. This is an important consideration for the grower of fresh produce, as it means that biological surface contamination is not automatically washed off the produce by the consumer. This fact is significant to farmers when creating a risk management plan. Physical or chemical contamination on the outside of the product can, in most situations, be washed off the product before distribution.

1.4 Conclusion

There will undoubtedly be more stories about people becoming ill from eating fruits and vegetables in future. The increase in consumption of 'ready-to-eat' fruit and vegetables calls for increased focus on minimizing food safety hazards.

To maintain high food safety in the primary production of fruit and vegetables, the farmer must have the necessary education on how to prevent hazards at the farm production level. A risk assessment is a valuable tool for recognizing and preventing hazards, and should be implemented as a basic procedure at all enterprises producing fruit and vegetables.

2

Niche farm fresh products

A. N. Jensen and D. L. Baggesen, Technical University of Denmark, Denmark

DOI: 10.1533/9781782420279.1.9

Abstract: There has been a general increase in demand for 'organic' or 'biodynamic' produce. As mineral nitrogen fertilizers are prohibited in organic farming, livestock manure often becomes the central fertilizer. Livestock manure is a known potential source of human pathogens, so it can be speculated that the observed increase in vegetable-associated foodborne outbreaks of pathogens is linked to its use. Legislation and guidelines vary regarding minimization of contamination risk in relation to use of livestock manure. An assessment of the effect of differences in management between organic and conventional production systems would help to provide more science-based recommendations for prevention of contamination.

Key words: organic, biodynamic, foodborne pathogens, contamination sources, livestock manure fertilizer.

2.1 Introduction

Consumption of fresh produce is increasingly recognized as an important part of a healthy lifestyle. Both organic and conventionally grown produce provide essential nutrients and vitamins as part of a balanced human diet. Non-conventional agricultural production systems adhering to organic or biodynamic principles appeal to many consumers because of the environmentally friendly and sustainable production methods used (Cottingham *et al.*, 2012; Paul and Rana, 2012). For example, the prohibition of pesticides helps to alleviate consumer concern about pesticide residues that might be present in fruit and vegetables. Furthermore, organic production systems promote a sound and natural ecological balance in agriculture without the use of mineral nitrogen fertilizers, and are designed to have minimal adverse effects on living systems (Rigby and Cáceres, 2001; Birkhofer *et al.*, 2008). On the other hand, the difference in microbial quality between conventional and organic produce seems to be uncertain as only a few direct comparisons have been made to date (Mukherjee *et al.*, 2004; Bohaychuk *et al.*, 2009).

Nevertheless, the concept of a less exploitative and more sustainable method of production has favored a general increase in organic agriculture around the world, although the market shares of organic produce in general, and predominant crops specifically, vary between regions and countries. The vegetable sector typically constitutes only a minor part of organic arable land. For example, only 1.4% of arable land in the EU in 2010 was being used for vegetable production (EC, 2010). However, organic fruit and vegetables are often the most popular organic products with shares in total organic sales between 15% and 36% in the four largest EU markets (EC, 2010), although the market for organic produce is minor relative to that for conventional produce. In the USA, 12% of all fruit and vegetables sold are organic, and in the EU this figure is 3.9% (EC, 2010; OTA, 2011).

2.1.1 Principles of organic production

Organic agriculture is based on four overall principles: health, ecology, fairness and care, as defined by the International Federation of Organic Agriculture Movements – IFOAM (http://www.ifoam.org/en/organic-landmarks/principles-organic-agriculture). The 'principle of health' implies that organic agriculture should sustain and enhance the health of soil, plants, animals, humans and the environment, because the health of individuals cannot be separated from the overall health of ecosystems. Accordingly, 'health' in organic agriculture refers to the 'wholeness and integrity' of all living systems involved in the process, rather than just the absence of disease.

The 'principle of ecology' means that organic agriculture should attain the ecological balance of living systems and cycles, and help to maintain genetic and agricultural diversity. Organic agricultural production should therefore be based on sound ecological processes and recycling. The 'principle of fairness' in organic agriculture entails providing a good quality of life for everyone involved, and managing any natural and environmental resources needed for production with respect for future generations. Likewise, the 'principle of care' states that organic agriculture should be managed in a precautionary and responsible way to protect the health and well-being of current and future generations and the environment. Precaution should be a central element in relation to management, development and technology choices in organic agriculture in order to prevent potential risks, such as the risks foreseen by some of genetic engineering.

2.1.2 Minimum standards for preventing contamination of produce with human pathogens

Standards have long been used within organic agriculture to create an agreement on what an 'organic' claim on a product really means. However, legislation for organic production is set by the governmental authority within each region and therefore varies worldwide. The European Union, for example, has provided the 'Council Regulation (EC) No 834/2007 on organic production and labeling of organic products and repealing Regulation (EEC) No 2092/91', whereas the USA has provided the Code of Federal Regulations (CFR)7, Part 205, National Organic Program (NOP) (Anonymous, 2000).

With respect to plant production, there is a common agreement on the prohibition of mineral nitrogen fertilizers, whereas the list of appropriate soil fertilizers and management techniques differ. According to the EU legislation, it is simply recommended (not mandatory) to compost livestock manure before applying it as fertilizer (Anonymous, 2007). The composting of livestock manure serves as a means to reduce the level of human pathogens such as *Salmonella*, *Campylobacter* and verotoxin-producing *Escherichia coli* (VTEC), which may have been shed into the manure by animals carrying these bacteria. These pathogens can persist in animal manure during storage for variable lengths of time, depending on the physical and chemical conditions.

Consequently, these pathogens may spread to the soil when the animal manure is applied as fertilizer and thereby constitute a potential crop contamination risk. Because of this contamination risk, some EU member states have specified further restrictions on the appropriate use of manure, as opposed to the non-compulsory recommendation on composting in the EU regulations. These restrictions involve mandatory composting of manure used for the growing of herbs or products intended for direct human consumption, and limitations in the use of manure of conventional origin.

However, some specifications mainly refer to registration and environmental concerns such as nitrogen load, rather than microbiological safety concerns. In the USA, NOP specifies that raw animal manure must be composted unless it is

- applied to land used for a crop not intended for human consumption; or
- incorporated into the soil not less than 120 days prior to the harvest of a product whose edible portion has direct contact with the soil surface or soil particles; or
- incorporated into the soil not less than 90 days prior to the harvest of a product whose edible portion does not have direct contact with the soil surface or soil particles (NOP, Section 205.203(c)(1)) (Anonymous, 2000).

The EC regulation for organic production (Anonymous, 2007) does not state a minimum time interval between application of manure and time of harvest, even though composting is non-mandatory.

2.1.3 Organic certification systems

To ensure consumer confidence in organic production systems and products, third party certification systems were developed, in order to guarantee that products stated to be produced organically actually comply with established organic standards. This certification started as a voluntary activity to facilitate trade, but is now required under many regulations if any kind of 'organic' claims are made on a product label. Governmental and private certification bodies operate according to organic standards that meet the minimum requirements enforced by law at the national level, but they may also certify to higher standards for a special logo with marketing advantages. Nevertheless, the varying organic standards and certification systems developed worldwide, including those originating from countries without organic food laws, means that 'organic' products are not directly comparable. A product may be recognized as organic in one country but not in another, which can be problematic for trade.

IFOAM has therefore developed the Organic Guarantee System (OGS), where two of the main components are the IFOAM Basic Standards and the Accreditation Criteria. Equivalence assessments of each standard and regulation have been conducted against one single international reference: the 'Common Objectives and Requirements of Organic Standards' (COROS), agreed upon by IFOAM members. This has resulted in the establishment of the IFOAM Family of Standards, which should provide a tool for drawing the line between organic and non-organic standards. Additionally, the US National Organic Program and the EU have signed a bilateral Equivalency Arrangement, which from June 2012 allows organic products certified in the USA to be sold as organic in the EU, and vice versa.

2.1.4 Concept of biodynamics

Biodynamic agriculture follows the regulations for organic farming, plus the overall principles and requirements specified in the Demeter International Production Standards (Demeter, 2011). These standards are typically formulated into guidelines by organizations for biodynamic agriculture in different countries to adjust for local conditions. Any product marketed with the Demeter trademark must have been produced in accordance with these standards.

Biodynamic agriculture is derived from Rudolf Steiner's 'Agricultural Course' of 1924, within the spiritual context of his anthroposophy philosophy. Biodynamic agriculture emphasizes the holistic development and interrelationships of the soil, plants and animals as a self-sustaining system, and has much in common with other organic approaches. Unique to the biodynamic approach is the regular use of biodynamic preparations made from fermented herbs and minerals as compost additives and field sprays, and the consideration of cosmic rhythms in plant production and animal husbandry. It is furthermore recommended that the 'inputs' needed for production are generated by the farm itself, and cattle should be kept for manure production. The use of composted livestock manure (particularly cattle manure) with added biodynamic preparations is essential, according to anthroposophy, in order to stimulate the fertility, life and humus content of soils, which in turn improves the yield, quality and vitality of plants. As composting is a necessary element of biodynamic agriculture, this system could help to reduce the pathogen contamination risk in biodynamically grown crops, because composting reduces or eliminates pathogens present in manure prior to its application as fertilizer.

2.2 Human pathogen contamination of 'niche products'

Besides norovirus contamination via human feces, the most common causes of foodborne gastrointestinal diseases are zoonotic pathogens such as *Salmonella*, *Campylobacter*, *E. coli* O157 and other vero–toxigenic *E. coli* (VTEC) that may be shed in animal feces. Raw produce is contaminated with these pathogens via numerous preharvest and postharvest routes, and can involve fecal contamination of animal as well as human origin. In addition to human pathogens, feces can contain high levels of non-pathogenic *E. coli*. Consequently, detection of *E. coli* bacteria can indirectly indicate levels of fecal contamination, which further indicates an increased risk of the presence of human pathogens. For *E. coli* to serve as a reliable fecal indicator for the presence of human pathogens, *E. coli* and the pathogen of interest must be almost equally persistent in the tested product. For example, it has been shown that a *Shigella sonnei* outbreak associated with baby corn correlated with high levels of generic *E. coli*. Likewise, the detection of *Salmonella* in fresh herbs coincided with *E. coli* numbers above 100 cfu/g, whereas the detection of *Campylobacter* in leafy greens and fresh herbs appeared not to correlate with the levels of *E. coli* (Anonymous, 2011). This shows that correlation between the fecal indicator *E. coli* and bacterial pathogens depends on the specific pathogen. For enteroviruses, this lack of correlation is even more pronounced (Ferguson *et al.*, 2012).

The number of worldwide gastrointestinal disease outbreaks in humans that have been associated with consumption of fresh produce is increasing (Doyle and Erickson, 2008; Lynch *et al.*, 2009; Berger *et al.*, 2010). However, some investigations into these outbreaks might have been based merely on epidemiological evidence, because traceback of suspected produce can often be difficult, and attempts to detect specific human pathogens usually fail. Difficulties in detecting human pathogens in fresh produce are a result of the rarity of their presence and typically low numbers (Anonymous, 2011). Extensive sampling is necessary, making it difficult to establish cost-efficient surveillance systems that can be used on a routine basis. Surveillance data concerning the occurrence of human pathogens in produce are therefore still relatively scarce. Nevertheless, sporadic surveys have been carried out to estimate the prevalence of the most common human pathogens (Harris *et al.*, 2003). Furthermore, existing data rarely specify whether the produce is of conventional or organic origin. Consequently, it is currently not at all clear whether produce from niche production systems differs from conventional produce with respect to pathogen contamination.

2.2.1 Potential contamination sources

In most cases where fresh produce has been linked to gastrointestinal disease outbreaks in humans, the specific source of contamination remains unidentified, because of the time and cost involved in investigating the many potential entrance points for contamination. Preharvest contamination in the production chain can occur directly or indirectly. Direct sources include untreated livestock manure, which is used as fertilizer, and contact with wildlife carrying pathogens. Indirect contamination occurs via irrigation and flooding with contaminated water, or through contaminated run-off water from nearby areas with fecal deposits.

Use of livestock manure as fertilizer

Use of livestock manure as fertilizer in agriculture provides an easy and effective way of meeting the nutrient requirements of crops, and it also helps to sustain the fertility and biological activity of the soil, which is an essential aspect of organic farming. Application of livestock manure in agriculture is a common practice, but it is especially important in organic plant production where mineral nitrogen fertilizers are prohibited.

Research results from the EU project PathOrganic showed that the application of livestock manure as fertilizer was common among organic farmers producing lettuce or carrots, but that fertilization practice for the producers differed between the five participating countries (Landau *et al.*, 2011). In all the countries investigated, the primary source of animal manure was cattle, while pig, chicken and horse manure were applied to variable degrees. Although the animal source will influence the potential presence of specific pathogens, the form of manure in terms of slurry or solid manure and the handling thereof will further affect the persistence of pathogens. Aeration of stored livestock manure by turning of manure heaps or stirring of slurry promotes decomposition and helps to eliminate or reduce pathogens. Similarly, the temperature rise during composting helps to reduce pathogens, although this depends on the temperature reached.

In addition, the preferred form (solid vs. slurry) of livestock manure varied between the participating countries in the 'PathOrganic' project. The length of storage and the time elapsing between the application of manure and planting also varied considerably between producers. As a result, the occurrence of pathogens in the manure at the time of use will vary significantly depending on the practice adopted. Although most producers used the livestock manure shortly before the crop was planted, thereby increasing the risk of pathogen contamination, the actual risk further depends on the type and prior treatment of the manure.

The potential presence of zoonotic pathogens in livestock manure depends on the animal species and the bacteriological status of the herd, which usually varies over time. Zoonotic pathogens shed into animal manure might, under microbiologically favorable storage conditions, persist for several months even without continuous input of fresh manure (Hutchison *et al.*, 2005). Moreover, once manure has been applied to the fields, the zoonotic pathogens may persist in the soil environment for up to several months (Ingham *et al.*, 2004; Islam *et al.*, 2004). This risk of spreading human pathogens via manure can be reduced by proper treatment of the manure before use, such as composting.

Composting is the microbial decomposition of organic matter, requiring oxygen and involving an increase in temperature. According to NOP, controlled composting ensures that temperature reaches 55–77 °C for three days in a static aerated system, or 15 days in a windrow system (windrow composting is the production of compost by piling organic matter or biodegradable waste, such as animal manure and crop residues, in long rows 'windrows') where the material is turned at least five times to reduce the pathogen levels satisfactorily (Anonymous, 2000). Processing methods in which all portions of the material reach 66–74 °C for at least one hour and the moisture level is reduced to below 12% are also accepted (NOP 5006). For instance, treatment of livestock slurry in a biogas plant will reduce or eliminate pathogens from the resulting fertilizer, depending on the combination of temperature and time.

2.2.2 International guidelines

In addition to the governmental regulations for organic farming, a number of guidelines on good agricultural practices (GAP) for fruit and vegetable production have been established, although these are more general guidelines and not specific to organic produce. There seems to be widespread acceptance in guidelines on good agricultural practice that the use of livestock manure as fertilizer may constitute a microbial contamination risk, although the magnitude of the actual risk is uncertain.

The Codex Alimentarius is a global reference point for ensuring the safety and quality of food that provides, among others, the Codex-code of Hygienic Practice for Fresh Fruits and Vegetables (CAC/RCP 53-2003, revised 2010). This codex recommends a number of precautions regarding the application of manure fertilizer, which are intended to limit the potential for microbial contamination. For example, proper treatment procedures, such as composting, pasteurization, heat drying, UV irradiation, alkali digestion, sun drying or combinations of these, should be adopted to reduce or eliminate pathogens in the manure. If untreated manure is to be used, appropriate corrective actions are needed, such as maximizing the time between the application of the manure and harvest of produce. Moreover, reducing direct or indirect contact between manure and fresh produce could lower the contamination risk, especially close to harvest time. Specific types of produce such as leafy greens are more at risk of contamination, as shown in human illness outbreak investigations (Harris *et al.*, 2003). This has formed the basis for 'Commodity specific food safety guidelines for the production and harvest of lettuce and leafy greens' (2007, by the US Food and Drug Administration). This guideline states several 'best practices' concerning soil amendments, particularly: 'DO NOT USE raw manure', or a soil amendment that contains un-composted, incompletely composted or non-thermally treated animal manure, on fields that will be used for lettuce and leafy green production.

The private sector body GLOBALG.A.P. (Good Agricultural Practices) also focuses on good hygiene. GLOBALG.A.P provides a practical manual for agriculture worldwide, and adherence to these practices could become necessary in order to access specific markets. One of the control points recommended by GLOBALG.A.P is the performance of a risk assessment before the use of organic fertilizers.

EU legislation

Although EU legislation concerning organic food production urges the use of organic animal fertilizer, a shortage of such fertilizer means that a certain proportion of manure originating from conventional sources is permitted. This practice was observed in three out of the five countries involved in the 'PathOrganic' project. However, the use of conventional manure in organic agriculture is not in agreement with the organic philosophy described above. Moreover, manure of conventional origin is likely to contain higher levels of antibiotic-resistant bacteria and other contaminants than manure of organic origin. In some EU member states, this has led to the development of more stringent standards for the use of conventional livestock manure, including restrictions on animal source, amount and treatment procedures, leading to markedly different practices between countries.

The PathOrganic project performed a field study on the presence of selected human pathogens; first in manure going to be used as soil fertilizer, and later in the organic lettuce or spinach grown in fields where the manure had been applied. The presence of the human pathogenic bacteria *Salmonella*, *Campylobacter*, *Staphylococcus aureus*, *Listeria monocytogenes* as well as *E. coli* virulence genes (*stx1*, *stx2*, *eae* and *rfbE*) was determined by PCR (not published). Examination of leafy greens from a total of 14 fields indicated that the occurrence of the pathogens varied considerably between the producers: 0, 2%, 4%, 11%, 41% for the *E. coli* virulence genes *stx1*, *stx2*, *rfbE* and *eae*, respectively; up to 35% for *Salmonella* and *Listeria*; up to 22% for *Staphylococcus aureus* and up to 7% for *Campylobacter*.

Although the PCR results could include non-viable bacteria incapable of causing illness, the relatively high number of PCR positives indicated that leafy greens are at risk of becoming contaminated when soil has been amended with livestock manure. It should be pointed out, however, that the vegetables tested were all grown on fields where the applied manure probably represented a worst case scenario in terms of pathogen contamination according to the results from the initial screening. Nevertheless, bacteriological culturing of lettuce samples from three Danish fields indicated that fecal contamination with *E. coli* was found in 45% of the lettuce samples (Jensen *et al.*, 2013). In 42% of these positive samples, the numbers of *E. coli* exceeded 100 cfu/g, which according to the EU regulation on microbiological hygiene criteria in foodstuff, is unsatisfactory for pre-cut fruits and vegetables (ready-to-eat) (EC, 2005). No equivalent criteria have yet been established for non-cut vegetables. Nevertheless, lettuce is probably close to the 'ready-to-eat' category as it is typically eaten minimally processed besides washing, which again may depend on the consumer. Interestingly, the field with the highest percentage of contaminated lettuce (54% vs. lowest contaminated field of 36%) concurred with the shortest interval between the application of animal slurry and harvest. This may indicate the importance of the length of the time gap between slurry fertilization and harvest when reducing the risk of fecal contaminating of lettuce, although the three fields differed also in other aspects. Although several previous surveys have resulted in a relatively low number of contaminated samples (Harris *et al.*, 2003), the conditions for fertilization, growth, etc., may have been different.

Despite the likelihood of livestock manure causing contamination of fresh produce, conclusive reports of vegetable-associated outbreaks are rare. A direct causal link between the use of livestock manure, and particularly untreated manure, for fertilization and the observed increase in produce associated human gastrointestinal outbreaks is still missing.

2.2.3 Environmental sources of contamination

An indirect point of preharvest contamination is the use of irrigation water contaminated with animal or human fecal matter. Water contaminated with feces also raises concern about the presence of enteric viruses (see Chapter 22) and parasites such as *Cryptosporidium* and *Giardia*. Many standards emphasize the importance of using drinking water for irrigation of crops for human consumption, but this may be unrealistic in arid regions.

The contamination risk from water varies according to the water source. The following sources are listed in the order of lowest to highest risk:

- groundwater from deep wells;
- groundwater from shallow wells (especially if they are badly maintained);
- surface water, particularly in the vicinity of animals, waste and habitation;
- raw or inadequately treated wastewater.

In addition to the water source, the irrigation method as well as the type of crop also influence the risk of contaminating crops. Irrigation methods that involve direct contact of contaminated water with the edible parts of crops constitute the greatest contamination risk. For example, drip irrigation systems appear to be safer than sprinkler irrigation for above-ground crops (such as lettuce and cabbage), whereas the drip irrigation method involves an increased risk for root crops (such as carrots).

Another indirect contamination risk, particularly for free-range crops, is via flooding with feces-contaminated water or via run-off water from nearby areas with animal fecal deposits (livestock or wildlife). Both wild birds and animals can act as reservoirs for human pathogens, or they may serve as passive carriers for varying lengths of time. For example, *Salmonella* has been isolated from various animal species, and larger birds are a common reservoir for *Campylobacter*, which are also commonly found in pet dogs and cats. Consequently, wildlife or pet animals may serve as a vehicle for the direct transmission of various human pathogens to crops if they forage on the fields. Insects are also suspected to play a role in the spread of bacteria.

In the PathOrganic project, detection of *E. coli* in lettuce in relatively high numbers indicated a fecal contamination and hence the presence of potential pathogens. The lettuce was grown in slurry-amended soils and therefore slurry seemed to be a likely source of *E. coli*, especially as no *E. coli* was detected in the soil before the slurry-amendment. However, the *E. coli* cfu found in the soil at the time of harvest was relatively low compared with the level of *E. coli* in the lettuce (Jensen *et al.*, 2013). A comparison of the similarity of *E. coli* sub-types found in the applied slurry, soil and lettuce showed that the majority of the *E. coli* sub-types differed. This finding indicates that the animal slurry fertilization was not the sole source of fecal contamination, but that the surrounding environment and wildlife were also involved.

Zoonotic pathogens may persist in extra-intestinal environments for several months, or even years, depending on physical and chemical conditions. For example, an experimental study with *Salmonella*-infected weaning pigs held on pastures, soil and water samples showed that *Salmonella* bacteria persisted for several weeks in the pasture environment (Jensen *et al.*, 2006). Moreover, several other *Salmonella* types were found in addition to the experimental one, which indicates the widespread occurrence of *Salmonella* in nature (Jensen *et al.*, 2004). Another study on the presence of *Salmonella* in wildlife (rodents, birds and insects) trapped in the surroundings of pig or cattle production sites with or without an ongoing *Salmonella* infection, indicated that *Salmonella* was transmitted from the production animals to the wildlife and not vice versa (Skov *et al.*, 2008). This also indicates that there is a potential risk of contaminating fruits and vegetables cultured in free-range farms within the range of wildlife migration. Furthermore, it emphasizes the importance of controlling *Salmonella* infections in livestock as well as protecting fields from wildlife access. The role of insects in the transmission of pathogens to produce is still uncertain, although wildlife has been recognized as an important vehicle in contamination of poultry houses. Finally, it is difficult (if not impossible) to prevent the access of insects into fields and this is also undesirable because of their beneficial role as pollinators or for biocontrol.

2.3 The difference in contamination risk for 'niche products'

Using livestock slurry as fertilizer is a common practice in organic production systems, and free-range production implies close contact with wildlife. These factors seemingly constitute an enhanced contamination risk for organic produce. However, organic farming may reduce the risk in certain circumstances, if appropriate soil management and fertilization regimes lead to more stable and diverse microbial communities in the soil. Similarly, the microbial communities of organic crops may differ from conventional crops, and could perhaps include antagonistic bacteria that compete with human pathogens, but this is not well-documented.

The potential occurrence of human pathogens in livestock manure differs depending on the animal species. Its organic or conventional origin might also influence pathogen levels because of different management practices. For example, feeding regimes in organic livestock production involve more roughage compared with conventional production. Feeding a greater proportion of roughage could induce a shift in the composition of the intestinal microbial community of the animal as a result of a prebiotic-like effect, and this shift might also affect the numbers of pathogens shed by the animals. The restricted use of antimicrobials in organic livestock production means that organic manure most likely contains fewer antibiotic-resistant pathogens than conventional manure, and this is important in the treatment of human infections.

The level of virulent human pathogens present at the time of consumption, as well as consumer behavior, are key factors in determining food safety. The potential presence of pathogens will depend not only on the initial pathogen load but also on the pathogens' ability to persist or grow on or in the plants after contamination (Harris *et al.*, 2003). The survival of human pathogens on plant surfaces is generally challenged by lack of nutrients, temperature and humidity fluctuations, as well as ultraviolet light. Human pathogens have also been found to reside inside plants, although the findings have been contradictory. Moreover, findings from experimental investigations may not apply to real-life conditions. Knowledge concerning the persistence of pathogens on fresh produce is limited, particularly in relation to the differences between conventional and organic farming.

2.4 Conclusion

Unfortunately, existing data on detected pathogens in fruits and vegetables rarely specify whether they are of conventional or organic origin. This makes it difficult to pinpoint the level of microbiological risk from organic produce. However, this uncertainty should by no means be allowed to undermine the overall health benefits of fresh produce. Organic produce can result in a lower intake of pesticides and other contaminants and provide an environmentally friendly alternative to conventional produce.

Despite an increase in gastrointestinal disease outbreaks associated with fresh produce, the sporadic occurrence of human pathogens in produce makes direct detection challenging, and the existing hygiene indicators such as *E. coli* often prove inadequate in identifying risk products. Preventing contamination in the first place is therefore a key issue, and a number of preventive measures have been suggested. However, in terms of the actual pathogen contamination risk to consumers, the relative importance of potentially contaminating behavior or conditions in each step of the organic production chain is still uncertain. Assessment of the relative importance of potentially risky behavior in vegetable production, including management differences between organic and conventional production systems, would help to provide more science-based recommendations for prevention of contamination. Finally, it would be interesting for the organic sector to elucidate whether production management, especially soil management and fertilization practices, provide any protective conditions as a consequence of their more diverse microbiota on the crop and in the soil.

2.5 Questions for discussion

1. What are the basic principles for organic farming?
2. What is the concept of biodynamic farming?
3. What characterizes a reliable indicator of pathogen contamination?

4. What are the potential contamination sources of fresh produce?
5. Does produce production on open land imply special contamination risks?
6. Which main factors will affect the level of pathogens in manure at the time of application to fields?
7. Is there a required minimum time interval between application of manure and harvest?
8. How can manure be treated to reduce the pathogen levels satisfactorily?
9. What are good agricultural practices (GAP) for fruit and vegetable production?
10. Is the contamination risk different for 'niche products'?

2.6 Acknowledgments

We would like to thank CORE Organic Funding Body Network, being former partners of the FP6 ERA-NET project, CORE Organic (Coordination of European Transnational Research in Organic Food and Farming, EU FP6 Project no. 011716). Special thanks to the assisting lab technicians and partners in the SafeOrganic project, as well as the participating organic farmers.

2.7 References

ANONYMOUS, 2000. *7 CFR 205.203: Soil fertility and crop nutrient management practice standard.* National Organic Program, United States Department of Agriculture.

ANONYMOUS, 2007. *Council Regulation (EC) No 834/2007 of 28 June 2007 on organic production and labelling of organic products and repealing Regulation (EEC) No 2092/91.* The Council of the European Union.

ANONYMOUS, 2011. *Annual Report on Zoonoses in Denmark 2010.* National Food Institute, Technical University of Denmark.

BERGER, C.N., SODHA, S.V., SHAW, R.K., GRIFFIN, P.M., PINK, D., *et al.*, 2010. Fresh fruit and vegetables as vehicles for the transmission of human pathogens. *Environmental Microbiology*, 12, 2385–97. doi: 10.1111/j.1462-2920.2010.02297.

BIRKHOFER, K., MARTIJN BEZEMER, T., BLOEM, J., BONKOWSKI, M., CHRISTENSEN, S., *et al.*, 2008. Long-term organic farming fosters below and aboveground biota: Implications for soil quality, biological control and productivity. *Soil Biology and Biochemistry*, 40, 2297–308.

BOHAYCHUK, V.M., BRADBURY, R.W., DIMOCK, R., FEHR, M., GENSLER, G.E., *et al.*, 2009. A Microbiological Survey of Selected Alberta-Grown Fresh Produce from Farmers' Markets in Alberta, Canada. *Journal of Food Protection*, 72, 415–20.

COTTINGHAM, M., ROSE, E., TWINE, J., COTTLE, F., and PERRETT, T. *Organic Market report 2012, UK Soil Association.* http://www.soilassociation.org/marketreport (accessed 29 September 2012).

DOYLE, M.P. and ERICKSON, M.C., 2008. Summer meeting 2007 – the problems with fresh produce: an overview. *Journal of Applied Microbiology*, 105, 317–30.

DEMETER 2011, *Production Standards for the Use of Demeter Biodynamic® and Related Trademarks*, Demeter-International Production Standards, pp. 1–41, July 2010.

EC (EUROPEAN COMMISSION), 2005. Commission Regulation (EC) No 2073/2005 on microbiological criteria for foodstuffs, amended by: EC No 1441/2007/EC of 5/12/2007. *Official Journal of the European Union.* http://eur-lex.europa.eu/LexUriServ/LexUriServ.do?uri=OJ:L:2005:338:0001:0026:EN:PDF and http://eur-lex.europa.eu/

LexUriServ/LexUriServ.do?uri=OJ:L:2007:322:0012:0029:EN:PDF (accessed 10 June 2012).

EC (EUROPEAN COMMISSION), 2010. *An analysis of the EU organic sector*. Directorate-General for Agriculture and Rural Development, Organic Farming. http://ec.europa.eu/agriculture/analysis/markets/organic_2010_en.pdf (accessed 29 September 2012).

FERGUSON, A.S., LAYTON, A.C., MAILLOUX, B.J., CULLIGAN, P.J., WILLIAMS, D.E., *et al.*, 2012. Comparison of fecal indicators with pathogenic bacteria and rotavirus in groundwater. *Science of the Total Environment* 431, 314–22.

HARRIS, L.J., FARBER, J.N., BEUCHAT, L.R., PARISH, M.E., SUSLOW, T.V., *et al.*, 2003. Outbreaks Associated with Fresh produce: Incidence, Growth, and Survival of Pathogens in Fresh and Fresh-Cut Produce. *Comprehensive Reviews in Food Science and Food Safety*, 2 (Supplement), Chapter III, 78–141.

HUTCHISON, M.L., WALTERS, L.D., AVERY, S.M., MUNRO, F. and MOORE, A., 2005. Analyses of livestock production, waste storage, and pathogen levels and prevalences in farm manures. *Applied and Environmental Microbiology* 71, 1231–6.

INGHAM, S.C., LOSINSKI, J.A., ANDREWS, M.P., BREUER, J.E., BREUER, J.R., *et al.*, 2004. *Escherichia coli* contamination of vegetables grown in soils fertilized with noncomposted bovine manure: garden-scale studies. *Applied and Environmental Microbiology* 70, 6420–7.

ISLAM, M., DOYLE, M.P., PHATAK, S.C., MILLNER, P. and JIANG, X., 2004. Persistence of enterohemorrhagic *Escherichia coli* O157:H7 in soil and on leaf lettuce and parsley grown in fields treated with contaminated manure composts or irrigation water. *Journal of Food Protection*, 67, 1365–70.

JENSEN, A.N., DALSGAARD, A. STOCKMARR, A., NIELSEN, E.M. and BAGGESEN, D.L., 2006. Survival and transmission of *Salmonella enterica* serovar Typhimurium in an outdoor organic pig farming environment. *Applied and Environmental Microbiology*, 72, 1833–42.

JENSEN, A.N., LODAL, J. and BAGGESEN, D.L., 2004. High diversity of Salmonella serotypes found in an experiment with outdoor pigs. *NJAS, Wageningen Journal of Life Sciences*, 52, 109–17.

JENSEN, A.N., STORM, C., BAGGESEN, D.L., FORSLUND, A. and DALSGAARD, A., 2013. *Escherichia coli* contamination of lettuce grown in soils amended with animal slurry. *Journal of Food Protection*, 76, 1137–44. doi:10.4315/0362–028X.JFP-13–011.

LANDAU, B., KOLLER, M. and MÄDER, P., 2011. *PathOrganic – Identification of Critical Control Points for organic vegetable crops*. Report, pp. 1–36, http://orgprints.org/20389/.

LYNCH, M.F., TAUXE, R.V. and HEDBERG, C.W., 2009. The growing burden of foodborne outbreaks due to contaminated fresh produce: risks and opportunities. *Epidemiology and Infection*, 137, 307–15.

MUKHERJEE, A., SPEH, D., DYCK, E. and DIEZ-GONZALEZ, F., 2004. Preharvest Evaluation of Coliforms, *Escherichia coli, Salmonella*, and *Escherichia coli* O157:H7 in Organic and Conventional Produce Grown by Minnesota Farmers. *Journal of Food Protection*, 67, 894–900.

NOP, 5006. *Guidance Processed Animal Manures in Organic Crop Production*, National Organic Program, United States Department of Agriculture.

ORGANIC TRADE ASSOCIATION (OTA), 2011. Organic Industry Survey.

PAUL, J. and RANA, J., 2012. Consumer behavior and purchase intention for organic food, *Journal of Consumer Marketing*, 29, 412–422. http://dx.doi.org/10.1108/07363761211259223.

RIGBY, D. and CÁCERES, D., 2001. Organic farming and the sustainability of agricultural systems, *Agricultural Systems*, 68, 21–40.

SKOV, M.N., MADSEN J.J., RAHBEK, C., LODAL, J., JESPERSEN, J.B., *et al.*, 2008. Transmission of *Salmonella* between wildlife and meat-production animals in Denmark. *Journal of Applied Microbiology*, 105, 1558–68.

3

Guidelines and protocols for safe practice in fresh produce production: FDA legislation

E. Snellman, FDA, USA

DOI: 10.1533/9781782420279.1.22

Abstract: As fresh produce consumption by US consumers increased, so too did outbreaks of illness related to consumption of those commodities. In response, the US Food and Drug Administration (FDA), together with the US Department of Agriculture (USDA), developed guidance setting out good agricultural practices (GAPs) to minimize contamination of produce, and began working with food safety partners to promote their adoption. In implementing GAPs, lessons learned during key outbreaks have led to enhanced efforts to minimize illness related to consumption of fresh produce through prevention, response, education and research. The Agency is now working to establish on-farm preventive controls that will be mandatory for many produce growers and packers.

Key words: FDA, produce safety, good agricultural practices (GAPs), Coordinated Outbreak Response & Evaluation (CORE), Food Safety Modernizaton Act (FSMA).

3.1 Introduction: a new strategy is tested

The conference call begins slowly as a large number of state public health experts, Food and Drug Administration (FDA) personnel from center and district offices and their federal partners dial in.[1] It's one of several that have taken place over the last few days. The call leader from the FDA's newly formed Coordinated Outbreak Response & Evaluation (CORE) team announces it will be just a few more minutes until the conference call starts as several additional agencies are expected to report in. They are waiting to hear the latest epidemiological information, case numbers and etiological data updates from various investigations taking place

[1] Conference call is for illustrative purposes only.

Published by Woodhead Publishing Limited, 2014

across the country. Beginning operations only one month prior, the CORE Signals and Surveillance, Response and Post-response teams, will be strenuously tested during this outbreak that will eventually result in 147 illnesses and 33 deaths across 28 states caused by one or more of five subtypes of *Listeria* (CDC, 2012a).

The week prior to the call, early signals of the emerging outbreak of listeriosis were being recognized by CORE's Signals and Surveillance team. They noted a Colorado Department of Public Health and Environment (CDPHE) press announcement concerning an unusually high number of monthly cases of listeriosis within the state (FDA, 2012a). Over the following days, CDPHE contacted the FDA Denver District Office and, working with the Centers for Disease Control and Prevention (CDC), its PulseNet database managers and epidemiologists, collectively began to piece together parts of the outbreak puzzle. Using PulseNet, an electronic bulletin board of DNA 'fingerprint' data entered by state, local and federal laboratories, and coordinated by the CDC, health officials defined the outbreak strains in ill Colorado patients and matched them to clinical strains of *Listeria* isolated from ill persons in neighboring states. An equally large piece of the puzzle remained in identifying the likely food vehicle contaminated by *Listeria*. Based on patient interviews using a standard CDC questionnaire, the likely suspect vehicle was identified as cantaloupe, even though *Listeria* had never been tied to whole cantaloupe before. Three ill persons recalled eating a particular cantaloupe variety called 'Rocky Ford', grown in the Rocky Ford region of southeastern Colorado. After confirming the suspect vehicle, CDPHE and FDA quickly mobilized teams of investigators who collected cantaloupe from retail locations where ill persons reported buying the product and performed *Listeria* testing on the samples. This conference call was to communicate the results of a supplemental questionnaire created by CDPHE to learn more about the cantaloupes eaten by case patients in the hope of identifying the source of the cantaloupes. The results of *Listeria* tests in cantaloupes collected by the CDPHE and FDA from grocery stores where several ill persons shopped were also to be announced.

FDA didn't traditionally invest its resources in mining data and information looking for an outbreak nor did it have a dedicated team of response experts (FDA, 2012a). In the past, FDA usually assembled groups of experts in response to outbreaks only after they were identified by CDC. However, the Agency found getting the right people assigned to the work and coordinating their efforts cost valuable response time. The new strategy of earlier involvement through signals monitoring was launched in mid-2011. The idea was integral to the design and organizational framework during the planning of CORE. The FDA hoped catching outbreak signals early in the chain of events that can lead to a multi-state outbreak would significantly increase the speed at which the Agency could respond. FDA also wanted to staff CORE with a pool of experts who had previous experience working with FDA district office field staff, state public health officials and experts in agriculture along with the CDC and other partners during outbreaks. These structural changes were expected to increase speed and consistency in response and minimize the overall number of foodborne illnesses by incorporating lessons learned from each outbreak into future responses, and into policies and other regulatory activities directed at preventing illness (FDA, 2012b; Karas, 2012).

3.2 Early produce safety policy

Early detection, improving outbreak response coordination and incorporating lessons learned into future policy decisions and other regulatory actions are consistent with the goals and objectives for produce safety set by the FDA, beginning in 1997. These were largely shaped by the disturbing rise in the number of foodborne disease outbreaks related to consumption of produce in the 1990s and food safety initiatives begun by the Clinton administration.

Throughout the 1990s, consumption of fresh fruits and vegetables (uncooked) increased as it widely became recognized that they are important components of a healthy diet and a source of minerals, fiber and antioxidants. The CDC estimated that during this period 76 million US citizens became sick, more than 325 000 people were hospitalized, and 5000 died each year from foodborne illness (Mead *et al.* 1999).[2] Further, they estimated that at least 12% of foodborne-outbreak-associated illnesses were linked to fresh produce commodities.

In his radio address on 25 January 1997, President Clinton announced that he would request $43 million in his 1998 budget to improve the safety of the nation's food supply using an interagency approach, and directed the Secretary of Agriculture, Secretary of Health and Human Services, and the Administrator of the Environmental Protection Agency to identify steps to improve the safety of the food supply. In what became the National Food Safety Initiative (FSI) report to the President, the FDA, USDA, USEPA and the CDC outlined actions to be taken to improve food safety, including enhanced surveillance and setting up an early warning system to help detect and respond to outbreaks earlier, improving responses to outbreaks, improving risk assessment, developing new research methods, improving inspections and compliance, and furthering food safety education (FDA, USDA, USEPA, CDC, 1997). During this period, specifically as a result of the increased numbers of outbreaks related to fresh produce consumption (CDC, 2000) including outbreaks associated with sprouted seeds, fresh and fresh-cut produce, the FSI was quickly expanded to include the Produce and Imported Food Safety Initiative (PIFSI) focusing on produce and microbial food safety (Beru, 2008). In 1998, as a part of the PIFSI, the FDA, in partnership with the USDA, issued guidance on GAPs focusing on the production and packing of fresh produce. This guidance for industry, entitled 'Guide to Minimize Microbial Food Safety Hazards for Fresh Fruits and Vegetables' (the GAPs guide), is broad based in scope and covers all fresh produce consumed in the USA that is produced domestically and abroad. It is intended to assist the US and foreign produce industries in enhancing the safety of domestic and imported produce during growing, harvesting, sorting, packing and distribution.

[2] More recent CDC estimates for the burden of foodborne illness show 48 million Americans get sick, 128 000 are hospitalized and 3000 deaths occur annually in the USA (CDC, 2012b). The reduction in numbers is best explained by innovations in methodology and improvements in data resulting in more accurate estimates, rather than true reduction in foodborne illness.

3.3 Challenges in good agricultural practices (GAPs) implementation

Despite wide availability and acceptance of the guidance in many industry segments, large outbreaks related to consumption of contaminated produce continued to occur. In November 2003, a large hepatitis A outbreak was identified among customers of a single Monaca, Pennsylvania Mexican-style restaurant. The outbreak eventually grew to unprecedented size involving over 600 patients, at least 124 hospitalizations and three deaths (Wheeler *et al.*, 2005). The size and seriousness of the outbreak quickly gathered plenty of media attention and placed the produce industry in general, but especially imported produce, under intense scrutiny (Burros, 2003; Conte and Fabregas, 2003). The vehicle identified by public health officials was green onions used in making mild salsa and more than 50 other menu items. The FDA traceback investigation revealed that the green onions were grown and packed on two farms in northern Mexico before being imported into the USA, passing through two or more distribution centers without repacking before delivery to the restaurant. Based on hepatitis A sequence comparisons and epidemiological information, it was deemed likely that the contamination occurred in Mexico during production (Calvin *et al.*, 2004; Wheeler *et al.*, 2005). Based on this information, and similar data collected during previous green onion-hepatitis A outbreaks in Tennessee and Georgia earlier that year, the FDA named the four Mexican farms (including the two that supplied the green onions to the Pennsylvanian restaurant) associated with the outbreaks and issued an import alert, instructing inspectors to refuse entry of green onions from these firms. Hepatitis A virus is transmitted by the fecal-oral route and is commonly transmitted by infected persons, often through insanitary hand contact, or by consumption or contact with contaminated water or food.

Production of green onions is labor-intensive, making them particularly vulnerable to contamination in the absence of appropriate controls. During harvest and postharvest handling, multiple workers are involved in peeling outer leaves, cutting roots, trimming stems and bundling prior to packing (Dentinger *et al.*, 2001); as many as nine different people may touch a green onion from planting to packing (Calvin *et al.*, 2004). In December 2003, FDA, CDC and Mexican officials jointly inspected the four green onion farms in Mexico. Although the FDA did not find the virus on the farms, several lapses in GAPs were identified including poor sanitation practices, inadequate hand-washing facilities and concerns of worker health and hygiene, along with the quality of water used in growing, packing and making ice (Calvin *et al.*, 2004). It is critically important when growing such a labor-intensive crop involving multiple persons handling the product, that strong worker health and hygiene and sanitation programs be in place and enforced by the grower. Further, understanding the potential routes of contamination associated with this pathogen (hand contact, poor water quality) and mitigation strategies directed toward decreasing the likelihood of contamination via these routes are also critical.

The 2003 hepatitis A outbreaks highlighted the fact that after nearly a decade of industry and FDA initiatives aiming to reduce the microbial contamination of fresh produce at the grower level, additional progress in achieving widespread adoption GAPs by growers was still needed. Although many growers had responded appropriately by adopting safer practices prior to the outbreaks, others fell behind in meeting government, consumer, and their own industry expectations for safer produce (Calvin *et al.*, 2004). Of these growers, the incentive for adopting these programs was missed while others were simply never reached or did not have the technical resources to follow GAPs. Slow progress in strengthening food safety programs by a few places all growers at economic risk when an outbreak occurs, as many consumers choose to avoid an implicated commodity entirely unless a specific grower, producer or shipper is identified (*Erie Times News*, 2003). In the case of Mexican green onions, consumer demand for green onions fell precipitously along with price after news of the outbreaks was announced. All Mexican growers suffered, but business for the firms named in the import alert completely collapsed as they largely could not sell green onions, or any other product, in any but the smallest, local markets (Calvin *et al.*, 2004).

Early in 2004, with the events of the previous year's hepatitis A outbreaks still in sharp focus, FDA began a retrospective look-back at its efforts in produce safety to assess what more could be accomplished to reduce the number of produce-related outbreaks (Beru, 2008). The review included not only assessing its progress in gaining industry adoption of GAPs, but also its communication and outreach efforts, the efficiency in responding to outbreaks and support of produce safety research. Critics of industry and governmental efforts to improve produce safety noted the rise in produce-related illnesses leading up to the 2003 hepatitis A outbreaks, and pointed out that lack of FDA regulatory authority and understaffing were partly to blame (Burros, 2003). In January 2004, the FDA and CDC, along with representatives from the produce industry and trade associations met to partner in evaluating produce food safety programs, government-industry communication strategies (public advisories, guidance, industry information), product sampling programs and advances in produce safety research in order to plan a way forward for produce safety.

In October 2004, FDA published the Produce Safety Action Plan (Action Plan) that addressed areas for improving produce safety identified in its internal review and those highlighted in discussions with its federal (CDC, USDA) and private sector (trade associations, industry and academia) food safety partners (FDA, 2004). The overarching goal of the Action Plan was to minimize the incidence of foodborne illness associated with the consumption of fresh produce. Four general objectives address ways to achieve the goal: 1) prevent the contamination of fresh produce with pathogens; 2) minimize the public health impact when contamination occurs; 3) improve communication with producers, preparers and consumers about fresh produce; and 4) facilitate and support research relevant to fresh produce (FDA, 2004). Each objective relies on a multi-layer approach and identifies specific steps that contribute to its achievement. Although this was an FDA Action Plan, the Agency acknowledged that achieving its objectives depends on collaborations among food safety partners in the federal government, industry (growers and trade associations), state and local public health officials, academia and consumers to improve produce safety. In this regard the Action Plan is much like the FSI, PIFSI and the GAPs guide—collaboration, communication and outreach in the absence of implementing regulations. At the time, little consensus existed for the idea of mandated produce safety regulations (Karst, 2004). Industry responded quickly to the Action Plan through its trade associations in developing and advocating for tighter supply-chain guidance for specific commodities, including lettuce, melons, tomatoes and green onions/herbs, along with better research into the cause of produce-related outbreaks before regulations were issued (Karst, 2004).

3.4 What more can we do?

FDA provided technical assistance to the produce industry in its development of commodity-specific guidance (CSG); and in 2005, began to adopt the CSGs as its own draft guidance with the addition of lessons learned by the Agency since publication of the GAPs Guide (FDA, 1998). The Agency believed this strategy would lend credibility to guidance recommendations and further promote implementation.

Although government and produce industry efforts had some success in reducing the number of produce-related illnesses, it was evident that full implementation of FDA recommendations in the GAPs guide and those in industry-led commodity-specific guidance were still not fully adopted. In 2005, prompted by a number of *Escherichia coli* O157:H7 outbreaks tied to fresh or fresh-cut lettuce (including one additional case related to fresh-cut spinach), the FDA issued a letter to California firms that grow, pack, process or ship fresh and fresh-cut lettuce. The letter outlined the Agency's serious concern with continued outbreaks in these commodities, citing 19 outbreaks responsible for over 400 illnesses and two deaths since 1995 (FDA, 2005). The letter pointed out that the efforts by the California Department of Health Services (CDHS) to actively engage the lettuce industry over the previous three years had yet to result in a comprehensive, collaborative plan by industry to address the issue of *E. coli* O157:H7 in lettuce. The letter encouraged the industry to begin or intensify efforts to improve communication between CDHS, California Department of Food and Agriculture (CDFA) and other industry partners to ensure active, widespread mitigation procedures to address the problem. It also urged the lettuce industry to expedite completion of lettuce and leafy green-specific supply-chain guidance, promote implementation of that guidance and establish a research coalition to characterize environmental reservoirs of *E. coli* O157:H7 and facilitate transfer of research findings to help mitigate the problem (FDA, 2005). In 2006, the FDA took additional action and developed the Lettuce Safety Initiative, partnering with CDHS and CDFA to assess the California lettuce industry by evaluating practices and processes in selected regions. An additional but important goal of the Initiative was to demonstrate FDA's commitment and concern about the safety of lettuce to the industry (FDA, 2006a). Leveraging this strategy, the following year the Agency similarly developed the Tomato Safety Initiative along with state health and agricultural departments of Florida and Virginia in response to recurring *Salmonella* outbreaks in fresh and fresh-cut tomatoes.

3.5 A pivotal outbreak prompts a policy shift

Against a backdrop of repeated outbreaks of illness caused by *E. coli* O157:H7 involving lettuce and other leafy greens, it seemed to many that a larger outbreak involving the pathogen was likely and in the fall of 2006 it arrived in a big way (Palmer, 2007). By early October, over 200 people were infected with the same outbreak strain, 102 were hospitalized, 31 suffered kidney failure and three died. Investigators quickly identified the suspect vehicle as raw spinach, primarily sold in bags but also in combination with other leafy greens (CDC, 2006). In its final report on the outbreak, the FDA reported that it found 13 bags of spinach obtained from sick persons contained the outbreak strain of *E. coli* O157:H7 (FDA, 2007). Using product codes from the bags and DNA fingerprinting on bacteria isolated from the spinach, investigators were able to match environmental samples of *E. coli* O157:H7 obtained from one production field to the strain that had caused the outbreak. CDHS investigators found samples of cattle manure collected from pastures surrounding the field were positive for the *E. coli* outbreak strain, and that they were a genetic match to *E. coli* isolated from bagged spinach and clinical isolates obtained from patients. They also found evidence that wild pigs had penetrated the fence surrounding the field and that samples obtained from a killed feral pig were also positive for the outbreak strain. All the spinach products implicated in the outbreak were traced back to Natural Selection Foods LLC of San Juan Bautista, California, who distributed and sold the produce under a variety of brand names (FDA, 2006b).

The consequences of the 2006 spinach outbreak were significant in terms of its economic impact on the produce industry and loss of consumer confidence in the safety of the nation's produce. Because spinach is a high-volume commodity that is used in several different types of products including mixed salads, spinach blends and baby spinach and sold under several different brand names, the product recalls initiated by Natural Selection Foods were broad in scope covering a diverse product mix distributed widely across the country. The tainted spinach caused a nationwide health scare as the number of states reporting illnesses quickly climbed (CNN, 2006). Consumers shunned products grown by the Western leafy green industry and sales of their products plummeted, costing the industry over $100 million in reduced leafy green sales (Blake, 2011). In addition, news of the outbreak caused Mexico to refuse shipments of lettuce from California (Russell, 2006). All growers suffered in the wake of the outbreak, even those who had strong food safety programs based on FDA and industry commodity-specific guidance recommendations for their products. Ultimately, the outbreak changed the culture of the leafy green industry—food safety became a top priority in order to restore consumer confidence and ensure sustainability of the industry (Blake,

2011). The outbreak pushed the Western leafy green industry towards tighter production and handling standards, ultimately giving rise to the California Leafy Green Marketing Agreement. It was also a catalyst that drove a shift in food safety compliance strategy from voluntary-based compliance to accepting and advocating for a federal regulatory oversight. Early in 2007, two large trade associations (United Fresh Produce Association and Produce Marketing Association) published letters to its members endorsing resolutions calling for a federal food safety regulatory framework to best protect public health and enhance consumer confidence in produce safety (Marshall and Miller, 2007; United Fresh, 2007).

3.6 New mandates: modernization of food safety

The Food Safety Modernization Act (FSMA) was signed into law in January 2011, representing the first major reform of the food safety provisions of the Food, Drug, and Cosmetic Act (FD&C Act) in over 70 years. It grants new powers to the FDA including mandatory recall authority and administrative detention of food only on 'reason to believe' it may be adulterated. Previously, detention required the Agency to prove that food was adulterated. FSMA provides the FDA a legislative mandate to require comprehensive, preventive control standards across the food supply chain, including those for the safe production, harvesting, packing and storage of raw agricultural commodities including produce. The preventive control standards for produce must include, with respect to growing, harvesting, science-based minimum standards related to soil amendments, hygiene, packaging, temperature controls, animals in the growing area and water. Further, the produce regulation must set procedures, processes and practices that the FDA determines are necessary to minimize the risk of serious adverse health consequences or death, including those determined to be reasonably necessary to prevent the introduction of hazards into fruits and vegetables. Clearly, the mandates included in FSMA will transform compliance with best practices (as they are promulgated in the rule-making) from a voluntary basis to mandatory compliance with federal enforcement. However, the greatest success in compliance and enforcement can only be achieved through continued partnerships with federal, state and local agricultural agencies along with support of the produce growers and their trade associations.

3.7 Future trends: back to the call

Everyone has joined now. The CORE lead on the call asks for an update from the Denver District Laboratory concerning the cantaloupe samples they collected at retail. The Denver microbiologists quickly announce nine out of ten cantaloupe samples tested were positive for *Listeria monocytogenes*. CDPHE follows with their report on epidemiological findings from the supplemental questionnaire they developed to learn more about the source of the suspect cantaloupe. They announce preliminary traceback and distribution data gathered by state officials show the cantaloupe collected at retail locations where several ill persons purchased cantaloupe may have come from Jensen Farms of Granada, Colorado, a grower and marketer of Rocky Ford cantaloupe. In addition, laboratory testing by CDPHE shows cantaloupe samples taken at two retail locations supplied by Jensen Farms yields a PFGE match to an outbreak strain of *Listeria* (CDC, 2011a). Another sample collected from the home of an ill person is positive for *Listeria*, but results from PFGE analysis are still pending. Colorado state officials explain that traceback information on these three samples also point to Jensen Farms as the likely source. They add some unsettling news—the cantaloupe were harvested in late August and September, and distributed widely across the USA where they still may be stored in consumers' homes or be available in grocery stores (CDC, 2011b). CORE, FDA field personnel, CDC and state public health experts carefully review the information presented and ask several questions to clarify the information. The epidemiological and analytical data are compelling, suggesting a strong link between Jensen Farms and the outbreak strains of *Listeria*. The national team of food safety and public health experts decide the next step in the investigation is to confirm the source of the outbreak and move to limit further public health impact.

3.8 Questions for discussion

1. Describe some difficulties in ensuring that imported produce is safe for US consumers.
2. With such large economic incentives present at the farm and industry levels, why might some growers not follow GAPs recommendations to minimize microbial contamination?
3. What role could industry associations, marketing agreements or commodity-specific groups play in advancing food safety initiatives to their members and other produce growers?
4. What are some of the recurrent themes in messaging that the FDA has pursued since 1998 with the produce industry, other federal and state public health agencies, and other food safety partners? Identify limitations in the approaches used by the FDA and provide alternative strategies that may have worked in increasing produce safety.
5. What tools did FSMA provide to allow the FDA to move quickly in response to suspect adulterated produce commodities?
6. In the *Listeria* outbreak scenario, what pieces of the outbreak puzzle do the food safety experts have and what is missing? Describe the next series of steps that could be taken to confirm the source of the outbreak.

3.9 Acknowledgment

I wish to thank Michelle Smith for her helpful reviews of this manuscript.

3.10 References

BERU, N. 2008. *FDA's Perspective on Global Produce Safety*. Conference for Food Protection Workshop, 12 April 2008. Accessed 22 September 2012 at: http://www.foodprotect.org/biennial-meeting/?id=2008-biennial-meeting.

BLAKE, C. 2011 'Leafy greens industry still reeling from *E. coli*-spinach outbreak.' *Western Farms Press* 2 March 2011. Accessed on 26 November 2012 at: http://westernfarmpress.com/vegetables/leafy-greens-industry-still-reeling-e-coli-spinach-outbreak.

BURROS, M. 2003 'Produce becomes a rising source of food-borne illnesses.' *New York Times* 23 November 2003. Accessed 12 October 2012 at: http://www.nytimes.com/2003/11/23/us/produce-becoming-increasing-source-for-food-illnesses.html.

CALVIN, L., AVENDAÑO, B. and SCHWENTESIUS, R. 2004. *The economics of food safety: the case of green onions and hepatitis A outbreaks*. Economic Research Service/USDA, VGS-305–01. Accessed 20 November 2012 at http://webarchives.cdlib.org/wayback.public/UERS_ag_1/20110914045200/http://www.ers.usda.gov/publications/vgs/nov04/VGS30501/VGS30501.pdf.

CDC. 2000. *Surveillance for Foodborne Disease Outbreaks—United States, 1993–1997*. Accessed 20 September 2012 at: http://www.cdc.gov/mmwr/preview/mmwrhtml/ss4901a1.htm#tab22.

CDC. 2006. *Update on Multi-State Outbreak of* E. coli *O157:H7 Infections From Fresh Spinach, October 6, 2006*. Accessed 16 November 2012 at: http://www.cdc.gov/ecoli/2006/september/updates/100606.htm.

CDC. 2011a. *Timeline of events multistate outbreak of listeriosis linked to whole cantaloupes from Jensen Farms in Colorado—United States, 2011*. Accessed 21 November 2012 at http://www.cdc.gov/listeria/outbreaks/cantaloupes-jensen-farms/110211/timeline.html.

CDC. 2011b. *Investigation update: multistate outbreak of listeriosis linked to Rocky Ford cantaloupes from Jensen Farms*, 19 September 2011. Accessed 21 November 2010 at http://www.cdc.gov/listeria/outbreaks/cantaloupes-jensen-farms/091911/index.html.

CDC. 2012a. *Multistate outbreak of listeriosis linked to whole cantaloupes from Jensen Farms, Colorado*, 27 August 2012 (final addendum). Accessed 18 November 2012 at: http://www.cdc.gov/listeria/outbreaks/cantaloupes-jensen-farms/082712/index.html.

CDC. 2012b. *CDC Estimates of Foodborne Illness in the United States: Question and Answers*. Accessed 20 September 2012 at http://www.cdc.gov/foodborneburden/questions-and-answers.html.

CNN. 2006. 'E. coli spinach scare increases to 21 states.' *CNN.com* 19 September 2006. Available at: http://www.cnn.com/2006/HEALTH/09/18/tainted.spinach/index.html.

CONTE, A. and FABREGAS, L. 'Many hands touch America's food.' *Pittsburgh Tribune Review* 26 November 2003. Print.

DENTINGER, C.M., BOWER, W.A., NAINAN, O.V., COTTER, S.M., MYERS, G., *et al.* 2001. 'An outbreak of Hepatitis A associated with green onions.' *Journal of Infectious Diseases* 183: 1273–6.

ERIE TIMES-NEWS. 2003. 'Stores, consumers responding cautiously to green onion concerns.' *Erie Times-News* 18 November 2003.

FDA, USDA, USEPA, CDC. 1997. *Food Safety from Farm to Table: A National Food Safety Initiative. Report to the President, May 1997*. Accessed 6 September 2012 at: http://www.cdc.gov/ncidod/foodsafe/report.htm.

FDA. 1998. *Guidance for Industry: Guide to Minimize Microbial Food Safety Hazards for Fresh Fruits and Vegetables*. Available at: http://www.fda.gov/Food/GuidanceRegulation/GuidanceDocumentsRegulatoryInformation/ProducePlantProducts/ucm064574.htm. Accessed 23 July 2013. Last modified 1998.

FDA. 2004. *Produce Safety From Production to Consumption: 2004 Action Plan to Minimize Foodborne Illness Associated with Fresh Produce Consumption*. Available at: http://www.fda.gov/Food/FoodborneIllnessContaminants/BuyStoreServeSafeFood/ucm129487.htm.

FDA. 2005. *Letter to California Firms that Grow, Pack, Process, or Ship Fresh and Fresh-cut Lettuce*. Available at: http://www.fda.gov/Food/GuidanceRegulation/GuidanceDocumentsRegulatoryInformation/ProducePlantProducts/ucm118911.htm. Accessed 23 July 2013. Last modified November 2005.

FDA. 2006a. *Lettuce Safety Initiative*. Accessed 31 August 2012 at: http://www.fda.gov/Food/FoodborneIllnessContaminants/BuyStoreServeSafeFood/ucm115906.htm.

FDA. 2006b. *FDA Announces Findings From Investigation of Foodborne* E. coli *O157:H7 Outbreak in Spinach*. FDA News Release, 29 September, 2006. Accessed 4 September 2012 at: http://www.fda.gov/NewsEvents/Newsroom/PressAnnouncements/2006/ucm108748.htm.

FDA. 2007. *FDA Finalizes Report on 2006 Spinach Outbreak*. FDA News Release, 23 March 2007. Accessed 12 November 2012 at: http://www.fda.gov/NewsEvents/Newsroom/PressAnnouncements/2007/ucm108873.htm.

FDA. 2012a. *Signals and surveillance: the detectors*. FDA Consumer Health Information accessed on 26 November 2012 at: http://www.fda.gov/ForConsumers/ConsumerUpdates/ucm314544.htm.

FDA. 2012b. *Response: stopping the outbreak*. FDA Consumer Health Information accessed on 26 November 2012 at: http://www.fda.gov/ForConsumers/ConsumerUpdates/ucm314851.htm.

KARAS, D. 2012. 'CORE: FDA foodborne illness response network.' *Food Safety* 18(4): 24–7.

KARST, T. 'Industry leaders weigh options for regulations.' *The Packer* 9 July 2004. Print.

MARSHALL, M. and MILLER, M. 2007. *A Letter From United Fresh Produce Association Co-Chairmen Maureen Marshall and Mark Miller*.

MEAD, P., SLUTSKER, L., DIETZ, V., MCCAIG, L.F., BRESEE, J.S., *et al.* 1999. 'Food-related illness and death in the United States'. *Emerg Infect Dis* 5(5): 607–25.

PALMER, S. 2007. 'The *E. coli* Outbreak — Lettuce Learn a Lesson.' *Today's Dietician* 9(1): 28.

RUSSELL, S. 'Spinach *E. coli* linked to cattle/manure on pasture had same strain as bacteria in outbreak.' *San Francisco Chronicle* 13 October 2006. Accessed 26 October 2012 at: http://www.sfgate.com/health/article/Spinach-E-coli-linked-to-cattle-Manure-on-2550111.php.

UNITED FRESH. 2007. *United Fresh Produce Association, Produce Marketing Association Endorse Common Principles for Federal Produce Safety Regulatory Oversight*. 23 May 2007. Accessed on 19 November 2012 at: http://www.unitedfresh.org/news/327/United_Fresh_Produce_Association_Produce_Marketing_Association_Endorse_Common_Principles_for_Federal_Produce_Safety_Regulatory_Oversight.

WHEELER, C., VOGHT, T.M., ARMSTRONG, G.L., VAUGHAN, G., WELTMAN, A., *et al.* 2005. 'An outbreak of hepatitis A associated with green onions.' *N Engl J Med* 353: 890–7.

4

Issues surrounding the European fresh produce trade: a global perspective

M. Uyttendaele, L. Jacxsens and S. Van Boxstael,
Ghent University, Belgium

DOI: 10.1533/9781782420279.1.33

Abstract: This chapter discusses food safety and trade challenges surrounding the global fresh produce supply chain. The chapter begins by highlighting the rise in fresh produce trade but also food safety concerns and problems such as foodborne outbreaks and EU border rejections. The chapter then describes consequences such as increasingly severe public and private standards and their impact on access to the high value EU market by developing countries. The chapter ends with a discussion on best practices to guarantee fresh produce safety and also on other relevant issues, for example the use of sanitizers, global water demand and the impact of climate change.

Key words: fresh produce, food safety, EU, trade, standards.

4.1 Introduction

Global consumption of fresh produce, including fruits and vegetables, has been increasing for several decades. This increasing demand is related to the numerous health benefits associated with eating fresh fruits and vegetables, including improved nutrition and reduced disease risk, and the increase in demand has been accompanied by a coinciding increase in trade. The global trade in fruits and vegetables is shaped by geographical proximity, trading arrangements, and historical and political elements, among other factors. Some regions have a comparative advantage for production of fruits and vegetables because of their optimal climate conditions and water and land availability. Mobility between countries and continents is also a key factor in the fruit and vegetable trade. European trade in commodities such as spices and cultivated tomatoes escalated in the fifteenth and sixteenth centuries, when pioneering explorers began spreading new commodities around the globe.

Several factors explain the emergence and globalization of trade in fresh produce: colonialism provided the means to install the infrastructure (e.g. roads) and thus support bilateral trade, and migration provided linguistic and cultural capital to support extensive trade. Once countries begin trading they install preferential trade relationships and trust in one another and continue to do so. The key element in these trade relationships is trust and the inherent confidence one has built in these partnerships and in attaining continuous food quality and safety. Although international trade has been present throughout the last 500 years, its economic, social and political importance has increased in recent centuries, mainly because of technological improvements that allow for better transport over longer distances and information technologies that enable rapid communication and data managing.

Over the last quarter of a century (1980–2004), the fruit and vegetable market has been one of the fastest growing of all agricultural markets. In particular, tropical fruits (bananas, melons, oranges), apples, tomatoes, chilies and peppers, as well as leafy vegetables, are top commodities according to production volume and trade value. In 2002–04 production of fruit and vegetables per year totaled roughly 1.35 billion tons of which 500 million tons were fruits and 850 million tons were vegetables. The average annual growth in the production of vegetables (4.2% p.a.) was almost double that of fruits (2.2% p.a.) during 1980–2004. The volume of fruit and vegetables traded as fresh products is more than 5% of the total production, with fruit exports heading for the 10% mark, while the trade in vegetables stagnates around 3%. The EU and the USA are among the biggest exporters and importers of fresh fruits and vegetables.

Table 4.1 presents the major imported fruits and vegetables (and the importing countries) to EU. The highest import volumes, with regard to fruits into EU-27, come from Costa Rica, Ecuador, Colombia and South Africa, which, combined, account for roughly half of the EU imports, while the four main third countries importing vegetables to the EU are Morocco, Turkey, Israel and Egypt, with a total share of more than 60% of the imported volume. The most imported vegetables into EU-27 include tomatoes, onions and peppers.

4.2 Challenges involved in the fresh produce trade

International agro-food trade flows between countries form a complex and dynamic web of interactions. It is the increasing flow of international trade that is usually meant by the term 'globalization'. Globalization provides new opportunities for economic development; however, the increase in produce production and consumption has been accompanied by a rise in the number of produce-associated foodborne disease outbreaks worldwide. For example, in the USA the number of outbreaks linked to fresh produce contamination increased almost 10-fold in the later part of the twentieth century, from 0.7% in the mid-1970s to 6% in the mid-1990s, even after adjusting for improved surveillance and reporting. In the EU the share of all foodborne outbreaks for vegetables and fruits increased from 4.4% in 2009 to 10% in 2010.

Table 4.1 Top 10 vegetables and fruits imported into the EU by third countries* in 2010

Rank	Product	Total imported (million tonnes (%))	Rank	Importing third country	Total imported by third country (million tonnes (%))
	Vegetables				
1	Tomatoes	0.48 (28.5)	1	Morocco	0.54 (32.3)
2	Onions	0.31 (18.3)	2	Turkey	0.26 (15.6)
3	Sweet peppers	0.23 (13.6)	3	Israel	0.17 (10.1)
4	Beans	0.19 (11.2)	4	Egypt	0.11 (6.9)
5	Carrots and turnips	0.08 (4.9)	5	New Zealand	0.09 (5.4)
6	Garlic	0.07 (4.3)	6	China	0.06 (3.7)
7	Asparagus	0.04 (2.2)	7	Kenya	0.06 (3.3)
8	Cucumber and gherkins	0.03 (2)	8	Chile	0.05 (3.1)
9	Peas	0.02 (1.1)	9	Argentina	0.04 (2.6)
10	Leeks	0.01 (0.8)	10	Peru	0.04 (2.2)
	Other	0.22 (13.1)		Other	0.25 (14.9)
	Total	**1.67 (100)**		**Total**	**1.67 (100)**
	Fruits				
1	Bananas, plantains	4.54 (43)	1	Costa Rica	1.58 (15)
2	Oranges	0.92 (8.7)	2	Ecuador	1.29 (12.2)
3	Pineapples	0.90 (8.5)	3	Colombia	1.21 (11.4)
4	Apples	0.59 (5.6)	4	South Africa	1.09 (10.3)
5	Grapes	0.56 (5.3)	5	Chile	0.60 (5.7)
6	Lemons	0.45 (4.2)	6	Turkey	0.57 (5.4)
7	Mandarins	0.39 (3.7)	7	Argentina	0.56 (5.3)
8	Grapefruit	0.36 (3.4)	8	Brazil	0.49 (4.7)
9	Melons	0.32 (3.1)	9	New Zealand	0.29 (2.8)
10	Pears and quinces	0.28 (2.6)	10	Cote d'Ivoire	0.29 (2.8)
	Other	1.26 (11.9)		Other	2.59 (24.5)
	Total	**10.57 (100)**		**Total**	**10.57 (100)**

Source: Freshfel (2011).

*Third countries are defined as countries who are not members of the EU-27.

Table 4.2 lists a number foodborne outbreaks in the EU and USA related to fresh produce products such as lettuce, onions, raspberries, watermelon and sprouted seeds. Bacterial pathogens (e.g. *Salmonella*, *Shigella*), viruses (e.g. Hepatis A, Norovirus) and also parasities (e.g. *Cyclospora*) were identified as causative agents of fresh produce related outbreaks. Some of these caused a relatively low number of disease cases, such as the *Cyclospora cayetensis* outbreak in 2009, which was a result of consumption of contamined sugar snaps imported

Table 4.2 Examples of foodborne outbreaks in the EU and USA related to imported fruits and vegetables

Causative agent	Year	Country	Number of human cases	Implicated food	Country of origin	Reference
Bacteria						
Salmonella Braenderup	2004	UK	>40	Iceberg lettuce	Import from Spain	Gajraj *et al.* (2012)
Salmonella Thompson	2004	Norway (and probably larger EU outbreak)	21	Rucola lettuce	Import from Italy	Nygard *et al.* (2008)
Salmonella Senftenberg	2007	UK	30	Fresh basil	Import from Israel	Pezzoli *et al.* (2008)
Salmonella Saintpaul	2008	USA	1500	Jalapeno and Serrano peppers	Import from Mexico	Behravesh *et al.* (2011)
Shigella sonnei	2009	Sweden, Norway	67	Sugar snap peas	Import from Kenya	Heier *et al.* (2009)
E. coli O104:H4	2011	Germany, France	3842	Sprouted seeds	Domestic production using seeds imported from Egypt	Buchholz *et al.* (2011); Altmann *et al.* (2011); Gault *et al.* (2011)
Shigella sonnei	2011	Norway	46	Fresh basil used in pesto	Import (country not specified)	Guzman-Herrador *et al.* (2011)
Yersinia enterocolitica O:9	2011	Norway	21	Bagged salad mix	Import from Italy	MacDonald *et al.* (2012)
Salmonella Newport	2011–2012	UK, Ireland, Germany	54	Ready-to-eat sliced watermelon	Import from Brazil	HPA (2012)
Viruses						
Hepatitis A virus	2003	USA	601	Green onions	Import from Mexico	Wheeler *et al.* (2005)
Norovirus	2005	Denmark	> 1000	Frozen raspberries	Import from Poland	Falkenhorst *et al.* (2005)
Norovirus and enterotoxi-genic *E. coli*	2010	Denmark and Norway	260	Lettuce used in sandwiches	Import from France	Ethelberg *et al.* (2010)
Norovirus	2012	Germany	10 950	Frozen strawberries	Import from China	RKI (2012)
Parasites						
Cyclospora cayetanensis	1996	Canada and US	> 1400	Raspberries	Import from Guatemala	Herwaldt *et al.* (1997)
Cyclospora cayetanensis	2009	Sweden	18	Sugar snaps	Import from Guatemala	Insulander *et al.* (2010)

from Guatemala (18 cases), while others caused a huge number, for example more than 10 000 children were reported to be ill during a norovirus outbreak in Germany as a result of eating school canteen desserts containing contaminated frozen strawberries that had been imported from China.

As well as microbiological hazards, chemical hazards such as pesticide residues, and, to a lesser extent, mycotoxins, also raise concerns in fresh produce. This can be seen in data from the EU rapid alert system for food and feed (RASFF). The RASFF system collects and disperses information on food safety problems (occurring on the EU market or with products entering EU) to the European Commission and its member states rapidly, so that effective measures such as rejections of products at borders can be undertaken. In order to evaluate whether the import of fruit and vegetable products are linked to food safety problems, the types of notifications for the category 'fruits and vegetables' within the European RASFF system were analyzed (Table 4.3). There are notifications linked to imported products that are controlled at the border inspection points (border

Table 4.3 Number of EU RASFF notifications during the period 2008–2011 allocated to fruits and vegetables, herbs and spices and nuts and nut products with indication of the share (%) of these alerts related to border controls

Hazard category	Fruits and vegetables	Herbs and spices	Nuts and nut products
Pesticide residues	838 (50.7)[a]	88 (85.2)[b]	10 (60)[c]
Mycotoxins	337 (88.1)	246 (76.8)	1992 (89.5)
Hygiene hazard/insufficient quality (e.g. spoilage)	222 (73.9)	40 (85)	256 (72.7)
Bacterial pathogens	160 (59.4)	164 (31.1)	82 (47.6)
Additives	146 (42.5)	3 (33.3)	13 (38.5)
Chemical hazard	84 (20.2)	4 (0)	13 (46.2)
Physical hazards	76 (5.3)	3 (66.7)	0 (0)
Heavy metals	32 (37.5)	2 (0)	1 (100)
Viruses	21 (4.8)	0 (0)	0 (0)
Migrants from food contact materials	17 (41.2)	1 (0)	0 (0)
Defective packaging hazard	15 (40)	0 (0)	0 (0)
Labeling hazard	14 (85.7)	0 (0)	15 (40)
Radioactivity	7 (85.7)	0 (0)	0 (0)
Genetically modified food	5 (80)	0 (0)	69 (4.3)
Allergens	4 (0)	4 (0)	0 (0)
Unauthorized irradiation	1 (100)	10 (60)	0 (0)
Process contaminants	0 (0)	0 (0)	0 (0)
Parasites	0 (0)	0 (0)	0 (0)
Antimicrobial resistance	0 (0)	0 (0)	0 (0)
Transport or storage hazard	0 (0)	0 (0)	3 (100)
Unauthorized color	0 (0)	78 (47.4)	7 (57.1)
Other	28 (53.6)	6 (33.3)	38 (92.1)
Total number of notifications	2007 (56.2)	649 (61.2)	2507 (82.8)

[a,b,c] Share of the notifications due to border controls.

controls) before entering the EU market, and other notifications are linked to controls on the EU market and reported outbreaks.

Table 4.3 shows that high shares of the notifications for fruits and vegetables, herbs and spices and nuts and nut products, respectively 56.2%, 61.2% and 82.8%, are related to border controls. For fruits and vegetables, the most reported notifications concern pesticide residues.

Mycotoxins are the most important notified hazard for nuts and nut products, and they represent 80% of all notifications for these products. Of these, 89.5% are border notifications. This high share can be expected, as production of nuts and nut products is very limited in the EU. As a result, they are largely imported from third countries. Mycotoxins are the second most important item for fruits and vegetables. However, one should take into account that the fruits and vegetables category in the RASFF system also contains dried products, and the mycotoxin notifications are entirely linked to these products (such as dried figs, raisins and apricots) and not to fresh vegetables, fruits and herbs.

Bacterial pathogens hold a limited share of the notifications for fruits and vegetables (8%), and a more significant one for herbs and spices (25.3%). Viruses are only notified for fruits and vegetables (4.8%) and not within the category of herbs and spices; however, despite the limited number of notifications, bacterial pathogens and viruses represent an important food safety concern in the fresh produce supply chain. This was also highlighted during a 2011 workshop on food safety of fresh produce held in the frame of the EU FP-7 project Veg-i-Trade. It asked experts from the EU fresh produce supply chain to rank fresh produce food safety issues and contextual factors that impact on food safety according to importance from their stakeholder point of view. Bacterial pathogens, foodborne viruses and pesticide residues were indicated as the most important issues, while globalization was identified as the most important contextual factor. One of the plausible explanations as to why experts classified bacterial pathogens and viruses as among the most important concerns is that in the event of a foodborne outbreak the consequences are often severe, from both a public health point of view (high number of cases and high severity of disease) and an economic point of view (loss of sales and trust).

4.3 Regulatory and economic aspects

4.3.1 Regulatory aspects

Emerging and developing countries face the stringent European legislative requirements on safe food production, which restricts opportunities for export. In the EU, fresh produce needs to comply with several safety and quality standards, and the basis for EU food safety public standards is laid down in the General Food Law or Regulation (EC) 178/2002. The main objective of this regulation[1] is to

---

[1] The European food safety law was strongly reformed to restore consumer trust after the BSE crisis in the mid-1990s. A white paper on food safety in 2000 served as the cornerstone for defining the General Food Law.

Table 4.4 Milestones of EU food law applying to food products such as fresh produce, herbs and nuts

Year	Content	Regulation (EC)
2002	General Food Law	178/2002
2004	Hygiene requirements	852/2004
2004	Official controls	882/2004
2005	Microbial hazards	2073/2005
2005	Pesticide residues	396/2005
2006	Contaminants	1881/2006
2009	Increased control for high-risk products of non-animal origin at borders	669/2009

secure a high level of protection for public health and consumer interests with regard to food products. The requirement of traceability ensures that products can be withdrawn from the market in the event of a problem. Concerning import/export of food products, the General Food Law specifies that products entering/leaving the EU market must comply with the relevant requirements of food law or equivalent requirements. After the General Food Law was established in 2002, several packages of new EU legislation followed. The relevant legislations for imported and exported fresh produce, nuts and herbs are summarized in Table 4.4.

The specifications of official controls of imported food products are defined by Regulation (EC) 882/2004, which stipulates that certain high-risk imported food products of non-animal origin (from a human health point of view) need to be inspected and analyzed at higher frequencies. A list containing these high-risk non-animal products is a part of (EC) Regulation 669/2009. This list is drawn up based on several types of information, such as notifications from the EU RASFF alert system (as also mentioned above), reports from the EC inspection service Food and Veterinary Office (FVO), scientific assessments from the European Food Safety Authority (EFSA) and exchange of information between EU member states.

Table 4.5 contains examples of fresh produce, herbs and nuts originating from third countries that are subject to a higher frequency of controls at the EU point of entrance. Although the most cited hazards refer to mycotoxins and pesticide residues, the microbial hazard *Salmonella* is also included. This list is revised every one to six months based on the latest available information.

Alongside governments' imposition of compliance to legal demands, private stakeholders in the fresh produce chain (retailers, branch organizations, multinational companies) increasingly require actors in the fresh (processed) produce chain to produce, process and trade according to their food safety requirements. They do this by imposing voluntary collective standards such as GLOBALG.A.P., British Retail Consortium (BRC) global standard for food safety and ISO 22000, which need to be certified by third parties.

GLOBALG.A.P. protocols have become the dominant standard for primary production in the fresh produce chain and are applied by nearly all major European retail chains as a prerequisite for procurement of produce. The number of

Table 4.5 Examples of vegetables, fruits, herbs and nuts imported from third countries with increased inspection frequency

Product	Country of origin	Hazard	Frequency of checks (%)
Sweet pepper, tomatoes	Turkey	Pesticide residues	10
Watermelon	Nigeria	Pesticide residues	50
Oranges, peaches, pomegranates, strawberries	Egypt	Pesticide residues	10
Pomelo	China	Pesticide residues	10
Coriander, basil, mint	Thailand	Salmonella	10
Dried grapes	Uzbekistan	Mycotoxins (Ochratoxin A)	50
Groundnuts, peanut butter	Brazil	Mycotoxins (Aflatoxin)	10
Groundnuts, peanut butter	South Africa	Mycotoxins (Aflatoxin)	10
Groundnuts, peanut butter	Ghana	Mycotoxins (Aflatoxin)	50

Source: EC (2009).

producers certified by GLOBALG.A.P. increased strongly from 18 000 in 2004 to 112 600 in 2011, and GLOBALG.A.P.. also actively promotes benchmarked national standards that are regarded as equivalent. Some individual companies, such as McDonald's, Dole and Chiquita, also impose their own private standards. Private standards are often perceived to be more stringent than the standards imposed by governments (public standards such as the EU standards, see Table 4.4), and the increasing demands by these private standards makes it increasingly difficult for farmers to sell their products to high-value markets.

4.3.2 Economic aspects

In order for farmers from third countries to be able to export their products to high-value markets such as the EU market they need to comply with the EU regulatory requirements; however, they also need to comply with voluntary private standards. These standards have evolved from regulatory requirements and consumer concerns, and appear as a means of competitive positioning in markets for high-value agricultural products. It is not compulsory for exporting farmers to comply with voluntary private standards; however, as the majority of buyers, in particular retailers, require compliance and certification before procurement they are virtually compulsory. As a result, private rather than public standards are becoming the predominant drivers of agri-food systems.

The 'bottom line' of EU regulation is that countries exporting to EU should meet the same requirements for producing food to meet the safety standards within the EU. This might be quite challenging for some of the exporting countries. As mentioned above, compliance testing may result in at-border rejections of products imported from third countries. Developing or transition countries accounted for more than 88% of all EU food and feed rejections between 2002 and 2008, and as roughly 70% of the imports of agricultural products originate from developing countries, it can be expected that the rejections are mainly related to products from developing countries. UNIDO calculated the losses of several food products by third countries as a result of EU border rejections, and the total value of the rejections between 2004 and 2008 for fruits and vegetables was calculated to be equal to 9 million dollars, or only 0.01% of the import value. Turkey accounts for 5.6 million dollars of these losses, followed by China (0.9 million dollars), Argentina (0.3 million dollars) and Iran (0.3 million dollars). As a result, it is not the lost value as a result of rejections of fruits and vegetables that represents the major cost for importing countries, but the difficulties they face in selling their products to high-value markets. According to UNIDO, the rejection data should be considered as indicators reflecting the challenges countries face to comply with EU food safety requirements and thus to obtain market access.

The economic consequences for complying with EU legislation and obtaining certifications for private standards by the exporting third countries are high. Several studies have demonstrated that investments in infrastructure, training and capacity building of workers and implementation of food safety management systems (FSMS) are demanding economical efforts from exporting countries (both by private companies and government, e.g. in airports or ports to ship the products). Additionally, any incidence where safety of fresh produce is associated with a severe impact on public health can have immediate direct economic costs as a result of a decrease in sales, import ban, food recalls, etc. However, indirect economic costs after an outbreak (such as loss of trust in the product or in the government) can also occur. This can be observed in the large enterohemorrhagic *E. coli* (EHEC) outbreak in May–July 2011 in Germany, which was associated with consumption of imported sprouted fenugreek seeds (from Egypt).

Until then, reported outbreaks relating to fresh produce in EU had not received such widespread media coverage or impacted so largely on the trust EU consumers have in fresh produce. The EHEC outbreak was mainly situated in Germany but affected citizens of other European countries (and some US citizens) who travelled to Germany. In total, 15 European countries in addition to Germany reported cases of EHEC infection (see Chapter 25 by Petra Luber). After the outbreak the entire supply chain was affected, including overcapacity in production and reduced retail sales. The costs for the fresh produce supply chain were estimated to be more than 1 billion Euros. The European Commission agreed on a rescue package of 230 million Euros, which covered products withdrawn from the market (but well below the production cost), and allocated an additional 15 million Euros in promotional funds for advertisement in the following three years to rebuild trust in safety of fresh produce for the EU consumer. In the meantime, increased monitoring for biological hazards in the fresh produce chain (mainly focused on *Salmonella* and EHEC) has been set up in the supply chain and by competent authorities in the EU, and food safety criteria related to EHEC in seeds for sprouting have been developed, as well as stricter rules for traceability of seeds for sprouting (EC Regulation 2073/2005 amendment (EC) 209/2013 of 11 March 2013). In fact, the EHEC crisis resulted in changes in governmental regulations affecting both domestic and/or imported food products.

Regarding private standards such as GLOBALG.A.P., many are of the opinion that developing countries, because of high compliance costs to meet food safety and other requirements, are increasingly excluded from high-value markets such as the European market. Others argue that these private standards can act as a catalyst for upgrading necessary processes and/or for enhanced competitive positioning in international markets. A study of Sub-Saharan Africa fresh produce exporters from 10 countries suggested that technical and/or financial assistance, as well as being in an established export country, are key discriminating factors for being GLOBALG.A.P. certified. They also showed that firms with certification have considerably higher export revenues, which suggests an appreciable return on the required investments; however, the economic impacts these types of standards have on developing countries remains an issue of debate.

4.4 Best practice in agriculture

The application of good agricultural practices (GAP) is widely recognized as the most important measure in assuring the safety of fresh produce, followed by the application of good hygienic practices (GHP) and the certification of food safety management systems (FSMS). GHP are distinguishable from GAP as they are applicable to the whole farm-to-fork continuum and not just primary production steps at the farms (i.e. cleaning and disinfection, pest control, water quality, etc.). Inspections are conducted by national competent food safety authorities and involve checking compliance to the legal demands. Certification of FSMS is an additional step in the verification of the application of GAP and/or GHP. The verification and certification of food safety systems by an outsider, a third party, is a driver to improve FSMS. In addition, the measured compliance with voluntary quality assurance standards such as ISO22000, BRC or GLOBALG.A.P. is, from a commercial point of view, very important for gaining buyer (and consumer) trust, where trust is a key factor in ensuring sustainable trade relationships, as discussed above.

However, audits and inspections are not enough. Whereas third party audits are a valuable tool to ensure food safety (although often based on extensive documentation), audit and inspection activities also have their limitations and represent only a snapshot in time. There is also a need for the in-house capacity to meaningfully assess the results of audits and inspections. The responsibility for the verification of the implementation of 'best practices' lies with the food business operator, whether a small- or large-scale farmer, processer, trader or retailer. Ideally, third party audits and regulatory compliance inspections should be supplemented with regular self-assessment studies and microbiological monitoring of products and environment. Moreover, as lessons learned from outbreak events show, it is necessary to develop a strong food safety culture to ensure adherence and actual implementation of best practices.

Within the EU FP7 Veg-i-Trade project (www.veg-i-trade.org) a self-assessment tool was developed and validated for fresh produce companies to enable them to perform an internal audit of their current implemented FSMS. Three tools were constructed, allowing self-assessment of businesses at primary production, processing and trade level. These self-assessment tools consist of 64 to 69 indicators and encompass a systematic analysis of core control and assurance activities in the FSMS of the company (to prevent and reduce microbiological, mycotoxin and pesticide residue contamination). Moreover, the self-assessment tool includes an assessment of the contextual situation wherein the FSMS operates in view of the decision-making activities. Each indicator uses grids with supporting information for self-assigning implemented FSMS to situation 1 (not present, not applicable), 2 (basic level), 3 (generic, average level) or 4 (advanced, tailored level). As such, a 'bird's-eye view' of the performance level of the current FSMS (the risk level of the context wherein the actor operates) and an indication of the food safety output is obtained, and can be presented in clear spider web diagrams. Using this self-assessment, farmers, processors or traders can also identify bottlenecks and weak points in their FSMS. A schematic structure of this approach for diagnosing performance of FSMS is presented in Fig. 4.1.

Furthermore, alongside producers, processors, traders and retailers, consumers also have a responsibility in their adherence to 'best practices'. Consumers can select and verify the overall quality of the fresh produce before buying, but they are also expected to keep the produce at low temperatures to avoid multiplication of microorganisms, to respect hygiene and good kitchen practices to avoid cross-contamination, and (sometimes) wash or peel the produce to reduce the number of microorganisms, although a residual risk may remain for fresh produce contaminated earlier in the food chain. As mentioned above, the remaining residual risk is to be minimized by the fresh produce supply chain actors (from farm to retail/catering). As a result, within this framework the adherence to 'best practices' and 'taking all reasonable precautions' is essential.

4.5 Troubleshooting approaches

Countries producing fresh produce work according to variable practices and product quality/safety standards and legislation. As a result, the product quality/safety parameters at the farm vary per country. Additionally, the logistic systems used to transport the products from the farm to the forks of the European consumers vary (e.g. air, truck or boat transport), thereby impacting the product quality/safety parameters. Typically, alongside biological variations, food quality and safety is determined by time and environmental conditions (such as temperature, humidity and the presence of contaminants), which might vary depending on the region or climate. These various contextual factors may present a challenge in ensuring the safety of fresh produce in the global supply chain.

In many countries (both EU and global) one of the major challenges lies in developing a policy framework that helps small and less developed businesses overcome constraints in applying 'best practices' programs. This is a prerequisite for supplying safe food as well as keeping competitiveness and gaining access to

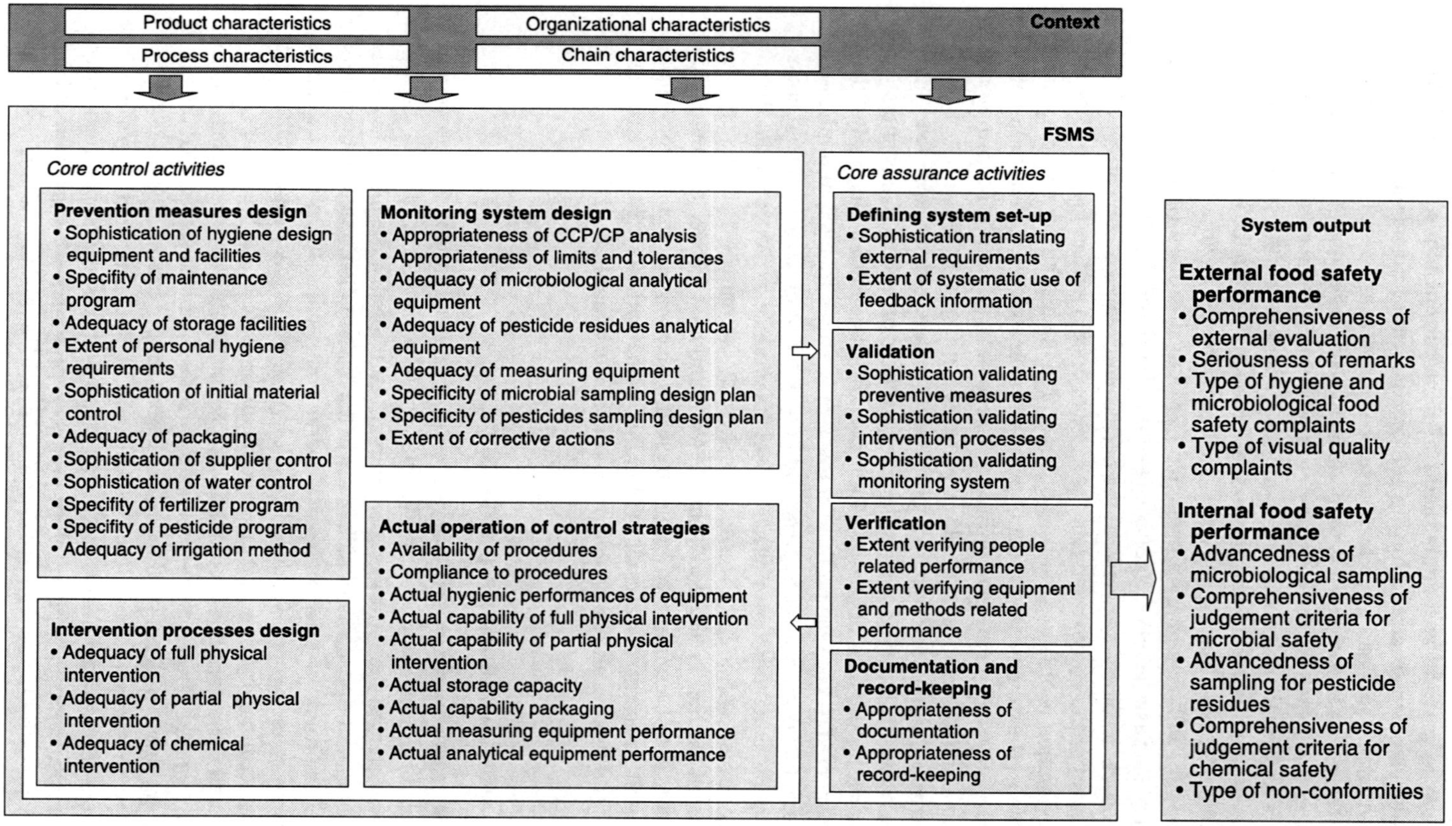

Fig. 4.1 Structure of Food Safety Management System (FSMS) consisting of core control and core assurance activities, in relation to the context of a company along the fresh produce chain, characterized by product and process requirements and characteristics of the organization and chain demands. An implemented FSMS is finally resulting in a system output defined by external and internal food safety performance indicators (after Kirezieva *et al.*, 2013).

profitable global food markets. Furthermore, cultural factors might affect production practices and usual consumption patterns of fresh produce, which may also reflect on the adequacy of understanding and implementation of demands on 'best practices'.

In this instance, training and capacity building are important. It is worth mentioning that in 2012 WHO published the manual *Five Keys to growing safer fruits and vegetables: promoting health by decreasing microbial contamination.* The manual is designed to support food safety education of rural workers who grow fresh fruits and vegetables for themselves, their families and for sale in local markets. It describes key practices in reducing microbial contamination and raises awareness of the links between the health of humans, animals and ecosystems and how failures in good hygienic practices in one sector can affect the others. The Five Keys to growing safer food are: 1 – Practice good personal hygiene; 2 – Protect fields from animal fecal contamination; 3 – Use treated fecal waste; 4 – Evaluate and manage risks from irrigation water; 5 – Keep harvest equipment and storage equipment clean.

In some countries food safety issues remain among the critical constraints in international trade of fresh produce with the EU. Increased communication between EU member states, as well as between EU and non-EU trade partners, will result in better understanding and two-way knowledge transfer in order to set up appropriate consensus-agreed demands to provide guarantees of consumer safety and avoid trade disruptions (which negatively affect productivity). With this goal in mind, the European Commission has set up the 'Better Training for Safer Food' program, covering food and feed law, animal health and welfare and plant health rules. Via this program, EU member states and candidate countries involved in official controls are trained, but this training is also organized specifically for third (particularly developing) country participants. The aim is to keep participants up-to-date with EU law and also to ensure more harmonized and efficient controls. Efficient controls are an essential factor in maintaining high levels of consumer protection, animal health and plant health. Additionally, in training participants from third countries the aim is to ease access for their products to the EU market.

In a global respect, the exchange of information (including the sharing of concerns and experiences in food safety issues) in an open and accessible manner between scientists, authorities and all actors (from farm to fork) in the fresh produce supply chain is important. Transparency, competency and efficiency of the precautionary systems and 'best practices' by suppliers to their purchasers (and consumers) are important for building and maintaining trust. Interaction and risk communication between stakeholders contribute to a better understanding of each other's concerns, constraints and interests in dealing with the food safety of the increasingly complex and globalized fresh produce supply chain.

Variability in supply chains and consumption patterns of fresh produce is a challenge for risk assessments. As much as resources allow, this variability should be factored into risk models. Data availability will play an important role and will differ across Europe and also on a global level. Grouping countries and selecting indicator food products should be considered, and, as such, generic risk assessment models can be developed to provide risk managers with an order of magnitude estimate for risk-ranking purposes. Identification and ranking of specific food/pathogen combinations via a risk-ranking tool was elaborated by the US-FDA and is aimed particularly at risk-ranking combinations of fresh produce commodities and specific pathogens. The FDA risk-ranking tool is based on an assessment of nine criteria (e.g. reported foodborne outbreaks in the USA), where each criterion is given a score from 1 to 4. At the end the scores of all criteria are summed to provide a total risk score for each food/pathogen combination. More recently, EFSA in the EU has adopted a similar approach for risk ranking. The latter risk-ranking tool was based on the strength of associations between food/pathogen combinations reported in EU Zoonoses monitoring of foodborne outbreaks (2007–11), together with estimates of the burden of illness, burden of disease, dose-response, consumption, prevalence of contamination and growth potential during shelf life. The top-ranking food/pathogen combination was *Salmonella* and leafy greens eaten raw followed by (in equal rank) *Salmonella* and bulb and stem vegetables, *Salmonella* and tomatoes, *Salmonella* and melons, and pathogenic *E. coli* and fresh pods, legumes or grains. The risk-ranking method used here undoubtedly underestimates the importance of diseases that appear to be of a more sporadic nature (such as those occurring from *Listeria*, *Campylobacter* and parasites) or of more general risks, such as antimicrobial resistance. Future fluctuations in the reported outbreaks are likely to impact on the current rankings, and, indeed, a drawback of this approach is that outbreak data are reliant on reporting that is incomplete, can vary between countries, may be greatly influenced by rare events occurring during the monitoring period, or have bias as a result of the preferential investigation of types of food perceived as higher risk.

4.6 Conclusion and future trends

Food safety is a major concern in the fresh produce supply chain and is one to which many uncertainties are linked, in particular regarding the microbial safety, as most fresh produce products in the EU are consumed without microbial killing steps such as heating. As such, the current focus remains on microbial food safety risk reduction and not risk elimination. Minimal processing of produce consists of a series of steps including harvesting, cold storage, trimming, shredding, washing/rinsing, draining, packaging, cold storage and distribution. Still a point of debate (in particular in the EU), it relates to the use of disinfectants for sanitation of postharvest processing water in the fresh-cut industry. The washing or rinsing step of (fresh-cut) produce is done to remove dirt, foreign materials, tissue fluids from cut surfaces and microorganisms. Chlorine is the most-used water disinfectant in general, and also specifically in the fresh-cut produce industry for wash, spray or flume waters, because of the low cost, the reliable availability and the high effectiveness against suspended vegetative bacteria and some enteric viruses, and because of the minimal impact on the nutritional and sensorial quality of the produce. Pros and cons of applying sanitizers to reduce microbial contamination of produce or processing water have been documented, but in the EU there is still debate about their use and each EU member state needs to approve their application. The use of chlorine in fresh-cut produce washing is prohibited altogether in some European countries, such as Germany, Switzerland, the Netherlands, Denmark and Belgium. However, in other countries in Europe, such as France, Spain, the UK, in the US and in most countries around the world, chlorine or another sanitizer is added to water in the fresh produce production and processing in order to reduce the risk of cross-contamination during the washing procedure. Nowadays there is a lot of pressure on proper water management and water (re)-use in a fresh produce production and processing industry. The global demand for water has tripled since the 1950s, and the supply of fresh water has been declining. This, together with the acknowledgement of water as a risk factor for introduction and spreading of contaminations on produce, both in primary production and processing, increases the need for clarity on which, if any, sanitizers might be used during fresh produce processing. Many fresh-cut processors in the EU are still bound by the lack of clarity in their national legislation on the re-use of water and application of sanitizers or the lack of microbial guidelines which have an impact on the intra-EU trade. In particular, the use of sanitizers in non-EU countries may potentially lead to unacceptable or unauthorized residues of chemical substance on the imported produce, which may lead to border rejections and trade dispute among countries.

An unresolved issue surrounding fresh produce is major uncertainty about the food safety implications under a changing climate. Some of the potential impacts of climate change on food safety might include: microbial evolution as a result of abiotic stress; new emerging pathogens; decreasing water availability influencing the quality and safety of fresh vegetables; and shifts of crop production areas and practices with impact on ecosystems. There is also a strong link expected between heavy rainfall (floods) and fecal pollution of water. Climate change as a driver will also have different effects on the various types of plant pests; it may affect pest development rates or host plant susceptibility to pests and may also increase the misuse or abuse of pesticide usage. This could contribute to environmental contamination and may lead to increased pest residues in fresh produce crops. As the fresh produce sector is highly variable in the characteristics of the individual commodities, their local production systems, such as variable climatic conditions, access to water resources for irrigation and postharvest handling, the safety of fresh produce is also very vulnerable to the effects of climate change.

There is a need for multidisciplinary research in order to tackle these major challenges. There is an increasing awareness that, in order to assure safety in the fresh produce chain and build a sustainable global fresh produce supply chain from a socio-economic point, as well as public health and well-being, part of the solution lies in international collaboration. Harmonization and understanding of legal demands and voluntary quality assurance standards to be implemented in food safety management systems, alongside the fresh produce supply chain in different regions, must be accompanied with respect and understanding of local practices, exchange of knowledge and building capacity to manage the understanding, acceptance and implementation (on a daily basis) of 'best practices'.

4.7 Questions for discussion

1. Why is globalization an important factor for the safety of fresh produce?
2. What is the impact of EU border rejections on third countries?
3. What are the main food safety challenges for fresh produce?
4. How can trade disputes between importing and exporting countries be resolved for fresh produce?
5. How can food safety management systems play a role in improving the safety of fresh produce?
6. Which key best practices can be identified along the fresh produce supply chain (farm, processing, trade and consumer level)?
7. What is the potential impact of climate change on food safety of fresh produce?

4.8 Acknowledgments

The information in the present chapter is a reflection of the research conducted for the Veg-i-Trade project 'Impact of Climate Change and Globalization on Safety of Fresh Produce – Governing a Supply Chain of Uncompromised Food Sovereignty' (www.veg-i-trade.org). Veg-i-Trade is funded under the Seventh Framework Program of the European Commission (Grant Agreement n. 244994). The authors would like to thank all Veg-i-Trade Consortium partners for interesting discussions and their contribution to shaping the way forward for the safety of fresh produce in a changing world.

4.9 Sources of further information and advice

ANDERSON, M., JAYKUS, L. A., BEAULIEU, S. and DENNIS, S. (2011). Pathogen-produce pair attribution risk ranking tool to prioritize fresh produce commodity and pathogen combinations for further evaluation (P(3)ARRT). *Food Control*, 22, 1865–72.

BEUCHAT, L. R. and RYU, J. H. (1997). Produce handling and processing practices. *Emerging Infectious Diseases*, 3, 459–65.

BRANDL, M.T. (2006). Fitness of Human Enteric Pathogens on Plants and Implications for Food Safety. *Annual Review of Phytopathology*, 44, 367–92.

BUZBY, J. (2001). Effects of Food-Safety perceptions on Food Demand and Global Trade. In: *Changing Structure of Global Food Consumption and Trade*. (Ed. Anita Regmi). Available at http://www.ers.usda.gov/publications/wrs-international-agriculture-and-trade-outlook/wrs01–1.aspx.

D'SOUZA, R. M., BECKER, N. G., HALL, G., and MOODIE, K. B. A. (2004). Does ambient temperature affect foodborne disease? *Epidemiology*, 15, 86–92.

EC. (2007). European Commission. Agricultural commodity markets past developments, fruits and vegetables: an analysis of consumption, production and trade based on statistics from the Food and Agriculture Organization (FAO). Available at http://ec.europa.eu/agriculture/analysis/tradepol/worldmarkets/fruitveg/072007_en.pdf.

EC. (2009). European Commission. Regulation (EC) No 669/2009 of 24 July 2009 implementing Regulation (EC) No 882/2004 of the European Parliament and of the Council as regards the increased level of official controls on imports of certain feed and food of non-animal origin and amending Decision 2006/504/EC. *Official Journal of the European Union*, L194, 11–20.

EFSA. (2013). EFSA Panel on Biological Hazards (BIOHAZ) Panel. Scientific Opinion on the risk posed by pathogens in food of non-animal origin. *EFSA Journal*, 11(1), 3025.

ERCSEY-RAVASZ, M., TOROCZKAI, Z., LAKNER, Z., and BARANYI, J. (2012). Complexity of the International Agro-Food Trade Network and its impact on food safety. *PLoS One*, 7, 1–7.

EU FP7 VEG-I-TRADE. (2013). Second project report. Impact of Climate Impact of Climate Change and Globalisation on Safety of Fresh Produce, Governing a Supply Chain of Uncompromised Food Sovereignty. Available at http://www.Veg-i-Trade.org.

FAO. (2008). Climate change: Implications for food safety. Available at http://www.fao.org/food/food-safety-quality/a-z-index/climate-change1/en/.

FAO/WHO. (2008). Microbiological hazards in fresh leafy vegetables and herbs: meeting report. Microbial Risk Assessment series No14. Rome. 151 pp. Available at ftp://ftp.fao.org/docrep/fao/011/i0452e/i0452e00.pdf.

FRESHFEL. (2012). European fresh produce association. Report on Freshfel Fruit and Vegetable Production, Trade, Supply & Consumption Monitor in the EU. Available at http://www.freshfel.org/asp/what_we_do/consumption_monitor.asp.

GIL, M., SELMA, M., SUSLOW, T., JACXSENS, L., UYTTENDAELE, M., and ALLENDE, A. (2013). Pre- and Post-harvest Preventive Measures and Intervention Strategies to Control Microbial Food Safety Hazards of Fresh Leafy Vegetables. *Critical Reviews in Food Science and Nutrition* (accepted for publication).

GLOBALG.A.P. (2012). Annual report 2012. Available at http://www.globalgap.org/export/sites/default/.content/.galleries/documents/130124_AR12_web_en.pdf.

HENSON, S., MASAKURE, O. and CRANFIELD, J. (2011). Do fresh produce exporters in sub-saharan exporters benefit from GLOBALG.A.P. certification? *World Development*, 39, 375–86.

ILIC, S., RAIJIC, A., BRITTON, C., GRASSO, E., WILKINS, W., *et al.* (2012). Review. A scoping study characterizing prevalence, risk factor and intervention research published between 1990 and 2010, for microbial hazards in leafy green vegetables. *Food Control*, 23, 7–19.

JACXSENS, L., LUNING, P. A., VAN DER VORST, J. G. A. J., DEVLIEGHERE, F., LEEMANS, R., and UYTTENDAELE, M. (2010). Simulation modeling and risk assessment as tools to identify the impact of climate change on microbiological food safety – the case study of fresh produce supply chain. *Food Research International*, 43, 1925–35.

KARIPIDIS, P., ATHANASSIADIS, K., AGGELOPOULOS, S., and GIOMPLIAKIS, E. (2009). Factors affecting the adoption of quality assurance systems in small food enterprises. *Food Control*, 20(2), 93–8.

LUNING, P. A., MARCELIS, W. J., VAN BOEKEL, M. A. J. S., ROVIRA, J., UYTTENDAELE, M., and JACXSENS, L. (2011). Assessment of context factors on riskiness within the Food Safety Management System-Diagnostic Instrument (FSMS-DI). *Trends in Food Science and Technology*, 22, S67–S79.

OLAIMAT, A. N. and HOLEY, R. A. (2012). Factors influencing the microbial safety of fresh produce: a review. *Food Microbiology*, 32, 1–19.

POWELL, D. A., ERDOZAIN, S., DODD, C., COSTA, R., MORLEY, K., and CHAPMAN, B. J. (2013). Audits and inspections never enough: a critique to enhance food safety. *Food Control*, 30, 686–91.

UNIDO. (2011). United Nations Industrial Development Organization. Meeting Standards, winning markets. Trade standards compliance 2010. 1–138. Available at http://www.unido.org/index.php?id=o723020.

UYTTENDAELE, M., GIL, M. I., ALLENDE, A., and SELMA, M. V. (2012). EU FP7 Veg-i-Trade investigates pre- and post-harvest practices influencing microbial quality and safety of leafy greens. *New Food*, 15, 66–70.

VAN BOXSTAEL, S., HABIB, I., JACXSENS, L., DE VOCHT, M., BAERT, L., *et al.* (2013). Food safety issues in fresh produce: bacterial pathogens, viruses and pesticide residues indicated as major concerns by stakeholders in the fresh produce chain. *Food Control*, 32, 190–7.

YAPP, C., and FAIRMAN, R. (2004). The evaluation of effective enforcement approaches for food safety in SMEs. Report of the Food Standards Agency. Available at http://www.foodbase.org.uk/results.php?f_report_id=206.

4.10 References

ALTMANN, M., WADL, M., ALTMANN, D., BENZLER, J., ECKMANNS, T., *et al.* (2011). Timeliness of Surveillance during Outbreak of Shiga Toxin-producing Escherichia coli Infection, Germany, 2011. *Emerging Infectious Diseases*, 17, 1906–9.

BEHRAVESH, C. B., MODY, R. K., JUNGK, J., GAUL, L., REDD, J. T., *et al.* (2011). 2008 Outbreak of Salmonella Saintpaul Infections Associated with Raw Produce. *New England Journal of Medicine*, 364, 918–27.

BUCHHOLZ, U., BERNARD, H., WERBER, D., BOEHMER, M. M., REMSCHMIDT, C., *et al.* (2011). German Outbreak of Escherichia coli O104:H4 Associated with Sprouts. *New England Journal of Medicine*, 365, 1763–70.

ETHELBERG, S., LISBY, M., BOTTIGER, B., SCHULTZ, A., VILLIF, A., *et al.* (2010). Outbreaks of gastroenteritis linked to lettuce, Denmark, January 2010. *Eurosurveillance*, 15, 2–4.

FALKENHORST, G., KRUSELL, L., LISBY, M., BO MADSEN, S., BÖTTIGER, S. and MØLBAK, K. (2005). Imported frozen raspberries cause a series of norovirus outbreaks in Denmark, September 2005, *Eurosurveillance*, 10, 1–2.

FRESHFEL. (2011). European fresh produce association. EU F&V Export and Import Trade Trends. In 2010 Fresh fruit and vegetable production, trade, supply and consumption monitor in the EU. Available at http://www.freshfel.org/asp/what_we_do/consumption_monitor.asp

GAJRAJ, R., POORANSINGH, S., HAWKER, J. I., and OLOWOKURE, B. (2012). Multiple outbreaks of Salmonella braenderup associated with consumption of iceberg lettuce. *International Journal of Environmental Health Research*, 22, 150–5.

GAULT, G., WEILL, F. X, MARIANI-KURKDJIAN, P., JOURDAN-DA SILVA, N., KING, L., *et al.* (2011). Outbreak of haemolytic uraemic syndrome and bloody diarrhoea due to Escherichia coli O104:H4, south-west France, June 2011. *Eurosurveillance*, 16, 5–7.

GUZMAN-HERRADOR, B., VOLD, L., COMELLI, H., MACDONALD, E., HEIER, B., *et al.* (2011). Outbreak of *Shigella sonnei* infection in Norway linked to consumption of fresh basil, October 2011. *Eurosurveillance*, 16, 2–3.

HEIER, B., NYGARD, K., KAPPERUD, G., LINDSTEDT, B., JOHANNESSEN, G., and BLEKKAN, H. (2009). Shigella Sonnei Infections in Norway Associated with Sugar Peas, May–June 2009. *Eurosurveillance*, 14.

HERWALDT, B. L., ACKERS, M. L., FARRAR, J., RICHARDSON, S., NELSON, R., *et al.* (1997). An outbreak in 1996 of cyclosporiasis associated with imported raspberries. *New England Journal of Medicine*, 336, 1548–56.

HPA. (2012). Press release of the Health Protection Agency on an outbreak of Salmonella Newport in watermelons. Available at http://www.hpa.org.uk/NewsCentre/NationalPressReleases/2012PressReleases/120202SalmonellaNewportoutbreak/.

INSULANDER, M., SVENUNGSSON, B., LEBBAD, M., KARLSSON, L., and DE JONG, B. (2010). A Foodborne Outbreak of Cyclospora Infection in Stockholm, Sweden. *Foodborne Pathogens and Disease*, 7, 1585–7.

KIREZIEVA, K., JACXSENS, L., UYTTENDAELE, M., VAN BOEKEL, M. and LUNING, P. (2013). Assessment of Food Safety Management Systems in the global fresh produce chain. *Food Research International*, 52, 230–42.

MACDONALD, E., HEIER, B. T., NYGARD, K., STALHEIM, T., CUDJOE, K. S., *et al.* (2012). Yersinia enterocolitica Outbreak Associated with Ready-to-Eat Salad Mix, Norway, 2011. *Emerging Infectious Diseases*, 18, 1496–9.

NYGARD, K., LASSEN, J., VOLD, L., ANDERSSON, Y., FISHER, I., *et al.* (2008). Outbreak of Salmonella Thompson infections linked to imported rucola lettuce. *Foodborne Pathogens and Disease*, 5, 165–73.

PEZZOLI, L., ELSON, R., LITTLE, C. L., YIP, H., FISHER, I., *et al.* (2008). Packed with Salmonella-Investigation of an International Outbreak of *Salmonella* Senftenberg Infection Linked to Contamination of Prepacked Basil in 2007. *Foodborne Pathogens and Disease*, 5, 661–8.

RKI. (2012). Robert Koch Institut. Großer gastroenteritis-ausbruch durch eine charge mit noroviren kontaminierter tiefkühlerdbeeren in kinderbetreuungseinrichtungen und Schulen in Ostdeutschland. *Epidemiologisches Bulletin*, 41, 415–17.

WHEELER, C., VOGT, T. M., ARMSTRONG, G. L., VAUGHAN, G., WELTMAN, A., *et al.* (2005). An outbreak of hepatitis A associated with green onions. *New England Journal of Medicine*, 353, 890–7.

5

Zoonotic transfer of pathogens from animals to farm products*

D. J. Bolton, Teagasc – Ashtown Food Research Centre, Ireland and T. S. Edrington, D. J. Nisbet and T. R. Callaway, US Department of Agriculture, USA

DOI: 10.1533/9781782420279.1.52

Abstract: Food animals contain a microbial population that lives on and within them, but this commensal microbial population can be penetrated by foodborne pathogenic bacteria that live asymptomatically in the animal. The presence of these pathogens in the live animal can contaminate water supplies and nearby crops, as well as be directly transmitted to humans via animal contact. Transmission to humans can be reduced via implementation of manure handling and spreading processes. Further pathogen reduction interventions can be grouped into antipathogen, pro-competition and management strategies. The rationale and effectiveness of these methods are discussed.

Key words: foodborne pathogens, animal production, preharvest, live animal, intervention strategies.

5.1 Introduction

Food animals are colonized by a complex microbiome on their skin/hide as well as in their gastrointestinal tract (Callaway *et al.*, 2004). Normally this population exists in an equilibrium that is beneficial to the host animal. However, this consortium can be penetrated by foodborne pathogenic bacteria that can be detrimental to the animal and to humans, and the presence of these pathogens in food animals often goes undetected because they are asymptomatically carried. Thus food animals are an important reservoir for pathogens that are found in meat products and via direct

---

* Mandatory Disclaimer: 'Proprietary or brand names are necessary to report factually on available data; however, the USDA neither guarantees nor warrants the standard of the product, and the use of the name by the USDA implies no approval of the product, or exclusion of others that may be suitable.'

Published by Woodhead Publishing Limited, 2014

contact with humans, but can also pose a threat via indirect transmission routes carried by manure, such as water and contaminated fruits and vegetables.

Foodborne pathogens are found on all types of animal production farms and in all stages of animal growth and production (LeJeune and Kersting, 2010). Foodborne pathogenic bacteria can survive in a variety of food production environments, including animals, and have been isolated from soil. Pathogenic bacteria can colonize produce by being taken up by plant roots or can be transferred during handling/processing. During harvesting and processing, contamination from a few plants may be spread horizontally to many via workers, hands, tools, crates, transport vehicles and washing (Franz and Van Bruggen, 2008). The latter is widely performed to remove soil and other debris and may facilitate dissemination of pathogens like *Salmonella*, Shiga-toxin-producing *Escherichia coli* (STEC), *Campylobacter* and *Listeria monocytogenes* if not performed correctly. However, many foodborne illnesses can be traced from animal production farms where pathogenic bacteria are often endemic.

Although postharvest pathogen-reduction strategies have been largely successful at reducing direct foodborne illness from meat products, these processing interventions have not been perfect. In order to reduce human illnesses and ensure a safe and wholesome food supply, research into preharvest pathogen reduction controls and interventions has grown in recent years (Callaway *et al.*, 2004; LeJeune and Wetzel, 2007). The logic underlying focusing on reducing foodborne pathogenic bacteria in live food animals is clear: reducing the amount of pathogens entering processing plants will reduce the burden on the plants and render the in-plant interventions more effective; reducing horizontal pathogens spread from infected animals on farms, in transport and lairage will reduce the pathogenic bacterial burden in the environment and wastewater streams, will reduce the spread of pathogens to fruits and vegetables from contact, run-off and contaminated irrigation water, and will reduce the direct risk to those (often city dwellers) in direct contact with animals via petting zoos and open farms. However, to date the impact of preharvest interventions has not been quantified. In order to understand how we can combat foodborne pathogenic bacterial illnesses, we must understand how the pathogens enter the food supply and how can we interrupt the transmission and reinfection cycles.

5.2 Zoonotic foodborne pathogenic bacteria in animals: prevalence estimates in food animal species

5.2.1 Ruminant animals

Cattle gastrointestinal tracts, especially the rumen, are occupied by a diverse, dense microbial consortium. The welcoming environment for anaerobic and facultative bacteria means that foodborne pathogenic bacteria can be found in many portions of the ruminant gastrointestinal tract. Cattle can be naturally colonized with enterohemorrhagic *E. coli* (EHEC; also known as STEC), *Salmonella enterica*, *Listeria* (although rarely *monocytogenes*) and *Campylobacter* in their gastrointestinal tract. Although all of these bacteria are pathogenic to human consumers, they do not necessarily cause illnesses in cattle.

Ruminant animals, in particular cattle, are the main reservoir for STEC. STEC-caused illnesses costs the American economy more than $1 billion each year in direct and indirect costs from more than 175 000 human illnesses (Scallan *et al.*, 2011; Scharff, 2010). Furthermore, since the emergence of STEC, more than $2 billion has been spent by the US cattle industry to combat STEC in processing plants. The reported prevalence rates for STEC-O157 in cattle are summarized in Table 5.1. However, over the past few years non-O157 cases have equaled or surpassed those caused by serotype O157:H7 in Europe and in the USA. Cattle are also the primary source of non-O157 STEC. Recent research has found that 40% of bovine fecal samples, 67% of hides and 27% of carcasses were non-O157 STEC positive. While bovine prevalence data for non-O157 STEC are limited beef carcass prevalence rates range from less than 2% to 54% (LeJeune and Wetzel, 2007). Hide prevalences of 4.5%, 11%, 28.2% and 56% have also been reported.

Salmonella serotypes have been isolated frequently from cattle under many production conditions and environments (Bolton *et al.*, 2012). *Salmonella* is the second most common cause of human foodborne bacterial illness, causing at least 1.2 million human illnesses at a cost of more than $14 billion a year in the USA (Scallan *et al.*, 2011; Scharff, 2010). These pathogenic bacteria survive in the gastrointestinal population, yet they are not dominant organisms, and colonization appears sporadic and transient. Populations of cattle colonized with *Salmonella* range up to 80% on some farms, but the predominant serotype or serogroup can vary widely within farm and between farms (Bolton *et al.*, 2012). Many of the most prominent serotypes that infect humans can be isolated from cattle. It is

Table 5.1 *E. coli* O157 prevalence in cattle

Country	Incidence	Comment
Denmark	5%	82 cattle tested
England and Wales	4%	2103 bovine rectal swabs
France	0.2%	471 fecal samples from cattle with a low sensitivity screening method
Ireland	3%	750 samples at slaughter
Italy	13%	Highest in feedlot cattle (17%) and dairy cows (16%). Highest in summer (18%) and lowest in winter (3%)
The Netherlands	0.8–22%	Excretion rates peaked in summer and were lowest in winter
Norway	0.3%	1970 animals tested using immunomagnetic separation (IMS) and PCR
Spain	2%	383 slaughter cattle by IMS
Sweden	14%	631 animals tested in 6 herds
USA	2%	11 881 fecal samples from cattle in 100 feedlots

Source: Bolton *et al.* (2009).

thought that *Salmonella* is spread between cattle via fecal-oral routes; however, fomites such as feed trucks and migratory birds also play a role and may disseminate *Salmonella* widely in feed troughs.

Campylobacter jejuni and *coli* are often isolated from cattle feces (Bolton *et al.*, 2012). The species isolated varies but the individual prevalence on farms ranges from 3% to 10%, and was less than 5% on Irish farms (Bolton *et al.*, 2012). In a longitudinal study in Irish feedlot cattle, it was found that over 50% of fecal samples contained *Campylobacter*. As feedlot cattle progress toward finishing, *Campylobacter* shedding incidence increased from 1% to over 60%, indicating the ease of horizontal spread in pens.

In general, foodborne pathogen colonization of cattle is asymptomatic, except for some *Salmonella* serotypes. Therefore, 'contaminated' cattle cannot be specifically isolated or addressed without significant diagnostic effort. Thus efforts to control foodborne pathogens in these animals must be economically feasible as well as amenable to application to a large number of animals.

5.2.2 Swine

Swine are also colonized with a complex intestinal consortium of bacteria in the gut that can be penetrated by foodborne pathogens, which can in turn be transmitted to pork. The primary pathogens that have been associated with swine include *Salmonella*, *Campylobacter*, *Listeria* and *Yersinia enterocolitica* (Davies, 2011). However, the role of live animals in carriage of *Listeria* and *Yersinia* is uncertain.

Live swine can contain *Salmonella*, which has been isolated from pigs and their environment through all stages of the pork production process (Rostagno *et al.*, 2003). Several *Salmonella* serotypes (such as the swine-adapted Choleraesuis) can cause severe illness in swine, but most serotypes are asymptomatic. The most common serotypes associated with swine include Derby, Typhimurium and Infantis. *Salmonella* can persist at a subclinical level within a farm or herd for years. The percentage of US swine colonized with *Salmonella* has been estimated at between 25% and 48% on the farm; however, fewer than 10% of marketed swine test positive for *Salmonella*. Recently, *Salmonella* was isolated from less than 4% of ground pork products in stores. As many as 66% of swine farms in Alberta had at least one positive *Salmonella* sample, and 20% of the on-farm environmental samples were positive for *Salmonella*. When swine are housed in outdoor pens, they can also harbor *Salmonella* isolates similar to those found in traditionally reared facilities.

Campylobacter is a common foodborne pathogen that is also commonly associated with swine (Davies, 2011). However, *Campylobacter* is not always found on farms and does not cause illness, or apparent morbidity in pigs (Bolton *et al.*, 2012). The incidence rate of *Campylobacter* varies on individual farms but increases from birth through finishing (Alter *et al.*, 2005). Several studies have found the incidence of *Campylobacter* in growing pigs to be approximately 45% (Alter *et al.*, 2005), yet other studies have found it to be less than 5% (Bolton *et al.*, 2012) depending on individual farm factors.

5.2.3 Poultry

Chickens and turkeys are frequently colonized by *Salmonella* (Vandeplas *et al.*, 2010) and *Campylobacter* (Adzitey and Nurul, 2011). When chicks are hatched, they have a naïve gastrointestinal tract that is colonized by bacteria from their environment, and this can often contain pathogens that survive fecal-oral passage. Unfortunately, these pathogens are also typically asymptomatic in poultry, thus detection is more difficult and the net must be cast wider to detect positive individuals or flocks. Because these pathogens can be spread via other reservoirs (such as insects and rodents), stringent biosecurity procedures are in place on most poultry farms (Bolton *et al.*, 2012).

In the USA, approximately 19% of broiler flocks were positive for *Salmonella*. The most common poultry-associated serotypes (e.g. Typhimurium and Enteritidis) account for one-third of the total human Salmonellosis cases in the USA. Thus it is apparent that much of the human *Salmonella*-burden is attributable to poultry sources. Because *Salmonella* can survive in the gut and tissues, it can be transmitted to consumers directly in meat and to the next generation of poultry via egg inclusion. Thus when a *Salmonella*-infected egg hatches, the chick can be immediately positive and then begin the horizontal spread of *Salmonella* to other birds. When 5% of birds were positive upon grower house entry, 72–95% of birds were *Salmonella* positive after three weeks. Therefore, strategies that inhibit this horizontal, within-house transfer are critical to reducing the number of birds containing foodborne pathogens from entering the food chain.

In poultry, *Campylobacter* is largely ubiquitous (Adzitey and Nurul, 2011). In some field observations, *Campylobacter* prevalence has been up to 85% within flocks and nearly 100% at the farm and processing plant level. Thus it is apparent that *Campylobacter* is endemic, and interventions to reduce this pathogen must be applicable at the flock level, rather than individually applied.

5.3 Survival, spread and transmission

5.3.1 Fecal survival and spread

Foodborne pathogens can survive for extended periods in the different environments encountered on farms, presenting considerable opportunity for recycling through farm environments, and cattle, with clear potential for subsequent transmission into the human food chain. For STEC, D-values (the time required to achieve a 1 log or 90% reduction in the population) of 8.79 and 8.16 days have been reported in soil stored at 4 °C and 14 °C, respectively. These soil D-values were similar to those reported in other studies. Another study examined the incidence of non-O157 STEC in farm pasture soils and investigated their survival in clay and sandy loam soils. Of samples from 20 farms recovered over a 12-month period, 27% were STEC positive. Inoculation studies with soil isolates of O26:H11, O113:H4, ONT:H4, O2:H27, O116:H28, O6:H8, ONT:H27 and O119:H5 and bovine fecal isolates of O145:H27, O20:H19,

O174:H21, O168:H8, O136:H2 and O86:H21 stored at 10 °C for up to 201 days, suggested greater persistence in the sandy-loam soil with D-values ranging from 50.26 to 75.60 days. The corresponding range in clay loams soils was 31.60 to 48.25 days.

STEC D-values in bovine feces stored at 4 °C of 7.56 and 9.79 days have been reported (O'Neill *et al.*, 2011; Himathongkham *et al.*, 1999). At 14 °C it took 9.79 days to reduce the STEC population by 1 log. These organisms also persist in bovine slurry where D-values of 3.9, 15.3 and 38.8 days have been reported. When combined with shedding data which suggest cattle may shed up to 10^8 STEC per gram of feces, the fecal and slurry persistence studies suggest persistence of up to approximately 80 days. Prolonged survival in farm water has also been reported. In a recent study, D-values of 8.63 and 3.59 days were reported in samples stored at 4 °C and 14 °C, respectively. As with soil, feces and slurry, these data suggest that contaminated water remains a potential source of STEC contamination for several months. *Salmonella* survival data on the farm are similar to those for STEC. D-values of 6.78, 15.32, 11.97 and 4.89 have been reported in soil, feces, slurry and water, respectively, stored at 4 °C and 9.43, 23.84, 10.51 and 5.25 days at 14 °C.

5.3.2 Transmission routes

Animal waste is the most significant source of STEC and *Salmonella*, and the application of manure and slurry to the land may result in contamination of fruit and vegetables directly or indirectly via contaminated water, tools, workers hands, birds, rodents, insects, etc. (Franz and Van Bruggen, 2008). Plants grown in STEC contaminated soils will become contaminated with these bacteria. Furthermore, some plants have been reported to internalize STEC from contaminated soils significantly increasing the survival of such pathogens, and the risks that they may not be removed from contaminated produce during harvesting and subsequent cleaning.

Conventional vegetable and fruit production does not use animal waste-based fertilizers so direct contamination should be avoided. Organic growers may also prevent direct contamination if the animal wastes are correctly composted and sufficient time is left between application to the soil and crop production. USDA guidelines, for example, require 120 days between application and product harvesting if parts of the plant to be consumed are in contact with the soil and 90 days if there is no direct contact.

Contaminated irrigation water is an important indirect route for the contamination of farm produce. Heavy rainfall is often associated with leaching of animal wastes into both surface and groundwater where bacterial pathogens such as STEC and *Salmonella* can be transported over large distances before being withdrawn to irrigate crops. The risk of human illness from consuming contaminated crops increases the closer the contaminated irrigation water is applied to harvest. Flood and spray irrigation systems that deposit the irrigation water directly onto the edible leaves of the crops increase the risk. Furthermore, repeated exposure results in pathogen build up on lettuce plants (Franz and Van Bruggen, 2008).

5.3.3 Outbreaks

Both STEC and *Salmonella* are commensal organisms carried in a range of livestock. STEC O157:H7, the serotype most often associated with outbreaks, is usually found in ruminants, especially cattle, where it colonizes the terminal rectum and is shed in the feces. *Salmonella* are asymptomatically carried by poultry, pigs and cattle. Both organisms may be spread person-to-person via a fecal-oral route. Primary contamination of farm produce occurs when feces or slurry from animals come in contact with fruit and/or vegetables (Franz and Van Bruggen, 2008). It has been suggested that the different pathogens are better adapted to survive and persist on different fruit and vegetables. *Salmonella* outbreaks, for example, have often been associated with tomatoes while lettuce has also been the vehicle in several O157:H7 outbreaks. However, *Salmonella* contaminated lettuce, cantaloupe, mung bean sprouts, alfalfa sprouts, fruit salad, peppers, basil citrus juice and watermelons have caused major outbreaks as have apple juice, sprouted seeds, spinach and watercress contaminated with *E. coli* O157 (Franz and Van Bruggen, 2008).

Within the EU, 14 member states tested for STEC in fruit and vegetables during the period 2004 to 2009. Only 11 of the 5910 samples (0.19%) tested were STEC positive, with eight of these being contaminated with O157 (0.14%). The food categories tested included fruit and vegetables (691 samples, one STEC positive, non-O157); vegetables (2019 samples, 10 positive, eight of which were O157); fruits (2774), juice (317), sprouts (104), spices and herbs (three) and ready-to-eat salads (two), all of which were STEC negative. However, in 2011 fenugreek seeds contaminated with O104:H4 caused an outbreak in Germany and France which resulted in 3911 confirmed cases and 47 deaths. The outbreak strain had a combination of STEC and enteroaggregative *E. coli* (EAggEC) virulence factors, carried a prophage encoded *stx*2a gene and had acquired a distinct set of additional virulence and antibiotic resistance factors by horizontal gene transfer.

5.4 How to stop pathogen transfer from and between live animals

Foodborne pathogenic bacteria are carried by livestock and shed directly into the farming environment where they persist for extended periods and can be spread or recycled (Bolton *et al.*, 2012; LeJeune and Kersting, 2010). However, animal waste (feces, urine, used bedding, etc.) represents a valuable nutrient source for grasslands. Farm management practices will therefore have a large impact on the dissemination of these pathogens within the farming environment and the potential from cross-contamination of other livestock and farm produce. Animal waste from animals housed indoors generally accumulates as slurry or farmyard manure.

Where animals are housed on slatted floors, feces and urine pass through the slats and accumulate as slurry (liquid waste with low dry matter content). Slurry can be stored in the pit underneath the slats or transferred to a separate slurry pit or overground tank. Where animals are housed on a solid floor with a bedding material (e.g. straw), high dry matter waste, consisting of bedding in addition to animal wastes, is periodically removed from housing and accumulates as farmyard manure. When animals are grazing outdoors on pastures, wastes are deposited directly onto grassland, including potential pathogens.

There is a serious potential food safety hazard if so called ready-to-eat (R-T-E) crops (i.e. vegetables and fruits not cooked prior to consumption) are exposed to animal waste. Furthermore, if animals graze on grass from animal waste treated fields, there is the possibility of the recycling of pathogens within a herd. Although it is imperative to prevent water contamination with animal wastes for environmental reasons, water contamination can also have food safety repercussions. Consumption of contaminated water by animals also contributes to the risk of pathogen recycling. Humans can indirectly come into contact with the contaminated water via water consumption, swimming, eating food cross-contaminated by the water, etc. In addition to run-off contamination of rivers, lakes and streams, pathogens can also pass through soil layers and contaminate groundwater. Thus, it is vital that sensible steps are taken to prevent pathogens in animal wastes from contaminating animals, crops and water. Pathogens are, however, sensitive to environmental stresses. Thus, pathogen numbers often decrease during storage and when spread onto pastures. Numerous factors influence this process: temperature, aeration, drying, freezing, thawing, sunlight, pH, composition of slurry/manure, indigenous soil microorganisms, soil type, chemical disinfection caused, for example, by the release of ammonia.

Risk assessment studies have investigated the rate of decline of various pathogens such that the time intervals required for the pathogens to be at an acceptable level can be determined or estimated. However, there are gaps in the scientific knowledge regarding the decline rates for specific pathogens in slurry and farmyard manure. Hence, recommendations are made based largely but not exclusively on the basis of scientific data. As further research is conducted and potential control measures investigated, recommendations are likely to be modified accordingly.

Present manure control measures to reduce the pathogen risk can be divided into pre-spreading, spreading and post-spreading controls.

5.4.1 Pre-spreading controls

- Proper storage facilities for animal waste are necessary to prevent leakage of waste into groundwater.
- Animals should be kept away from slurry pits/dung heaps.
- From a food safety point of view, extended storage of slurry prior to spreading is an important control measure. Storage is recommended for at least one month or preferably for three months. This ideally requires that farmers employ a

batch system with two separate waste storage units so that one is allowed to age without the addition of new material, while the second one takes new slurry. However, this is not always practical because of the cost implications involved.

- Farmyard manure undergoes a natural composting process. Active aerobic or anaerobic digestion processes may also be applied. All of these processes generate sufficient heat to destroy non-spore forming bacteria. In order to gain maximum benefit from this process, steps should be taken to ensure that the compost heap is mixed well (at least twice) so that all parts reach the high temperatures required to kill pathogens. This also requires that no new waste is added to the compost heap, and recommended storage for one month where farmyard manure is mixed to ensure complete composting or for three months otherwise.
- Other treatments should be used when recommended by the competent authorities in the case of specific disease outbreaks. These include the addition of lime to slurry to raise the pH and thus kill pathogens.
- Aerobic digestion and anaerobic digestion are potential treatments to remove pathogens from slurry. However, cost implications limit their use at present.

5.4.2 Spreading controls

- Spreading should be avoided in conditions where contamination of a water course is more likely to occur, for example during wet weather, on wet land or on land sloping steeply towards a water course. Animal waste should never be spread in the vicinity of a domestic well or public water supply. A filter strip (area left untreated around the water course) can significantly reduce the risk of run-off into water courses.
- The use of injection systems and splash plate type slurry spreaders with downward splash (as opposed to splash plate type slurry spreaders with upward splash) are recommended to minimize aerosols. Aerosols contribute to the loss of nitrogen from animal waste and to odor problems. Also viruses and bacteria can be spread over a wide area as aerosols. Seals and fittings on spreaders should be fully functional to prevent animal waste dripping onto the farmyard, roadways, etc. There are, however, conflicting views regarding the implications of recommended equipment on pathogen survival. Some investigators believe that the use of injector systems may enhance pathogen survival as pathogens have reduced exposure to drying and sunlight.
- Where possible, animal waste should be spread immediately postharvest of the last crop, thus maximizing the time interval between spreading and harvesting of the next crop.

5.4.3 Post-spreading controls

Following the application of animal waste to land, it is reasonable to expect that a time interval should be adhered to prior to the use of that land for grazing by animals or the harvesting of crops.

- It is recommended that pastures should be left ungrazed for at least one month or until all visible signs of animal waste have disappeared.
- In the case of ready-to-eat crops, a six-month interval has been recommended between animal waste application/animal grazing and crop harvest. In the UK, future guidelines are likely to recommend a 12-month interval between the application of fresh, untreated animal waste and R-T-E crop harvest, with this interval reduced to two months in the case of stored or treated waste.
- In the case of animal waste application to grass selected for silage production, the time interval before grass is deemed suitable for silage production is usually sufficient to eliminate any concerns regarding the animal waste.

5.5 Pathogen control strategies in live animals: novel solutions

The ecological niche occupied by, and ubiquity of these foodborne pathogens in food animals makes the development of global strategies difficult, if not impossible. However, it is nevertheless critical for public health to reduce the prevalence of these pathogens generally on the farm. Unfortunately, there is no 'silver bullet' that will eliminate all pathogens from the food supply for all food animals under all production conditions. Rather a continuum of good production practices and interventions at targeted critical control points during production (e.g. at time of hatch in poultry) is critical to erecting multiple, complementary, tailored hurdles to the entry of pathogens to the food supply in a comprehensive fashion. With that in mind, potential interventions can be broadly divided into three categories: 1 directly antipathogen, 2 pro-competition and 3 management.

5.5.1 Antipathogen strategies

Antipathogen strategies specifically target the pathogen or pathogens most frequently isolated from each food animal species (Table 5.2). Although antibiotics

Table 5.2 Antipathogen strategies for use in food animals

Strategy	Pathogen	Food animal species	Outlook
Bacteriophage	*Salmonella* STEC	Cattle Swine Poultry	Good, some commercial products on market as hide sprays
Bacteriocins/ colicins	STEC *Salmonella*	Poultry Swine	Good, as peptide production increases
Vaccination	STEC *Salmonella*	Swine Cattle	Excellent, various products in marketplace
Sodium chlorate	STEC *Salmonella*	Swine Cattle Poultry	Good, awaiting approval by FDA in USA

are the most well-known strategy, because of the concerns surrounding dissemination of antibiotic resistance this is not recommended for implementation or even further examination.

Bacteriophages are viruses that prey solely on bacteria and have rather narrow target spectra, which allows phages to be used as 'smart bombs' against specific microorganisms in a mixed microbial population without perturbing the overall ecosystem. Phages are a normal component of the gastrointestinal ecosystem and farm waste microbial ecosystems (Niu *et al.*, 2009). Phage treatment reduced disease that impacted production efficiency or health in swine, sheep and poultry, including enterotoxigenic *E. coli* (ETEC)-induced diarrhea and splenic ETEC colonization. Bacteriophages have been used experimentally to control foodborne pathogenic bacteria in ruminant animals, especially *E. coli* O157:H7 (Niu *et al.*, 2009). Further studies have found that *Salmonella* populations in swine and poultry can be controlled by phage addition (Wall *et al.*, 2010). Additional studies have demonstrated that phage sprays could be useful on the skin of live animals and carcasses. Recently, the FDA approved the use of phages as surface cleaning agents as well as in use on live animal hide sprays to reduce *E. coli* O157:H7 on cattle hides before they enter the slaughter plant.

Vaccines have long been used in food animal production, primarily to reduce animal disease causing bacteria and viruses. To date, there have been vaccines targeted at reducing foodborne pathogenic bacteria (Gyles, 1998). *Salmonella* strains do cause some diseases in swine and cattle, and vaccines have been developed against them. Vaccination has also been successfully used to combat post-weaning *E. coli* edema diarrhea (PWECD) in young pigs, and to reduce *Salmonella* colonization in poultry and swine. A vaccine has been developed for use in feedlot cattle that significantly reduces fecal *E. coli* O157:H7 shedding. Preliminary experimental results indicated that this vaccine reduced *E. coli* O157:H7 shedding in feedlot cattle from 23% to less than 9% (Moxley *et al.*, 2003). Further studies demonstrated that vaccinated animals shed *E. coli* O157:H7 at 14% compared with 46% of untreated control animals. Other specific vaccines have been developed as well that target STEC proteins (siderophores), and the future development of these vaccines appears promising.

Salmonella and *E. coli* bacteria can respire under anaerobic conditions by reducing nitrate to nitrite via a dissimilatory nitrate reductase. The intracellular bacterial enzyme nitrate reductase does not differentiate between nitrate and its analog, chlorate which is reduced to chlorite in the cytoplasm; chlorite accumulation kills bacteria equipped with nitrate reductase. Chlorate treatment reduced experimentally inoculated *Salmonella* and *E. coli* O157:H7 populations in swine and sheep intestinal tracts and poultry feces and intestinal contents. Other studies indicated that soluble chlorate administered via drinking water significantly reduced ruminal, cecal and fecal *E. coli* O157:H7 populations in cattle and populations of *E. coli* on their hides. The use of chlorate to reduce foodborne pathogens has been suggested to be included in the last meal (approximately 24 hours) before animals are shipped for processing, thus chlorate could work against these pathogens during the transport process. The use of chlorate in food animals to reduce foodborne pathogenic bacteria is presently under review by the US Food and Drug Administration.

Some bacteria produce antimicrobial proteins as they engage in biochemical warfare against bacteria that occupy the same environmental niche, and these antimicrobial proteins are classified as bacteriocins; those that specifically target *E. coli* species are known as colicins. Feeding of probiotic bacteria that produce colicins/bacteriocins has been suggested and successfully utilized against animal and foodborne pathogens in experimental models (Stahl *et al.*, 2004). Only recently have purified bacteriocins, including colicins, been produced in quantities great enough for examination as feed additives to reduce foodborne pathogens in food animals (Stahl *et al.*, 2004), and potentially for use on carcasses and ready-to-eat foods to prevent foodborne illness.

5.5.2 Pro-competition strategies

In the opposite manner to directly antipathogen strategies, pro-commensal strategies seek to promote the growth of groups of beneficial bacteria that are competitive with, or antagonistic to, pathogens. The use of native or novel microflora introduced to reduce pathogenic bacteria in the gut has been termed as a 'probiotic' strategy (Fuller, 1989). Providing a limiting nutrient ('prebiotic') to the microbial population that allows an existing commensal microbial population a competitive advantage in the gastrointestinal tract is another pro-commensal approach. The overall goal of pro-competition strategies is to fill all ecological niches in the gut and thereby prevent the establishment of, or cause the displacement of, a pathogenic bacterial population in the gut.

Probiotics are a 'live microbial feed supplement which beneficially affects the host animal by improving intestinal microbial balance' that are included in animal rations to enhance performance, or to reduce zoonotic pathogens (Fuller, 1989). In general, probiotics aim to improve animal growth and performance, but some have been reported to reduce foodborne pathogens in food animals. In swine, the addition of a probiotic culture comprising *Streptococcus (Enterococcus) faecium* reduced enterotoxigenic *E. coli* (ETEC) colonization and subsequent diarrhea. Inclusion of a *Saccharomyces* fermentation feed product did not reduce populations of *Salmonella* in experimentally infected pigs, but did reduce the negative effects of *Salmonella* infection in these animals (Price *et al.*, 2010). Other studies have found that *Lactobacillus* and *Streptococcus* cultures reduced *Salmonella* populations in poultry.

Research studies in ruminants have demonstrated that a probiotic containing *Lactobacillus acidophilus* cultures fed to finishing cattle reduced *E. coli* O157:H7 shedding by more than 50%. Further studies have confirmed that this product can reduce *E. coli* O157 populations in cattle (Moxley *et al.*, 2003). Currently, this probiotic is fed to beef cattle to reduce pathogens and improve growth efficiency; thus ensuring that the pathogen reduction economically pays for its inclusion in the cattle ration, a critical factor in any pathogen reduction strategy to be implemented in the food production industry.

Prebiotics are substrates that are unavailable to or indigestible by the host animal, but are digestible by a portion of its microbial population. Prebiotics provide energy or other limiting nutrients to the intestinal mucosa as well as substrates for the intestinal bacterial fermentation, resulting in enhanced production of vitamins and antioxidants that further directly benefit the host animal (Kim *et al.*, 2011). Some prebiotics can provide a competitive advantage to members of the native microflora that can help to exclude pathogenic bacteria from the intestine via direct competition for nutrients or for binding sites through the production of 'blocking factors', or antimicrobial compound (e.g. bacteriocins, volatile fatty acid (VFA) production) (Price *et al.*, 2010). Other research has indicated that specific carbohydrate prebiotics reduced colitis type symptoms and improved mucosal immunity in swine. In poultry, the use of prebiotics has been found to increase immune measures and performance (Vandeplas *et al.*, 2010), as well as to reduce *Salmonella* and *Clostridium perfringens* populations in broilers and models of *in vitro* intestinal infection (Kim *et al.*, 2011).

Competitive exclusion (CE) involves the addition of a (non-pathogenic) bacterial culture to the intestinal tract of food animals in order to reduce colonization or decrease populations of pathogenic bacteria in the gastrointestinal tract (Fuller, 1989). A CE culture may be composed of one or more strains or species of bacteria, but it should be derived from the animal of interest and thus attempt to exploit relationships developed during co-evolution of host and microbiome. *Salmonella* colonization in young chickens was reduced by administration of a preparation of gut bacteria originating from healthy adult chickens. Other researchers have demonstrated that a swine mucosal CE culture could reduce *Salmonella* populations in young pigs. Recent studies demonstrated that a swine CE culture derived from the cecal contents of healthy pigs reduced the incidence of *Salmonella cholerasuis* and enterotoxigenic *E. coli*. The use of true CE products in ruminants has been limited because of the complexity of the ruminant gastrointestinal microbial population, and the length of time involved in cattle production (up to 18 months).

The simultaneous application of probiotics or CE with prebiotics is known as 'synbiotics', and could yield a synergistic effect in reduction of foodborne pathogenic bacterial populations in food animals prior to slaughter. Furthermore, synbiotic approaches increased glucose transport in broilers and increased the availability of B-vitamins. Several studies have demonstrated that synbiotic strategies could reduce postweaning *E. coli* diarrhea in swine and necrotic enteritis in poultry. In a variation on synbiotics, researchers included bacteriophages phages along with a competitive exclusion culture to successfully reduce *Salmonella* population in poultry.

5.5.3 Animal management

Good animal management is critical for efficient animal production, but to date no typical 'management' procedures have been shown to affect colonization or shedding of foodborne pathogenic bacteria. Lairage and handling have been demonstrated to play a role in *Salmonella* transmission in swine (Rostagno *et al.*, 2003), but that is largely a function of fomite exposure and stress increasing susceptibility. However, although no management practices have been shown to directly improve food safety (LeJeune and Wetzel, 2007), some practices may reduce horizontal transmission.

Generally speaking, the most effective 'management' strategy that can be used to alter the microbial population of the gut involves animal ration composition. The effects of dietary shifts and forage inclusion on *E. coli* O157:H7 shedding in cattle are too complex for detailed inclusion in this chapter. However, it appears that there is some effect of feeding fresh high-quality forage on the fecal shedding of *E. coli* O157:H7, but that forage quality is intimately involved.

Some dietary ingredients alter the microbial population through direct impact of ingredients. For example, organic acids reduce the pH or alter other environmental factors of the gut, similar to the hypothesized mode of action of probiotic/CE cultures. The inclusion of organic acids in animal feedstuffs has been demonstrated to impact pathogen populations. Other feedstuffs include essential oils, which are toxic to some bacterial species. Thymol and other organic acids from spices have demonstrated their ability to alter the intestinal microbiome, especial in regards to populations of *Campylobacter*. Feeding of citrus peels to cattle have been shown to reduce intestinal populations of *Salmonella* and *E. coli* O157:H7; however, this is apparently related simply to the concentrations of essential oils in the peels. Phenolic compounds and tannins also provide a natural antimicrobial effect in animal rations. Thus there are many avenues of exploration and exploitation available to attack pathogens directly or indirectly by manipulating the competitive microbiome via dietary manipulations.

5.6 Conclusion

Although the North American and European food supplies are generally safe, foodborne illnesses still occur and are frequently associated with animal agriculture. The increasingly integrated and global nature of food production has increased our awareness and sensitivity to foodborne illness outbreaks that originate with food animals. Live animals are often colonized by foodborne pathogenic bacteria which can be spread directly through food, but also through human contact and via manure exposures. Manure controls and procedures can be implemented to reduce the horizontal spread within a herd, but can also be utilized to reduce the spread to human consumers of fruits and vegetables fertilized with, or exposed to manure. The use of antipathogen and pro-competition strategies, diet and proper animal management can potentially reduce the incidence of foodborne pathogenic bacteria that enter the food chain. Further research into intervention strategies, diet and management procedures that specifically target 'critical control points' are crucial to improving overall food and environmental safety. All intervention strategies and production practice changes cannot be implemented without an economic cost. The increased economic burden must be shared by all concerned parties, because without an economic incentive to implement these pathogen reduction procedures, pathogen intervention strategies will remain economically unfeasible.

5.7 Questions for discussion

1. Which are the sources of, and routes taken by, foodborne pathogens that can directly enter your life?
2. Are there any of the alternative pathogen exposure routes (e.g. animal contact, water) with which you come into contact?
3. Which foods that you eat carry the greatest risk of farm pathogen contamination?
4. Can pathogenic bacteria truly be eliminated from all environments?
5. Is sterilization of foods the solution to improving food safety? Via processes such as cold pasteurization/irradiation?
6. Are foods with no bacteria on them a good idea for humans? Keep in mind the concept behind the 'hygiene hypothesis' and immunostimulation.
7. What critical control points can you devise or apply to reducing pathogen entry into your own food production system at home/work?
8. What kind of manure control/disposal/utilization system could produce the greatest benefit to food safety? How can this figure into the organic/processed foods debates?
9. Which type of pathogen reduction intervention intuitively appeals most to you? What are some potential drawbacks to it?
10. Can a 'silver bullet' be found that reduces all foodborne pathogens in live animals?

5.8 References

ADZITEY, F. and H. NURUL. 2011. *Campylobacter* in poultry: Incidences and possible control measures. *Res. J. Microbiol.* 6(2):182–92.

ALTER, T., F. GAULL, S. KASIMIR, M. GÜRTLER, H. MIELKE, *et al.* 2005. Prevalences and transmission routes of *Campylobacter* spp. strains within multiple pig farms. *Vet. Microbiol.* 108(3–4):251–61.

BOLTON, D. J., G. DUFFY, C. J. O'NEILL, C. L. BAYLIS, R. TOZZOLI, *et al.* 2009. *Epidemiology and Transmission of Pathogenic Escherichia coli*, ISBN 1 84170 535 7, Teagasc, Dublin.

BOLTON, D. J., C. J. O'NEILL, and S. FANNING. 2012. A Preliminary Study of *Salmonella*, Verocytotoxigenic *Escherichia coli/Escherichia coli* O157 and *Campylobacter* on Four Mixed Farms. Zoono. *Pub. Health* 59(3):217–28.

CALLAWAY, T. R., R. C. ANDERSON, T. S. EDRINGTON, K. J. GENOVESE, R. B. HARVEY, *et al.* 2004. Recent pre-harvest supplementation strategies to reduce carriage and shedding of zoonotic enteric bacterial pathogens in food animals. *Anim. Health Res. Rev.* 5(1):35–47.

DAVIES, P. R. 2011. Intensive swine production and pork safety. *Foodborne Pathog. Dis.* 8(2):189–201.

FRANZ, E. and A. H. C. VAN BRUGGEN. 2008. Ecology of *E. coli* O157:H7 and *Salmonella* enterica in the primary vegetable production chain. *Crit. Rev. Microbiol.* 34(3–4): 143–61.

FULLER, R. 1989. Probiotics in man and animals. *J. Appl. Bacteriol.* 66:365–78.

GYLES, C. L. 1998. Vaccines and shiga toxin-producing *Escherichia coli* in animals. Pages 434–444 in *Escherichia coli O157:H7 and other shiga toxin-producing E. coli strains.* J. B. KAPER and A. D. O'BRIEN, eds. *Amer. Soc. Microbiol. Press,* Washington, DC.

HIMATHONGKHAM, S., S. BAHARI, H. RIEMANN, and D. CLIVER. 1999. Survival of Escherichia coli O157:H7 and Salmonella Typhimurium in cow manure and cow manure slurry. *FEMS Microbiol. Lett.* 178:251–7.

KIM, G. B., Y. M. SEO, C. H. KIM, and I. K. PAIK. 2011. Effect of dietary prebiotic supplementation on the performance, intestinal microflora, and immune response of broilers. *Poult. Sci.* 90(1):75–82.

LEJEUNE, J. and A. KERSTING. 2010. Zoonoses: An occupational hazard for livestock workers and a public health concern for rural communities. *J. Agric. Safe. Health* 16(3):161–79.

LEJEUNE, J. T. and A. N. WETZEL. 2007. Preharvest control of *Escherichia coli* O157 in cattle. *J. Anim. Sci.* 13 (Suppl):85–97.

MOXLEY, R. A., D. SMITH, T. J. KLOPFENSTEIN, G. ERICKSON, J. FOLMER, *et al.* 2003. Vaccination and feeding a competitive exclusion product as intervention strategies to reduce the prevalence of *Escherichia coli* O157:H7 in feedlot cattle. Page 23 (Abstr.) in *Proc. 5th Int. Symp. on Shiga Toxin-Producing Escherichia coli Infections*, Edinburgh, UK.

NIU, Y. D., T. A. MCALLISTER, Y. XU, R. P. JOHNSON, T. P. STEPHENS, and K. STANFORD. 2009. Prevalence and impact of bacteriophages on the presence of *Escherichia coli* O157:H7 in feedlot cattle and their environment. *Appl. Environ. Microbiol.* 75:1271–8.

O'NEILL, C. J., D. J. BOLTON and S. FANNING. 2011. Comparative studies on the survival of Verocytotoxigenic Escherichia coli and Salmonella in different farm environments. *Agric. Food Analyt. Bacteriol.* 1(2):116–22.

PRICE, K. L., H. R. TOTTY, H. B. LEE, M. D. UTT, G. E. FITZNER, *et al.* 2010. Use of *Saccharomyces cerevisiae* fermentation product on growth performance and microbiota of weaned pigs during *Salmonella* infection. *J. Anim. Sci.* 88(12):3896–908.

ROSTAGNO, M. H., H. S. HURD, J. D. MCKEAN, C. J. ZIEMER, J. K. GAILEY, and R. C. LEITE. 2003. Preslaughter holding environment in pork plants is highly contaminated with *Salmonella enterica. Appl. Environ. Microbiol.* 69(8):4489–94.

SCALLAN, E., R. M. HOEKSTRA, F. J. ANGULO, R. V. TAUXE, M.-A. WIDDOWSON, *et al.* 2011. Foodborne illness acquired in the United States—major pathogens. *Emerg. Infect. Dis.* 17:7–15.

SCHARFF, R. L. 2010. *Health-related costs from foodborne illness in the United States*. The Project Safety Project at Georgetown University, Washington, DC.

STAHL, C. H., T. R. CALLAWAY, L. M. LINCOLN, S. M. LONERGAN, and K. J. GENOVESE. 2004. Inhibitory activities of colicins against *Escherichia coli* strains responsible for postweaning diarrhea and edema disease in swine. *Antimicrob. Ag. Chemother.* 48(8):3119–121.

VANDEPLAS, S., R. DUBOIS DAUPHIN, Y. BECKERS, P. THONART, and A. THEWIS. 2010. *Salmonella* in chicken: Current and developing strategies to reduce contamination at farm level. *J. Food Prot.* 73(4):774–85.

WALL, S. K., J. ZHANG, M. H. ROSTAGNO, and P. D. EBNER. 2010. Phage therapy to reduce preprocessing *Salmonella* infections in market-weight swine. *Appl. Environ. Microbiol.* 76(1):48–53.

Part II

Environmental issues impacting the potential safety of fresh produce

6

Postharvest washing as a critical control point in fresh produce processing: alternative sanitizers and wash technologies

K. Warriner and A. Namvar, University of Guelph, Canada

DOI: 10.1533/9781782420279.2.71

Abstract: The postharvest wash process in fresh produce processing is considered a critical control point (CCP) for removing field-acquired contamination. However, it is now acknowledged that hypochlorite washes are relatively ineffective at decontaminating fresh produce. This chapter will provide an outline of chlorine alternatives along with advantages and disadvantages. Factors that affect the wash process and how contamination can be readily spread between batches will be outlined. Parameters that can be used to monitor the wash process will be reviewed. Recent advances in postharvest decontamination methods will be discussed with respect to establishing a true CCP in fresh produce processing.

Key words: fresh produce, washing, hypochlorite, decontamination, hazard analysis and critical control point (HACCP).

6.1 Introduction

The fresh-cut produce industry that includes leafy greens, in addition to other vegetables and fruit, has experienced rapid growth since the first introduction of pre-packed produce back in 1986. The increase in the fresh-cut sector has been mirrored by numerous high-profile foodborne illness outbreaks implicating leafy greens, tomatoes, cantaloupe and soft fruit, among other product types (Olaimat and Holley, 2012). To enhance food safety there has been a sustained effort to introduce hazard analysis and critical control point (HACCP) into the fresh-cut sector (Francis *et al.*, 2012). One of the key foundations of HACCP is to have an intervention step that can remove field-acquired contamination that in the current context is the postharvest wash step. In the following chapter the limitations and challenges of postharvest wash step as CCP will be identified. The underlying reasons why hypochlorite is still widely applied in the industry will be outlined and currently available alternatives described. Emerging technologies to improve, or as an alternative to postharvest washing in terms of pathogen reduction will be described. The challenges and potential future directions to enhance the efficacy of postharvest washing will be discussed in the context of designating the wash step as a true CCP.

6.1.1 Objectives and challenges of postharvest washing

The first bagged salad was introduced by the British retailer Marks and Spencer in 1986. Consumers appreciated the convenience of pre-prepared fresh produce and as a consequence the market underwent rapid growth. In the early development of the market sector there was a belief that the postharvest wash would effectively remove field-acquired contamination, thereby rendering the product microbiologically safe (Gonzalez *et al.*, 1987). However, it was soon realized that aqueous-based washes could only achieve 1–2 log cfu reduction even under laboratory conditions (Beuchat, 1996). More recently an additional hazard of contamination being distributed between batches has been proven, thereby making postharvest wash a cross-contamination point rather than a decontamination step (Johnston *et al.*, 2005; Allende *et al.*, 2008; Lopez-Galvez *et al.*, 2010; Nou *et al.*, 2011; Buchholz *et al.*, 2012a). Here, pathogens released from one batch of produce can contaminate the water thereby becoming distributed onto subsequent batches. As a consequence, more recent innovations in postharvest washing have focused on inactivating pathogens released into the water as opposed to decontaminating produce directly (Olaimat and Holley, 2012).

With the limitations and potential cross-contamination events that could occur during the wash process, it has been debated whether washing of fresh produce should actually be performed. However, it should be noted that washing is not only applied for decontaminating produce but also removal of field-soils and also for transportation through the processing line. Consequently, contact with water is inevitable in processing with the inherent risk of cross-contamination between batches. Therefore, the wash process cannot be considered a CCP with respect to removing or controlling the hazard, but is a CCP by virtue that there is no subsequent step to reduce the food safety risk.

6.2 When postharvest washing goes wrong

Although there are limitations to the postharvest wash process, there is potential to cause major food safety issues if cross-contamination events are not controlled. The consequences of improper washing can be summarized by two high-level outbreaks linked to fresh produce. The case studies to be described are true events but the names of companies and individuals have not been identified.

6.2.1 *Escherichia coli* O157:H7 outbreak linked to lettuce in 2008

In May 2008 an outbreak of *E. coli* O157:H7 involving 10 individuals was reported and traced to a batch of fresh-cut lettuce processed within Washington State. The lettuce was produced by Company A, who obtained the product from Supplier A, who obtained the batch from Grower A, who in turn sub-contracted to Grower B who harvested the lettuce. The processing of lettuce followed operations commonly applied by industry. Specifically, the lettuce heads were harvested and field heat removed by vacuum cooling. The Romaine lettuce was chopped/shredded then transported via water flumes containing sodium hypochlorite and pH adjusted with citric acid. Under the process description as written in the HACCP plan, the free chlorine and pH level within the water flume was automatically controlled by an oxidation reduction potential (ORP) feedback loop to maintain free chlorine levels. However, the ORP unit had been delivered the previous month and had not been installed. Therefore, instead of ORP, the chlorine levels were measured using a handheld meter with manual addition of hypochlorite and citric acid.

The investigation to trace the sources of the *E. coli* O157 did not yield any positive samples in soil or well samples. There was a cattle ranch located 0.5 miles from the field, although there was no evidence that a contamination event had occurred. Inspection of the auditor's report for the facility did not yield any major infractions, and, relevant to the case, the ORP monitoring and set-point (750 mV) were appropriate for lettuce processing. However, the auditor either neglected to check the line or was misinformed with regards to dosing of chlorine into the tank. As mentioned, the ORP control unit had been delivered but not installed. The FDA inspectors questioned the management of the facility about chlorine dosing into the tank. The response of the management was that Quality Assurance technicians manually measured the chlorine content and pH every 30–60 minutes. The chlorine levels (target 25 ppm total chlorine) were adjusted by the addition of hypochlorite and pH adjusted with citric acid to a value of pH 7.5. When asked to provide documents on the frequency of checks, there were no specific times provided to verify that the monitoring had taken place at appropriate times. Of equal concern were the monitoring forms, which recorded an ORP 750 mV at every sampling point taken, which was unexpected given that manual dosing was performed.

There are several noteworthy aspects to the case. First is the relative complexity of sourcing fresh produce. It is common practice in the industry to use contractors who in turn work with growers and other sub-contractors. Indeed, in certain operations the whole process of sourcing and processing is under a sub-contract. Although contracting and sub-contracting is the most efficient means of producing fresh-cut products, it can raise problems when tracking back to the raw material source.

The role of third party auditor reports was also highlighted as a concern. Despite issues with the wash process, the report did not raise concerns. This may have been because of a focus on documentation that defined the process and how it was monitored, rather than actual observation of the plant during operating hours. Also, there is the question of whether third party auditors are independent, given that they are contracted by the company under inspection (Dawson, 2011). Yet, with the advent of the Global Food Safety Initiative (GFSI), more consistency between audits is anticipated (Percy, 2012).

Focusing on the wash process, there were several issues that could have contributed to the *E. coli* O157:H7 outbreak. The most obvious is that, although an ORP system was in place it was not being used and the chlorine added was low compared with that usually applied in industry (150–200 ppm, pH 6.5). Furthermore, because of the instability of hypochlorite, the free chlorine levels would have readily been depleted despite the monitoring records indicating that the ORP of the water remained at 750 mV. In effect, the lack of chlorine would mean any *E. coli* O157:H7 released from the lettuce would have cross-contaminated over batches of product. It should be noted that the specific source of the *E. coli* O157:H7 was never determined nor the contribution of the wash process conclusively proven.

A further notable factor of the incident was that the monitoring records indicated that the process was under control (i.e. 750 mV) and in accordance with the written HACCP plan. Clearly, this was not the case in the facility, which illustrates that what is written in the plan was not necessarily put into practice.

6.2.2 *Listeria monocytogenes* outbreak linked to cantaloupes in 2011

One of the most deadly foodborne illness outbreaks in the USA occurred in the fall of 2011, and resulted in over 140 confirmed cases of listeriosis and 33 deaths across 28 states. The source of the *Listeria monocytogenes* was traced to a family farm in Colorado that produced and distributed cantaloupes. The farmers were experienced and had produced cantaloupes on the farm for some 30 years.

The cultivation and postharvest handling of cantaloupes is a relatively straightforward process. Once harvested, the cantaloupes are transferred to the grading and packing plant. Within the facility, the cantaloupes can be transported via conveyors or water flumes.

Cantaloupes are especially susceptible to contamination given that the fruit is grown on the ground and has a rough 'netted' surface that accumulates debris. Therefore, the fruit undergo a vigorous washing process that typically includes air-agitation to dislodge debris from the cantaloupe surface. The cantaloupes are then inspected, packed into cartons and cooled prior to distribution.

The farm implicated in the outbreak made critical errors of judgment that ultimately proved disastrous to both the company and those stricken by listeriosis. From the investigation it was hypothesized that the *L. monocytogenes* (five strains were isolated) originated from the field and also conveyor equipment that had been brought in from a potato processing operation. *Listeria* naturally inhabits decaying organic material and can survive over extended periods by virtue of its psychrotrophic growth (Warriner and Namvar, 2009).

In addition to introducing the conveyor, the processor also made the decision to remove chlorine from the postharvest wash. The motivation for removing chlorine from the wash was not disclosed, but is likely to be concern over the generation of potentially carcinogenic disinfection byproducts (DBP) by the reaction of hypochlorite with organic matter. Trihalomethanes (chloroform, bromoform) and haloacetic acids form the predominant classes of DBP encountered in water of food. The precursors for DBPs include natural organic matter (e.g. humic acid) and bromide (Chowdhury *et al.*, 2011; Cardador and Gallego, 2012; Hsiao *et al.*, 2012; Jeong *et al.*, 2012). With the concerns linked to DBPs, there has been a trend for those processing organic produce to remove chlorine from wash tanks. In addition, in 2011 the US Environmental Protection Agency (EPA) also recommended that alternatives to chlorine should be sought to reduce the environmental impact of DBP.

With the lack of chlorine in the wash tank there was no barrier to prevent the widespread distribution of *L. monocytogenes* throughout the plant. Indeed, *Listeria* was recovered from equipment, pools of standing water, coolers and the cantaloupes. Although there were numerous infractions within the facility such as rusting equipment and lack of temperature control, the lack of sanitizer was highlighted as a significant contributory factor in the foodborne illness outbreak.

The listeriosis outbreak linked to cantaloupes caused major disruption across the entire melon sector with multi-million dollar losses. In addition, the outbreak placed further focus on food safety standards within the cantaloupe industry. In this respect it was a disappointment to the industry when a further cantaloupe outbreak occurred in 2012 linked to *Salmonella* Typhimurium and Newport. The outbreak resulted in 261 confirmed salmonellosis cases with three deaths. Among other factors, the processor failed to monitor chlorine levels within the flume and dunk tank water with a suspicion that the sanitizer was not applied on a regular basis.

The aforementioned outbreaks clearly illustrate the importance of postharvest washing as a food safety intervention. Although it is established to be limited in terms of reducing actual levels on produce, sanitizers in wash water contribute to minimizing cross-contamination events in the course of processing.

6.3 Approved sanitizers for fresh-cut processing

Sanitizers applied in the fresh produce wash require to be granted approval or letter-of-no-objection from the appropriate regulatory authority. Within Canada, sanitizer approval is sought through Health Canada and the Canadian Food Inspection Agency. In the USA, the approval process is performed by the Food and Drugs Administration in conjunction with the Environmental Protection Agency. Within Europe, the approval process is undertaken by food safety agencies in the member states with approved sanitizers differing between countries. For example, chlorine is approved as a sanitizer for washing fresh produce in the majority of countries, although is banned in Germany, Holland, Denmark and Belgium because of the risk from DBP (Nguyenthe and Carlin, 1994).

There are different categories of sanitizers with different applications and targets. With respect to produce washes, any sanitizer with a no residue claim requires to pass the AOAC Germicidal and Detergent Sanitizers Method. The target of 5 log cfu reduction of the appropriate pathogen requires to be achieved with a 30 second contact period in the present of organics (e.g. serum albumin). There is also a need to provide evidence that no residues are left on the product and generation of DBP is within permit limits. To minimize the risk of residue carry over, there is a stipulation for certain sanitizers (e.g. acidified sodium chlorite) that a potable wash rinse is applied following the biocidal wash.

Within the USA, chemical sanitizers for fresh produce must meet the requirements specified in 40 CFR 180.940 and 21 CFR 173.315 with ozone being covered in 21 CFR 173.368. A selection of approved sanitizers for fresh produce is provided in Table 6.1 along with permitted levels. It should be noted that specific sanitizer concentrations differ between produce types. For example, cantaloupes can be washed in up to 1000 ppm hypochlorite.

6.3.1 Washing of organic produce

Organic fruit and vegetables are governed by certification organizations such as the Soil Association. Although the philosophy of organic produce is to avoid chemicals, there is a relatively wide range of sanitizers that can be applied. For example, chlorine can be used in wash water provided that the residual levels in discharge or effluent are below 4–10 ppm in accordance with the EPA Safe Drinking Water Act. To be effective, hypochlorite is more commonly added at 50–200 ppm free chlorine, and consequently there is a need to perform water treatment (e.g. holding in lagoons) prior to disposal to meet the 4 ppm residual limit.

The Soil Association, along with others, deter the use of chlorine and encourage processors to use alternative sanitizers such as ozone, acetic acid, hydrogen peroxide and peroxyacetic acid. Quaternary ammonium salts (QAS) are permitted

Table 6.1 List of sanitizers approved for use in washing fresh-cut produce

Sanitizer	Maximum concentration (ppm)
Hypochlorite	200 ppm
Organic acids	GRAS
Aliphatic acids	1.0%
Potassium bromide	0.2%
HEDP	13 ppm
Peroxyacetic acid	80 ppm
Acidified sodium chlorite	1200 ppm
Hydrogen peroxide	59 ppm
Trisodium phosphate	GRAS
Chlorine dioxide	3 ppm
Alkylene oxide adducts	0.2%
Polyacrylamide	10 ppm
Naphthalene sulfonates	0.2%
Ozone	GRAS

Source: taken from Anon. (2012).

HEDP, hydroxyethylidene diphosphonic acid. GRAS, Generally Regarded As Safe.

but produce needs to undergo a postwash rinse in potable water to remove any sanitizer residues. Although a wide range of sanitizers can be used, it is evidently not applied across the entire organic fresh produce industry (Olmez and Kretzschmar, 2009).

6.4 Current postharvest decontamination methods

6.4.1 Hypochlorite

Chlorine is the most commonly applied sanitizer in fresh produce washing. In many ways, chlorine represents an ideal sanitizer in terms of range of microbes affected, inability of microbes to generate significant resistance, cheap, stable, ease of handling, monitoring and application (Olaimat and Holley, 2012). Yet, there are disadvantages with regards to the rapid reaction (sequestering) of chlorine with organic (e.g. proteins, phenols) materials to generate DBP that can be potentially carcinogenic and also leaves swimming pool taints on produce (Cardador and Gallego, 2012; Hsiao *et al.*, 2012; Jeong *et al.*, 2012). The formation of DBP can rapidly deplete free chlorine, thereby reducing antimicrobial activity and necessitates continuous introduction of hypochlorite in the presence of organic loading encountered in fresh-cut produce processing (Olaimat and Holley, 2012).

The chemistry of chlorine is strongly linked to pH. Under alkaline conditions (pH >8) the hypochlorite anion dominates, which, although stable, has negligible antimicrobial activity. As the pH is decreased to pH 6.0–7.5, the hypochlorous acid dominates and represents the most antimicrobial form of chlorine. Yet, the hypochlorous acid readily reacts with organics to produce the DBP (Huang and Batterman, 2009). The extent of DBP formation is not only dependent on the pH but also the chlorine concentration and produce type. With regards to the latter, haloacetic acid concentrations associated with Iceberg lettuce are generally higher than Romaine (Table 6.2). Although not studied to any great extent, it is thought the generation of DBPs is related to the tryptophan concentration of the component proteins (Huang and Batterman, 2009).

As the acidity increases further, the formation of DBP decreases with chlorine becoming increasingly volatile. At pH 4 the chlorine converts to chlorine gas that is highly toxic and represents a hazard to worker safety.

Given that hypochlorous acid is the most active form, the pH of wash tanks is poised at pH 6.5–7.0 using an organic acid such as citrate, although inorganic acids such as phosphoric acid have also been applied. Although not studied to any great extent, the merits of inorganic acid pH control are that it does not contribute to the organic loading of water that needs to be removed during wastewater treatment.

Free chlorine levels can be monitored intermittently using test strips, then hypochlorite added to reach the target concentration (50–200 ppm). However, feedback loops based on measuring the ORP are more commonly encountered in fresh-cut processing because of the high organic loading content of water that

Table 6.2 Haloacetic acids (HAA) disinfection byproducts recovered from bagged fresh-cut produce purchased at retail

Produce type	HAA concentrations (μg/kg)				
	Dichloroacetic (DCAA)	Trichloroacetic (TCAA)	Bromochloroacetic (BCAA)	Dibromoacetic (DBAA)	Dibromochloroacetic (DBCAA)
Carrot	6.3	13	1.2	0.9	<2
Green pepper	ND	0.7	ND	ND	ND
Iceberg lettuce	15	18	3.9	1.3	3.2
Romaine lettuce	3.5	5.2	ND	<0.8	<2
Chicory	<0.4	1.6	ND	ND	ND
Spinach	ND	0.5	ND	ND	ND
Mixed salad	22	24	3.2	1.4	4.8

Source: adapted from Cardador and Gallego (2012).
ND, not detected.

rapidly depletes free chlorine concentration. Here, the ORP provides an estimation of the oxidizing species in water that in the case of hypochlorite solutions will be free chlorine. The ORP is set at a designated value (650–750 mV) that is attained through the addition of hypochlorite. There is narrow linear range between free chlorine and ORP with interference from other oxidizing agents, in addition to the influence of pH (Fig. 6.1). The benefit of the ORP system is that the free chlorine levels are maintained at a constant, thereby providing a degree of control over the process. Limitations of ORP include that it is an indirect measure of chlorine level, sensor response is non-linear to chlorine and it has a slow response time. Consequently, it has been proposed to apply the ORP feedback loop system but also monitor the free chlorine concentration periodically to verify the wash process is under control. It should also be noted that the addition of increasing amounts of hypochlorite in the presence of accumulating organics leads to a greater proportion of disinfectant byproduct formation. Although the health risk posed by DBP is lower than that of microbiological hazards occurring in the absence of chlorine, there is a sustained effort to find an alternative sanitizing agent in the fresh-cut sector (Artes and Allende, 2005).

An advance in the area of chlorine has been the introduction of stabilizing agents that function to maintain antimicrobial activity in the presence of high organic loading. Cyanuric acid is an example of a stabilizing agent and reacts with hypochlorite to produce chloromide that exhibits antimicrobial activity but is more stable in the presence of organics or photo-oxidation. With respect to

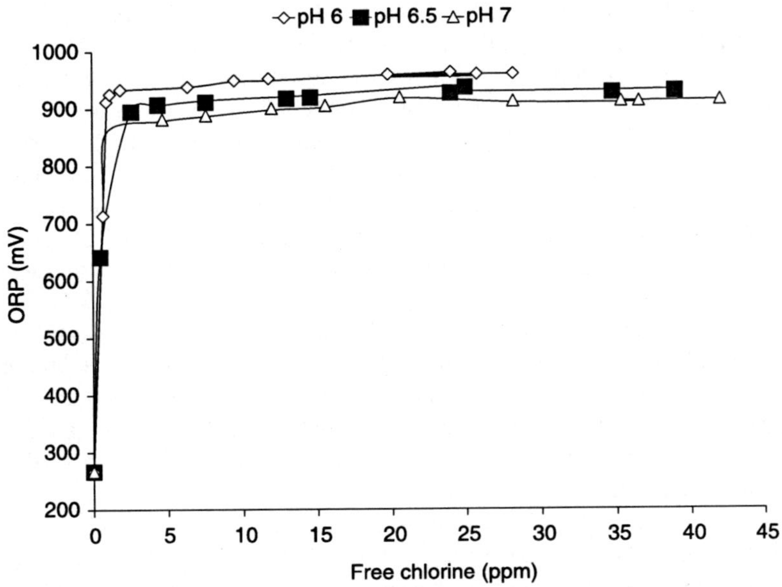

Fig. 6.1 Relationship between free chlorine (ppm) and measured oxidation reduction potential (ORP) at different pH. Taken from Warriner (unpublished data).

produce washing, the chlorine stabilizer designated T-128 has been demonstrated to maintain a high ORP in the presence of soil or homogenated lettuce (Nou *et al.*, 2011). It was noted that the addition of T-128 did not enhance the log reduction of *E. coli* O157:H7 that was the same as for chlorine wash (1 log cfu) without stabilizing agent. Yet, the inclusion of T-128 did support a 2 log cfu reduction of *E. coli* O157:H7 and *Salmonella* in wash water supplemented with lettuce homogenate (Nou *et al.*, 2011). Consequently, the stabilizing agent can provide an extra degree of protection to minimize the transfer of pathogens between batches.

6.4.2 Electrolyzed water

Electrolyzed water is produced by the electrolysis of NaCl to produce an acidic hypochlorite fraction (pH 2.45) at the anode and basic NaOH (pH >9) at the cathode (Graca *et al.*, 2011). As a result of the instability/reactivity of the hypochlorite, it is common to generate electrolyzed water on-site then store for a brief time (<24 hours). There are several manufacturers of electrolyzed water generators ranging from domestic units through to systems that can produce over 400 liters/hour. Yet, the cost of generators can be expensive (>$100k), corrosion of equipment/resin coatings can occur and units require constant maintenance (Ayebah and Hung, 2005). However, the chemistry of electrolyzed water does provide certain advantages over bleach. For example, the alkaline fraction can be used in sanitation to remove protein deposits and also has a degree of antimicrobial activity (Rahman *et al.*, 2011). The biggest advantage of electrolyzed water is the lower generation of DBP compared with hypochlorite (Olmez and Kretzschmar, 2009).

The hypochlorite generated by electrolyzed water is no different to that found in bleach. However, the key difference relates to pH that in turn affects the extent to which DBPs are formed. As previously discussed, bleach is essentially an alkaline solution of chlorite anion that is converted to hypochlorite by adjusting the pH to 6.5–7.5. The neutral pH promotes generation of DBP. With respect to the electrolyzed water, the pH of the acidic fraction is typically adjusted to pH 6 to enhance stability compared with when in the highly acidic form (Issa-Zacharia *et al.*, 2011). Consequently, the formation of DBP is less likely and the hypochlorite simply reverts back to sodium chloride, although can volatize into chlorine gas if high acid (pH <4) preparations are applied.

The antimicrobial properties of electrolyzed water are comparable with bleach, with the added effect of controlling pH through combining the appropriate concentrations of acidic and alkaline fractions. In the main, the highest antimicrobial activity is observed in the slightly acidic region (pH 3.5–5.5) compared with when in the strong acidic form or neutralized (Graca *et al.*, 2011; Issa-Zacharia *et al.*, 2011).

Electrolyzed water has been evaluated as an alternative to bleach for produce washing. The acidic fraction (pH 2.45, 16 free chlorine ppm) supported a 1 log cfu reduction of total aerobic count on non-inoculated cilantro (Wang *et al.*, 2004). Park *et al.* (2008) reported a >5 log cfu reduction of *E. coli* O157:H7, *Salmonella* and *Listeria* inoculated onto green onions then treated with acidic electrolyzed water (pH 2.06, 37.5 ppm free chlorine). When applied in slightly acidic form (pH 5.6, 104 ppm free chlorine), electrolyzed water decreased levels of *E. coli* O157:H7 and *Salmonella* on lettuce by 2.7–2.9 log cfu (Issa-Zacharia *et al.*, 2011).

Spinach inoculated with *E. coli* O157:H7, *Salmonella*, *Staphyloccocus aureus*, *L. monocytogenes* and *Enterococcus faecalis* could be decreased by a 10 minute soak in electrolyzed water (pH 6.5, 100–200 ppm free chlorine) (Guentzel *et al.*, 2008). The same authors reported a decrease of only 0.24 log cfu of *E. coli* O157:H7 when the same treatment was applied to lettuce with the other bacterial levels being decreased by 2.43–3.81 log cfu. Graca *et al.* (2011) applied basic electrolyzed water (pH 8.06, 50 ppm free chlorine) to decontaminate lettuce inoculated with *E. coli* O157:H7, *Salmonella* and *L. monocytogenes*. The log reduction was restricted to 1–2 log cfu reduction and comparable to standard chlorine washes with no improvement in the disinfection process.

As with many sanitizer validation studies, comparing the results from one study with another is problematic because of differences in methodology such as pH and free chlorine levels, in addition to produce type. One of the main omissions of studies is the lack of a bleach control making direct comparisons between standard chlorine washes and those based on electrolyzed water problematic.

Electrochemical disinfection of wash water is a novel extension of the electrolyzed water process (Lopez-Galvez *et al.*, 2012). As the term suggests, electrochemical disinfection involves the electrolysis of water to generate hydroxyl radicals, ozone and hypochlorite. The main point to note is that with electrochemical disinfection the aim is to reduce the microbial loading of water as opposed to disinfection of product. Therefore, the process should be viewed as a means of reducing cross-contamination events as opposed to removing field acquired contamination.

Electrochemical disinfection essentially recirculates water over polarized boron doped diamond electrodes (Lopez-Galvez *et al.*, 2012). The electrochemical reaction not only generates free radicals and hypochlorite, but also decreases the chemical oxygen demand (COD) that aids in water remediation prior to disposal. From studies performed to date it has been reported that electrochemical disinfection could support >5 log cfu reduction of *E. coli* O157:H7 with a short treatment time (<1 minute). Yet, in the presence of organic matter (COD of 800 mg/l) the exposure time to support of 5 log cfu was increased to >100 minutes (Lopez-Galvez *et al.*, 2012). The presence of organic matter presumably reacts with the generated radicals thereby sequestering antimicrobial activity. The treatment time to achieve the 5 log cfu reduction in *E. coli* O157:H7 in the presence of organic matter could be decreased to 25 minutes by the addition of sodium chloride, which likely generates hypochlorite with less contribution from hydroxyl radicals.

6.4.3 Chlorine dioxide

Chlorine dioxide can be generated *in situ* by combining sodium chlorite with chlorine (gas, hypochlorite or hydrochloric acid), or derived from acidification of a stabilized sodium chlorite solution. As with hypochlorite, the chemistry of chlorine dioxide is dependent on pH. The addition of acid (organic or inorganic) to stabilized sodium chlorite produces a green-yellow solution referred to as acidified sodium chlorite. In the case of chlorine dioxide generators, the pH of the chlorite solution is decreased to below pH 4, thereby generating chlorine dioxide gas that is unstable and volatile. As a consequence, the chlorine dioxide gas partitions from the water or decomposes back to chlorite or chlorate. Importantly, chlorine dioxide has 2.5 times more oxidizing power than hypochlorite, is relatively stable in the presence of organic material and does not generate significant DBP (Lopez-Galvez *et al.*, 2010; Sharma and Sohn, 2012).

In terms of disinfection potential, 2.5 ppm chlorine dioxide can effectively decrease counts in wash water but does not result in greater log reductions of pathogens inoculated onto fresh produce such as tomatoes (Lopez-Velasco *et al.*, 2012; Tomas-Callejas *et al.*, 2012a, b). Although stable in the presence of organics, chlorine dioxide is lost to the atmosphere, degradation back to its constituent part and dilution by water recharge. Consequently, a feedback loop based on ORP monitoring is commonly undertaken to maintain chlorine dioxide levels within the wash tank.

The decontamination efficacy of acidified sodium chlorite is the same when compared with hypochlorite-based washes (Elano *et al.*, 2010). Although the antimicrobial activity of acidified sodium chlorite within spent wash water has yet to be reported, it can be considered to be similar to chlorine dioxide. Therefore, acidified sodium chlorite and chlorine dioxide can be considered to share the same antimicrobial efficacy with respect to produce disinfection. Acidified sodium chlorite has the advantage of not requiring expensive *in situ* generators, but the disadvantages include hazards in handling chlorite and the need to manually introduce the sanitizer into the wash tank.

6.4.4 Ozone

The antimicrobial activity of ozone was identified around the same time as that of chlorine, although the latter became popular because of cost, stability and efficacy. Ozone can be generated by passing air (oxygen) across an electrical arc or UV (174 nm) lamps (Perry and Yousef, 2011). Ozone has 30 times the oxidizing power of hypochlorite but is essentially insoluble in water (saturates at 4 ppm) and rapidly decomposes in the presence of organic matter. Consequently, ozone requires to be constantly introduced into wash water via an online generator.

Aqueous ozone does not result in higher log reductions on fresh produce compared with hypochlorite, although does have the benefit of not producing DBP (Rosenblum *et al.*, 2012). Yet, because of the poor solubility of ozone, there can be issues with off-gassing into the environment, in addition to corrosion of wash tanks and negative impact on the sensory characteristics of the product (Kim *et al.*, 2003).

Advances in ozone treatment of fresh produce include the delivery of the gas in nanobubbles, which increases antimicrobial efficacy through greater surface area, in addition to enhancing stability (Sakamoto, 2011). Yet, studies performed using ozone nanobubbles report the same efficacy for leafy green decontamination as hypochlorite-based washes (Inatsu *et al.*, 2011).

There continues to be interest in applying ozone when washing fresh produce, although the cost of generators, instability in the presence of organics and equivalency in decontaminating efficacy compared with hypochlorite are clear limitations.

6.4.5 Organic acids

Organic acids such as lactic, citric, malic and acetic, among others, have advantages over hypochlorite in terms of being stable in high organic loading and are Generally Regarded As Safe (GRAS). However, their antimicrobial activity is generally low and requires long contact times, combined with high concentrations (i.e. parts per thousand as opposed to parts per million for chlorine), to reduce microbial loading on fresh produce (Olaimat and Holley, 2012). For example, Park *et al.* (2011) reported the efficacy of different organic acids in decontamination of sliced apples and lettuce. Treatments using 1–2% organic acids with a 10 minute contact time supported a 2 log cfu reduction in pathogen (*E. coli* O157:H7, *Salmonella* and *Listeria*) levels with malic and citric acids being notably more effective compared with the other acid types tested (Park *et al.*, 2011).

Peroxyacetic acid is an organic acid that has found increasing utility in the fresh produce sector despite the relatively high cost relative to bleach (Vandekinderen *et al.*, 2009). The sanitizer is essentially produced from the reaction between hydrogen peroxide and acetic acids with a working concentration of 20–80 ppm being applied in fresh produce washing. The main advantage of peroxyacetic acid over chlorine is the enhanced stability in the presence of organic matter (Olmez and Kretzschmar, 2009). The log reductions of pathogens (*E. coli* O157:H7, *Listeria monocytogenes and Salmonella*) on fresh produce achieved with peroxyacetic acid vary between 0.4 and 3.5 log cfu. Although bacterial pathogens do not vary in resistance to peroxyacetic acid, there is variation in reductions achieved when different product types are evaluated. For example, highest log reductions of pathogens inoculated onto cabbage are higher compared with leeks (Vandekinderen *et al.*, 2009; Ho *et al.*, 2011). The efficacy of peroxyacetic acid (70 ppm) can be enhanced through combination with lactic acid (2000 ppm) (commercially marketed as Freshrinse) to increase the log reduction of the endogenous microflora on lettuce from 0.5 to 1.5 log cfu compared with chlorine washes (pH 7, 10 ppm) (Ho *et al.*, 2011). However, the main antimicrobial effect was observed in wash water whereby pathogens could be decreased by >7 log cfu in spent wash water (Ho *et al.*, 2011). Therefore, peroxyacetic acid can be considered an effective sanitizer to reduce cross-contamination events during the washing process (Ho *et al.*, 2011; Ho, 2012).

6.4.6 Surfactants

Surfactants act as wetting agents that reduce the surface tension of liquids and physically remove contamination from surfaces. Surfactants can be subdivided into anionic (e.g. dodecyl sulfate; SDS), cationic (quaternary ammonium compounds) or non-ionic (e.g. Tween, Triton X-100). Surfactants alone are ineffective antimicrobial agents, although they enhance the antimicrobial action of sanitizers (Predmore and Li, 2011). A combination of 50 ppm of anionic and non-ionic surfactants applied in combination with 200 ppm chlorine supported 3 log reduction of murine norovirus 1 inoculated onto strawberries (Predmore and Li, 2011). This compares with 1 log cfu reduction in virus levels with chlorine alone.

Combinations of anionic surfactants and short chain fatty acids (70% caprylic acid, 10% PE-surfactant and 20% lactic acid) applied with aqueous chlorine (200 ppm) or chlorine dioxide (100 ppm) solution were used to decontaminate lettuce inoculated with *E. coli* O157:H7 (Keskinen and Annous, 2011). The combined methods supported >5 log cfu reduction of *E. coli* O157:H7, with treatments applied individually resulting in <1 log cfu decrease in levels (Keskinen and Annous, 2011).

Disadvantages of surfactants include the negative effects on sensory properties of fresh produce and excessive foaming when applied in commercial operations (Ho, 2012).

6.4.7 Hot water

Thermal treatment of fresh produce offers a chemical-free decontamination process. Yet, because of detrimental changes in texture, there are a limited number of produce types that are compatible with the treatment. In commercial terms, melons (watermelon, cantaloupe) are the most relevant type of fresh produce treated using hot water. A treatment of 85 °C for 60 seconds supported a 4.7 log cfu reduction in *Salmonella* levels without causing excessive softening to the cantaloupe (Solomon *et al.*, 2006). However, there is evidence that hot water treatment of melons can induce changes in the texture that make the product more susceptible to rapid spoilage during subsequent storage (Fan *et al.*, 2008). Consequently, there is interest in using combined treatment to enable the time/temperature to be reduced or enhance the decontamination efficacy through inactivation of spoilage microflora. Hot water (90 °C) in combination with 5% acetic acid did not enhance decontamination of cantaloupes compared with when the organic acid was omitted (Fouladkhah and Avens, 2010). A combination of hot water (75 °C for 60 seconds) followed by ozone gas treatment (10 000 ppm for 30 minutes) reduced endogenous microflora by 3.8 log cfu with no detrimental changes in shelf-stability (Selma *et al.*, 2008). A further successful treatment involved initially treating melons with hot water at 60 °C for 90 seconds followed by a dip in paracetic acid (80 ppm 60 seconds) (Silveira *et al.*, 2011). Hot water temperature can also be reduced by incorporating physical brushing of the produce during treatment. A 3 log cfu reduction of *E. coli* on melons was achieved using a combination of brushing and hot water at 58 °C (Fallik *et al.*, 2007).

A further currently available hot water-based treatment combines UV light (254 nm) to enhance shelf-life of mango, cantaloupe, onion, garlic, leek, celery and broccoli (Siddiqui *et al.*, 2011). The system has been commercialized by Freshappeal based in Australia and consists of a reactor constructed from quartz tubes surrounded by UV lamps. The primary application is for inactivating enzymes implicated in browning, in addition to reducing endogenous microflora on fresh-cut fruit. The hot water (50 °C) acts to transport, separate the fruit and contributes to the lethality of the process. Although the system is novel, no independent trials have been undertaken to assess the efficacy for pathogen inactivation.

6.5 Best practices in postharvest washing

6.5.1 Monitoring the efficacy of the postharvest wash process

Given that the postharvest wash process is a designated CCP, there is a need to establish a validated process and also designate a parameter that can be monitored to ensure the step is under control. The difficulty arises in defining the CCP given that there is no clear definition on what expectations can be placed on the wash process in terms of human pathogen control. Specifically, is the objective of the postharvest wash to achieve a designated log reduction of human pathogens or to prevent cross-contamination between batches? The majority of research to date has focused on achieving maximal log reductions of pathogens (*E. coli* O157:H7, *Salmonella*, *Listeria monocytogenes*).

Although a standardized protocol for undertaking validation trials has been proposed, it has not been widely adopted (Burnett *et al.*, 2004). Consequently, comparing the results between trials is difficult and partly explains the wide variation encountered in efficacy data. Regardless of this fact, there has yet to be a designated target log reduction to be attained by a decontamination process. A reduction of 5 log cfu has been proposed given that this is the level of inactivation required for pasteurization-based methods. The performance of any sanitizer or process can be compared with that of standard chlorine washes, although many studies do not include comparative controls.

Although laboratory-based validation trials on pathogen reduction are informative, the performance of sanitizers rarely can be extrapolated to predict efficacy within commercial processing lines. One obvious effect is the method of introducing pathogens that are invariably on the surface of produce and at greater levels than encountered under natural conditions. Naturally contaminated produce harbors lower levels with a proportion located at cut-edges and sub-surface structures (Warriner *et al.* 2009). However, the main factor to explain the inconsistency between laboratory trials and commercial scale processing is the dynamic nature of the industrial produce washing process. Within commercial washes the microbial counts on incoming raw material, accumulation of microbes and organic matter within the wash water, in addition to active concentration of sanitizer, are constantly changing. Consequently, there is a need to establish which parameter(s) impact on the wash efficacy, then attempt to set limits to define a process that is under control or at least operating at an acceptable level.

To establish critical limits, there is a need to determine which of the parameters have an impact on the wash process. In this respect there are challenges with respect to recreating the conditions of the wash water process in the laboratory. Consequently, the majority of studies to date have been performed within commercial processing facilities.

Barrera *et al.* (2012) reported on a study performed in three facilities, two of which processed salad spinach and the other shredded lettuce. Throughout the processing activity, samples were taken of pre- and postwash product, in addition to spent wash water. Microbiological counts were made in produce samples and water, determined along with water parameters (temperature, conductivity, sanitizer concentration and turbidity). The results of the study confirmed that limited reductions occurred during washing of leafy greens (<1 log cfu). In general, the log reductions in total aerobic counts and coliforms could be positively correlated to bacterial counts of prewashed leafy greens (range 4.98 to >7.20 log cfu/g) and conductivity (solids content; range 49–540 μS/cm) of the wash water (Table 6.3).

Table 6.3 Range and averages of processing parameters measured in spinach (Facility A and B) and shredded lettuce (Facility C) processors

	Range	Average	Correlation coefficient* ACC LCR	Correlation coefficient* coliform LCR
Facility A				
Prewash spinach ACC (log cfu/g)	4.78 to –>7.40	6.44 ± 0.77	0.40 (P<0.05)	
Prewash spinach coliforms (log cfu/g)	0.89–4.54	2.90 ± 1.10		0.40 (P<0.05)
Prewash spinach temperature (°C)	6.6–16.3	11.19 ± 2.94	0.18 (P>0.05)	−0.22 (P<0.05)
Postwash spinach temperature (°C)	8.1–18.2	13.50 ± 1.16	0.27 (P>0.05)	−0.19 (P<0.05)
Wash water HPC (log cfu/100 ml)	0.95 to >3.34	2.73 ± 0.82	−0.04 (P>0.05)	
Wash water coliforms (log cfu/100 ml)	<0.90–4.45	3.12 ± 1.26		−0.30 (P>0.05)
Wash water temperature (°C)	7.6–18.7	13.49 ± 3.23	0.22 (P>0.05)	−0.15 (P>0.05)
Temperature difference (prewash spinach – water) (°C)	−5.5 to 0.8	−2.30± 1.84	−0.23 (P>0.05)	−0.12 (P>0.05)
Water conductivity (μS/cm)	32.4–480	83.81 ± 27.41	0.34 (P<0.05)	0.02(P>0.05)
pH	4.79–8.36	7.33 ± 2.19		0.15 (P>0.05)
Turbidity	0.008–0.171	0.058 ± 0.053	0.08 (P>0.05)	0.25 (P>0.05)
Peroxyacetic acid concentration (ppm)	<0.1–18	6.10 ± 5.22	−0.04 (P>0.05)	−0.18 (P>0.05)

(*Continued*)

Table 6.3 Continued

Facility/parameter	Range	Average	Correlation coefficient* ACC LCR	Correlation coefficient* coliform LCR
Facility B				
Prewash spinach ACC (log cfu/g)	4.88–6.93	5.98 ± 0.53	0.13 (P<0.05)	
Prewash spinach coliforms (log cfu/g)	0.89–4.54	1.45 ± 0.74		0.53 (P<0.05)
Prewash spinach temperature (°C)	4.0–10.2	6.74 ± 1.40	−0.14 (P>0.05)	0.30 (P>0.118)
Postwash spinach temperature (°C)	5.5–11.5	8.71 ± 1.50	−0.52 (P<0.05)	0.02 (P>0.05)
Wash water HPC (log cfu/100 ml)	<0.9–3.74	1.49 ± 0.66	−0.19 (P>0.05)	
Wash water coliforms (log cfu/100 ml)	0.6–2.44	1.42 ± 0.81		−0.03 (P>0.05)
Wash water temperature (°C)	9.4–17.11	12.75 ± 2.47	−0.41 (P>0.05)	−0.37 (P=0.05)
Temperature difference (prewash spinach – water) (°C)	−12.5 to −0.9	−6.01 ± 2.78	−0.12 (P>0.05)	0.57 (P<0.05)
Water conductivity (µS/cm)	49–260	119 ± 14	−0.17 (P>0.05)	0.38 (P<0.05)
pH	6.74–8.15	7.53 ± 0.11	0.22 (P>0.05)	0.45 (P<0.05)
Turbidity	0.002–0.153	0.036 ± 0.036	0.056 (P>0.05)	0.34 (P>0.05)
ORP (mV)	247–748	383 ± 127	−0.19 (P>0.05)	0.20 (P>0.05)
Facility C				
Prewash lettuce ACC (log cfu/g)	4.59–6.50	5.34 ± 0.59	−0.46 (P>0.05)	
Prewash lettuce coliforms (log cfu/g)	0.60–4.13	1.30 ± 0.57		0.90 (P<0.05)
Prewash lettuce temperature (°C)	3.4–7.0	5.30 ± 0.98	0.30 (P>0.05)	0.08 (P>0.05)
Postwash lettuce temperature (°C)	1.0–4.0	2.24 ± 0.90	0.02 (P>0.05)	−0.33 (P>0.05)
Wash water HPC (log cfu/100 ml)	<0.9–3.74	1.58 ± 0.65	0.07 (P>0.05)	
Wash water coliforms (log cfu/100 ml)	0.65–3.52	0.91 ± 0.33		−0.037 (P>0.05)
Wash water temperature (°C)	0.2–2.1	1.08 ± 0.57	−0.18 (P>0.05)	−0.43 (P<0.05)
Temperature difference (prewash lettuce – water) (°C)	2.3–5.4	4.22 ± 1.10	0.41 (P>0.05)	0.38 (P>0.05)
Water conductivity (µS/cm)	69–540	240 ± 168	0.16 (P>0.05)	0.021 (P>0.05)
pH	6.97–8.11	7.47 ± 0.26	0.08 (P>0.05)	0.078 (P>0.05)
Turbidity	0.001–0.141	0.123 ± 0.27	0.036 (P>0.05)	0.142 (P>0.05)
ORP (mV)	344–890	598 ± 152	−0.09 (P>0.05)	0.48 (P<0.05)

Source: taken from Barrera *et al.* (2012). *Significance was set at 5% confidence level. Sample sets of fresh produce and wash water were taken at different intervals during the processing activity.

Spearman's coefficient was then applied to determine which processing parameters significantly correlated to the LCR achieved in ACC and coliform counts.

Abbreviations: ACC, aerobic colony counts; HPC, heterotrophic plate counts; LCR, log count reduction; ORP, oxidation reduction potential.

A negative correlation was found between LCR and water temperature (range 0.2–18.7 °C). Interestingly, within the ranges measured the log reduction was independent of the bacterial loading of the water (range 0.95–>3.34 log cfu/100 ml) and temperature differential between produce and water (Table 6.3). The results from the study illustrated that solids content (conductivity) and low water temperature enhanced the log count reductions of the wash process. The results were counterintuitive given that warm water temperatures are considered necessary to prevent ingress of contamination within produce (Xia *et al.*, 2012). In addition, the presence of solids (inorganic or organic) decreased the efficacy of the process through sequestering sanitizers. Yet, it is known that solids such as clays can bind microbial cells that could account for positive effects on log reductions obtained in the washing process (Stevik *et al.*, 2004). A further noteworthy finding of the Barrera *et al.* (2012) study was that, despite the differences in how the facilities operated in terms of size of operation, sanitizer type and water recharge rates (i.e. the rate at which fresh water was introduced), the log reductions obtained by the wash process were the same (Barrera *et al.*, 2012). It was also observed that apart from microbial loading on prewash leafy greens, there was no single parameter that could be used to predict the log reduction obtained by the wash process. Given the multiple variables of the wash process, it is likely that interaction of different parameters needs to be taken into account and that this will be the focus of future research.

Given the limitation of postharvest to remove contamination from fresh produce, there is starting to be a collective view that the process or critical point should be to avoid cross-contamination between batches. Prevention of cross-contamination represents a different challenge to decontaminating produce directly, given that the water parameters (sanitizer concentration, pH, temperature, solids content) require to be considered. Approaches to minimizing cross-contamination within produce wash tanks have not received significant attention to date. Yet studies have been performed on relatively large scales to illustrate how pathogens (specifically *E. coli* O157:H7) transfer between batches during the washing process. A study described by Buchholz *et al.* (Buchholz *et al.*, 2012b) illustrated how *E. coli* O157:H7 can be transferred across batches during a processing period. Here, *E. coli* O157:H7 was inoculated into 22 kg of lettuce at levels ranging from 2 to 6 log cfu/g. The inoculated product (75 kg) was run through the shredding line and followed with non-inoculated product. The results of the study illustrated that, in the absence of sanitizer, *E. coli* O157:H7 could be recovered on 90 kg of product when introduced at the higher inoculation level. With lettuce inoculated at the level of 2 log cfu/g, the initial 21 kg of product run after inoculated lettuce harbored that pathogen. The study confirmed that in the absence of sanitizer contamination can persist and transfer throughout the line even when pathogen levels are low (Buchholz *et al.*, 2012b, a). The same conclusions were reported in a comparable study performed by Luo *et al.* (Luo *et al.*, 2012), who illustrated the transfer of *E. coli* O157:H7 between lettuce batches in a large-scale study (45 kg/min). The wash water was supplemented with 20 ppm free chlorine, and inoculated lettuce passed through the wash tank.

The researchers illustrated how free chlorine concentration was totally depleted after passage of 540 kg of lettuce (12 minutes of processing). The study went on to illustrate that a critical free chlorine level of >1.0 ppm at 30 seconds exposure could inactivate *E. coli* O157:H7 present at levels of 1–8 MPN/ml (Luo *et al.*, 2012). Yet, it has been proposed to ensure a minimum of 10 ppm free chlorine to provide assurance that cross-contamination between batches is negligible (Luo *et al.*, 2011). Therefore, by monitoring the free chlorine levels and ensuring levels do not decrease below 10 ppm, a CCP can be put in place. With non-hypochlorite-based sanitizers, the target concentration is typically derived from the Use-Dilution tests, which essentially determine the extent of microbial inactivation over a 30 second period (Cole *et al.*, 1988). Although the test can be performed in the presence of organic matter such as albumin, it is more informative to determine the required concentration and contact time in the spent wash water of interest.

6.5.2 Reducing the organic loading of wash water

Although a degree of organic matter in water appears to beneficial with respect to enhancing the log reductions of the wash process, it is well established that it can also result in rapid depletion of hypochlorite. Although the depletion of free chlorine can be compensated by simply introducing more hypochlorite into the wash tank via an ORP feedback loop, this would lead to an increase in DBP accumulation. Therefore, it is critical to keep to a minimum the amount of organic accumulation within the wash tank. With respect to lettuce, it has been demonstrated that washing whole heads prior to shredding releases less latex into the wash water thereby minimizing the accumulation of organic material and ingress of contamination via cut edges (Nou and Luo, 2010). A further approach is to spray lettuce with sanitizer solution during the shredding process to remove both contamination and exudate from cut surfaces that would otherwise accumulate in the wash tank (Toivonen *et al.*, 2011).

A more common approach to reduce organic loading of wash tanks is to perform recharging/replacing with fresh water over the course of processing. The recharge rate varies between facilities and is rarely based on scientific data. In certain facilities the water recharge rate is sufficient to ensure a constant level within the tank, thereby having a negligible effect on organic loading. Yet in other facilities the recharging rate can be up to 50% every hour, which represents a significant investment in water resource, in addition to increasing costs of wastewater treatment.

With pressure on water resources and increasing costs of wastewater treatment, there is increasing interest in recycling technologies in the fresh-cut industry. To date, wash water treatment has focused on removing organics and non-organic material prior to disposal (Kern *et al.*, 2006). In the case of wastewater treatment there is interest in reducing such parameters as biological oxygen demand (BOD), chemical oxygen demand (COD), turbidity, total soluble solids (TSS) and in the case of organic production, residual chlorine (Table 6.4). The characteristics of wastewater leaving wash tanks are dependent on the type of produce, product:water

Table 6.4 Wash water parameters of significance in treatment and recycling

Parameter	Description	Method of determination	Acceptable limit	Spent wash water*
Biological oxygen demand (BOD)	Biological available organic matter	Oxygen consumed by mixed culture inoculated into the water sample	5 mg/l	500
Chemical oxygen demand (COD)	Oxidizable organic matter	Permanganate titration	10 mg/l	1500
Conductivity	Estimate of charged organic and inorganic species	Conductivity meter probe	400 μS/cm	1500 μS/cm
Total solids	Organic and inorganic soluble and insoluble solids	Evaporation and weighing residue Soluble solids determined using filtered samples	500 mg/l	800
Turbidity	Cloudiness of water caused by suspended solids	Nephelometric turbidity unit (NTU) meter	<5 NTU	3.5

Source: taken from Luo (2007).

*Spent water from processing 18 kg shredded Romaine lettuce in 40 liters chlorinated (150 ppm) water.

ratio and recharge rate. In relative terms, water derived from washing leafy greens has a low loading (Table 6.4) compared with root vegetable processing where the BOD can be in the order of 20 000 mg/l.

Many large fresh produce processing facilities have on-site wastewater processing facilities, although the water is simply disposed of or used for irrigation rather than returned to the plant. Yet, with pressures on water resources it can be envisaged that recycling will become increasingly important.

There are different stages of water recycling treatment that can be defined as physical or chemical removal of solids. The initial stage is to remove the vegetable debris that can be achieved through bag filtration, centrifugation, cyclone separators and parabolic screens. The latter is preferred with regards to cost and low maintenance needs (Kentish and Stevens, 2001). The remaining solid particles <3 mm can be removed by cross-flow filtration, although the process is costly, readily fouled and cannot remove soluble solids unless ultrafiltration is applied. Therefore, a coagulation step can be undertaken to increase particle sizes to ease filtration and solubilize low molecular weight constituents. The nature of soluble and insoluble solids is dependent on the product type. For example, root vegetables contain a heavy loading of soils and clay. In contrast, leafy greens contain low molecular weight constituents and polymeric materials such as pectin, in addition to latex. Consequently, the wash water from leafy greens has a tendency to be colloidal in nature and can represent a challenge to coagulation. To induce coagulation, there is a need to add an agent to facilitate the aggregation of solids to facilitate removal (Badawy *et al.*, 2012). There are a range of coagulating agents available principally based on aluminum or ferric salts or polymers (Pacala *et al.*, 2012). The coagulation process essentially is the formation of aggregates by forming ionic bonding between solids that insolubilize to trap organic or inorganic constituents. Because of the range of constituents in wastewater, the type of coagulating agent and conditions require to be optimized by using a technique referred to as the jar test. In its most basic form, the jar test takes a representative sample of the wastewater into which the coagulating agent is added. After a 30 minute period of rapid stirring, the sediments are allowed to settle over time and the change in the relevant water parameter (e.g. turbidity, BOD, COD, dissolved organic carbon (DOC) or TSS) is recorded. Specialized units are available in which to perform the jar test, although it can be performed using a simple beaker and magnetic stirrer. The variables to be tested include the coagulant type and concentration, aeration, pH, temperature, mixing rate and supplements such as activated carbon. The criteria for success are the conditions that achieve optimum removal of solids following the downstream filtration step.

The next step following coagulation is a filtration step to remove the aggregates. There are several options, with sand filters representing the low cost and maintenance option. However, it should be noted that sand filters require a high volume of water (2 liters per kg sand) to perform backwashing, and hence are not always an appropriate choice. Alternatives are cyclone separators, bag filters, cross-flow filtration or centrifugation.

The final stage of water recycling is a disinfection step, which in the case of wash water could be undertaken by UV or injection of sanitizer such as chlorine prior to returning to the wash tank.

A small-scale (20 l/min) water recycling system is depicted in Fig. 6.2. The coagulation agent is introduced into the wastewater stream, which then passes through a static mixer to a retention or holding tank. The coagulated materials sediment to the bottom of the tank, with residual material being removed by mesh filters. The filtrate is passed through a carbon filter primarily to remove non-reacted coagulant, which can be potentially toxic at high levels. The final UV disinfection step is to inactivate microbes prior to returning to the wash tank.

Process water recycling technologies are in their infancy and are only now being developed. Yet, there are offline systems available, although these represent a significant investment in terms of hardware, energy, footprint and maintenance. Therefore, when selecting water treatment systems it is advisable to undertake a cost-benefit analysis.

6.6 Future trends

Given the poor efficacy of current fresh produce washes to ensure pathogen reduction, it can be envisaged that future research will be directed at finding more effective intervention strategies. A number of these technologies currently exist but will require further development to be compatible with the high throughput,

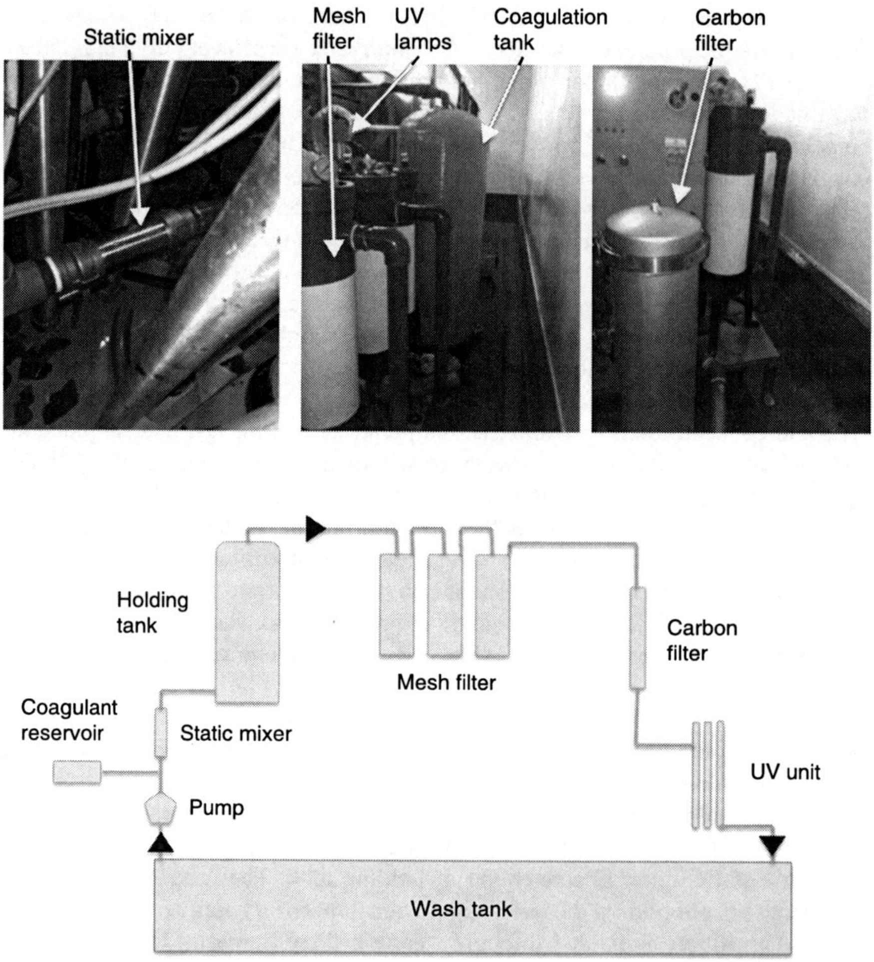

Fig. 6.2 Images and a schematic of a small-scale wash water recycling unit. Taken from Warriner (unpublished).

broad product range and limited financial resources of the fresh-cut produce sector.

Perhaps the most notable postwash intervention is ionization radiation, which has proven an effective method for inactivating pathogens and, notably, is independent of produce type (Moreira *et al.*, 2012). Among other factors, negative consumer perception of irradiation and cost are the underlying reasons for restricting irradiation within the fresh produce sector. To address such issues it has been proposed to apply low-dose e-beam irradiation, and research has demonstrated efficacy for inactivating *Listeria monocytogenes*, *Salmonella* and *E. coli* O157:H7 on a wide range of produce types with D values being in the region of

0.10–0.26 kGy (Moreira *et al.*, 2012). Enteric viruses exhibit greater resistance to irradiation, with D values in the order of 1.03–2.24 kGy being observed.

A further intervention technology showing promise is gas phase decontamination interventions. Unlike aqueous-based washes, the risk of cross-contamination is negligible and greater penetration of the antimicrobial species into the produce sub-surface is possible (Olaimat and Holley, 2012). However, gas phase treatments require specialized treatment chambers to contain noxious gases and, as a consequence, are compatible with batch treatments rather than the continuous flow commonly encountered in commercial processing operations. The latter has been addressed to a degree by using antimicrobial gases during the postharvest vacuum cooling treatment used to remove field heat and also modifying the headspace in stores/transporters. For example, ozone gas (935 ppm for 30 minutes) introduced into the vacuum-cooling chamber (prior to applying the vacuum) supported a 2.4 log cfu of *E. coli* O157:H7 on spinach. When the spinach was then treated with 10 ppm ozone over 3 days (mimic transport and storage), a further 1.4 log reduction of *E. coli* O157:H7 was observed (Vurma *et al.*, 2009).

Other gas-based treatments use ethyl pyruvate, chlorine dioxide and hydrogen peroxide vapor, all of which have demonstrated a high level of efficacy to inactivate pathogens (4 log cfu) (Durak *et al.*, 2012) although have yet to be applied commercially.

A variation of gas phase treatment is advanced oxidative process (AOP), which uses a combination of UV (254 nm) and hydrogen peroxide spray. Here, the UV degradation of hydrogen peroxide generates free radicals that can inactivate microbes both on the surface and sub-surface. Hadjok *et al.* (Hadjok *et al.*, 2008) used a combination of UV and 1.5% hydrogen peroxide to decontaminate lettuce inoculated with *Salmonella* and *E. coli* O157:H7. Log reductions >3 log cfu were obtained on a variety of fresh produce types, including leafy greens and brassica (Table 6.5). Li *et al.* (2011) demonstrated that a combination of UV and hydrogen peroxide (2.5% v/v) supported a 1.4 log reduction of MNV-1 on inoculated lettuce, which was significantly higher compared with UV or hydrogen peroxide applied alone.

A further intervention technology based on free radical disinfection is atmospheric gas plasma. Plasma is generated by passing a gas (e.g. air, oxygen, argon or nitrogen) through a high voltage field. The resultant ionized gas is highly antimicrobial, but short-lived and leaves no residual protection. Gas plasma has been demonstrated to reduce *Salmonella* levels by 2.72, 1.76 and 0.74 log cfu on lettuce, strawberry and potato respectively, with a 15 minute treatment time (Fernandez *et al.*, 2013).

An interesting application of gas plasma is the treatment of prepacked fresh-cut produce. Here, the product is packed within a pillow pack of a high dielectric material (LPDE, HDPR) filled with air or oxygen. The electrodes are then contacted with the pack and 12 kV is applied to generate the plasma within the pack. With oxygen as the gas phase, a >5 log reduction of *E. coli* O157:H7 was reported for a 5 minute treatment time (Klockow and Keener, 2010).

Table 6.5 Inactivation of model bacteria on and within different produce types using a combination of UV (37.8 mJ/cm^2) and hydrogen peroxide (1.5% v/v, 50 °C).

Produce type/organism	Initial loading (cfu g^{-1})		Log count reductions			
			UV:hydrogen peroxide		Calcium hypochlorite	
	Surface	Internal	Surface	Internal	Surface	Internal
Iceberg lettuce						
Salmonella	5.77±0.04	2.84±0.34	4.12±0.45^{Aa}A	2.84±0.34^{Xa}A	1.85±0.02^{Ba}A	0.20±0.10^{Ya}AB
Pseudomonas fluorescens	5.16±0.06	2.63±0.04	4.21±0.40^{Aa}	0.80±0.23^{Xb}	2.10±0.28^{Ba}	−0.40±0.22^{Ya}
E. coli O157:H7	5.52±0.04	3.07±0.15	3.87±0.96^{Aa}AC	0.48±0.16^{Xb}A	1.55±0.32^{Ba}A	0.2±0^{Ya}A
Romaine lettuce						
Salmonella	5.23±0.10	2.54±0.17	3.75±0.48^{a}Ac	0.52±0.21^{a}B	NT	NT
Pseudomonas fluorescens	4.97±0.21	3.01±0.16	4.01±0.15^{a}	0.77±0.31^{a}	NT	NT
E. coli O157:H7	5.19±0.07	2.83±0.14	4.94±0.31^{b}C	0.71±0.37^{a}	NT	NT
Pectobacterium carotovora	4.10±0.08	2.13±0.42	3.79±0.31^{a}	0.53±0.27^{a}	NT	NT
Spinach						
Salmonella	6.35±0.05	3.89±0.20	3.65±0.12^{A}AC	0.89±0.18^{X}B	0.48±0.08^{B}B	−0.34±0.34^{Y}A
E. coli O157:H7	6.14±0.05	4.01±0.12	4.75±0.85^{A}C	0.63±0.15^{X}A	0.46±0.07^{B}B	−0.05±0.05^{Y}A
Cauliflower						
Salmonella	3.06±0.09	3.76±0.04	2.02±0.55^{Aa}BC	0.70±0.21^{Xa}B	0.28±.09^{Ba}B	0.45±0.17^{Xa}B
Pseudomonas fluorescens	2.97±0.23	2.12±0.25	3.32±0.15^{Ab}	0.63±0.45^{Xa}	0.23±0.41^{Ba}	0.26±0.24^{Xa}
E. coli O157:H7	4.20±0.10	3.64±0.08	3.11±0.43^{Aab}A	0.68±0.13^{Xa}A	0.77±0.15^{Ba}B	0.28±0.07^{Ya}A
Pectobacterium carotovora	3.76±0.33	2.92±0.30	2.67±0.74^{Aa}	0.84±0.73^{Xa}	0.36±0.40^{Ba}	0.05±0.49^{Xa}

Broccoli						
Salmonella	4.64±0.38	3.59±0.11	2.18±0.65^{Aa}BC	0.79±0.13^{Xa}B	0.25±0.17^{Ba}B	0.31±0.29^{Xa}B
Pseudomonas fluorescens	3.45±0.51	2.94±0.64	2.97±0.25^{Aa}	0.85±0.45^{Xa}	0.45±0.23^{Ba}	0.54±0.12^{Ya}
E. coli O157:H7	4.73±0.20	3.46±0.21	2.56±0.82^{Aa}A	0.61±0.19^{Xa}A	0.44±0.25^{Ba}B	0.34±0.27^{Xa}A
Spanish onion						
Salmonella	4.21±0.08	3.76±0.05	3.66±0.30^{Aa}AC	0.97±0.17^{Xa}B	0.34±0.10^{Ba}B	0.05±0.06^{Ya}B
Pseudomonas fluorescens	3.70±0.10	2.59±0.17	3.06±0.51^{Aa}	1.40±0.28^{Xa}	0.21±0.19^{Ba}	0.05±0.20^{Ya}
E. coli O157:H7	4.61±0.09	3.76±0.04	2.09±0.49^{Ab}B	0.70±0.21^{Xa}A	0.28±0.09^{Ba}B	0.45±0.17^{Xa}A
Tomato						
Salmonella	3.87±0.38	2.64±0.31	2.22±1.51^{a}BC	1.95±0.11^{a}C	NT	NT
E. coli O157:H7	3.82±0.27	1.88±0.55	3.55±1.35^{a}B	0.89±0.54^{b}A	NT	NT

Source: taken from Hadjok *et al.* (2008).

The different produce types were inoculated with the individual bacterial types and treated with UV:hydrogen peroxide for 60 seconds. For comparison, samples were washed for 3 minutes in calcium hypochlorite solutions (200 ppm).

Means within columns of the individual produce types followed by the same superscript lower case letter are not significantly different. Means within rows followed by the same superscript capital letter are not significantly different. Means for reductions in *Salmonella* and *E. coli* O157:H7 levels followed by the same capital letter are not significantly different.

NT, not tested.

Ultrasound is an example of a physical intervention approach that is based on applying high frequency waves (20–100 kHz), which detaches microbes from surfaces and inactivates via cavitation forces. When applied in isolation, ultrasound has been demonstrated to achieve 2.0–2.5 log cfu reduction of *Listeria monocytogenes*, *E. coli* O157:H7 or *Salmonella* on lettuce with an extended treatment time of 60 minutes, although negligible reductions are observed at exposure times of 5 minutes or less. Yet, ultrasound can significantly enhance the efficacy of sanitizers in enhancing the decontamination of fresh produce. For example, ultrasound combined with 2% organic acid (malic, citric or lactic) could increase the reduction of endogenous microflora by 0.8–1.0 log cfu compared with acid alone (Sagong *et al.*, 2011). Ultrasound in combination with 40 ppm peracetic acid was demonstrated to reduce *Salmonella* levels on cherry tomatoes by 3.9 log cfu (Jose and Vanetti, 2012).

6.7 Conclusion

Ongoing food safety issues related to fresh-cut produce have led to intensive research on approaches to reduce the microbiological risk associated with such products. Despite twenty years of research, there is still the notion that the postharvest wash step is considered a CCP in fresh-cut processing. However, it is now becoming well established that the best that can be achieved is to minimize cross-contamination events within the wash tank as opposed to produce decontamination. Consequently, rather than focusing on actual log reductions achieved by the wash process, more critical attention should be paid to conditions that inactivate pathogens in the wash water. In this respect, sanitizers that are active in the presence of organics would be the preferred choice.

A further consequence is that the quality of the raw material entering the line needs to be controlled. Although there are active on-farm options, the open nature of the cultivation process is problematic. Therefore, there is a need for a postwash intervention step to ensure pathogen inactivation. From the technologies currently available, the obvious choice would be irradiation given that the process is effective and independent on produce type. Nevertheless, alternative treatment based on gas phase interventions may also be a viable option.

6.8 Questions for discussion

1. Why is postharvest washing ineffective at ensuring elimination of human pathogens from fresh-cut produce?
2. What is the major disadvantage of chlorine-based postharvest washes, given that despite the limitation the sanitizer remains the most widely used in the fresh-cut sector?
3. Some organic processors elect not to add sanitizers to wash water. Why is this a dangerous practice and what food safety risk does it pose?

4. From the list of currently available alternative sanitizers, which have demonstrated enhanced efficacy compared with chlorine and why?
5. Which water parameters have the most significant impact of the log reductions achieved by postharvest washing of leafy greens?
6. Why is the organic loading of water significant, and what approaches are available to maintain low levels in wash tanks?
7. Which of the emerging decontamination technologies are most promising for fresh-cut produce, and what barriers exist in commercial implementation?
8. Can current postharvest wash processes be considered a critical control point in fresh produce process? Provide reasons for your view.

6.9 Acknowledgments

The authors wish to thank Atosa Namvar for her assistance in preparing the manuscript. Appreciation also goes out to the Center for Produce Safety (CPS) and Ontario Ministry of Agriculture, Food and Rural Affairs (OMAFRA) for financial support, in addition to their dedication to enhance the safety of fresh produce.

6.10 Sources of further information and advice

CFIA Guidelines for the Use of Food Additives and/or Processing Aids Intended for Fresh Fruits and Vegetables http://www.inspection.gc.ca/english/fssa/frefra/safsal/additivese.shtml

FAO Processing of Fresh-Cut Tropical Fruits and Vegetables: A technical guide. http://www.fao.org/docrep/014/i1909e/i1909e00.pdf

FDA Food Safety of Fruits, Vegetables and Juices http://www.fao.org/ag/agn/CDfruits_en/others/docs/maryland_manual.pdf.

FDA Guidance for Industry: Guide to Minimize Microbial Food Safety Hazards of Fresh-cut Fruits and Vegetables. http://www.fda.gov/Food/GuidanceRegulation/GuidanceDocumentsRegulatoryInformation/ProducePlantProducts/ucm064574.htm.

FDA Guidance for Industry: Guide to minimize microbial food safety hazards of melons http://www.fda.gov/Food/GuidanceRegulation/GuidanceDocumentsRegulatoryInformation/ProducePlantProducts/ucm174171.htm.

6.11 References

ALLENDE A, SELMA MV, LOPEZ-GALVEZ F, VILLAESCUSA R, GIL MI. 2008. Impact of Wash Water Quality on Sensory and Microbial Quality, Including Escherichia coli Cross-Contamination, of Fresh-Cut Escarole. *Journal of Food Protection* 71:2514–2518.

ANON. 2012. FDA CFR-Code of federal regulation titel 21. At: http://www.accessdata.fda.gov/scripts/cdrh/cfdocs/cfcfr/cfrsearch.cfm?fr=173.315 (January 2013).

ARTES F, ALLENDE A. 2005. Processing lines and alternative preservation techniques to prolong the shelf-life of minimally fresh processed leafy vegetables. *European Journal of Horticultural Science* 70:231–245.

AYEBAH B, HUNG YC. 2005. Electrolyzed water and its corrosiveness on various surface materials commonly found in food processing facilities. *Journal of Food Process Engineering* 28:274–264.

BADAWY MI, GAD-ALLAH TA, ALI MEM, YOON Y. 2012. Minimization of the formation of disinfection by-products. *Chemosphere* 89:235–240.

BARRERA MJ, BLENKINSOP R, WARRINER K. 2012. The effect of different processing parameters on the efficacy of commercial postharvest washing of minimally processed spinach and shredded lettuce. *Food Control* 25:745–751.

BEUCHAT LR. 1996. Present and emerging control measures for fresh-cut packaged vegetables. *Journal of Food Protection* 59:204–216.

BUCHHOLZ AL, DAVIDSON GR, MARKS BP, TODD ECD, RYSER ET. 2012a. Quantitative Transfer of Escherichia coli O157:H7 to Equipment during Small-Scale Production of Fresh-Cut Leafy Greens. *Journal of Food Protection* 75:1184–1197.

BUCHHOLZ AL, DAVIDSON GR, MARKS BP, TODD ECD, RYSER ET. 2012b. Transfer of Escherichia coli O157:H7 from Equipment Surfaces to Fresh-Cut Leafy Greens during Processing in a Model Pilot-Plant Production Line with Sanitizer-Free Water. *Journal of Food Protection* 75:1920–1929.

BURNETT AB, ITURRIAGA MH, ESCARTIN EF, PETTIGREW CA, BEUCHAT LR. 2004. Influence of variations in methodology on populations of Listeria monocytogenes recovered from lettuce treated with sanitizers. *Journal of Food Protection* 67:742–750.

CARDADOR MJ, GALLEGO M. 2012. Effect of the chlorinated washing of minimally processed vegetables on the generation of haloacetic acids. *Journal of Agricultural and Food Chemistry* 60:7326–7332.

CHOWDHURY S, RODRIGUEZ MJ, SADIQ R. 2011. Disinfection byproducts in Canadian provinces: Associated cancer risks and medical expenses. *Journal of Hazardous Materials* 187:574–584.

COLE EC, RUTALA WA, SAMSA GP. 1988. Disinfectant testing using a modified use-dilution method - collaborative study. *Journal of the Association of Official Analytical Chemists* 71:1187–1194.

DAWSON F. 2011. Third-party audits under attack. *Food Manufacture* 86:33–34.

DURAK MZ, CHUREY JJ, GATES M, SACKS GL, WOROBO RW. 2012. Decontamination of Green Onions and Baby Spinach by Vaporized Ethyl Pyruvate. *Journal of Food Protection* 75:1012–1022.

ELANO RR, KITAGAWA T, BARI ML, KAWASAKI S, KAWAMOTO S, INATSU Y. 2010. Comparison of the Effectiveness of Acidified Sodium Chlorite and Sodium Hypochlorite in Reducing Escherichia coli. *Foodborne Pathogens and Disease* 7:1481–1489.

FALLIK E, RODOV V, HOREV B, SELA S, ALKALAI-TUVIA S, VINOKUR Y. 2007. Hot water rinsing and brushing technology for the fresh-cut industry. *Proceedings of the International Conference on Quality Management of Fresh Cut Produce*: 229–235.

FAN X, ANNOUS BA, BEAULIEU JC, SITES JE. 2008. Effect of hot water surface pasteurization of whole fruit on shelf life and quality of fresh-cut cantaloupe. *Journal of Food Science* 73:M91–M98.

FERNANDEZ A, NORIEGA E, THOMPSON A. 2013. Inactivation of Salmonella enterica serovar Typhimurium on fresh produce by cold atmospheric gas plasma technology. *Food Microbiology* 33:24–29.

FOULADKHAH A, AVENS JS. 2010. Effects of Combined Heat and Acetic Acid on Natural Microflora Reduction on Cantaloupe Melons. *Journal of Food Protection* 73:981–984.

FRANCIS GA, GALLONE A, NYCHAS GJ, SOFOS JN, COLELLI G, *et al.* 2012. Factors Affecting Quality and Safety of Fresh-Cut Produce. *Critical Reviews in Food Science and Nutrition* 52:595–610.

GONZALEZ B, FERNANDEZ-CUESTA D, LEON MP, MORENO E. 1987. Microbiological study on lettuce influence of washing in water with and without the addition of commercial sodium. *Alimentaria* 24:45–48.

GRACA A, ABADIAS M, SALAZAR M, NUNES C. 2011. The use of electrolyzed water as a disinfectant for minimally processed apples. *Postharvest Biology and Technology* 61:172–177.

GUENTZEL JL, LAM KL, CALLAN MA, EMMONS SA, DUNHAM VL. 2008. Reduction of bacteria on spinach, lettuce, and surfaces in food service areas using neutral electrolyzed oxidizing water. *Food Microbiology* 25:36–41.

HADJOK C, MITTAL GS, WARRINER K. 2008. Inactivation of human pathogens and spoilage bacteria on the surface and internalized within fresh produce by using a combination of ultraviolet light and hydrogen peroxide. *Journal of Applied Microbiology* 104: 1014–1024.

HO KLG. 2012. *Peracid and 2-hydroxy organic acid compositions and methods for treating produce*. US patent 8 263 151 B2.

HO KLG, LUZURIAGA DA, RODDE KM, TANG S, PHAN C. 2011. Efficacy of a Novel Sanitizer Composed of Lactic Acid and Peroxyacetic Acid against Single Strains of Nonpathogenic Escherichia coli K-12, Listeria innocua, and Lactobacillus plantarum in Aqueous Solution and on Surfaces of Romaine Lettuce and Spinach. *Journal of Food Protection* 74:1468–1474.

HSIAO A, KOMAKI Y, IMAAD SM, MARINAS BJ, PLEWA MJ, LIU GL. 2012. Cytotoxicity analysis of water disinfection byproducts with a micro-pillar microfluidic device. *Lab on a Chip* 12:3891–3900.

HUANG AT, BATTERMAN S. 2009. Formation of trihalomethanes in foods and beverages. *Food Additives and Contaminants Part A-Chemistry Analysis Control Exposure and Risk Assessment* 26:947–957.

INATSU Y, KITAGAWA T, NAKAMURA N, KAWASAKI S, NEI D, *et al.* 2011. Effectiveness of Stable Ozone Microbubble Water on Reducing Bacteria on the Surface of Selected Leafy Vegetables. *Food Science and Technology Research* 17:479–485.

ISSA-ZACHARIA A, KAMITANI Y, MIWA N, MUHIMBULA H, IWASAKI K. 2011. Application of slightly acidic electrolyzed water as a potential non-thermal food sanitizer for decontamination of fresh ready-to-eat vegetables and sprouts. *Food Control* 22: 601–607.

JEONG CH, WAGNER ED, SIEBERT VR, ANDURI S, RICHARDSON SD, *et al.* 2012. Occurrence and Toxicity of Disinfection Byproducts in European Drinking Waters in Relation with the HIWATE Epidemiology Study. *Environmental Science and Technology* 46: 12120–12128.

JOHNSTON LM, JAYKUS LA, MOLL D, MARTINEZ MC, ANCISO J, *et al.* 2005. A field study of the microbiological quality of fresh produce. *Journal of Food Protection* 68:1840–1847.

JOSE J, VANETTI MCD. 2012. Effect of ultrasound and commercial sanitizers in removing natural contaminants and Salmonella enterica Typhimurium on cherry tomatoes. *Food Control* 24:95–99.

KENTISH SE, STEVENS GW. 2001. Innovations in separations technology for the recycling and re-use of liquid waste streams. *Chemical Engineering Journal* 84:149–159.

KERN J, REIMANN W, SCHLUTER O. 2006. Treatment of recycled carrot washing water. *Environmental Technology* 27:459–463.

KESKINEN LA, ANNOUS BA. 2011. Efficacy of adding detergents to sanitizer solutions for inactivation of Escherichia coli O157:H7 on Romaine lettuce. *International Journal of Food Microbiology* 147:157–161.

KIM J-G, YOUSEF AE, KHADRE MA. 2003. Ozone and its current and future application in the food industry. *Advances in Food and Nutrition Research* 45:167–218.

KLOCKOW PA, KEENER KM. 2010. Safety and quality assessment of packaged spinach treated with a novel ozone-generation system (vol 42, pg 1047, 2009). *Lwt-Food Science and Technology* 43:1471–1471.

LI D, BAERT L, DE JONGHE M, VAN COILLIE E, RYCKEBOER J, *et al.* 2011. Inactivation of Murine Norovirus 1, Coliphage phi X174, and Bacillus fragilis Phage B40–8 on Surfaces and

Fresh-Cut Iceberg Lettuce by Hydrogen Peroxide and UV Light. *Applied and Environmental Microbiology* 77:1399–1404.

LOPEZ-GALVEZ F, ALLENDE A, TRUCHADO P, MARTINEZ-SANCHEZ A, TUDELA JA, *et al.* 2010. Suitability of aqueous chlorine dioxide versus sodium hypochlorite as an effective sanitizer for preserving quality of fresh-cut lettuce while avoiding by-product formation. *Postharvest Biology and Technology* 55:53–60.

LOPEZ-GALVEZ F, GIL MI, TRUCHADO P, SELMA MV, ALLENDE A. 2010. Cross-contamination of fresh-cut lettuce after a short-term exposure during pre-washing cannot be controlled after subsequent washing with chlorine dioxide or sodium hypochlorite. *Food Microbiology* 27:199–204.

LOPEZ-GALVEZ F, POSADA-IZQUIERDO GD, SELMA MV, PEREZ-RODRIGUEZ F, GOBET J, *et al.* 2012. Electrochemical disinfection: An efficient treatment to inactivate Escherichia coli O157:H7 in process wash water containing organic matter. *Food Microbiology* 30: 146–156.

LOPEZ-VELASCO G, TOMAS-CALLEJAS A, SBODIO A, ARTES-HERNANDEZ F, SUSLOW TV. 2012. Chlorine dioxide dose, water quality and temperature affect the oxidative status of tomato processing water and its ability to inactivate Salmonella. *Food Control* 26: 28–35.

LUO, YG. 2007. Fresh-cut produce wash water reuse affects water quality and packaged product quality and microbial growth in Romaine lettuce. *Hortscience* 42:1413–1419.

LUO Y, NOU X, MILLNER P, ZHOU B, SHEN C, *et al.* 2012. A pilot plant scale evaluation of a new process aid for enhancing chlorine efficacy against pathogen survival and cross-contamination during produce wash. *International Journal of Food Microbiology* 158:133–139.

LUO Y, NOU X, YANG Y, ALEGRE I, TURNER E, *et al.* 2011. Determination of Free Chlorine Concentrations Needed To Prevent Escherichia coli O157:H7 Cross-Contamination during Fresh-Cut Produce Wash. *Journal of Food Protection* 74:352–358.

MOREIRA RG, PUERTA-GOMEZ AF, KIM J, CASTELL-PEREZ ME. 2012. Factors Affecting Radiation D-Values (D10) of an Escherichia Coli Cocktail and Salmonella Typhimurium LT2 Inoculated in Fresh Produce. *Journal of Food Science* 77:E104–E111.

NGUYENTHE C, CARLIN F. 1994. The microbiology of minimally processed fresh fruits and vegetables. *Critical Reviews in Food Science and Nutrition* 34:371–401.

NOU X, LUO Y, HOLLAR L, YANG Y, FENG H, *et al.* 2011. Chlorine Stabilizer T-128 Enhances Efficacy of Chlorine against Cross-Contamination by E. coli O157:H7 and Salmonella in Fresh-Cut Lettuce Processing. *Journal of Food Science* 76:M218–M224.

NOU XW, LUO YG. 2010. Whole-Leaf Wash Improves Chlorine Efficacy for Microbial Reduction and Prevents Pathogen Cross-Contamination during Fresh-Cut Lettuce Processing. *Journal of Food Science* 75:M283–M290.

OLAIMAT AN, HOLLEY RA. 2012. Factors influencing the microbial safety of fresh produce: A review. *Food Microbiology* 32:1–19.

OLMEZ H, KRETZSCHMAR U. 2009. Potential alternative disinfection methods for organic fresh-cut industry for minimizing water consumption and environmental impact. *Lwt-Food Science and Technology* 42:686–693.

PACALA A, VLAICU I, RADOVAN C. 2012. Comparative studies on the efficiency of aluminium coagulants. *Environmental Engineering and Management Journal* 11:427–432.

PARK EJ, ALEXANDER E, TAYLOR GA, COSTA R, KANG DH. 2008. Fate of foodborne pathogens on green onions and tomatoes by electrolysed water. *Letters in Applied Microbiology* 46:519–525.

PARK SH, CHOI MR, PARK JW, PARK KH, CHUNG MS, *et al.* 2011. Use of Organic Acids to Inactivate Escherichia coli O157:H7, Salmonella Typhimurium, and Listeria monocytogenes on Organic Fresh Apples and Lettuce. *Journal of Food Science* 76:M293–M298.

PERCY B. 2012. GFSI compliance and the Integrated Management System create a holistic, harmonized solution for the enterprise. *Food Quality* Feb/March.

PERRY JJ, YOUSEF AE. 2011. Decontamination of Raw Foods Using Ozone-Based Sanitization Techniques. *Annual Review of Food Science and Technology* 2:281–298.

PREDMORE A, LI J. 2011. Enhanced Removal of a Human Norovirus Surrogate from Fresh Vegetables and Fruits by a Combination of Surfactants and Sanitizers. *Applied and Environmental Microbiology* 77:4829–4838.

RAHMAN SME, JIN Y-G, OH D-H. 2011. Combination treatment of alkaline electrolyzed water and citric acid with mild heat to ensure microbial safety, shelf-life and sensory quality of shredded carrots. *Food Microbiology* 28:484–491.

ROSENBLUM J, GE C, BOHREROVA Z, YOUSEF A, LEE J. 2012. Ozonation as a clean technology for fresh produce industry and environment: sanitizer efficiency and wastewater quality. *Journal of Applied Microbiology* 113:837–845.

SAGONG HG, LEE SY, CHANG PS, HEU S, RYU S, *et al.* 2011. Combined effect of ultrasound and organic acids to reduce Escherichia coli O157:H7, Salmonella Typhimurium, and Listeria monocytogenes on organic fresh lettuce. *International Journal of Food Microbiology* 145:287–292.

SAKAMOTO Y. 2011. *Ozone microbubble-containing water producing device, ozone microbubble-containing water producing method, article washing device, article washing method, culture method for marine product and hydroponic culture method.* JP patent 2011240206 A.

SELMA MV, IBANEZ AM, ALLENDE A, CANTWELL M, SUSLOW T. 2008. Effect of gaseous ozone and hot water on microbial and sensory quality of cantaloupe and potential transference of Escherichia coli O157: H7 during cutting. *Food Microbiology* 25:162–168.

SHARMA VK, SOHN M. 2012. Reactivity of chlorine dioxide with amino acids, peptides, and proteins. *Environmental Chemistry Letters* 10:255–264.

SIDDIQUI MW, CHAKRABORTY I, AYALA-ZAVALA JF, DHUA RS. 2011. Advances in minimal processing of fruits and vegetables: a review. *Journal of Scientific and Industrial Research* 70:823–834.

SILVEIRA AC, AGUAYO E, ESCALONA VH, ARTES F. 2011. Hot water treatment and peracetic acid to maintain fresh-cut Galia melon quality. *Innovative Food Science and Emerging Technologies* 12:569–576.

SOLOMON EB, HUANG LH, SITES JE, ANNOUS BA. 2006. Thermal inactivation of Salmonella on cantaloupes using hot water. *Journal of Food Science* 71:M25–M30.

STEVIK TK, AA K, AUSLAND G, HANSSEN JF. 2004. Retention and removal of pathogenic bacteria in wastewater percolating through porous media: a review. *Water Research* 38:1355–1367.

TOIVONEN PMA, LU CW, DELAQUS P, BACH S. 2011. Slicing under Chlorinated Water Improves the Disinfection of Fresh-Cut Romaine Lettuce. *Journal of Food Protection* 74:2142–2147.

TOMAS-CALLEJAS A, LOPEZ-GALVEZ F, SBODIO A, ARTES F, ARTES-HERNANDEZ F, SUSLOW TV. 2012a. Chlorine dioxide and chlorine effectiveness to prevent Escherichia coli O157: H7 and Salmonella cross-contamination on fresh-cut Red Chard. *Food Control* 23:325–332.

TOMAS-CALLEJAS A, LOPEZ-VELASCO G, VALADEZ AM, SBODIO A, ARTES-HERNANDEZ F, *et al.* 2012b. Evaluation of Current Operating Standards for Chlorine Dioxide in Disinfection of Dump Tank and Flume for Fresh Tomatoes. *Journal of Food Protection* 75:304–313.

VANDEKINDEREN I, DEVLIEGHERE F, DE MEULENAER B, RAGAERT P, VAN CAMP J. 2009. Optimization and evaluation of a decontamination step with peroxyacetic acid for fresh-cut produce. *Food Microbiology* 26:882–888.

VURMA M, PANDIT RB, SASTRY SK, YOUSEF AE. 2009. Inactivation of Escherichia coli O157:H7 and Natural Microbiota on Spinach Leaves Using Gaseous Ozone during Vacuum Cooling and Simulated Transportation. *Journal of Food Protection* 72:1538–1546.

WANG H, FENG H, LUO YG. 2004. Microbial reduction and storage quality of fresh-cut cilantro washed with acidic electrolyzed water and aqueous ozone. *Food Research International* 37:949–956.

WARRINER K, HUBER A, NAMVAR A, FAN W, DUNFIELD K. 2009. Recent Advances in the Microbial Safety of Fresh Fruits and Vegetables, pp. 155–208. In Taylor SL (ed.), *Advances in Food and Nutrition Research*, vol. 57. Elsevier Academic Press Inc, 525 B Street, Suite 1900, San Diego, Ca 92101–4495 USA.

WARRINER K, NAMVAR A. 2009. What is the hysteria with Listeria? *Trends in Food Science and Technology* 20:245–254.

XIA XD, LUO YG, YANG Y, VINYARD B, SCHNEIDER K, MENG JH. 2012. Effects of Tomato Variety, Temperature Differential, and Post-Stem Removal Time on Internalization of Salmonella enterica Serovar Thompson in Tomatoes. *Journal of Food Protection* 75:297–303.

7

Preventing cross-contamination during produce wash operations

Y. Luo, USDA, USA, D. T. Ingram, FDA, USA and K. Khurana, Pulse Instruments, USA

DOI: 10.1533/9781782420279.2.103

Abstract: Produce washing is a common commercial practice intended to enhance product quality and shelf-life, and improve food safety through the removal of foreign materials, plant tissue exudates and the inactivation of both human pathogens and spoilage microorganisms. However, in the absence of sufficient sanitizer presence, washing also has the potential to allow pathogen survival and spread to uncontaminated produce in the wash water. Maintenance of effective sanitizer concentrations in the wash water is a critical control point to reduce risk by preventing wash water itself from becoming a major source of pathogen cross-contamination.

Key words: post-harvest, produce, wash water, sanitizers, foodborne pathogens.

7.1 Introduction: commercial produce wash operation, water quality and sanitizer concentration

There are many different instruments and protocols employed by industry for commercial produce wash, but the process-flow has been standardized to the following order: 1 cut, 2 wash, 3 de-water. Fresh-cut leafy green vegetables are usually washed in double-flume or tank systems, followed by a final rinse and de-watering process. Fig. 7.1 illustrates a typical double-wash system currently used in the industry. This wash system consists of two separate water tanks: primary and secondary, each connected to a refrigerated chiller system. Air is continuously injected into each wash tank to create turbulence that washes and moves product along the process. Fresh-cut produce, especially leafy greens, are typically buoyant so the wash tanks are equipped with rotating screens to facilitate product submersion. After the washing stage (usually 30 seconds), the product exits the tank on a conveyor and is placed into a de-watering system, which removes excess

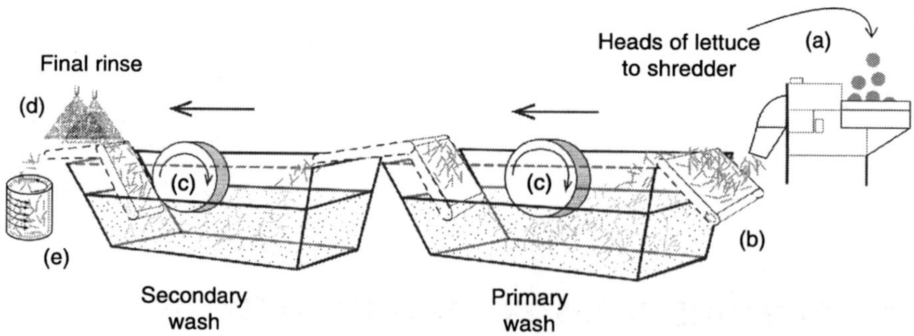

Fig. 7.1 Illustration of commercial two-stage fresh-cut washing operation of shredded lettuce. (a) Lettuce heads passed through slicer. (b) Shredded lettuce added to transfer belt. (c) Submersion drums facilitate contact of lettuce pieces with wash water. (d) Final rinse. (e) Product collecting bins for subsequent de-watering/drying. Luo *et al.* (2012). Reprinted with permission from Elsevier.

water through centrifugation. The process water is constantly screened and re-circulated into stage-specific reservoir tanks, allowing continuous removal of debris (soil, plant material, fines, etc.) from the system. Operational water loss in the primary wash tank is replenished from the secondary tank, which in turn is replenished from the water chiller.

Water reuse and re-circulation is a common practice in the fresh and fresh-cut produce industry for the purposes of water conservation and reducing operational cost (Gil *et al.*, 2009). Throughout the washing process, water quality rapidly and continuously deteriorates as the result of accumulation of soil, debris and plant exudates in the washing system, and is manifested by a measured increase in turbidity (NTU), chemical oxygen demand (COD) and decrease in effective sanitizer concentration. In a pilot plant study, Luo *et al.* (2012) found that before the introduction of shredded iceberg lettuce and spinach into the commercial washing system, the initial turbidity and COD were 0.5–0.6 NTU and 301–366 mg/l, respectively. As shredded lettuce was continuously added into the wash system, the COD and turbidity values increased rapidly and proportionately with the increase in added produce. At the end of the 36 minutes, or 1620 kg of lettuce washed, the water turbidity had reached 24 NTU and the COD reached 1374 mg/l (Fig. 7.2). Under the operational conditions in this experiment, the increases in both water turbidity and COD linearly corresponded to the amount of lettuce washed and hence the amount of tissue exudates released into the wash system. Wash water turbidity was negatively affected by the presence of free chlorine, with a steep decrease in measured NTUs occurring immediately following chlorine replenishment (Fig. 7.2, arrows indicate chlorine addition).

The commercial washing process for tomatoes is similar to that of leafy greens, and chlorinated water is the mainstay disinfectant for industrial packing house operations. During the washing process, tomatoes are emptied from bins into dump tanks where chlorinated water is used as a cushion to prevent mechanical

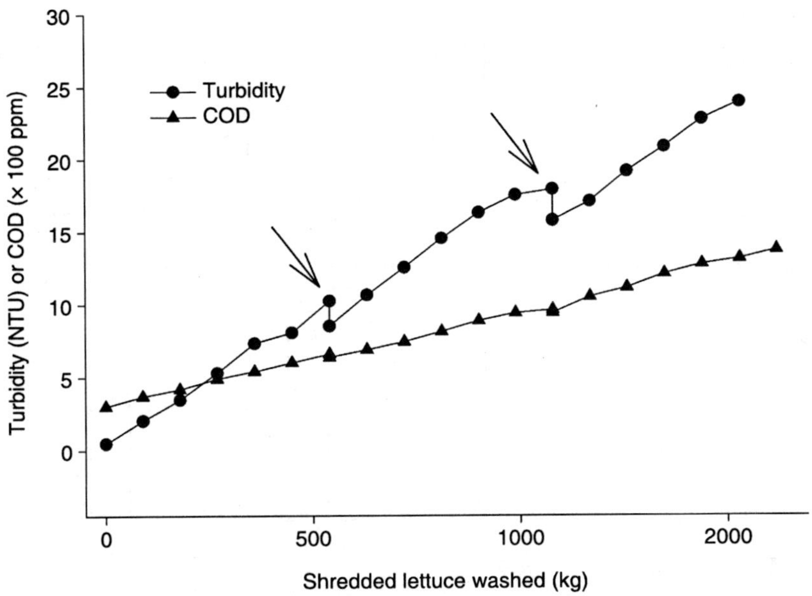

Fig. 7.2 Dynamic changes in wash water chemical oxygen demand (COD) and turbidity of the wash water. Lettuce shreds were continuously fed into the wash system at a rate of 45 kg/min. Arrows indicate when the process was paused and sodium hypochlorite was replenished in the wash solution. Data are the average of three replicate experiments. Adapted from Luo *et al.* (2012). Reprinted with permission from Elsevier.

damage. Studies by Schneider *et al.* (2011) indicate that most packinghouses also reuse and re-circulate dump-tank water. As seen with leafy green operations, tomato dump-tank water quality also rapidly and proportionately deteriorates with accumulation of waxes, dust, pesticide residues, leaf debris, soil and fruit exudates of damaged fruits in the process water (Bartz, 2001; Tomas-Callejas, 2012). The increasing organic load in the process wash water results in a dramatic increase in chlorine demand, which correlates with a steady decline in water quality and free available chlorine. Data from a Florida survey indicate that water quality continuously declined during packinghouse operations, and the fruit-loading-rates correlated well with a significant increase in total dissolved solids (TDS), turbidity and COD over time (Schneider *et al.*, 2011).

7.2 Changes in sanitizer concentration during wash

The concentration of free available chlorine in wash water is considered the most important critical control factor for preventing cross-contamination in produce washing systems (Hurst, 2002). However, maintaining a constant chlorine concentration during commercial fresh-cut processing is a difficult technical

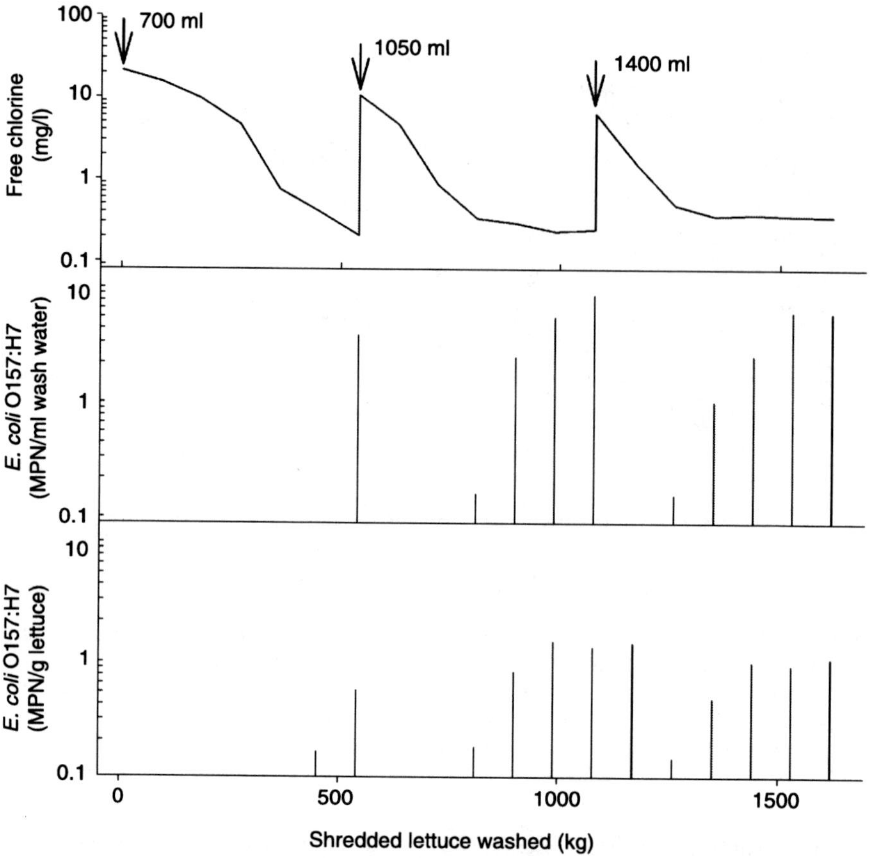

Fig. 7.3 Recovery of *E. coli* O157:H7 in uninoculated iceberg lettuce. Uninoculated shredded lettuce and inoculated baby spinach (1000:2, gram basis, respectively) were continuously introduced into the wash system at a rate of 45 kg/min. Continuous lines indicate free chlorine level in the wash water, and bars represent either the recovery of inoculated *E. coli* O157:H7 in process wash water or on uninoculated lettuce. Arrows indicate points when aliquots of 700 ml, 1050 ml and 1400 ml of concentrated sodium hypochlorite were added to the primary wash system. MPN, most probable number. Modified from Luo *et al.* (2012). With permission from Elsevier.

challenge because of the rapid accumulation of organic matter in the wash solution (Adams *et al.*, 1989). This process has two effects on chlorine. First, as chlorine oxidizes all organic matter, an increasing organic load will dramatically reduce the residual chlorine concentrations in the wash solution. Second, sodium or calcium hypochlorite solutions are basic solutions (pH 10.5–11.5), and the addition of these solutions tends to increase the pH of wash water. The most biologically reactive species of chlorine solution is hypochlorous acid (HOCl), the stability of which is highly pH dependent (Connell, 1996). When solution pH is above 7.5, the hypochlorous acid converts to hypochlorite ion (ClO^-), which has limited antimicrobial efficacy (Erkmen, 2010). However, when the pH is too low, the formation of chlorine off-gas could lead to immediate termination of the wash process and evacuation of the processing plant as a result of associated health issues. Therefore, the pH of the produce wash solution needs to be maintained at 6–7 in order to achieve the highest concentration of hypochlorous acid and thus highest antimicrobial efficacy, without off-gas risks.

Pilot plant studies from Luo *et al.* (2012) show that as shredded lettuce was continuously introduced into the washing system, the rapid increases in COD and turbidity were accompanied by a rapid decline in residual chlorine concentration (Fig. 7.3). Free chlorine decreased from the initial concentration of ~20 mg/l to near depletion at the end of the first wash segment, when approximately one-sixth of the weight equivalent of shredded lettuce was washed in the solution (540 kg in 3200 l). Adding fresh sodium hypochlorite increased the free chlorine levels in the wash water; however, this concentration decreased alarmingly quickly to below effective concentrations. This trend was also seen in a survey of commercial tomato dump tank operations (Schneider *et al.*, 2011). Although it is technically possible for free chlorine to be replenished to the initial set point, more and more chlorine is required during each replenishment event to maintain effective bactericidal concentrations of chlorine (Fig. 7.3). In summary, constant monitoring of wash water quality (pH, oxidation reduction potential (ORP), COD and turbidity) is necessary to determine the frequency and amount of chlorine input to maintain effective bactericidal properties of the wash water. Each type of produce presents a unique challenge to the wash industry. In general, the chlorine demand for various produce washing systems is listed as chopped onions > shred cabbage > shred carrots > shred lettuce > spinach > broccoli > cauliflower.

7.3 Factors affecting pathogen survival and cross-contamination during wash

Most of the common industrial sanitizers are oxidizers, and work by the principle of increasing the oxidation potential in the water system. A number of sanitizers have been approved by the US FDA for fresh and fresh-cut produce wash application, including chlorine, ozone, chlorine dioxide and peroxyacetic acid (FDA 2009; Selma *et al.*, 2008). Among these, chlorine continues to be the most commonly used sanitizer in the produce industry. The prevalence of commercial chlorine usage is attributed to its long established capacity for killing human pathogens in solution, minimal impact on product quality and low cost.

Pathogen inactivation in a chlorine solution is a complex function of the concentration of chlorine in solution, the association with its most reactive and biocidal species (HOCl) and exposure time. The antimicrobial effect of chlorine varies among pathogen strains. In pure water (no organic residue), when chlorine concentration equaled or exceeded 0.5 mg/l, and exposure time equaled or exceeded 5 seconds, all of the tested pathogens, with the exception of *Salmonella spp.*, were reduced from 4.5 log cfu/ml to below the detection limit. These results suggest that a minimum of 0.5 mg/l free chlorine and a contact time of 30 seconds are necessary to inactivate most pathogens in suspension. Recent studies from Luo *et al.* (2011) also reported that no *E. coli* O157:H7 survival was detected (>4 log reduction) in solutions after 30 seconds when the chlorine concentration was greater than or equal to 0.4 mg/l. Similar results on *E. coli* O157:H7 inactivation by chlorinated solutions were reported by others (Li *et al.*, 2001; Rice *et al.*, 1999; Zhao *et al.*, 2001). *Salmonella* strains as a group are known to be more resistant than *E. coli* strains to low levels of chlorine in solution, and non-O157 STEC strains were generally more susceptible than *E. coli* O157:H7 strains to chlorine solution.

Although 200 mg/l free chlorine is the maximum allowable concentration (FDA) in commercial wash water systems, the chlorine levels in most commercial systems often fluctuate within the single digit range (0–10 mg/l). When free chlorine level is this low, the introduction of additional loads of cut produce into the wash system could easily cause a temporary but complete depletion of free chlorine in the entire system, thus resulting in the possibility that fresh products will be washed in water without sufficient protection. In such a system, human pathogens from contaminated lettuce could result in widespread cross-contamination and increase the potential for foodborne disease. Luo *et al.* (2011) recently investigated the minimum chlorine concentration necessary in commercial washing systems to prevent this cross-contamination scenario for fresh-cut produce wash. It was determined that, although ≤1 mg/l of free chlorine was sufficient to prevent pathogen survival in pure water after 30 seconds of exposure, ≥10 mg/l of chlorine was needed to prevent pathogen cross-contamination during actual produce wash (Luo *et al.*, 2011). By maintaining ≥10 mg/l of free chlorine in the wash solution throughout the entire washing event, pathogens that are removed from contaminated produce into the wash solution will be quickly killed or injured beyond repair, thus reducing the potential for cross-contamination.

7.4 Common industrial practices: process flow and re-wash

In an attempt to mitigate pathogen cross-contamination of produce washed in water with inadequate free chlorine, many fresh-cut operations have prescribed re-wash as the corrective action for their Hazard Analysis and Critical Control Point (HACCP) Program (Hurst, 2002). This approach is partially built on the practice and theory from the 1990s that 1 chlorine wash is a terminal kill step for fresh-cut produce, and 2 pathogens that have 'loosely' attached to produce from wash water through the aforementioned 'cross-contamination' events may not have had time to attach securely to the fresh-cuts, and thus the subsequent wash would completely remove or inactivate these pathogens.

The effectiveness of this practice for ensuring produce safety has not been validated. A recent study reported that when lettuce is contaminated during prewash, a subsequent wash was not able to completely eliminate the pathogen population. Luo *et al.* (2011) reported that re-washing with 50 mg/l free chlorine for 30 seconds significantly reduced *E. coli* O157:H7 population on freshly contaminated lettuce; however, small numbers of *E. coli* O157:H7 cells remained on lettuce samples from all three experimental trials. This suggests that, contrary to the current practice, re-washing is not an effective means for correcting a failure in process control, such as insufficient sanitizer presence. In the washing solution, planktonic *E. coli* O157:H7 cells may have only a few seconds to effectively adhere and attach to the surfaces, cut edges and crevasses on fresh-cut produce. Once human pathogens, for example *E. coli* O157:H7, have firmly attached to fresh produce in the wash water, they are considered to be effectively protected and inaccessible to the chlorine disinfectant (FAO/WHO 2008; Lopez-Galvez *et al.*, 2009; Takeuchi and Frank, 2000).

Recent research has evaluated the impact of cut-wash sequence on pathogen survival and cross-contamination during produce wash. Studies show that the process of cutting leafy greens and other vegetables, indeed, increases the release of plant exudates, latex, waxes and other organic materials from the tissue (Luo, 2007; Garg *et al.*, 1990). When this freshly cut material is then constantly introduced into the washing system, the chlorine demand is dramatically increased, which requires proportionately more amounts of added chlorine to maintain bactericidal concentrations. Furthermore, the smaller fresh-cut pieces exhibit a larger surface area that can be potentially cross-contaminated with any trace amounts of surviving pathogens. Current research has shown that reversing this current practice by first washing the produce prior to cutting, maintains the ability to effectively reduce pathogens on the produce surfaces while greatly mitigating the potential for cross-contamination events in wash water (Nou and Luo, 2010).

Produce washing systems are highly dynamic, continuously changing systems that include much natural variation. Dynamic control of many factors is needed to maintain a steady-state level of disinfectant in wash water (Suslow, 2012). Major sources of variation include wash tank/flume size and water volume, rate of fresh water replenishment, and product type and throughput (this will in turn determine the changes in organic load, and water quality). These factors together will determine the consumption rate of sanitizers during the wash process. In order to create a stable process that produces predictable results, it is important to create a dynamic process control that may manage normal variations and maintain sanitizer concentration constantly above the target levels.

7.5 Conclusion

Effective produce washing operations are one critical step in the maintenance of the farm-to-fork food safety continuum. Washing effectively removes foreign materials and tissue exudates, and can reduce human pathogens and spoilage microorganisms present on produce surfaces. However, in the absence of sufficient process controls, the amount of sanitizers in washing solutions can be quickly degraded. This could increase the widespread dissemination of pathogens to previously uncontaminated products, which is especially a concern when water is both reused and re-circulated through the entire system. Therefore, maintaining effective sanitizer concentrations that can kill pathogens within a brief contact time in the wash solution is the key to ensuring food safety and preventing the potential of wash water becoming a source of contamination. Understanding how different sanitizers work and how they are measured, monitored and effectively distributed is an important element in every produce wash/sanitation system and the maintenance of an effective and cost-efficient food safety program.

7.6 Questions for discussion

1. What is the main purpose of employing washing systems in the fresh-cut produce industry?
2. What are the main chemical sanitizers used in the industry for produce wash systems?
3. What are the technical, practical and environmental limitations associated with each of these sanitizers?
4. What are the major differences and problems associated with tomato and leafy green commercial washing operations?
5. What is the effective pH range of chlorine disinfectant in wash water and why?
6. Describe the process flow of current industrial fresh-cut operations and how this could be improved.
7. What operational parameters or critical control points should be included in every industrial produce wash HACCP plan?
8. What is the currently recommended operational chlorine concentration in industrial wash systems?
9. What improvements can you suggest to improve the current produce wash work-flow?
10. What are the limitations associated with oxidizing disinfectants in high-throughput industrial produce washing systems?

7.7 References

ADAMS, M. R., A. D. HARTLEY, and L. J. COX. 1989. Factors affecting the efficacy of washing procedures used in the production of prepared salads. *Food Microbiol* 6:69–77.

BARTZ, J. C. 2001. Chlorine concentration and the inoculation of tomato fruit in packinghouse dump tanks. *Plant Disease* 85 (8):885–9.

CONNELL, G. 1996. *The chlorination/chloramination handbook.* American Water Works Association. Denver, CO.

ERKMEN, O. 2010. Antimicrobial effects of hypochlorite on *Escherichia coli* in water and selected vegetables. *Foodborne Pathog Dis* 7:953–8.

FAO/WHO. 2008. Microbiological hazards in fresh leafy vegetables and herbs. Expert panel report. *Microbiological Risk Assessment Series 14.* WHO Press, 138 pp.

FDA. 2009. *Chapter 5. Methods to Reduce/Eliminate Pathogens from Produce and Fresh-Cut Produce, Analysis and Evaluation of Preventive Control Measures for the Control and Reduction/Elimination of Microbial Hazards on Fresh and Fresh-Cut Produce.* Available at: http://www.fda.gov/Food/FoodScienceResearch/SafePracticesforFoodProcesses/ucm091363.htm. Accessed 17 December 2012.

GARG, N., J. J. CHUREY, and D. F. SPLITTSTOESSER. 1990. Effect of processing conditions on the microflora of fresh-cut vegetables. *J Food Prot* 53:701–3.

GIL, M. I., M. V. SELMA, F. LÓPEZ-GÁLVEZ, and A. ALLENDE. 2009. Fresh-cut product sanitation and wash water disinfection: problems and solutions. *Int J Food Microbiol* 134:37–45.

HURST, W. 2002. Safety aspects of fresh-cut fruits and vegetables, pp. 45–90. In O. Lamikanra (ed.), *Fresh-cut Fruits and Vegetables: Science, Technology, and Market.* CRC Press. Boca Raton, FL.

LI, Y., R. E. BRACKETT, J. CHEN, and L. R. BEUCHAT. 2001. Survival and growth of *Escherichia coli* O157:H7 inoculated onto cut lettuce before or after heating in chlorinated water, followed by storage at 5 or 15 °C. *J Food Prot* 64:305–9.

LOPEZ-GALVEZ, F., A. ALLENDE, M. V. SELMA, and M. I. GIL. 2009. Prevention of *Escherichia coli* cross-contamination by different commercial sanitizers during washing of fresh-cut lettuce. *Int J Food Microbiol* 133:167–71.

LUO, Y. 2007. Wash operation affects water quality and packaged fresh-cut romaine lettuce quality and microbial growth. *HortScience* 42:1413–19.

LUO, Y., X. NOU, Y. YANG, I. ALEGRE, E. TURNER, *et al.* 2011. Determination of free chlorine levels needed to prevent *Escherichia coli* O157:H7 cross-contamination during fresh-cut produce wash. *J Food Prot* 74:352–8.

LUO, Y., X. NOU, P. MILLNER, B. ZHOU, C. SHEN, *et al.* 2012. A pilot plant scale evaluation of a new process aid for enhancing chlorine efficacy against pathogen survival and cross-contamination during produce wash. *Int J Food Microbiol* 158:133–9.

NOU, X., and LUO, Y. 2010. Whole-leaf wash improves chlorine efficacy for microbial reduction and prevents pathogen cross-contamination during fresh-cut lettuce processing. *J Food Sci* 75:M283–90.

RICE, E. W., R. M. CLARK, and C. H. JOHNSON. 1999. Chlorine inactivation of *Escherichia coli* O157:H7. *Emerg Infect Dis* 5:461–3.

SELMA, M. V., A. ALLENDE, F. LÓPEZ-GÁLVEZ, M. Á. CONESA, and M. I. GIL. 2008. Disinfection potential of ozone, ultraviolet-C and their combination in wash water for the fresh-cut vegetable industry. 2008. *Food Microbiol* 25:809–14.

SCHNEIDER, K., Y. LUO, S. SARGENT, and R. SCHNEIDER. 2011. *Evaluation and optimization of postharvest intervention strategies for the reduction of bacterial contamination on tomatoes*. Available at: https://cps.ucdavis.edu/researchproject/258/awards/Evaluation_and_optimization_of_postharvest_intervention_strategies_for_the_reduction_of_bacterial_contamination_on_tomatoes.html. Accessed on 17 December 2012.

SUSLOW, T. 2012. *Water Preventive Controls for definition of Validation and Verification.* Available at http://www.pma.com/preventative_controls_and_wash_water_fs12. Accessed 17 December 2012.

TAKEUCHI, K., and J. FRANK. 2000. Penetration of *Escherichia coli* O157:H7 into lettuce tissues as affected by inoculum size and temperature and the effect of chlorine treatment on cell viability. *J Food Prot* 63:434–40.

TOMAS-CALLEJAS, A. G.-V.-H. (2012). Evaluation of current operating standards for chlorine dioxide in disinfection of dump tank and flume for fresh tomatoes. *J Food Prot* 75 (2):304–13.

ZHAO, T., M. P. DOYLE, P. ZHAO, P. BLAKE, and F. M. WU. 2001. Chlorine inactivation of *Escherichia coli* O157:H7 in water. *J Food Prot* 64:1607–9.

8

Organic environmental chemical contaminants in fresh produce and fruits

T. L. Cederberg, Technical University of Denmark, Denmark

DOI: 10.1533/9781782420279.2.112

Abstract: Plant uptake of organic environmental contaminants is an important process when considering chemical food safety and risk assessments of fresh produce. Understanding of the principal pathways involved in transfer and accumulation of contaminants in plants is the basis for developing mathematical models for plant uptake and human exposure. Risk associated with crop growth on contaminated sites can in this way be described and more effective risk management actions can be taken, for example with regards to soil quality standards.

Key words: organic environmental contaminants, pathways for plant uptake, mathematical models, human exposure.

8.1 Introduction

Consumption of fruit and vegetables is an essential part of a healthy human diet and consumers expect that fresh produce is safe to eat. Chemical food safety of fresh produce is partly about the assigned use of chemicals in the production of crops, for example pesticides and biocides, and partly about the unwanted exposure of plants to environmental chemical pollutants. It is the latter scenario which is covered in this context.

Risk assessments related to human exposure to fresh produce contaminated by organic environmental compounds can be attributed to different situations, whether the crop has been produced for private or commercial purposes. A crop may be inadvertently exposed to environmental pollution before harvest by a growth condition in contaminated soil or atmosphere. After harvest the exposure can occur under storage or transport by, for example, engine exhaust fumes, cross-contamination in shipping containers or other contaminated postharvest treatment (Biedermann, 1995). Another situation is growth on contaminated sites or the use of recycled materials, for example sewage sludge or decontaminated soils. Mathematical models have been developed for calculating the transfer of contaminants into food and estimating human exposure (Elert *et al.*, 2011). For growth on contaminated sites, one of the goals of the models is to be able to use them as decision-support tools for risk assessments and establishment of cut-off values in soil quality criteria (Samsøe-Petersen *et al.*, 2002).

It should be noted that organic food can be expected to contain lower levels of agrochemical residues than conventional grown food, but this is not true for environmental contaminants. For these compounds the risk of increased levels is the same for foods of both origins (Magkos *et al.*, 2003).

8.2 Regulatory aspects

For commercial production of fresh produce, international rules and guidelines are in place in order to protect public health, maintain consumer confidence and ensure fair trade. The Codex Alimentarius (from Latin: Food Law or Code) Commission implements the Joint FAO (Food and Agricultural Organization of the United States)/WHO (World Health Organization) Food Standards Programme, the purpose of which is to protect the health of consumers and to ensure fair practices in the food trade. The Codex Alimentarius is a collection of internationally adopted food standards presented in a uniform manner. It also includes provisions of an advisory nature in the form of codes of practice, guidelines and other recommended measures to assist in achieving the purposes of the Codex Alimentarius (Codex Alimentarius, 2003, 2010). In the European Union (EU) general and specific hygienic rules are in place to ensure a high level of consumer protection (European Community Regulation No 852, 2004). Both Codex and the EU recommend food safety procedures to be implemented by the food business operators based on hazard analysis and critical control point (HACCP) principles. On a small number of specific organic environmental contaminants maximum levels have been established by national and international regulation, but for the majority of possible contaminants no maximum levels exist. In this case hazard identification, critical limits, corrective actions, etc., must rely on the knowledge available at food authorities, scientific communities or producer associations.

8.3 Modelling of uptake

Before harvest fruit and vegetables may be exposed to organic environmental contaminants from polluted air and soil. Plants grow by uptake of water and nutrients from the soil, and by exchange of carbon dioxide and oxygen from the air. The large surface areas of roots and leaves can potentially exchange gasses and fluids other than those needed by the plant, and this may cause uptake of organic environmental contaminants and other non-needed components. Many factors contribute to the concentration of chemical contaminants in crops. The contaminants can potentially be transported by the air over long distances before the exposure takes place. From contaminated soil the plant can uptake the contaminants directly from the soil particles, from the water in the soil or from the gas phase in the soil. Plant parts above the ground such as leaves, fruits and shoots can also be exposed to soil contaminants as a result of evaporation from the soil, from dust particles under dry conditions or soil splash in the rain.

Understanding of the processes involved during uptake of organic compounds in plants has advanced in recent years. There are four principal uptake pathways: passive and active uptake through the root system, gaseous and particulate deposition to above-ground shots, and direct contact between soil and plant tissues. Other factors in the uptake process are growth dilution, plant lipid content and plant metabolism (Collins *et al.*, 2011). Crop-specific models have been developed for roots, potatoes, leaves and fruits (Trapp and Legind, 2011). The models are based on the same physicochemical principles and involve basic processes such as advective uptake into plants, diffusive uptake, chemical equilibrium, transport inside plants in xylem and phloem tissues, dilution by growth, and particle deposition from soil and air. In Fig. 8.1 the various crop-specific plant uptake processes are depicted. Active uptake processes, which include energy or enzymes of the plant, have not yet been shown to be significant for organic environmental contaminants (Trapp and Legind, 2011).

Review of experimental data and model simulation results have revealed some common relationships between the chemical properties of contaminants, pathways for uptake in plants and the potential for accumulation in various parts of the plant (Collins *et al.*, 2011; Trapp and Legind, 2011). As shown in Table 8.1, polar and

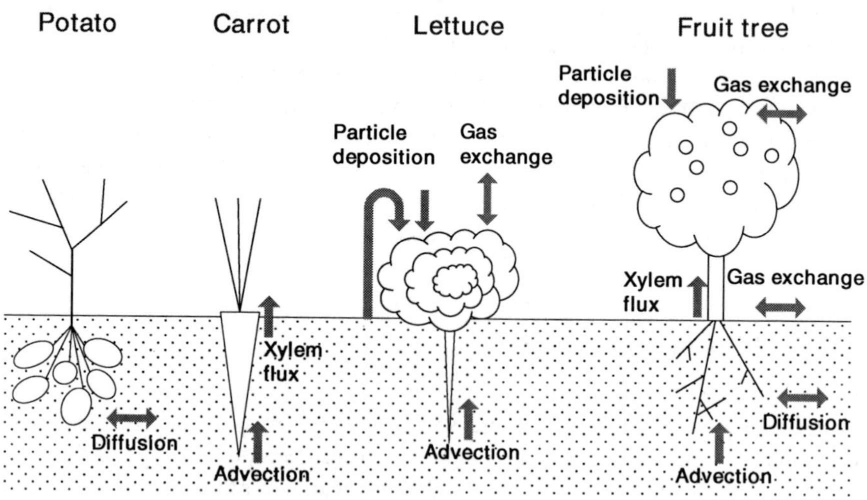

Fig. 8.1 Exposure of vegetables and fruits to organic environmental contaminants. Crop-specific uptake models for organic environmental contaminants and the selected pathways involved (adapted from Legind and Trapp, 2009).

Table 8.1 Overview of relationship between chemical property of contaminants, pathway for uptake and potential for accumulation in plants

Chemical property of contaminant	Pathway for uptake of contaminants in vegetables and fruits
Polar (log K_{OW} < 3) and non-volatile (K_{AW} <10^{-6})	High potential for transfer from soil Accumulation in leaves and fruits
Lipophilic (log K_{OW} >3)	Is retained in soil, roots and lower stem Transport to leaves by soil particles or from air
Volatile	Low potential for accumulation Rapid evaporation into the air

Source: from Trapp and Legind (2011).

K_{OW}, partition coefficient octanol–water; K_{AW}, partition coefficient air–water.

non-volatile compounds exhibit high potential for transfer from soil to plant and for further accumulation in leaves and fruits. Lipophilic compounds have low potential for this pathway but can be transported to leaves by soil particles or through the air. Volatile contaminants have low potential for accumulation because of rapid evaporation to the air. Although the models for plant uptake have improved, the predictions are uncertain because of the great complexity of the plants involved and the large variability in environmental conditions.

8.4 Contaminated sites and risk assessments

The use of chlorinated chemicals such as polychlorinated biphenyls (PCBs) and organochlorine pesticides (OCPs) has been banned in most countries because of their persistence and accumulation in the global environment. PCBs and OCPs are hydrophobic and lipophilic and are subject to biomagnification in the food chain. In the terrestrial environment uptake of PCBs and OCPs at a heavily contaminated site from soil and air into radishes has been studied by applying bioconcentration factors (BCP) (Mikes *et al.*, 2009). Bioconcentration factor (BCF) is defined as the ratio between concentration of a contaminant in the plant and the concentration in the soil. BCF values for roots were constant and were not correlated to octanol-water partition coefficient (log K_{OW}). A negative correlation was found for BCF versus log K_{OW} for edible radish bulbs. Shoot BCF values were constant. Elevated concentrations in shoots of radishes grown in control soil indicated that the uptake from air was more significant for shoots than uptake from soil.

Transfer of perfluorooctanoic acid (PFOA) and perfluorooctane sulfonate (PFOS) from soil to different plant compartments in carrots, potatoes and cucumbers has been investigated (Lechner and Knapp, 2011). PFOA and PFOS are widely used surfactants and are found in sewage sludge. The use of sewage sludge as fertilizer can contaminate the soil with these fluorinated compounds. PFOA and PFOS are water-soluble ionic surfactants and were suspected to behave differently to, for example, chlorinated persistent organic pollutants which accumulate in plant roots by diffusion and sorption onto the lipophilic root solids. Transfer of PFOA and PFOS were measured to the edible compartment of the studied crops as well as to the vegetative part. The concentration in the vegetative part was found to be particularly higher than in the edible parts. Toxicological risk assessments on carrots or cucumbers grown in heavily contaminated soil showed that a man of 60 kg would have to eat 3 kilos of the vegetables every day in order to exceed the tolerable daily intake (TDI) for PFOA. This amount of possible contaminated vegetables was not considered applicable. However, vegetative parts of the crops are used for feeding stuffs and the high levels found in these plant tissues could possibly account for the main uptake of PFOA and PFOS into the food chain via plants.

Polycyclic aromatic hydrocarbons (PAH) are potential contaminants in soils, particularly in urban areas. For non-smokers the main contributor to human exposure to PAH is food. Crop-specific models for leafy vegetables, root vegetables, potatoes and tree fruits have been applied for estimation of BCF for PAH in edible plant tissues (Kulhánek *et al.*, 2005). BCF were compared with results derived from empirical regression. The crop-specific models resulted in lower BCF values in fruits, potatoes and leafy vegetables, particularly for high molecular weight PAH. Higher BCF values were obtained with the crop-specific models for root vegetables and leafy vegetables if 1% attached soil particles were considered. Average daily intake of benzo(*a*)pyrene (BaP) by an adult person was calculated to be 190 ng BaP with the results from the crop-specific models and 460 ng from the empirical regression results. With the use of a toxicological safe dose below 4.2–35 ng per person per day, an acceptable soil concentration of BaP could be estimated to be below 0.02–0.18 mg kg^{-1} (wet weight) with the crop-specific models.

Human exposure to contaminants through consumption of vegetables and fruits is potentially significant. The relevance of the exposure pathways depends on the land use, that is residential sites, vegetable gardens and agricultural sites. Calculation of human exposure through consumption of vegetables and fruits must use a representative concentration. (Elert *et al.*, 2011). The representative concentration is determined by selection of the most relevant vegetables and fruits. The relevant vegetables may vary in different parts of the world, and the selection can be important as uptake and accumulation of contaminants differs significantly between plants.

8.5 Conclusion and future trends

Organic environmental contaminants in soil or air can be taken up by plants. Uptake pathways depend on the chemical properties of the compounds, the type of plants and the physicochemical conditions where the plants are grown. The developing of mathematical models for simulation of the uptake aids in understanding this complex relationship and in the risk assessment of the human exposure through consumption of vegetables and fruits. Predictions of uptake of contaminants in plants are uncertain because of variability in the parameters involved, and harmonisation of the procedures for estimating humane exposure is required.

8.6 Questions for discussion

1. How does contamination of vegetables and fruits with environmental chemical pollutants differ from residues of, for example, pesticides or disinfecting agents in these products?
2. How will organic grown produce compare with conventional grown products when considering the risk for content of environmental contaminants?
3. Which chemical properties affect the uptake of chemicals in plants?
4. What are the principal pathways for uptake of chemical compounds in plants?
5. What roles do plant composition and edible parts play in risk assessments of human exposure with environmental contaminants through uptake in plants?
6. How can model simulations help in risk assessments of environmental contaminants in vegetables and fruits?
7. What preventive actions can be taken in order to reduce the risk of uptake of contaminants in plants grown on contaminated soils?
8. What can the consumer do in order to reduce the content of environmental contaminants in fresh produce and what are the limitations?
9. What are the sources of uncertainties in estimating human dietary exposure to contaminants through consumption of vegetables and fruits?
10. Can uptake of chemical compounds in plants be of benefit in some situations?

8.7 References

BIEDERMANN, M., K. GROB and G. MORCHIO. 1995. On the origin of benzene, toluene, ethylbenzene and xylene in extra virgin olive oil. *Z. Lebensm. Unters. Forsch.* 200: 266–72.

CODEX ALIMENTARIUS. 2010. *Code of hygienic practice for fresh fruits and vegetables.* CAC/RCP 53–2003.

CODEX ALIMENTARIUS. 2003. *General principles of food hygiene*. CAC/RCP 1–1969.

COLLINS, C. D., I. MARTIN and W. DOUCETTE. 2011. Plant Uptake of Xenobiotics, pp. 3–16. In P. Schröder and C.D. Collins (eds.), *Organic Xenobiotics and Plants: From Mode 3 of Action to Ecophysiology*, Plant Ecophysiology 8, Springer Science+Business Media B.V., Dordrecht, Netherlands.

ELERT, M., R. BONNARD, C. JONES, R. A. SCHOOF and F. A. SWARTJES. 2011. Human Exposure Pathways, pp. 455–515. In F.A. Swartjes (ed.), *Dealing with Contaminated Sites*, Springer Science+Business Media B.V., Dordrecht, Netherlands.

EUROPEAN COMMUNITY. 2004. Regulation (EC) No 8521 2004 of the European Parliament and of the Council of 29 April 2004 on the hygiene of foodstuffs. *Official Journal of the European Union* L139 30.4.2004, p. 1.

KULHÁNEK, A., S. TRAPP, M. SISMILICH, J. JANKU and M. ZIMOVÁ. 2005. Crop-specific human exposure assessment for polycyclic aromatic hydrocarbons in Czech soils. *Sci. Total Environ.* 339:71–80.

LECHNER, M. and H. KNAPP. 2011. Carryover of Perfluorooctanoic Acid (PFOA) and Perfluorooctane Sulfonate (PFOS) from Soil to Plant and Distribution to the Different Plant Compartments Studied in Cultures of Carrots (*Daucus carota* ssp. *Sativus*), Potatoes (*Solanum tuberosum*), and Cucumbers (*Cucumis Sativus*). *J. Agric. Food Chem.* 59:11011–18.

LEGIND, C. H. and S. TRAPP. 2009. Modeling the exposure of children and adults via diet to chemicals in the environment with crop-specific models. *Environ. Pollut.* 157:778–85.

MAGKOS, F., F. ARVANITI and A. ZAMPELAS. 2003. Putting the safety of organic food into perspective. *Nutr. Res. Rev.* 16:211–21.

MIKES, O., P. CUPR, S. TRAPP and J. KLANOVA. 2009. Uptake of polychlorinated biphenyls and organochlorine pesticides from soil and air into radishes (*Raphanus sativus*). *Environ. Pollut.* 157:488–96.

SAMSØE-PETERSEN, L., E. H. LARSEN, P. B. LARSEN and P. BRUUN. 2002. Uptake of Trace Elements and PAHs by Fruit and Vegetables from Contaminated Soils. *Environ. Sci. Technol.* 36:3057–63.

TRAPP, S. and C. N. LEGIND. 2011. Uptake of Organic Contaminants from Soil into Vegetables and Fruits, pp. 369–408. In F.A. Swartjes (ed.), *Dealing with Contaminated Sites*, Springer Science+Business Media B.V., Dordrecht, Netherlands.

9

Water: waste, recycling and irrigation in fresh produce processing

C. Chaidez, Centro de Investigacion en Alimentacion y Desarrollo, Mexico, M. Soto, The University of Arizona, USA and M. Jimenez, Universidad Autonoma de Sinaloa, Mexico

DOI: 10.1533/9781782420279.2.119

Abstract: Irrigation water can be contaminated with pathogenic microorganisms which cause foodborne outbreaks worldwide. The extent of contamination depends on the degree of treatment provided and the design of the irrigation system. The highest possibility of contamination is presented by the flooding irrigation and sprinkler irrigation technique, whereas dripping irrigation shows the lowest. Also, water used for fertilizers, pesticides and herbicides has to meet certain chemical and microbiological requirements before its use. Agricultural irrigation is crucial for improving the quality and quantity of food production. Therefore, there is a need to identify the possible sources and problems involved with its use.

Key words: agriculture, irrigation water, irrigation system, pathogenic microorganism.

9.1 Introduction

Water-related problems are a serious environmental threat to human health and food security. Lack of access to a safe water supply and sanitation affects the health of billions of people annually. Such problems may be attributed to many factors. Inadequate water management is accelerating the depletion of surface water and groundwater. Wastewater reuse has become increasingly important in water resource management for both environmental and economic reasons. There is a long tradition of reuse of sewage effluent for agricultural practices, primarily in agriculture, and additional applications including industrial, household and urban are becoming more prevalent. Two-thousand years ago, water and sewerage systems, irrigation and navigation were strategically designed to supply the growing population with water for villages and agriculture (Angelakis and

Table 9.1 Worldwide volume of wastewater used for irrigation

Country	Wastewater used for irrigation (m^3/d)	Country	Wastewater used for irrigation (m^3/d)
Mexico	449 300	Iran	422 000
Egypt	1 918 000	Chile	380 000
China	1 239 000	Jordan	225 000
Syria	1 182 000	UAE	200 000
Spain	932 000	Turkey	137 000
USA*	911 000	Argentina	130 000
Israel	767 000	Tunisia	118 000
Italy	741 000	Libya	110 000
Saudi Arabia	595 000	Qatar	80 000
Kuwait	432 000	Cyprus	68 000

Source: Jiménez and Asano (2008).

*California and Florida.

Spyridakis, 1996). The Minoans may have collected wastewater for reuse at Knossos, Crete, some 4000 years ago (Angelakis, 2005). In the sixteenth and seventeenth centuries, the first documented sewage farms (in which wastewater was applied to land for disposal and for use in agriculture) were operated in Bunzlau, Germany and Edinburgh, Scotland (Shuval *et al.*, 1986). Recently, a global estimation on wastewater use indicated that about 20 million hectares of agricultural land is irrigated with (treated and untreated) wastewater (Jiménez and Asano, 2008). Agriculture is therefore the largest sector using wastewater, and this is expected to increase further, particularly in developing countries. In lower income countries and in arid and semi-arid high-income countries, wastewater irrigation is one of the main options being considered as a new source of water (World Bank, 2010). Table 9.1 summarizes the volume of wastewater used for irrigation worldwide.

9.2 Technological challenges

Wastewater treatment and reuse are potential solutions to the problem of poor quantity and quality of water, because this could satisfactorily reduce the levels of pollutants and can be used for irrigation or recreational activities. However, reused wastewater needs to meet some microbial standards before any application. In recent years, both the European Union and State legislators in the USA have recognized the benefits of wastewater reuse (Asano and Levine, 1996; Rogers *et al.*, 2004). However, responsibility for municipal wastewater collection and treatment primarily lies with the local public health authorities and wastewater treatment facilities. These public services are regulated by state and federal environmental protection agencies, which are guided by state and federal legislation.

Currently, water reuse practices are not governed at a federal level, but individual states have water reuse guidelines and regulations (EPA, 2012). Certain parameters are imposed on wastewater facilities to ensure the production of good-quality water. The most common microbial and chemical parameters monitored under state water reuse regulations include biological oxygen demand (BOD), total suspended solids (TSS), turbidity, and total and fecal coliforms. For this reason, the required treatment types are used to remove pathogens, nutrients, trace organics, trace metals, total dissolved solids (TDS) and micro-constituents (Koivunen *et al.*, 2003; AAWA and WEF, 2008). For higher-level uses, such as irrigation of public-access lands or vegetables to be consumed without processing, requirements are higher for the treatment process (Bell, 2011).

The 1972 Amendments to the Federal Water Pollution Control Act (known as the Clean Water Act) established the foundation for wastewater discharge control. The objective was to restore and maintain the chemical, physical and biological integrity of the USA's waters (EPA, 2004). The recommended sewage treatment for this use comprises three major steps: primary, secondary and tertiary treatment (EPA, 2004; Maier *et al.*, 2009). Tertiary treatment of wastewater is an advanced level of treatment used for the protection of wildlife after its discharge into rivers and lakes, especially if the water is to be reused for irrigation, recreational purposes or for drinking water. Examples of these treatments are physicochemical, such as coagulation, filtration and activated carbon, adsorption of organics, reverse osmosis and additional disinfection; however, filtration is usually the most commonly selected (Maier *et al.*, 2009). In the last decade there has been an increased interest in the use of natural systems for the treatment of municipal wastewater as a form of tertiary treatment. Appropriate technologies such as a slow sand filter for tertiary treatment have shown promising results for disinfection of wastewater to allow safe reuse (Langenbach *et al.*, 2009). Constructed wetlands may also be an excellent technology for treating domestic wastewater generated by small communities (Belmont *et al.*, 2004). Wastewater treatment is, in any case, recognized as a practice that may reduce risks to public health, not only to farm-workers but also the nearby population and consumers.

9.3 Significant factors in environmental challenges to food safety

9.3.1 Significance of pathogens or chemical agents

Wastewater used in agriculture has considerable benefits, but can also carry risks to public health, especially when untreated wastewater is used for produce irrigation (World Bank, 2010). The fundamental precondition for water reuse is that applications will not cause unacceptable public health risks. The major risks to public health are microbial and chemical contamination. However, wastewater use in agriculture can also create environmental risks in the form of soil and groundwater pollution.

Untreated sewage can contribute to a high concentration of waterborne pathogens, especially viruses (Fong *et al.*, 2010).The efficient removal of pathogens varies depending on the treatment process type, retention time, biological flora present in the activated sludge, oxygen concentration, pH, temperature and the efficiency in removing suspended solids (Koivunen *et al.*, 2003; EPA, 2004; Maier *et al.*, 2009). Municipal wastewater can transport diverse microorganisms such as *Escherichia coli*, *Salmonella*, *Shigella*, *Listeria*, *Cryptosporidium parvum*, *Giardia lambia*, *Cyclospora cayetanensis*, *Toxoplasma gondi*, norovirus, hepatitis A, among others at the levels indicated in Table 9.2 and will survive in the environment for long periods (Table 9.3). Pathogenic bacteria are present in wastewater at much lower levels than the coliform group of bacteria,

Table 9.2 Levels of pathogens in wastewater

Pathogen		Concentration per liter in municipal wastewater*
Viruses	Enteroviruses†	5000
Bacteria	Pathogenic *E. coli*‡	?
	Salmonella spp.	7000
	Shigella spp.	7000
	Vibrio cholerae	1000
Protozoa	*Entamoeba histolytica*	4500
Helminths	*Ascaris lumbricoides*	600
	Hookworms§	32
	Schistosoma mansoni	1
	Taenia saginata	10
	Trichuris trichiura	120

Source: Feachem *et al.* (1983).

*Based on 100 litres per capita per day (lpcd) of municipal sewage and 90% inactivation of excreted pathogens.
†Includes polio-, echo- and coxsackie viruses.
‡Includes enterotoxigenic, enteroinvasive and enteropathogenic *E. coli.*
§*Anglostoma duedenale* and *Necator americanus.*
?: uncertain.

Table 9.3 Survival of excreted pathogens (at 20–30 °C)

	Survival times in days			
Type of pathogen	In feces, nightsoil and sludge	In fresh water and sewage	In the soil	On crops
Viruses Enteroviruses	<100 (<20)	<120 (<50)	<100 (<20)	<60 (<15)
Bacteria				
Fecal coliforms	<90 (<50)	<60 (<30)	<70 (<20)	<30 (<15)
Salmonella spp.	<60 (<30)	<60 (<30)	<70 (<20)	<30 (<15)
Shigella spp.	<30 (<10)	<30 (<10)	–	<10 (<5)
Vibrio cholerae	<30 (<5)	<30 (<10)	<20 (<10)	<5 (<2)

Sources: Feachem *et al.* (1983); Pescod (1992).

Figures in brackets show the usual survival time.

which are much easier to identify and enumerate (as total coliforms/100 ml) (Pescod, 1992). In this sense, fecal coliforms are selected as the indicator bacteria most commonly used in fecal pollution, *E. coli* being the preferred adopted indicator bacteria (Edberg *et al.*, 2000).

Chemical health risks are indicated in the presence of heavy metals (e.g. cadmium, lead and mercury) and many organic compounds (e.g. pesticides, fertilizers and herbicides) in treated water reuse. These contaminants are derived from domestic and industrial sources. Even if toxic materials are not present in concentrations likely to affect humans, they might well be at phytotoxic levels, which would limit their agricultural use (Pescod, 1992). On the other hand, the chemical constituents of concern for agricultural irrigation are salinity, sodium, trace elements and excessive chlorine residual. However, treated wastewater tends to have higher concentrations of these constituents than groundwater or surface water sources (EPA, 2004). Therefore, the monitoring of its use must be considered in estimating the capital and operating costs of the reuse system, and a complete understanding of monitoring requirements is needed as part of a cost-benefit analysis (EPA, 2004).

Improving wastewater irrigation and minimizing public health and environmental risks, as well as integrating wastewater into the water resources management context, will fully integrate wastewater into a water resources management system.

9.3.2 Geographical and regional significance of wastewater

Wastewater use in agriculture indicates important patterns between a country's economic status and the types of wastewater issues it faces. In low-income countries, wastewater treatment plants are minimal or function poorly, and basic sanitation is the primary focus. Policymakers and consumers in these countries face financial, institutional and technical challenges when considering improvements in sanitation and wastewater management (World Bank, 2010). In upper-middle-income countries, wastewater management is becoming an increasingly important concern, and more complex regulatory structures are being developed, along with the political and the financial resources to implement them. For example, the Latin American and the Caribbean (LAC) regions have the highest rate of sanitation coverage among the developing regions. WHO and UNICEF (2000) global assessment reported that 49% of the LAC population was connected to a sewer system in 2000 compared with 42% in 1990; this means that 68 million people have gained access to a sewer connection over a single decade.

In high-income countries, wastewater is generally treated. In North America, over 90% of wastewater is treated, generally to secondary, and in many cases, tertiary levels (WHO/UNICEF 2000). The high-income countries have made substantial investments in wastewater treatments over the past two decades and have implemented more complex and comprehensive regulatory frameworks. However, a number of constraints need to be addressed when planning wastewater irrigation, such as technical, institutional, regulatory, financial and social issues.

Improperly managed, it imposes significant health risks on farmers and consumers leading to a costly burden of diarrheal and parasitic diseases, and environmental degradation. Furthermore, the Intergovernmental Panel on Climate Change (2007) predicts that global warming will alter precipitation patterns around the world; the subtropics and mid latitudes, where much of the world's poorest populations live, are expected to become substantially drier, resulting in heightened water scarcity (Meehl *et al.*, 2007). In this sense, innovative approaches and new methodologies for protecting water resources from pollution are necessary. Although developed countries have addressed the concerns of wastewater sludge management, ±2.6 billion of the world's people lack basic sanitation. They live in areas where the development of robust systems for excreta management that fully protect public health are non-existent or in their earliest stages (UN-HABITAT, 2008).

Cultural beliefs have a strong impact on sanitation, and even on the possibility of talking about sanitation. In many cultures, the handling of excreta is considered a taboo and viewed as disgusting or a dangerous nuisance. These perceptions block the safe recovery of valuable agricultural resources from human wastes (WHO/UNICEF 2000). Without the general acceptance and involvement of the stakeholders – including regulatory authorities, water and sanitation companies, farmers and consumers alike – wastewater treatment and irrigation projects are unlikely to succeed. The active participation of farmers and consumers is of particular importance. Farmers need to be educated on safe irrigation and postharvest practices and consumers need to be informed on safe handling and preparation of food crops irrigated with treated wastewater.

9.4 Regulatory and economic aspects

9.4.1 Regulatory aspects

The WHO guidelines continue to represent a benchmark target for decision makers in developing the wastewater-recycling sector. The fecal coliform guideline (that is, =1000 FC/100 ml for food crops eaten raw) intends to protect against risks from bacterial infections, and the intestinal nematode egg guideline was intended to protect against helminth infections (WHO, 1989). In contrast, the US Environmental Protection Agency and the US Agency for International Development (1992) have recommended the use of much stricter guidelines for wastewater use (Hespanhol and Prost, 1994). For irrigation of crops likely to be eaten uncooked, no detectable fecal coliforms/100 ml are allowed, and for irrigation of commercially processed crops, fodder crops, the guideline is ≤200 FC/100 ml (where only a nematode egg guideline is set by WHO). For unrestricted irrigation of food crops, the range is from 10 to 1000 fecal coliform bacteria/100 ml for surface irrigation to 2.2 to 200 fecal coliform bacteria/100 ml for spray irrigation. California has some of the strictest standards, requiring <2.2 total coliform bacteria/100 ml for irrigation of food crops (to be achieved through secondary treatment followed by filtration and disinfection) and <23 total coliform bacteria/100 ml for irrigation of pasture and landscaped areas (through secondary treatment and disinfection) (State of California, 1978). This essentially means that fecal contamination should be absent and there should be no potential risk of infection present (although low coliform levels do not necessarily equate to low pathogen levels). European Union countries have also been influenced by WHO (1989) Guidelines (Bontoux, 1998). Conversely, Mexico follows the NOM-001-ECOL-1996 (NORMA Oficial Mexicana, 1996) standard governing wastewater quality, as well as NOM-003-ECOL-1997 (NORMA Oficial Mexicana, 1997), which establishes wastewater quality for public services reuse, but not for agricultural purposes. Table 9.4 shows the Mexican standard governing wastewater reuse in agriculture.

In the past two decades, recycling of urban wastewater for agricultural use has been receiving increasing attention from decision makers, planners and external support agencies, largely as a result of the rapid dwindling of easily accessible freshwater sources (groundwater in particular) and the consequently sharp rise in cost of procuring irrigation water (Havelaar *et al.*, 2001). Reduction in environmental pollution caused by wastewater disposal was seen as a benefit of recycling of human waste. With this change of paradigm in urban water resources management, a renewed need for informed guidance on health protection has arisen. As a result, stakeholders such as WHO, United Nations Development Programme (UNDP), the World Bank, United Nations Environment Program (UNEP), Food and Agriculture Organization (FAO), and bilateral support agencies have commissioned reviews of credible epidemiological literature related to the health effects of excreta and wastewater use in agriculture. The results are documented in Blum and Feachem (1985) and Shuval *et al.* (1986). The above stakeholders, along with independent academic institutions and experienced scientists, aimed to develop a rational basis for the updated health guidelines. To this end, Blumenthal *et al.* (2000) reviewed established guidelines for the microbiological quality of treated wastewater used in agriculture. This guideline limit for fecal coliform bacteria in unrestricted irrigation (≤1000 fecal coliform bacteria/100 ml) is valid, but for restricted irrigation $=10^5$ fecal coliform bacteria/100 ml is recommended when adult farm-workers are exposed to spray irrigation. A limit of $\leq 10^3$ fecal coliform bacteria/100 ml is recommended if flood irrigation is used or children are exposed. The guideline limit for nematode eggs

Table 9.4 Mexican standard governing wastewater reuse in agriculture

Irrigation	FC/100 ml (MPN)	Helminthes ova/liter
Restricted	1000_m–2000_d	≤5
Unrestricted	1000_m–2000_d	≤1

Source: Blumenthal *et al.* (2000).

Unrestricted irrigation is defined as permitting irrigation of all crops; whereas restricted irrigation excludes salad crops and vegetables that are eaten raw (individual crops are no longer specified in the standard).

FC, fecal coliform; m: monthly mean; d: daily mean; MPN: most probable number.

for both types of irrigation except when conditions favor the survival of nematode eggs and where children are exposed; in these cases it should be reduced from ≤1 egg/l to ≤0.1 egg/l.

9.4.2 Economic aspects

Assessment and management of the microbial risks to human health caused by wastewater use in agriculture are essential to avoid any excessive additional burden of disease for those who work in wastewater-irrigated fields or consume wastewater-irrigated foods (World Bank, 2010).

Financial costs of reclaimed wastewater use make it an expensive water resource; the financial costs mainly include the investment and operational costs related to collection and treatment of influent and conveyance and distribution of treated effluents. However, in practice, there are economic impacts involved that make reclaimed wastewater irrigation reasonably competitive compared with other water sources. This stems from added value to water resources and agriculture, and the costs avoided related to reclaimed wastewater utilization; the avoided costs are those of developing new water resources, prevention of health risks and of environmental degradation.

Three aspects have been analyzed for the successful reuse of water: an increase in the quantity of wastewater treated; technical improvements for the production of high-quality water at affordable costs; and the institutional focus on water reuse regulations to help reduce health and environmental risks (Molinos-Senante *et al.*, 2012). However, sustainable use of water resources must be economically justified and Urban Agriculture (UA) may provide that economic justification because food production requires personnel and therefore contributes to the economy by retaining or creating jobs. Moreover, treated wastewater used in UA can provide organic nutrients and soil conditioners for optimal agricultural production. In fact, the use of reclaimed wastewater in agriculture has been reported in around 50 countries and this practice is winning increased attention worldwide (FAO, 2010). In the case of American wastewater treatment utilities, the benefits amount to less than 25% of the operating costs, which is not desirable and requires corrective action. Even when water reuse is completely justified, the cost-benefit balance is not always equilibrated (AWWA and WEF, 2008).

9.5 Market issues

International trade in fresh produce has considerably increased in recent decades and expansion of fresh produce production, particularly from Latin American countries, has the potential to meet most of the growing global demand for fruit and vegetable products. The need to increase food production to feed an ever-growing world population is more urgent now than ever, as FAO has predicted that global food production will need to increase by 50% over current levels by 2050. The complexity of the food chain leading to human consumption has increased and globalization of the food supply has presented new challenges for food safety contributing to the international public health problem of foodborne disease. To initiate and sustain efforts aiming to prevent foodborne disease, the magnitude of the problem needs to be determined. Although several countries have conducted studies to determine the burden of foodborne diseases, global estimates are fragmentary and inconsistent. Foodborne pathogens cause a range of diarrheal and flu-like illnesses. An estimated 1.8 million childhood deaths are associated with disease-causing organisms acquired via food consumption, with the greatest number of cases occurring in developing countries (WHO, 2008). Developed countries also suffer from a significant burden of diarrheal illnesses (Scallan *et al.*, 2005). More than 130 million people from European countries suffer from foodborne diseases and diarrhea is the most common symptom, while in Australia around 17 457 098 cases occurred in 2003 (WHO, 2008).

Documented foodborne outbreaks associated with the consumption of minimally processed or uncooked produce has increased in recent years, raising concerns regarding the quality of fruit and vegetables as they are processed and transported from preharvest to the consumer (Bean *et al.*, 1997; Mead *et al.*, 1999). It has been speculated that the recent produce-related outbreaks in the USA originated with crops harvested outside the USA, such as in Latin American countries.

The USA imports millions of pounds of fruit and vegetables each year (CIAD, 2006; Florkowski, 2006), and during the 2004–2005 growing season alone, more than 200 000 trucks crossed the USA-Mexico border to deliver fresh produce (CIAD, 2006). In 2008, nearly 3 million metric tons of fresh vegetables were imported from Mexico to the USA, representing more than one-half of all revenues from US vegetable imports (ERS, 2009; FAS, 2009). This technological and organizational change has influenced the methods of growing crops; most agriculture in Central America is still not developed and most of the production is dedicated to domestic markets.

9.6 Critical factors in using irrigation water

The risk of contamination of fruit and vegetables when water is in contact with them depends on the microbiological quality of the water source. Water has multiple uses in diverse agricultural activities, including the application of pesticides and fertilizers. This water has to meet chemical and microbiological requirements before its use (Siller-Cepeda *et al.*, 2002). There are critical factors that need to be monitored in order to ensure safe water supply: all water sources must be submitted periodically to chemical and microbiological examination, the results must be recorded and existing problems corrected; bathing and grazing animals near to water sources should be prohibited as this practice prevents fecal contamination and reduces the risks to human health from consuming contaminated fresh produce.

Frequent contamination is associated with the irrigation systems and the type of crop. Studies have proved that flooding irrigation represents the greatest risk of contamination if it is used on produce having direct contact with the soil, while the sprinkler irrigation technique provides a rapid means to contaminate the product if the water is contaminated. The drip irrigation technique has the lowest risk of contamination of produce.

Microbiological quality has always been low in Central America because wastewater from sewage has sometimes been used for irrigation of crops intended for the domestic market. Sadovski *et al.* (1978) noted that the use of wastewater was of greatest concern if applied immediately before harvest rather than during the early stages of the production cycle. Among the types of irrigation used in Mexico and Central American countries, drip irrigation is the most common type used on agriculture dedicated to exportation, particularly in Northwest Mexico, reducing risk of contamination; however, this technique is used for around 15% only of the irrigated land in this area. Overhead irrigation with sprinklers is most commonly utilized to irrigate fruit crops, especially citrus and apple crops; however, most tropical crops in all of these regions also depend on seasonal rain periods for irrigation. Flooding irrigation is the type of irrigation most extensively used in countries where water is conducted from dams to fields. This type of irrigation has higher possibility of contaminating produce, especially when it is applied on creeping crops and where water drives through contaminated soil. In crops staked and raised from soil, such as those vegetables mostly grown for exportation, the risk of contamination is low.

9.7 Troubleshooting and best practice

In recognizing that water-related problems are one of the most important and immediate challenges to the environment and public health, it is important to act now. Water scarcity and water pollution are some of the crucial issues that must be addressed within both local and global perspectives. One of the ways to reduce the impact of water scarcity as well as minimizing water pollution is to expand water and wastewater reuse. In this regard, it is vital to implement Good Agricultural Practices (GAPs) in wastewater reuse and recycling. So far, wastewater reuse is most significant in the agriculture sector, and stabilization ponds and artificial wetlands based on phytotechnologies are commonly used in developing countries to improve the water quality. Other methods including reservoirs and soil aquifer treatments are also in use in some places. The increasing awareness of food safety, and the influence of countries that import food produce, is influencing policymakers and agriculturists to improve the standards of wastewater reuse in agriculture. Fresh fruit and vegetable growers from different countries are implementing programs to prevent and reduce produce contamination, such as GAPs and Good Manufacturing Practices. However, their success and sustainability depend on political will, public awareness and active support from national and international agencies to create an enabling environment for the promotion of environmentally sustainable technologies.

9.8 Conclusion and future trends

9.8.1 Public health risks of foodborne pathogens in foods (fresh produce)

- Fresh produce is highly associated with foodborne pathogen outbreaks.
- Even low levels of particular foodborne pathogens (i.e. viruses) are a considerable public health problem.
- This low level of prevalence may pose human health risks if fresh produce is consumed raw, but even when low levels are present, cooking or adequate disinfection will eliminate the risk.

9.8.2 Pathways of contamination of fresh produce environments

- There are a variety of pathways for foodborne pathogens to enter the food chain environment, ranging from wild animals, domestic stock, poor sanitation and inappropriate disposal of human and animal wastes. Control of such pathways poses major challenges, such as land run-off during rain and control of wild animals in the farm environment.
- Several foodborne pathogens have been reported to survive for long periods of time in the aquatic tropical environment where a large proportion of fresh produce production takes place.
- There is currently disagreement in scientific literature as to whether some foodborne pathogens (i.e. *Salmonella* species) are a part of the normal flora in the tropical aquatic environment. Although long-term survival of some *Salmonella* serovars has been reported in these environments, it is unclear if these species are growing and multiplying and have found an ecological niche. In some studies, the serotypes isolated from the environment are not those associated with humans and animals.

9.8.3 Risk management

- Good Agricultural Practices, Good Manufacturing Practices and Good Hygiene Practices during food production and biosecurity measures can minimize but not eliminate foodborne pathogens in agricultural products.
- Good hygienic practices will minimize the opportunity for cross-contamination in processing environments.
- Currently there are insufficient data to carry out quantitative risk assessment for several foodborne pathogens.

9.9 Questions for discussion

1. Is wastewater suitable for fresh produce production?
2. List the waterborne pathogens commonly present in reused wastewater.
3. Explain the benefits of reusing wastewater in fresh produce production.

4. Explain how Good Agricultural Practice prevents waterborne pathogens entering fresh produce in field production systems.
5. What are the most common microbial and chemical parameters monitored in wastewater?
6. Describe the recommended treatment for sewage.
7. What are the WHO categories of treated wastewater for agricultural use?
8. Is HACCP a good system for using treated wastewater in fresh produce production?
9. Is risk assessment a good system for using treated wastewater in fresh produce production?
10. Should consumers be notified of fresh produce being irrigated by treated wastewater?
11. What are the benefits of implementing water and wastewater reuse initiatives?
12. Which cases and approaches of wastewater reuse result in efficient and successful management of resources?

9.10 References

AMERICAN WATER WORKS ASSOCIATION (AWWA) and WATER ENVIRONMENT FEDERATION (WEF). 2008. *Using Reclaimed Water to Augment Potable Water Resources* 2nd ed. pp. 4–6.

ANGELAKIS AN. 2005. *Waste and wastewater technologies in ancient Greece with emphasis on Minoan era.* IWA SG on Water and Wastewater in Ancient Greece and Nation Foundation For Agricultural Research. Iraklion, Greece.

ANGELAKIS AN, SPYRIDAKIS S. 1996. The status of water resources in Minoan times: A preliminary study. In: *Diachronic Climatic Impacts on Water Resources in Mediterranean Region.* (Angelakis AN and Issar A, Editors) 8: 161–91. Springer-Verlag, Heidelberg, Germany.

ASANO T, LEVINE A. 1996. Wastewater reclamation, recycling and reuse: Past, present and future. *Wat. Sci. Tech.* 10:1–14.

BEAN NH, GOULDING JS, DANIELS MT, ANGULO FJ. 1997. Surveillance for foodborne disease outbreaks – United States, 1988–1992. *Journal of Food Protection* 60: 1265–86.

BELL KY. 2011. Water reuse – Treatment technology advances and applications. *Water Environmental Federation* 50: 2914–21.

BELMONT MA, CANTELLANO E, THOMPSON S, WILLIAMSON M, SANCHEZ A, METCALFE CD. 2004. Treatment of domestic wastewater in a pilot-scale natural treatment system in central Mexico. *Ecological Engineering* 23: 299–311.

BLUM D, FEACHEM RG. 1985. *Health aspects of nightsoil and sludge use in agriculture and aquaculture – part III: An epidemiological perspective,* IRCWD Report No. 04/85, SANDEC, CH-8600, Dubendorf, Switzerland.

BLUMENTHAL UJ, MARA DD, PEASEY A, RUIZ-PALACIOS G, STOTT R. 2000. Guidelines for the microbiological quality of treated wastewater used in agriculture: recommendations for revising WHO guidelines. Environment and Health. *Bulletin of the World Health Organization* 78: 1104–16.

BONTOUX L. 1998. The regulatory status of wastewater reuse in the European Union. In: *Wastewater reclamation and reuse* (ed. T. Asano). Appendix 1. pp. 1463–76. Lancaster, PA: Technomic Publishing Co.

CENTRO DE INVESTIGACIÓN EN ALIMENTACIÓN Y DESARROLLO (CIAD). 2006. *Overview of the Mexican fresh produce industry.* Food Safety Workshop, Culiacán Research Station, February 2006.

ECONOMIC RESEARCH SERVICE (ERS). 2009. *Foreign agricultural trade of the United States*. Available at: http://www.ers.usda.gov/Data/FATUS/. Accessed 13 March 2009.

EDBERG SC, RICE EW, KARLIN RJ, ALLEN MJ. 2000. *Escherichia coli*: the best biological drinkingwater indicator for public health protection. *Journal of Applied Microbiology* 88: 106S–16S.

EPA (U.S. ENVIRONMENTAL PROTECTION AGENCY). 2004. *Primer for municipal wastewater treatment systems*. Document no. EPA 832-R-04–001. Washington, DC.

FAO. 2010. *The Wealth of Waste: The Economics of Wastewater Use in Agriculture*. ISBN 978–92-5–106578-5. Rome, Italy.

FEACHEM RG, BRADLEY DJ, GARELICK H, MARA DD. 1983. *Sanitation and disease: Health aspects of excreta and wastewater management*. John Wiley, Chichester.

FLORKOWSKI WJ. 31 October 2006. U.S. food imports of produce and seafood. *Report for the Book Project on Microbial Contamination. Economic Research Service*. 2009. Foreign agricultural trade of the United States. Available at: http://www.ers.usda.gov/Data/FATUS/. Accessed 13 March 2009.

FONG TT, PHANIKUMAR MS, XAGORARAKI I, ROSE JB. 2010. Quantitative detection of human adenoviruses in wastewater and combined sewer overflows influencing a Michigan river. *Applied and Environmental Microbiology* 76: 715–23.

FOREIGN AGRICULTURAL SERVICE (FAS). 2009. U.S. Trade Internet System. Available at: http://www.fas.usda.gov. Accessed March 2009.

HAVELAAR A, BLUMENTHAL UJ, STRAUSS M, KAY D, BARTRAM J. 2001. Guidelines: the current position. World Health Organization (WHO). *Water Quality: Guidelines, Standards and Health*. Edited by LORNA FEWTRELL and JAMIE BARTRAM. Published by IWA Publishing, London, UK. ISBN: 1 900222 28 0.

HESPANHOL I, PROST AME. 1994. WHO guidelines and national standards for reuse and water quality. *Wat. Res.* 28: 119–22.

JIMÉNEZ B, ASANO T. (eds.) 2008. *Water Reuse: An International Survey of Current Practice, Issues and Needs*. London: IWA Publishing.

KOIVUNEN J, SIITONEN A, HEINONEN-TANSKI H. 2003. Elimination of enteric bacteria in biological-chemical wastewater treatment and tertiary filtration units. *Water Research* 37: 690–8.

LANGENBACH K, KUSCHK P, HORN H, KASTNER M. 2009. Slow sand filtration of secondary clarifier effluent for wastewater reuse. *Environmental Science and Technology* 43: 5896–901.

MAIER RM, PEPPER IL, GERBA CP. 2009. *Environmental microbiology*. Second edition. Academic Press.

MEAD PS, SLUTSKER L, DIETZ V, MCCAIG LF, BRESEE JS, *et al.* 1999. Food-related illness and death in the United States. *Emerg. Infect. Dis.* 5: 607–25.

MEEHL GA, STOCKER TF, COLLINS WD, FRIEDLINGSTEIN P, GAYE AT, *et al.* 2007. Global Climate Projections. In: Solomon S, Qin D, Manning M, Chen Z, Marquis M, *et al.* (eds.) *Climate Change 2007: The Physical Science Basis. Contribution of Working Group I to the Fourth Assessment Report of the Intergovernmental Panel on Climate Change*. Cambridge, UK: Cambridge University Press.

MOLINOS-SENANTE M, HERNÁNDEZ-SANCHO F, SALA-GARRIDO R. 2012. Economic feasibility study for new technological alternatives in wastewater treatment processes: a review. *Water Sci & Tech* 65:898–906.

NORMA OFICIAL MEXICANA. 1996. *NOM-001-ECOL-1996, Que establece los límites máximos permisibles de contaminantes en las descargas de aguas residuales: En aguas y bienes nacionales*. Diario Oficial de la Federación. Junio 1996.

NORMA OFICIAL MEXICANA. 1997. *NOM-003-ECOL-1997, Que establece los límites máximos permisibles de contaminantes para las aguas residuales tratadas que se reusen en servicios al público*. Diario Oficial de la Federación. Enero 1998.

PESCOD MB. 1992. *Wastewater treatment and use in agriculture – FAO irrigation and drainage paper 47*. Food and Agriculture Organization of the United Nations. Technology and Engineering.

ROGERS PP, LLAMAS RM, MARTINEZ-CORTINA L. 2004. *Water crisis: myth or reality?* Marcelina Biotin Water Forum 2004. Taylor and Francis plc. London, UK.

SADOVSKI A, FATTAL YB, GOLDBERG D. 1978. Microbial contamination of vegetables irrigated with sewage effluent by the drip method. *J. Food Prot.* 41: 336–40.

SCALLAN E, MAJOWICZ SE, HALL G, BAJERJEE A, BOWMAN CL, *et al.* 2005. Prevalence of diarrhea in the community in Australia, Canada, Ireland and the United States. *International Journal of Epidemiology* 34: 454–60.

SHUVAL HI, ADIN A, FATTAL B, RAWITZ E, YEKUTIEL P. 1986. *Wastewater Irrigation in Developing Countries – Health Effects and Technical Solutions Integrated Resource Recovery*, UNDP Project Management Report No. 6, World Bank Technical Paper No. 51.

SILLER-CEPEDA JH, BAEZ-SAÑUDO MA, SANUDO-BARAJAS A, BAEZ-SANUDO R. 2002. *Manual de buenas practicas agricolas. Guia para el agricultor.* Centro de Investigacion en Alimentacion y Desarrollo, A.C. SAGARPA. 1ra (ed). 62 pp.

STATE OF CALIFORNIA. *Wastewater reclamation criteria. Berkeley, CA, Department of Health Services, 1978.* California administrative code, Title 22, Division 4, Environmental Health.

UN-HABITAT. 2008. *Global Atlas of Excreta, Wastewater, Sludge, and Biosolids Management: Moving Forward the Sustainable and Welcome Uses of a Global Resource.* Edited by: LEBLANC RJ, MATTHEWS P, RICHARD RP. ISBN: 978–92-1–132009-1.

US ENVIRONMENTAL PROTECTION AGENCY/US AGENCY FOR INTERNATIONAL DEVELOPMENT. *Guidelines for water reuse.* Washington, DC, Environmental Protection Agency, Office of Wastewater Enforcement and Compliance, 1992 (technical report no. EPA/625/R-92/004).

WORLD BANK. 2010. *Improving wastewater use in agriculture: An emerging priority.* Washington, DC: World Bank.

WORLD HEALTH ORGANIZATION (WHO). 2008. *Foodborne disease outbreaks: guidelines for investigation and control.* WHO Library ISBN 978 924 1547222.

WHO. 1989. *Health Guidelines for the Use of Wastewater in Agriculture and Aquaculture.* Report of a WHO Scientific Group, Technical Report Series No. 778, WHO, Geneva.

WHO. 2006. *WHO Guidelines for the Safe Use of Wastewater, Excreta and Greywater. Vol. I: Policy and Regulatory Aspects. Vol. II: Wastewater Use in Agriculture. Vol. III: Wastewater and Excreta Use in Aquaculture. Vol. IV: Excreta and Greywater Use in Agriculture.* Geneva: World Health Organization.

WHO/UNICEF JOINT MONITORING PROGRAMME FOR WATER SUPPLY AND SANITATION. 2000. *Global Water Supply and Sanitation Assessment 2000 Report.* Geneva: World Health Organization.

10

Maintaining sustainable and environmentally friendly fresh produce production in the context of climate change

D. Pimentel and M. Burgess, Cornell University, USA

DOI: 10.1533/9781782420279.2.133

Abstract: This chapter compares conventional and organic agricultural systems for making agriculture more sustainable in terms of conserving soil, water, energy and biological resources. Organic agricultural technologies provide numerous benefits including abundant soil organic matter and nitrogen, reduced fossil energy inputs, increased crop yields, decreased soil erosion and conservation of soil water – especially beneficial under drought conditions. Conventional agriculture can be made more sustainable and ecologically sound by adopting some of the traditional organic farming technologies.

Key words: organic, agriculture, conventional, conservation, water, soil, nutrients.

10.1 Introduction

Global climate change is a serious and growing problem with the heavy increase in fossil energy use worldwide (Pimentel, 2011). After the publication of the fourth report of the Intergovernmental Panel on Climate Change (IPCC), the arguments on the security aspects of climate change have greatly intensified (Sheffran *et al.*, 2012). Investigations in some areas have made progress but remain highly controversial. There is a clear need to reduce all the energy inputs especially in corn and soybean production, the largest crops produced in the USA and the crops that have the largest energy inputs (USDA, 2011; USCB, 2011). More than 95% of all corn and soybeans rely on herbicides for weed control (USDA, 2011). Insecticides are also heavily used in corn and soybean production (USDA, 2011).

Other inputs for conventional corn and soybean production, such as nitrogen fertilizer, labor and machinery, are expensive in terms of costs as well as energy consumption, which is predominantly from fossil fuels that relates directly to greenhouse gas emissions, specifically carbon dioxide (Pimentel, 2011). There is also serious pollution of water and soil from the heavy use of pesticides, fertilizers and other inputs (Frankenberger and Turco, 2003; Pimentel *et al.*, 2005). The National Academy of Sciences (NAS, 2003) reports that excessive fertilizer inputs in corn and soybeans costs more than $2.5 billion per year. The estimated annual public health and environmental health losses related to soil erosion, especially in corn and soybean production, may exceed $45 billion per year (Pimentel *et al.*, 1995).

Organic farming systems can reduce the reliance on agro-chemical inputs as well as make commercial agriculture environmentally and economically sound. Pimentel and Pimentel (2008) and NAS (2003) have confirmed that sound management practices can reduce pesticide and fertilizer inputs while maintaining high crop yields, and improve farm economics. Some government programs in Indonesia and Sweden have demonstrated that pesticide use can be reduced by 65% to 68% without sacrificing high crop yields and quality foods while continuing to practice conventional agricultural systems (Oka, 1991; Eckstrom and Bergkvist, 2008).

If organic and conservation agriculture can be implemented on a global scale, fossil energy inputs will be greatly reduced, soil and water resources will be conserved and the impacts of global climate change will be lessened (Pimentel *et al.*, 2005; Pimentel, 2011). The fastest growing system of agriculture is organic agriculture.

10.2 Experimental design

Field investigations were conducted from 1981 to 2002 at the Rodale Institute in Kutztown, Pennsylvania. The rainfall averaged 1105 mm per year. The experimental design included three cropping systems: one conventional and two different organic systems – an animal-based system and a legume-based system. The main plots were 18 × 92 meters, which were split into three 6 × 92 meter subplots. The subplots were large enough to be worked with regular farm equipment. Each of the three cropping systems was replicated eight times.

10.2.1 Conventional cropping system

The conventional cropping system followed Pennsylvania State University Cooperative Extension recommendations for corn and soybean production. Thus, during the growing season, the conventional systems had no more exposed soil than the organic system. However, the organic systems had cover crops during the non-growing season, whereas the conventional system did not.

10.2.2 Organic cropping systems

Like the conventional system, both organic systems produced a cash crop every year. However, neither organic system used commercial synthetic fertilizers, relying instead on nitrogen-fixing green manure crops as the nitrogen source. The organic rotation system included hairy vetch as a winter cover crop, planted after the corn or the soybean crops were harvested. The total nitrogen added to this system per ha per year averaged 49 kg (or 140 kg per ha for any given year with a corn crop). Weed control was practiced without herbicides. The energy input in the conventional corn system totaled 5.2 million kcal, whereas the energy input for the organic system was only 3.5 million kcal (Pimentel *et al.*, 2005), 33% less than the conventional system. The economic costs of the organic systems were 15% less than the conventional system costs.

A new no-till system was implemented later on both organic systems which allowed for the winter vetch to be rolled down in the spring right before no-till planting. This helped reduce the need for plowing and reduced the need for fossil fuel. This no-till system was estimated to require only 3.3 million kcal or 37% less energy than the conventional system.

The production costs per hectare for the conventional system are $538, whereas the production costs for the organic system were slightly lower or $457 per hectare (Pimentel *et al.*, 2005). Early in the rotations the return per hectare was higher in the conventional or $127 per ha, but only $109 per ha for the organic. Later on the returns for the organic system increased to $162 per ha and remained at this level for the remainder of the experiment, as did the returns for the conventional system (Pimentel *et al.*, 2005).

The organic system required 35% more manual labor, but because the labor was spread out during the growing season, the hired labor costs per hectare were about equal between the two systems. Each system was allowed 250 hours of 'free' family labor per month. When labor requirements exceeded this level, labor was hired at $13 per hour. With the organic system, the farmer was busy throughout the summer with the hairy vetch cover crop and mechanical weed control (but worked less than the 250 hours per month). In contrast, the conventional farmer had large labor inputs in the spring and fall, planting and harvesting, but little during the summer months. This may have implications for the growing number of part-time farmers for whom the availability of family farms is severely limited.

Organic farming systems other than the ones practiced in this experiment, on average, require about 15% more labor than conventional farming systems (Sorby, 2002; Granatstein, 2003). Over the 21-year period, organic corn (without price premiums) was 25% more profitable than the conventional corn ($221 per ha versus $178 per ha). This was possible because the organic corn yields were only 3% less than the conventional yields (5843 per ha versus 6011 kg per ha), while costs were 15% less ($351 per ha versus $412 per ha).

10.3 Soil content in the organic and conventional farming systems

10.3.1 Carbon

Soil carbon, which correlates with the soil organic matter levels, was measured in 1981 and 2002. In 1981, the soil carbon levels were about the same (about 2%) in both the organic and conventional systems (Pimentel *et al.*, 2005). In 2002, however, the soil organic matter in the organic systems was about 5%, whereas in the conventional system, it remained about 2%. The higher soil organic matter levels found in the organic systems cause the soil to retain more water that results in better crop yields during droughts, reduces soil erosion, increases plant nutrient retention and increases biological diversity. There was about 4000 kg per ha of soil carbon in the conventional system but about 10 000 kg per ha of soil carbon in the organic system in 2002. The higher water content accounted for the higher corn and soybean yields during the drought years faced by these systems (Lotter *et al.*, 2003).

Higher soil organic matter levels cause greater soil nitrogen retention, greater microbial biodiversity, and promote the presence and growth of arbuscular mycorrhizal fungi that penetrate the roots of crops and facilitate the movement of plant nutrients from the soil into the crop plants resulting in better crop growth and yields (Pimentel *et al.*, 2005). Higher soil carbon levels also hold soil particles together so less soil erosion occurs (Pimentel *et al.*, 2005).

10.3.2 Nitrogen

Soil nitrogen levels at the start of the experiment in 1981 averaged about 0.31% (Pimentel *et al.*, 2005). After 21 years, the conventional system remained at 0.31% but the organic system increased to 0.33%. The accumulated soil organic matter in the organic system accounted for the increase in soil nitrogen.

10.3.3 Herbicide leaching

Four herbicides were applied in the conventional system: atrazine (to corn), pendimethalin (to corn), metolachlor (to corn and soybeans) and metribuzin (to soybeans) (Pimentel *et al.*, 2005). From 2001 to 2003, atrazine and metolachlor were only detected in water leachate samples collected from the conventional system. No metribuzin or pendimethalin were detected after application (Pimentel *et al.*, 2005).

10.3.4 Soil biology

Among the natural biological processes on which the organic rotation depends is symbiosis of arbuscular mycorrhize and the crop plants which were investigated in these experiments (Pimentel *et al.*, 2005). Arbuscular mycorrhize fungi are indigenous to most soils. The fungi colonize the roots of most plants, forming mutualistic symbiosis (the mycorrhiza). The fungus receives sugars from the root of the host plant, and the plant benefits primarily from enhanced nutrient uptake from soil through the fungus. The extraradical mycelia of the arbuscular mycorrhize act, in effect, as extensions of the root system. Arbuscular mycorrhize have been shown to enhance disease resistance, improve water relations and increase soil aggregation (Miller and Jastrow, 1990; Hooker *et al.*, 1994; Wright *et al.*, 1999; Auge, 2000). This symbiotic relationship between crop and arbuscular mycorrhize contributes to the success of the organic production systems (Pimentel *et al.*, 2005).

10.4 Soil organic matter and biodiversity

Soil organic matter provides the basis for productive organic farming and sustainable agricultural systems. After 22 years of separate management, soil carbon (soil organic matter) was significantly higher in the organic systems than in the conventional system. Soil carbon increased 15.1% in the organic system compared with only 8.6% in the conventional system (Pimentel *et al.*, 2005).

The amount of organic matter in the upper 15 cm of soil in the organic system was approximately 110 000 kg per ha (Pimentel *et al.*, 2005). The soil in the 15 cm weighed about 2.2 million kg per ha. Approximately 41% of the volume of the organic matter consisted of water, compared with only 35% in the conventional system (Sullivan, 2002). The amount of water held in the organic system is calculated to 816 000 liters per ha. The large amount of soil organic matter and water in the organic system made it more tolerant to droughts, like that which occurred in 1999 and other drought years (Pimentel *et al.*, 2005).

Large amounts of biomass (soil organic matter) are expected to significantly increase soil biodiversity (Pimentel *et al.*, 1992; Troeh and Thompson, 1993; Lavelle and Spain, 2001; Mader *et al.*, 2002; Pimentel *et al.*, 2006). Microbes and invertebrate species can number 21 000 per hectare (Pimentel *et al.*, 2006). The number of arthropods per hectare can range from 2 million to 5 million, and earthworms 1 million to 5 million (Lavelle and Spain, 2001; Gray, 2003). The microarthropods and earthworms were reported to be twice as abundant in organic versus conventional agricultural systems in Denmark (Hansen *et al.*, 2001). The weight of earthworms per ha in agricultural soils can range from 2000 to 4000 kg/ha (Lavelle and Spain, 2001). There can be as many as 1000 earthworm and insect holes per square meter of agricultural land. Earthworms and insects are particularly helpful in constructing large holes in the soil that increase the percolation of water into the soil and decrease run-off.

Soil organic matter is an important source of nutrients and can help increase biodiversity which provides vital ecological services, increasing crop protection and recycling nutrients (Pimentel *et al.*, 2005, 2006). For example, adding compost and other organic matter to the soil reduces the incidence of crop diseases (Cook, 1988; Hoitink *et al.*, 1991), and increases the number of species of microbes in the ecosystem (van Elsen, 2000; Pimentel *et al.*, 2006). The organic systems, in avoiding the use of toxic pesticides and commercial fertilizers, minimize the harmful effects of these chemicals on the soil organisms (Pimentel *et al.*, 2005, 2006).

10.5 Conclusion

Overall, environmental damage from agricultural chemicals was reduced in the organic systems because of the use of cover crops and because no commercial fertilizers or pesticides were employed. As a result, overall public health and ecological integrity could be improved through the adoption of these practices, which decrease the quantities of pesticides and commercial fertilizers applied in agriculture (NAS, 2003; Pimentel *et al.*, 2005, 2006).

10.6 References

AUGE, R.M. 2000. Stomatal behavior of mycorrhizal plants. Pages 201–238. In Kapulnik, Y. and Douds, D.D., Eds. *Arbuscular Mycorrhizas: Physiology and Function*. Kluwer Academic. Dordrecht, Netherlands.

COOK, R.J. 1988. Biological control and holistic plant-health care in agriculture. *American Journal of Alternative Agriculture* 3: 51–62.

EKSTROM, G. and BERGKVIST, P. 2008. Swedish pesticide reduction: 1986–2006. Pages 1–7. In D. Pimentel, Ed. *Encyclopedia of Pest Management*. Taylor and Francis. Boca Raton, Florida. http://dx.doi.org/10.1081/E-EPM-120044723

FRANKENBERGER, J. and TURCO, R. 1999. Hypoxia in the Gulf of Mexico: A Reason to Improve Nitrogen Management. *Purdue Animal Issues Briefing AI-6.* (3 December 2012) http://www.ansc.purdue.edu/anissue/AI6.pdf.

GRANATSTEIN, D. 2003. Tree Fruit Production with Organic Farming Methods. Wenatchee (WA): Center for Sustainable Agriculture and Natural Resources, Washington State University. (29 August 2012). http://modern-prepper.com/download/plants/Fruit%20tree%20Production%20with%20Organic%20Farming%20Methods.pdf.

GRAY, M. 2003. Influence of agricultural practices on earthworm populations. *The Bulletin: Pest Management and Crop Development*. (29 August 2012). http://bulletin.ipm.illinois.edu/pastpest/articles/200305d.html.

HANSEN, B., ALROE, H.F., and STEEN, K.E. 2001. Approaches to assess the environmental Impact of organic farming with particular regard to Denmark. *Agriculture, Ecosystems and Environment* 83: 11–26.

HOITINK, H.A.J., INBAR, Y. and BOEHM, M.J. 1991. Status of compost-amended potting mixes naturally suppressive to soilborne diseases of floricultural crops. *Plant Diseases* 75: 869–73.

HOOKER, J.E., JAIZME-VEGA, M. and ATKINSON D. 1994. Biocontrol of plant pathogens using arbscular mycorrhizal fungi. Pages 191–200. In S. Gianinazzi, and H. Schuepp. Eds. *Impact of Arbuscular Mycorrhizas on Susceptible Agricultural and Natural Ecosystems*. Birkhauser Verlag, Basel, Switzerland.

LAVELLE, P. and SPAIN A.V. 2001. *Soil Ecology*. Kluwer Academic, Dordrecht.

LOTTER, D.W., SIDEL, R. and LIEBHARDT, W. 2003. The performance of organic and conventional cropping systems in extreme climate year. *American Journal of Alternative Agriculture* 18: 146–54.

MADER, P., FLIEBACH, A., DUBOIS, D., GUNST, L., FRIED, P. and NIGGLI, U. 2002. Soil fertility and biodiversity in organic farming. *Science* 296: 1694–7.

MILLER, R.M. and JASTROW, J.D. 1990. Hierarchy of root and mycorrhizal fungal interactions with the soil aggregation. *Soil Biology and Biochemistry* 22: 579–84.

NAS. 2003. *Frontiers in Agricultural Research: Food, Health, Environment, and Communities*. National Academy of Sciences, Washington, DC (13 January 2011) Briefs from the book at the following website: http://dels.nas.edu/Report/Frontiers-Agricultural-Research-Food-Health/10585.

OKA, I.N. 1991. Success and challenges of the Indonesia National Integrated Pest Management Program in the rice-based cropping system. *Crop Protection* 10(3): 163–5.

PIMENTEL, D. 2011. *Global Climate Change: Its Impacts on Food and the Environment*. Rachel Carson Council, Inc., Silver Springs, MD.

PIMENTEL, D. and PIMENTEL, M. 2008. *Food, Energy and Society*. 3rd Edition. CRC Press (Taylor and Francis Group): Boca Raton, FL. 380 pp.

PIMENTEL, D., STACHOW, U., TAKACS, D.A., BRUBAKER, H.W., DUMAS, A.R., *et al.* 1992. Conserving biological diversity in agricultural/forestry systems. *BioScience* 42:354–62.

PIMENTEL, D., HARVEY, C., RESOSUDARMO, P., SINCLAIR, K., KURTZ, D., *et al.* 1995. Environmental and Economic Costs of Soil Erosion and Conservation Benefits. *Science* 267: 1117–23.

PIMENTEL, D., HEPPERLY, P., HANSON, J., DOUDS, D. and SEIDEL, R. 2005. Environmental, energetic, and economic comparisons of organic and conventional farming systems. *Bioscience* 55(7): 573–82.

PIMENTEL, D., PETROVA, T., RILEY, M., JACQUET, J., NG, V., *et al.* 2006. Conservation of Biological Diversity in Agricultural, Forestry, and Marine Systems. Pages 151–73 in Burk, A.R. (ed.) *Focus on Ecology Research*. Nova Science Publishers: New York. (Also, 2007. Pages 1–25 in Schwartz, J. (ed.) *Focus on Biodiversity Research*. Nova Science Publishers: New York.)

SHEFFRAN, J., BRZOSKA, M., KOMINEK, J., LINK, P.M. and SCHILLING, J. 2012. Climate change and violent conflict. *Science* 336: 869–71.

SORBY, K. 2002. *What is Sustainable Coffee?* Background paper to the World Bank Agricultural Technology Note 30. Washington (DC): World Bank.

SULLIVAN, P. 2002. *Drought resistant soil. Fayetteville, (AR): Appropriate Technology Transfer for Rural Areas (ATTRA)*. Agronomy Technical Note (August 29, 2011). http://www.plant-trees.org/resources/infomaterials/english/agroforestry_technologies/Drought%20Resistant%20Soils.pdf.

TROEH, F.R. and THOMPSON, L.M. 1993. *Soils and Soil Fertility*. Oxford University Press, New York.

USCB. 2011. *Statistical Abstract of the United States – 2011*. U.S. Census Bureau, Washington, DC

USDA. 2011. *Agricultural Statistics 2011*. U.S. Department of Agriculture, Washington, DC

VAN ELSEN, T. 2000. Species diversity as a task for organic agriculture in Europe. *Agriculture, Ecosystems and the Environment* 77: 101–9.

WRIGHT, S.F., STAR, J.L. and PLALTINEAU I.C. 1999. Changes in aggregate stability and concentration of glomalin during tillage management transition. *Soil Science Society of America Journal* 63: 1825–9.

11

Reducing waste in fresh produce processing and households through use of waste as animal feed

J. Zentek, F. Knorr and A. Mader, Freie Universität Berlin, Germany

DOI: 10.1533/9781782420279.2.140

Abstract: Byproducts of the food industry are important sources of animal feed. Along the traditional food chain, many valuable energy-or protein-rich byproducts are generated during the processing of animal- and plant-derived foods. Technical developments are providing opportunities for innovative materials, such as byproducts from bio-ethanol production or new fermentation technologies. The composition of such novel byproducts must be characterized prior to implementation in feed, and their potential impact on animal health, the environment, and the quality and safety of the final product should be evaluated.

Key words: byproducts, animal nutrition, sustainability, feed value.

11.1 Introduction

According to the Food and Agriculture Organization (FAO), approximately one-third of the edible portions of food produced for human consumption is lost or wasted globally, which is estimated at 1.3 billion tons per year (Gustavsson *et al.*, 2011). Although global contention for food and feed resources has in recent years furthered the effort to recycle and reutilize food waste as fuel, fertilizer, carbohydrate sources for microbial fermentation processes (Boucqué and Fiems, 1988) or livestock feed, much of the food waste accrued by food industries is still being disposed of in landfill, incinerators or by land application. This indicates an enormous potential for exploitable byproducts.

Feeding food waste to animals has long been important in livestock production and provides an alternative to conventional feed sources (Westendorf, 2000; Minkler, 1914). Food scraps were fed to domestic pigs living in close proximities to rural as well as urban settlement areas and as supplements to cereal-based diets on pig farms (Kamphues, 2001). Recent technical advancements in food processing and biotechnology have enabled the recovery of an even wider array of residual materials from the production, processing, distribution and consumption of food, including cull beans and potatoes, dairy processing waste, bakery waste and garbage. Some byproducts or co-products of other industries, such as corn gluten meal, wheat middlings and brewers' and distillers' grain, are no longer considered food waste byproducts but rather normal commodities used in livestock nutrition (Westendorf, 2000).

11.2 Legal aspects for using food waste and byproducts for animal feed

The main goal of feed regulations at national or European levels is to ensure the supply of food of animal origin at a high level of quality, safety and efficacy. From the European perspective, the objectives can be summarized as follows: Protection of animal and human health, protection of the ecosystem and, finally, protection against fraud. The EU Hygiene Regulation (EC) No 1774/2002 has set basic rules for the usage of animal byproducts not intended for human consumption. The feeding of food residues to livestock has been prohibited since the end of 2002 or 2006 (Austria and Germany, respectively). EC Regulation No 1069/2009 repealed EC Regulation No 1774/2002 as of 21 October 2009.

Catering waste from facilities where food is prepared for immediate consumption, including restaurants, catering facilities, as well as central kitchens and household kitchens, may not be used as a feed constituent for livestock. The EU member states have decided on this feed ban, as the traceability of catering waste, including used cooking oil from restaurants and catering facilities cannot be guaranteed. The EU authorities see a lack of tracking mechanism as the key to incidents in the past, such as the dioxin crises, and foot and mouth disease and swine fever outbreaks. Whether or not these regulations will be modified in the future depends on several factors, considering rapidly changing economic, ecologic and safety aspects.

The basis for the standardization and the labeling of feed materials is the Community Catalogue of Feed Materials, including the descriptions of each feed and information on the manufacturing process. The catalogue is supplemented by a separate register, in which feedstuffs, which are not listed in the catalogue and that are placed on the market for the first time, are registered. The register is continuously updated and published at www.feedmaterialsregister.eu/.

National lists are available, for instance the 'positive list', which is supported by the Standards Commission for feed ingredients of the German Agricultural Society. However, this is not legally binding, but is based on an economic agreement. In addition, various quality assurance systems such as QS ('quality and safety') are relevant for the consideration of feed materials. Compound feeds, either complete feeds or feed supplements, are regulated for instance by the

Regulation (EC) No 767/2009, which regulates placing on the market and the use of feed, and defines compound feed and labeling rules.

Feed additives must be considered separately from other feed ingredients and are regulated in Regulation (EC) No 1831/2003. Feed additives are substances, microorganisms or preparations, other than feed material or pre-mixtures, which are added deliberately to feed or water to meet one or more of the following functions: To positively affect feed characteristics, quality of animal products, animal production, animal performance, animal well-being; to meet nutritional requirements; to induce coccidiostatic or histomonostatic effects. Feed additives need to be authorized according to the principles in Regulation (EC) No 429/2008 and No 1831/2003. The authorized feed additives are published in the European Community Register, which is continuously updated: http://ec.europa.eu/food/food/animalnutrition/feedadditives/registeradditives_en.htm

11.3 Feedstuffs from catering waste

The nutritional composition of food waste is, by nature of origin, subject to variation (Waller, 2011), depending primarily on the raw constituents (growing and harvesting conditions), postharvest processing, and storage and handling. Catering waste in particular comprises a wide array of dietary constituents, limiting the potential for incorporation into practical feeding programs. A thorough survey of studies on the nutrient contents of catering waste originating from restaurants and catering facilities, as well as central kitchens and household kitchens has been provided by Schlüter (2002).

In addition to the legal restriction for using catering waste as feed (Regulation (EC) No 1069/2009), these materials have widely varying nutrient contents. The main limitation associated with catering waste is reported to be the high moisture content, which requires immediate processing in order to limit spoilage. Kornegay *et al.* (1970) demonstrated that in catering waste collected from hotels and restaurants, institutions such as nursing homes, hospitals and prisons, military bases and municipalities, crude protein and crude fat contents were sufficient for feeding swine, if not abundant in most cases. The mineral and vitamin contents were adequate also when compared with traditional diets, although calcium and phosphorus were either borderline or deficient in some sources. The digestibility of the samples varied considerably, and it was found that supplementation with grain proved beneficial for swine performance.

Outbreaks of disease in livestock have often been associated with the feeding of uncooked food residues. The animal diseases of concern include hog cholera, foot and mouth disease, African swine fever and swine vesicular disease. There is also concern regarding domestic pathogens of public significance such as *Salmonella*, *Campylobacter*, *Trichinella* and *Toxoplasma* (Westendorf, 2000). As a response to the need to prevent the spread of Transmissible Spongiform Encephalopathy (TSE), several regulations within the EU presently restrict the use of animal-derived byproducts in feed for livestock. Findings have linked the spread of Bovine Spongiform Encephalopthy (BSE) to the consumption of feed contaminated by processed ruminant protein infected with prions (Collee and Bradley, 1997).

Based on these reports, a ban on the feeding of mammalian processed animal protein to cattle, sheep and goats was introduced in 1994. The ban was expanded by Regulation (EC) 999/2001, which prohibits the feeding of mammalian protein to ruminants with limited exceptions. These include the use of fish meal, some blood products and dicalcium phosphate as feed for non-ruminants, as described in Regulation (EC) 1292/2005. Commission Regulation (EC) 956/2008 provides a deviation from the general prohibition of feeding animal proteins to ruminants by allowing the implementation of proteins derived from fish meal in milk replacers to non-weaned ruminants. Regulation (EC) No 1774/2002, which was repealed by Regulation (EC) No 1069/2009 that regulates animal byproducts, bans the feeding of animals with proteins derived from the same species. Only so-called category 3 material, which is deemed fit for human consumption, can be used to feed farm animals.

The recommendation by the EU Scientific Steering Committee and member states' national advisory scientific bodies that the ruminant intra-species recycling ban be extended to non-ruminant animals subsequently led to the introduction of an extended feed ban for all livestock by amending Regulation (EC) 999/2001 and through Commission Regulation (EC) 1234/2003. This was furthered by the previous lack of analytical methods capable of species-specific identification of used materials. Therefore, catering waste cannot be used as a feed ingredient for food producing animals. This includes waste from premises such as food factories that contain or have been in contact with animal byproducts. The ban is particularly relevant in the prevention of further crises as the traceability of catering waste cannot be guaranteed. The EU authorities regard this lack of traceability as the instigator of incidents in the past, such as the dioxin crises, as well as outbreaks of foot and mouth disease and swine fever. In the future, these regulations may be modified in accordance with changing economic and ecologic aspects and feed safety considerations.

11.4 Feedstuffs from the processing of fruits and vegetables

11.4.1 Byproducts of fruit processing

Apple

The highly variable composition of apple pomace has been reviewed by Kennedy *et al.* (1999). Production of pectin is considered to be the most reasonable method for utilization both from economic and ecologic viewpoints. In comparison with citrus pectins, apple pectins are considered to have superior gelling properties. Apple pomace has also been shown to contain substantial amounts of polyphenols, which are predominantly located in the peels and extracted into the juice to a small amount (Schieber *et al.*, 2001). The phenolic constituents have been shown to exhibit strong antioxidant activities *in vitro* (Lu and Foo, 2000), suggesting possibilities for commercial exploitation (Schieber *et al.*, 2001). Apple pomace has been utilized as animal feed after ensiling or drying. Rumsey and Lindahl (1982) demonstrated that apple pomace supplemented with urea impaired weight gain and lambing performance in ewes. These adverse effects could be mitigated by the inclusion of straw in the diet of cows (Rumsey *et al.*, 1979). Ghoreishi *et al.* (2007) reported that ensiled apple pomace can be included into diets up to 30% without a negative effect on milk yield and milk composition.

Citrus fruit

As a result of the large amounts of citrus fruits being processed into juice, a byproduct industry has evolved that utilizes the residual peels, membranes, seeds and other compounds, as reviewed by Braddock (1995). Residues include fresh and dried pulp, molasses, fiber-pectin, cold pressed oils, juice pulps and pulp wash, ethanol, seed oil, pectin, limonoids and flavonoids (Bampidis and Robinson, 2006; Siliha *et al.*, 2000; Braddock, 1995; Martinez and Carmona, 1980). Fiber-pectins are characterized by high fiber contents. Citrus seeds and peels have been found to possess a high antioxidant activity (Bocco *et al.*, 1998). Citrus byproduct feedstuffs are generally suitable for inclusion in ruminant diets as ruminants are able to ferment fiber very efficiently. Citrus pulp is usually dehydrated before being fed to animals, and should be introduced gradually into a ration to allow the animals to become accustomed to its distinctive smell and taste (Bampidis and Robinson, 2006). Fresh citrus is readily consumed by dairy cattle, although transportation, storage and handling issues limit its use (Lundquist, 1995). The nutritionally unbalanced Ca:P ratio of citrus byproducts can result in higher incidences of milk fever in cattle near parturition (Bath *et al.*, 1980). Partial or total substitution of corn or barley grain by dried orange pulp or dried lemon pulp in concentrated form fed to dairy cattle had no effects on feed uptake (Lanza, 1984). Migwi *et al.* (2001) suggested that the level of citrus pulp ensiled with wheat straw and poultry litter should be maintained at 15–20% dry matter in sheep rations as higher levels may result in lower palatability.

Grapes

Apart from oranges, grapes are the world's largest fruit crop, with approximately 80% of the total crop being used in the production of wine (Mazza and Miniati, 1993). Pomace represents about 20% of the weight of the grapes processed (Schieber *et al.*, 2001). A wide range of products including ethanol, citric acid, grape seed oil and dietary fiber can be recovered from grape pomace (Bravo and Saura-Calixto, 1998; Valiente *et al.*, 1995; Hang, 1988). Grape skins and seeds are rich sources of flavonoids, which have been shown to act as antioxidants by scavenging free radicals and terminating oxidative reactions (Brenes *et al.*, 2008). This may be adversely affected by drying of the pomace at high temperatures (Larrauri *et al.*, 1998). Grape peels and skins have also been reported to be high in polyphenols as well as dietary fiber. Grape seed oil has high levels of unsaturated fatty acids (Llobera and Canellas, 2007) as well as polyphenolics (Jayaprakasha *et al.*, 2001). A study by Goni *et al.* (2007) demonstrated that increased concentrations of grape pomace in chicken diets did not negatively affect growth performance compared with animals fed non-supplemented diets or diets supplemented with vitamin E. Dorri *et al.* (2012) concluded that grape pomace could be used up to 6% in broiler diets without affecting growth performance.

Mango

The tropical mango fruit and mango products have recently gained popularity in the European market. The processing yields peels and stones, amounting to 35–60% of the total fruit weight (Larrauri *et al.*, 1996). Mango peels have been reported to be a good source of high-quality pectin (Sudhakar and Maini, 2000) and dietary fiber, containing high amounts of extractable polyphenolics (Larrauri *et al.*, 1996). The peel is palatable to cattle, but its use in animal feed is limited on account of the high moisture and acidity content. Diarra and Usman (2008) replaced 20% of dietary maize in broiler chickens with raw or boiled mango kernel meal and observed no significant differences in performance between the control and the boiled kernel meal diets. In another study, it was concluded that up to 60% of the maize in broiler finisher diets can be replaced with boiled mango kernel meal without adverse effects on the growth, health and carcass parameters (Diarra *et al.*, 2010).

11.4.2 Byproducts of vegetable and potato processing

Tomato

Tomato juice is the most important vegetable juice with regard to human consumption, followed by carrot juice (Schieber *et al.*, 2001). Approximately 3–7% of the raw material is lost as waste during juice processing (Otto and Sulc, 2001). Tomato pomace consists of dried and crushed skin as well as seeds. The seeds comprise about 60% of the total waste and are a source of 35% protein and 25% fat. Bordowski and Geisman (1980) reported that tomato seed proteins contain approximately 13% more lysine than soy protein. Tomato seed oil is rich in unsaturated fatty acids, particularly in linoleic acid. Skin extracts are especially rich in lycopene, the principal carotenoid present causing the red hue of tomatoes. Baysal *et al.* (2000) indicated that large quantities of carotenoids are lost as waste during tomato processing. A study by Al-Betawi (2005) reported that tomato pomace can be used at a level of up to 10% in broiler diets.

Carrot

A steady increase in carrot juice consumption has resulted in considerable improvements in processing techniques. Carrot pomace is usually disposed as feed or fertilizer (Schieber *et al.*, 2001). Sims *et al.* (1993) reported that up to 80% of carotenes in carrots may be lost with the pomace. High levels of cull carrots from the vegetable industry in livestock diets can result in an off-colored fat in finishing lambs and cattle, and levels should be limited to 20% or less in finishing diets (Lardy and Anderson, 2009).

Pumpkin

Food processing techniques create large quantities of waste in the form of pumpkin peels and seeds that can be utilized as feed ingredients in a similar way to other waste stream products and fiber-rich components. Pumpkin pulp is a good source of carotenoids (e.g. beta-carotene, lutein and zeaxanthin) (Scherz and Senser, 2000). Although animal trials with pumpkin pulp have shown high digestibility in pigs, feeding high amounts can have adverse effects on fat consistency. Dried pumpkin pulp has also been successfully implemented in swine feed; however, drying the raw material is time-consuming and costly (Becker and Nehring, 1969).

Potato

Peels are the major waste of potato processing. According to Putz (1991), losses range from 15% to 40%, with their amount depending on the technique applied. Aqueous peel extracts have been shown to be a source of phenolic acids (Rodriguez de Sotillo *et al.*, 1994). As reviewed by Schlüter (2002), starch utilization is crucial for the use of potatoes in swine feed, making steaming necessary to improve its ileal digestibility. According to Stalljohann *et al.* (2000), 1.5–3 kg steamed potatoes, or 10–20% of the total ration at 88% dry matter, can be fed to swine in the finishing period per day. As a result of their low energy density and viscosity-increasing properties, the use of potatoes and their byproducts should be limited to 20–25% in the total ration. The steroidal glycoside solanine further limits their use, whereby 150 mg/kg solanine in feed material has been reported as acceptable (Jeroch *et al.*, 1993). Other antinutritional factors including inhibitors of digestive enzymes (trypsin, chymotrypsin) and lectins can be reduced by heat processing (Kaiser, 1970).

Red beet

The majority of red beet produced is consumed as vegetable, and the remainder is processed into juice and food colorant. The pomace from the juice industry accounts for 15–30% of the raw material (Otto and Sulc, 2001), and can be disposed of as feed or fertilizer (Schieber *et al.*, 2001).

Olive

The byproducts resulting from olive oil extraction are vegetation water, as well as the olive husk including skin and stones (Vitolo *et al.*, 1999). The husk can be reprocessed for the recovery of olive oil or to yield husk oil. Dried husk can be used for fuel or animal feed (Gasparrini, 1999).

Onion

Major byproducts resulting from industrial peeling of onion bulbs are skin, the outer two fleshy leaves and the top and bottom bulbs. However, because of their strong characteristic aroma and their susceptibility to phytopathogens, onion wastes are generally not suitable as fodder. The feeding of high levels of cull onions to ruminants can cause anemia as a result of the presence of sulfur compounds, which may result in hemolysis of red blood cells (Lardy and Anderson, 2009).

11.5 Feedstuffs from other food processing systems

11.5.1 Byproducts of sugar production

Sugar beet

Sugar cane and sugar beet are the primary crops for the production of sugar. Molasses represents the run-off syrup obtained at the final stage of crystallization, which can be used as feed. Depending on the process, exhausted beet pulp has a dry matter content of 8–15%, and therefore requires dewatering, which is usually performed by mechanical pressing and thermal drying (Schieber *et al.*, 2001). The shelf-life of pressed pulp, an energy-rich feed, can be extended considerably by ensiling (Heller, 1998). Damaged beets and roots obtained during washing can also be used in animal feeding (Mirzaei-Aghsaghali and Maheri-Sis, 2008). An overview of the applicability of sugar beet pulp in livestock feeding is provided by Kelly (1983).

11.5.2 Oil seed and oil fruit byproducts

Vegetable oils are produced by pressing the oil from the seeds and refining it to remove free fatty acids and other impurities. Processing of the seeds for oil production provides protein meal and cake as byproducts. In the case of soybean, the feed demand for soybean meal has driven soybean production growth (FAO, 2006). In oil extraction, soybeans yield 18–19% oil and 73–74% meal (Schnittker, 1997), the remainder being waste. Soybean meal is used primarily in the diet of monogastric species, particularly chickens and pigs. Oil palm vegetable oil production can provide diverse animal feedstuffs. By chopping, drying, cubing and pelletizing, oil palm fronds can be implemented as a source of ruminant feed, while oil palm trunks constitute a readily available source of fiber. Oil palm fronds, used either alone or combined with other ingredients such as palm kernel cake and palm oil mill effluent, have been transformed into feed as pellets or cubes for ruminants (Prieler and Fischer, 2009). The use of rapeseed cake in pig nutrition depends on its glucosinolate content (Schöne *et al.*, 1996). Sunflower cake should be from dehulled or semi-dehulled seeds, otherwise its fiber content is too high for pig nutrition. In some regions of central Europe, seeds of specific pumpkin varieties are processed into oil and oilcake, which is very rich in protein, and may be used as a protein source for pigs (Wetscherek-Seipelt *et al.*, 1991). Variable nutritional values have been reported for other oilseeds, including flax, poppy seed and false flax, as well as their cakes (Flachowsky *et al.*, 1998; Salewski, 1994; Kling and Wöhlbier, 1983).

11.5.3 Starch production, fermentation and distillation industry byproducts

Distillers' and brewers' grains

Distillers' grains are primary fermentation products resulting from the fermentation of cereal grains by yeast into alcohol. Brewers' grains are extracted residues of barley malt alone or in a mixture with other cereal grains or grain products resulting from the manufacture of wort or beer. The spent grains are dried and sold as feed. Solubles left over from fermentation usually are added to the distillers' grains before drying, resulting in the product distillers' dried grains with solubles. These products may be pressed to remove a part of the water and fed with a dry matter content of 30–40% or dried to a moisture content of 8–12% before feeding (Clark *et al.*, 1987). Wet and dried distillers' grains with solubles have protein contents between 27% and 30%, and have been used as a protein supplement in feedlot cattle diets (Klopfenstein *et al.*, 2008a). Although rumen escape protein has been shown to be variable among distillers' grains sources, protein in dried distillers' grains has about 2.4 times greater protein value and DDGS has 1.8 times greater protein value than soybean meal. As a result of the highly digestible and rapidly fermentable fiber in DDGS, it is also being used as a high energy and protein source in diets for feedlot finishing cattle (Klopfenstein *et al.*, 2008b). Farlin (1981) reported that wet distillers' grains with solubles had more energy per kg of dry matter than corn. Brewers' grains contain about 26% crude protein (Arosemena *et al.*, 1995), but their protein quality is poor. Therefore, higher amounts of dried brewers' grains are not suitable for growing-finishing or lactation diets, but can be used in gestation diets for pigs (Miller *et al.*, 1987). Brewers' yeast contains high amounts of crude protein, essential amino acids and minerals; however, it was recommended to limit the level of inclusion to approximately 5% in pig diets (Boucqué and Fiems, 1988). In dairy cattle, feeding up to 30% of the dry matter as wet brewers' grain was not found to reduce milk production in dairy cattle (West, 1994). Inclusion rates up to 20% of brewers' grains did not depress gains or feed conversion in broilers during early growth, and up to 30 % was not observed to decrease performance (Deltoro-López and Fernandez-Carmona, 1981). An extensive review of cereal byproducts and spent grains was conducted by McKee and Latner (2000).

Corn gluten feed

Corn gluten feed is a product of the corn wet milling industry, which produces high fructose corn syrup and starch. It is relatively high in protein, moderately high in fiber and moderate in energy, whereas the starch content is low because most of the starch is converted enzymatically to syrup glucose, fructose and dextrose. In cattle growth trials, corn gluten feed protein was more degradable than soybean meal. As less corn gluten feed protein escapes ruminal digestion, performance of growing calves may be reduced compared with calves fed soybean meal. As an energy source in roughage rations, corn gluten feed provided similar gains and feed efficiencies to corn grain. In high-grain finishing diets, dry corn gluten feed reduced feed efficiency 5–15% (Kubik and Stock, 1990).

11.6 Conclusion

The use of byproducts of the food industry that cannot be used directly for human consumption increases the sustainability of animal production. Interesting raw materials can be used for animal nutrition and thus remain in the food chain. Future developments should take into consideration technological processes to improve nutritional characteristics and to characterize specific effects on animal and human health and safety, performance and environmental aspects from the whole food chain perspective. Research activities are required to characterize nutritional value, hygiene and the potential impact on food safety. In view of the impending shortage of conventional feed materials, it seems imperative to identify byproducts not only as sources of energy and nutrients for livestock, but also to explore their potential functional impact on the animal organism.

11.7 Questions for discussion

1. What legal aspects in Europe have to be considered when residues of the food industry are fed to livestock?
2. Can any food byproduct be used as animal feed?
3. Which fresh produce byproducts of the food industry are used as animal feed?
4. Which further possible fresh produce residues could be of interest?
5. List the reasons why the use of residues from the food industry in animal feed is of importance.
6. State the physical, chemical and biological problems that may prevent the usage of fresh produce wastes for animal feed.
7. What nutritional aspects have to be considered when using fresh produce byproducts from food in animal nutrition?
8. List past cases in which outbreaks of diseases in livestock have been associated with the feeding of non-thermally treated food residues.

11.8 References

AL-BETAWI NA (2005). Preliminary study on tomato pomace as unusual feedstuff in broiler diets. *Pakistan Journal of Nutrition*, 4 (1): 57–63.

AROSEMENA A, DEPETERS EJ, FADEL JG (1995). Extent of variability in nutrient composition within selected byproduct feedstuffs. *Animal Feed Science and Technology*, 54: 103–20.

BAMPIDIS VA, ROBINSON PH (2006). Citrus byproducts as ruminant feeds: A review. *Animal Feed Science and Technology*, 128: 175–217.

BATH DL, DUNBAR JR, KING JM, BERRY SL, LEONARD RO, OLBRICH SE (1980). *Byproducts and unusual feedstuffs in livestock rations*. Western Regional Extension Publication No. 39. USDA-ARS, Washington, DC, USA.

BAYSAL T, ERSUS S, STARMANS DAJ (2000). Supercritical CO_2 extraction of ß-carotene and lycopene from tomato paste waste. *Journal of Agriculture and Food Chemistry*, 48: 5507–11.

BECKER M, NEHRING K (1969). *Handbuch der Futtermittel*. Hamburg, Berlin: Parey.

BOCCO A, CUVELIER ME, RICHARD H, BERSET C (1998). Antioxidant activity and phenolic composition of citrus peel and seed extracts. *Journal of Agricultural and Food Chemistry*, 46: 2123–9.

BORDOWSKI DL, GEISMAN JR (1980). Protein content and amino acid composition of protein of seeds from tomatoes at various stages of ripeness. *Journal of Food Science*, 45: 228–35.

BOUCQUÉ CV, FIEMS LO (1988). Vegetable byproducts of agro-industrial origin. *Livestock Production Science*, 19: 97–135.

BRADDOCK RJ (1995). Byproducts of citrus fruit. *Food Technology*, 49: 74–7.

BRAVO L, SAURA-CALIXTO F (1998). Characterization of dietary fiber and the in vitro indigestible fraction of grape pomace. *American Journal of Enology and Viticulture*, 49: 135–41.

BRENES A, VIVEROS A, GONI I, CENTENO C, SAYAGO-AYERDY SG, *et al.* (2008). Effect of grape pomace concentrate and vitamin E on digestibility of polyphenols and antioxidant activity in chickens. *Poultry Science*, 87: 307–16.

CLARK JH, MURPHY MR, CROOKER BA (1987). Supplying the protein needs of dairy cattle from byproduct feeds. *Journal of Dairy Science*, 70: 1092–109.

COLLEE JG, BRADLEY R (1997). BSE: A decade on – Part 1. *The Lancet*, 349 (9052): 636–41.

DELTORO-LÓPEZ J, FERNANDEZ-CARMONA J (1981). Evaluation of brewers dried grains in the diets of broiler chickens. *Animal Feed Science and Technology*, 6 (2): 179–88.

DIARRA SS, USMAN BA (2008). Growth performance and some blood parameters of broiler chickens fed raw or cooked mango kernel meal. *International Journal of Poultry Science*, 7 (4): 315–18.

DIARRA SS, USMAN BA, IGWEBUIKE JU (2010). Replacement value of boiled mango kernel meal for maize in broiler finisher diets. *Journal of Agricultural and Biological Science*, 5 (1): 47–52.

DORRI S, TABEIDIAN SA, TOGHYANI M, JAHANIAN R, BEHNAMNEJAD F (2012). *Effect of different levels of grape pomace on performance broiler chickens*. 1st International and 4th National Congress on Recycling of Organic Waste in Agriculture. Isfahan, Iran.

FAO (2006). *Livestock's Long Shadow – Environmental Issues and Options*. Food and Agriculture Organization, Rome, Italy. Available online at: ftp://ftp.fao.org/docrep/fao/010/a0701e/a0701e.pdf.

FARLIN SD (1981). Wet distillers grains for finishing cattle. *Animal Nutrition and Health*, 36: 35.

FLACHOWSKY G, LANGBEIN T, BOHME H, SCHNEIDER A, AULRICH K (1998). Effect of false flax expeller combined with short-term vitamin E supplementation in pigs feeding on the fatty acid pattern, vitamin E concentration and oxidative stability of various tissues. *Journal of Animal Physiology and Animal Nutrition*, 78 (4–5): 187–95.

GASPARRINI R (1999). Treatment of olive oil processing residues. *Oils and Fats International*, February: 32–3.

GHOREISHI SF, PIRMOHAMMADI R, YANSARI AT (2007). Effects of ensiled apple pomace on milk yield, milk composition and DM intake of Holstein dairy cows. *Journal of Animal and Veterinary Advances*, 6 (9): 1074–8.

GONI I, BRENES A, CENTENO C, VIVEROS A, SAURA-CALIXTO F, *et al.* (2007). Effect of dietary grape pomace and vitamin E on growth performance, nutrient digestibility, and susceptibility to meat lipid oxidation in chickens. *Poultry Science*, 86: 508–16.

GUSTAVSSON J, CEDERBERG C, SONESSON U, OTTERDIJK RV, MEYBECK A (2011). *Global food losses and food waste: Extent, causes and prevention*. Food and Agriculture Organization, Rome, Italy. Available online at: http://www.fao.org/docrep/014/mb060e/mb060e00.pdf.

HANG YD (1988). Recovery of food ingredients from grape pomace. *Process Biochemistry*, 23: 2–4.

HELLER C (1998). Preservation of pressed pulp by ensiling. In: van der Poel PW, Schiweck H, Schwartz T (Eds.), *Sugar Technology*. Berlin: Bartens.

JAYAPRAKASHA GK, SINGH RP, SAKARIAH KK (2001). Antioxidant activity of grape seed (*Vitis vinifera*) extracts on peroxidation models in vitro. *Food Chemistry*, 73: 285–90.

JEROCH H, FLACHOWSKY G, WEISSBACH F (1993). *Futtermittelkunde*. Stuttgart: Gustav Fischer Verlag.

KAISER KP (1970): *Trypsin- und Chymotrypsininhibitoren in Kartoffeln.* Dissertation, Technische Universität München.

KAMPHUES J (2001). Utilization of byproducts in animal nutrition – from its beginning up to the white paper of food safety. *Landbauforschung Völkenrode*, Sonderheft 223: 5–20.

KELLY P (1983). Sugar Beet Pulp – A Review. *Animal Feed Science and Technology*, 8: 1–18.

KENNEDY M, LIST D, LU Y, FOO LY, NEWMAN RH, *et al.* (1999). Apple pomace and products derived from apple pomace: Uses, composition and analysis. In: Linskens H-F, Jackson JF (Eds.), *Modern Methods in Plant Analysis*, Volume 20. Berlin: Springer.

KLING M, WÖHLBIER W (1983). Handelsfuttermittel. Volume 2A + 2B. Ulmer Verlag, Stuttgart, Germany.

KLOPFENSTEIN TJ, ERICKSON GE, BREMER VR (2008a). Use of distillers byproducts in the beef cattle feeding industry. *Journal of Animal Science*, 86: 1223–31.

KLOPFENSTEIN TJ, ERICKSON GE, BREMER VR (2008b). Use of Distillers Co-Products in Diets Fed to Beef Cattle. In: Babcock BA, Hayes DJ, Lawrence JD (Eds.), *Using Distillers Grains in the U.S. and International Livestock and Poultry Industries*. Midwest Agribusiness Trade Research and Information Center, Center for Agricultural and Rural Development, Iowa State University, Ames, Iowa, USA.

KORNEGAY ET, VAN DER NOOT GW, BARTH KM, GRABER G, MACGRATH WS, *et al.* (1970). *Nutritive evaluation of garbage as a feed for swine*. Bulletin No. 829. College of Agriculture and Environmental Science. New Jersey Agricultural Experiment Station. Rutgers, State University of New Jersey, New Brunswick.

KUBIK DJ, STOCK R (1990). G90–978 Byproduct feedstuff for beef and dairy cattle. *Historical Materials from University of Nebraska-Lincoln Extension.* Paper 311.

LANZA A (1984). *Dried citrus pulp in animal feeding*. Proceedings of the International Symposium on Food Industries and the Environment. Budapest, Hungary.

LARDY G, ANDERSON V (2009). *Alternative feeds for ruminants*. North Dakota State University Agriculture. Available online at: http://www.ag.ndsu.edu/pubs/ansci/livestoc/as1182.html.

LARRAURI JA, RUPEREZ P, BORROTO B, SAURA-CALIXTO F (1996). Mango peels as a new tropical fiber: Preparation and characterization. *Lebensmittel-Wissenschaft und – Technologie*, 29: 729–33.

LARRAURI JA, SANCHEZ-MORENO C, SAURA-CALIXTO F (1998). Effect of temperature on the free radical scavenging capacity of extracts from red and white grape pomace peels. *Journal of Agriculture and Food Chemistry*, 46: 2694–7.

LLOBERA A, CANELLAS J (2007). Dietary fiber content and antioxidant activity of *Manto Negro* red grape (Vitis vinifera): Pomace and stem. *Food Chemistry*, 101: 659–66.

LU Y, FOO L (2000). Antioxidant and radical scavenging activities of polyphenols from apple pomace. *Food Chemistry*, 68: 81–5.

LUNDQUIST R (1995). *Current uses of traditional co-products*. Proceedings of the Second National Alternative Feed Symposium. St. Louis, MO, USA.

MARTINEZ, PJ, CARMONA JF (1980). Composition of citrus pulp. *Animal Feed Science and Technology*, 5: 1–10.

MAZZA G, MINIATI E (1993). Grapes. In: Mazza G, Miniati E (Eds.). *Anthocyanins in Fruits, Vegetables, and Grains*. Boca Raton, FL, USA: CRC Press.

MCKEE LH, LATNER TA (2000). Underutilized sources of dietary fiber: A review. *Plant Foods for Human Nutrition*, 35: 341–71.

MIGWI PK, GALLAGHER JR, VAN BARNEVELD RJ (2001). The nutritive value of citrus pulp ensiled with wheat straw and poultry litter for sheep. *Australian Journal of Experimental Agriculture*, 41: 1143–8.

MILLER ER, HOLDEN PJ, LEIBBRANDT VD (1987). Byproducts in swine diets. *Pork Industry Handbook, No 108*. Purdue University, Cooperative Extension Service, West Lafeyette, Illinois, USA.

MINKLER FC (1914). *Hog cholera and swine production.* Circular No. 40. New Jersey Agricultural Experiment Station. Trenton, New Jersey.

MIRZAEI-AGHSAGHALI A, MAHERI-SIS N (2008). Nutritive value of some agro-industrial byproducts for ruminants – a review. *World Journal of Zoology*, 3 (2): 40–6.

OTTO K, SULC D (2001). Herstellung von Gemüsesäften. In: Schobinger U (Ed.), *Frucht- und Gemüsesäfte*. Stuttgart: Ulmer.

PRIELER S, FISCHER G (2009). *Agricultural byproducts associated with biofuel production chains. Assessment of biofuel policy impacts in food and feed markets*. Available online at: http://www.elobio.eu/fileadmin/elobio/user/docs/DeliverableE_5.1.pdf.

PUTZ B (1991). *Kartoffeln*, 1. Auflage. Hamburg: Behrs Verlag.

RODRIGUEZ DE SOTILLO D, HADLEY M, HOLM ET (1994). Phenolics in aqueous potato peel extract: Extraction, identification and degradation. *Journal of Food Science*, 59: 649–51.

RUMSEY TS, KERN DL, SLYTER LL (1979). Ruminal microbial population, movement of ingesta from the rumen and water intake of steers fed apple pomace diets. *Journal of Animal Science*, 48: 1202–8.

RUMSEY TS, LINDAHL IL (1982). Apple pomace and urea for gestating ewes. *Journal of Animal Science*, 54: 221–34.

SALEWSKI A (1994). Raps- und Leinkuchen im Schweinefutter. *Deutsche Geflügelwirtschaft und Schweineproduktion*, 15: 19–21.

SCHERZ H., SENSER F (2000). *Food composition and nutrition tables*. Stuttgart: Medpharm.

SCHIEBER A, STINTZIG FC, CARLE R (2001). Byproducts of plant food processing as a source of functional compounds – recent developments. *Trends in Food Science and Technology*, 12: 401–13.

SCHLÜTER R (2002). *Auswirkungen des Einsatzes von einem Flüssigfutter aus thermisch vorbehandelten Speiseresten und Kartoffeln auf die Mastleistungen, Gesundheitsstatus, Schlachtkörperqualität und Ökonomie in der Schweinemast*. Dissertation, Tierärztliche Hochschule Hannover.

SCHNITTKER J (1997). *The history, trade, and environmental consequences of soybean production in the United States*. Report to the World Wide Fund for Nature. World Wildlife Fund: Washington DC, USA.

SCHÖNE F, KIRCHHEIM U, SCHUMANN W, LÜDKE H (1996). Apparent digestibility of high-fat rapeseed press cake in growing pigs and effects on feed intake, growth and weight of thyroid and liver. *Animal Feed Science and Technology*, 62: 97–110.

SILIHA H, EL-SAHY K, SULIEMAN A, CARLE R, EL-BADAWY A (2000). Citrus-wastes: Composition, functional properties and utilization. *Obst-, Gemüse- und Kartoffelverarbeitung*, 85: 31–6.

SIMS CA, BALABAN MO, MATTHEW RF (1993). Optimization of carrot juice color and cloud stability. *Journal of Food Science*, 58: 1129–31.

STALLJOHANN G, ORLOWSKI K, KRÄMER M (2000). *Rechenmeister 2000*. Landwirtschaftskam mer Westfalen-Lippe, Münster.

SUDHAKAR DV, MAINI SB (2000). Isolation and characterization of mango peel pectins. *Journal of Food Processing and Preservation*, 24: 209–27.

VALIENTE C, ARRIGONI E, ESTEBAN RM, AMADO R (1995). Grape pomace as a potential food fiber. *Journal of Food Science*, 60: 818–20.

VITOLO S, PETARCA L, BRESCI B (1999). Treatment of olive oil industry wastes. *Bioresource Technology*, 67: 129–37.

WALLER JC (2011). Byproducts and unusual feedstuffs. *Feedstuffs*, 74: 18–23.

WEST JW (1994). Wet brewers grains for lactating dairy cows during hot, humid weather. *Journal of Dairy Science*, 77 (1): 196–204.

WESTENDORF ML (2000). *Food Waste to Animal Feed*. Iowa State University Press: Ames, Iowa.

WETSCHEREK-SEIPELT G, WETSCHEREK W, ZOLLITSCH W (1991). Einsatzmöglichkeiten von Kürbiskernkuchen in der Schweinemast. *Die Bodenkultur*, 42: 277–289.

12

Risk assessment of microbial and chemical contamination in fresh produce

S. Koseki, National Food Research Institute, Japan

DOI: 10.1533/9781782420279.2.153

Abstract: Outbreaks of foodborne illnesses related to the consumption of fresh produce occur all over the world. Quantitative microbial risk assessment (QMRA) for fresh produce can be used to enhance microbiological safety in the production, distribution and consumption of fresh produce. This chapter describes the prevalence of contamination of fresh produce, possible routes of pathogen contamination and pathogen behaviour on fresh produce during distribution. The information described here will be useful for QMRA as supporting knowledge.

Key words: fresh produce, foodborne pathogens, quantitative microbial risk assessment (QMRA), predictive microbiology, growth modelling.

12.1 Introduction

There has been a substantial amount of research into outbreaks of foodborne illnesses related to the consumption of fresh produce (Heaton and Jones, 2008; Sivapalasingam *et al.*, 2004). In a recent risk prioritization study, leafy green vegetables were identified as the commodity group of highest concern from a microbiological safety perspective (Harris *et al.*, 2003). *Escherichia coli* O157:H7, *Salmonella enterica* and *Listeria monocytogenes* are among the bacterial pathogens most frequently associated with foodborne disease resulting from the consumption of fresh produce.

Salmonella spp. is the most common cause of disease outbreaks linked to fresh produce, occurring in, among others, sprouted seeds, cantaloupe melons, tomatoes, unpasteurized citrus juices, rocket and other lettuce varieties (Heaton and Jones, 2008). Greene *et al.* (2008) reported that a recurrent multistate outbreak of *Salmonella* Newport had been associated with tomatoes. In 2002, tomatoes grown and packed on the eastern shore of Virginia were contaminated with a pan-susceptible *S.* Newport strain, causing illness in 510 patients in 26 states. In July–November 2005, the same strain caused illness in at least 72 patients in 16 states. Munnoch *et al.* (2009) reported that a multi-state outbreak of *Salmonella* Saintpaul in Australia associated with cantaloupe consumption occurred in 2006. An outbreak of 26 cases of *Salmonella* Litchfield infection also occurred in the states of Western Australia and Queensland between October 2006 and January 2007 (Gibbs *et al.*, 2009).

Escherichia coli O157:H7 is another documented cause of produce-related foodborne disease outbreaks. Although fresh produce was not considered a significant vector for the transmission of *E. coli* O157:H7 until the mid-1990s, a series of outbreaks associated with minimally processed produce clearly showed that contamination can occur (Harris *et al.*, 2003). The largest *E. coli* O157:H7 outbreak occurred in 1996, when more than 6000 schoolchildren in Japan were infected with *E. coli* O157:H7 from white radish seed sprouts (Itoh *et al.*, 1998). Between 1993 and the year 2006, 26 reported outbreaks of *E. coli* O157:H7 infection have been traced to contaminated lettuce and leafy green vegetables (Maki, 2006). A recent multi-state outbreak in the USA linked to bagged fresh spinach caused approximately 205 confirmed illnesses, 31 cases of haemolytic uremic syndrome and three deaths (California, 2007; Doyle and Erickson, 2008). Between 14 September and 20 October 2007, an outbreak of Shiga-toxin-producing *Escherichia coli* (STEC) O157 simultaneously occurred in the Netherlands and Iceland. The most probable cause of this international outbreak was contaminated lettuce, shredded and pre-packed in a Dutch food processing plant (Friesema *et al.*, 2008). A multi-state outbreak of *E. coli* O157:H7 infection associated with consumption of packaged spinach was also reported in August–September 2006 (Wendel *et al.*, 2009). More recently, a multi-national outbreak in Europe linked to the consumption of sprouts caused approximately 47 deaths by *E. coli* O104 (Frank *et al.*, 2011).

The Gram-positive bacterium *Listeria monocytogenes* is another significant foodborne pathogen in terms of both public health and food safety. Although there is a substantial amount of literature concerning the isolation, attachment, survival and growth of *L. monocytogenes* on produce (Lianou and Sofos, 2007), only two fresh-cut produce-related listeriosis outbreaks have been reported in the USA (Ho *et al.*, 1986; Schlech *et al.*, 1983) until 2010. However, a multi-state outbreak of listeriosis linked to whole cantaloupes occurred in 2011. The outbreak of listeriosis sickened more than 146 individuals in 28 states. The infection was eventually linked to contaminated cantaloupe, and the outbreak has been considered responsible for at least 30 deaths (CDC, 2011).

As mentioned above, fresh produce is one of the critical vectors of foodborne illness. In this chapter, the relevant information for a quantitative microbial risk assessment on fresh produce is described. The information presented in this chapter can be applied to the appropriate risk assessment of consumption of fresh-cut produce. Please note that this chapter does not detail the technical aspects of calculation methods for risk estimation; however, the information given here can be incorporated into risk assessment procedures.

12.2 The frequency with which bacterial pathogens contaminate fresh produce

Although many outbreaks caused by contaminated fresh produce have been documented, various field surveys of fresh produce have indicated extremely low or no occurrence of pathogenic bacteria such as *Salmonella* spp., *E. coli* O157:H7 and *L. monocytogenes* (Abadias *et al.*, 2008; Arthur *et al.*, 2007; Bohaychuk *et al.*, 2009; Johannessen *et al.*, 2002; Johnston *et al.*, 2006, 2005; Mukherjee *et al.*, 2004, 2006; Sagoo *et al.*, 2001, 2003; U.S. Department of Agriculture, 2007, 2008, 2009). For example, our study (Koseki *et al.*, 2011a) showed that no pathogenic bacteria, including *Salmonella* spp., *L. monocytogenes* and *E. coli* O157:H7, was detected in any of the 419 samples analyzed by a highly sensitive and validated procedure using PCR. Other studies have also noted similar results for the prevalence of pathogens in fresh produce. In a study of locally grown produce collected from farm markets in Alberta (Canada) in June to October 2007, no pathogenic bacteria were isolated from 128 lettuce samples (Bohaychuk *et al.*, 2009). The U.S. Department of Agriculture (USDA) Microbiological Data Program analyzed a total of 7646 samples collected from produce terminals and wholesale distribution centres in 2008 and did not find any samples positive for *E. coli* O157:H7 or *Salmonella* (U.S. Department of Agriculture, 2009). Furthermore, no *Salmonella*, *Campylobacter*, or *E. coli* O157:H7 were detected in fresh produce from various regions including Spain (Abadias *et al.*, 2008), Norway (Johannessen *et al.*, 2002; Loncarevic *et al.*, 2005), Ireland (McMahon and Wilson, 2001) or the UK (Sagoo *et al.*, 2001, 2003). Many other studies have also reported the microbiological quality of fresh produce and found no or very low levels of contamination with pathogenic bacteria (Johnston *et al.*, 2006, 2005; Mukherjee *et al.*, 2004; U.S. Department of Agriculture, 2007, 2008, 2009).

From these numerous studies we can only conclude that cases of pathogens being detected were very rare, if detected at all, in limited worldwide surveys of fresh produce. However, because of the frequency of outbreaks of foodborne illness, it is reasonable to assume that there are always opportunities for pathogen contamination to occur. To prevent pathogen contamination and assess the risk of foodborne illness induced by bacterial pathogens, we need to know how bacterial pathogens contaminate fresh produce in the real field and/or during distribution.

12.3 How bacterial pathogens contaminate fresh produce

12.3.1 Possible contamination routes

Outbreaks of foodborne illness associated with the consumption of leafy vegetables are a result of contamination of the edible portions of plant leaves with foodborne pathogens. The potential cause of preharvest contamination of the edible portion of leaves has been considered to be direct contact with the pathogen through contaminated irrigation water and/or soil (Brandl and Amundson, 2008; Delaquis *et al.*, 2007; Harapas *et al.*, 2010; Lapidotand, 2009; Solomon *et al.*,

2003), and uptake from the root system through the vasculature (Bernstein, 2007; Bernstein *et al.*, 2007; Erickson *et al.*, 2010; Mahbub, 2004; Mitra *et al.*, 2009; Solomon *et al.*, 2002; Wachtel *et al.*, 2002; Warriner *et al.*, 2003). However, few comparative studies of these two contamination routes have been conducted. It is unclear which of these routes is the principal source of contamination of edible aerial leaves. If the main route of contamination of leafy green vegetables was clarified, more effective and critical measures for controlling microbial contamination during production would be possible. Thus, elucidation of the contamination route will play a key role in ensuring the safety of leafy vegetables for consumption.

12.3.2 Case study on a hydroponic system

In this section, a case study on a hydroponic cultivation system is discussed as an example (Koseki *et al.*, 2011c). To clarify the translocation of the bacterial pathogens from the roots to the aerial leaves, we used a hydroponic growing system in a small plastic pot for directly assessing the uptake of bacterial pathogens. The contamination from the leaf itself was also investigated by inoculating bacterial pathogens directly onto the leaves. In addition, because the contamination by bacterial pathogens through both of the possible contamination routes would be affected by the contamination level (inoculum size) (Bernstein *et al.*, 2007; Erickson *et al.*, 2010; Solomon *et al.*, 2002), the effect of inoculum size was also investigated. Finally, the effects of the contamination route (from roots versus from leaves), inoculum level and kind of pathogens on the probability of contamination of the leave were analyzed using logistic regression.

Transmission of the pathogens to the growing leaves through the roots was confirmed in the hydroponic system used in this study. Although the higher pathogen concentration ($>10^6$ cfu/ml) contaminated growing leaves at a high rate, the lower concentration ($\sim10^3$ cfu/ml) did not transmit pathogens from roots to the growing leaves. This result was consistent with those of other reports (Bernstein *et al.*, 2007; Erickson *et al.*, 2010; Solomon *et al.*, 2002). Although there have been some reports regarding the translocation of pathogens (Bernstein, 2007; Bernstein *et al.*, 2007; Erickson *et al.*, 2010; Mahbub, 2004; Mitra *et al.*, 2009; Solomon *et al.*, 2002; Wachtel *et al.*, 2002; Warriner *et al.*, 2003), no probabilistic comparative analysis of pathogen contamination has been conducted. Koseki *et al.* (2011c) have elucidated that the probability of the contamination of the growing leaves from the roots indicated a 6.93 times higher risk of contamination from the roots than from the leaf surfaces. Furthermore, the statistical analysis indicated that the contamination was dependent on the inoculum level, which represents the contamination level. Although pathogen concentrations $>10^6$ cfu/ml would not be present in a hydroponic solution in a real situation, the results of the present study showed a low probability of contamination from the root system even when the concentration of the pathogen was low. Thus, sanitary control of the solution directly in contact with the roots must play a key role in ensuring the safety of hydroponically grown leafy greens.

In most of the previous studies, the translocation of pathogens through the root system was investigated in soil-grown plants. As Warriner *et al.* (2003) noted, there might be a different response of pathogenic bacteria to the growing roots in the soil environment versus hydroponic systems. Because the roots of spinach in a hydroponic system are more accessible to pathogenic bacteria than those in a soil environment, translocation of pathogens from the roots to the leaves might occur relatively easily in a hydroponic system. Thus, the results of the present study might be difficult to extrapolate to a soil environment. However, because the fundamental mechanism underpinning the internalization and translocation of pathogenic bacteria through the roots of leafy vegetables would be common regardless of the environment, a similar trend can be observed in hydroponic systems and soil environments. The results presented here are of great significance to the expanding hydroponics industry, and suggest that root uptake of bacteria must be considered in soil-grown plants.

In this investigation, there were few cases of contamination through the leaf surface, even when the inoculum level was high. In cases where contamination was confirmed, the number of pathogens was at most ca. 3.0 log cfu/leaf (Koseki *et al.*, 2011c). This result was consistent with previous studies (Aruscavage *et al.*, 2008; Erickson *et al.*, 2010). Although a different trend in the contamination of growing leaves was confirmed in some reports (Brandl and Amundson, 2008; Harapas *et al.*, 2010; Lapidotand, 2009; Solomon *et al.*, 2003), the experimental conditions in the present study differed from those in previous reports. Pathogens could survive on leaves after inoculation onto the leaves under high relative humidity in a closed experimental chamber (Brandl and Amundson, 2008; Harapas *et al.*, 2010; Lapidotand, 2009; Solomon *et al.*, 2003). In contrast, pathogens could not survive on the leaves under relatively low humidity (<55% RH), as in this study and in field studies (Aruscavage *et al.*, 2008; Erickson *et al.*, 2010). This information suggests that the relative humidity in the environment of the growing plant plays an important role in the survival of the inoculated bacteria on the leaf surface. However, maintaining a high relative humidity is not likely in a real field cultivation environment. If the plants were cultivated in a field, weather conditions could not be controlled and high relative humidity would not be expected to last throughout the cultivation period. Thus, in terms of relative humidity conditions, the results of the present study might be relevant to field cultivation conditions (Aruscavage *et al.*, 2008; Dreux *et al.*, 2007; Erickson *et al.*, 2010; Stine *et al.*, 2005). Furthermore, as the relative humidity in a real hydroponic green house cultivation would be higher than that of a field, such an environment would assist the survival of pathogens on leaves. However, pathogen contamination from leaf surface via irrigation water and/or soil would rarely occur in a hydroponic closed environment.

A significantly lower contamination rate was observed for *L. monocytogenes* than for *Salmonella* spp. and *E. coli* O157:H7. The reason for this result remains unclear. However, the lower rates of contamination by *L. monocytogenes* are consistent with the fact that only a few fresh-cut produce-related listeriosis outbreaks have been documented in the USA (Ho *et al.*, 1986; Schlech *et al.*, 1983). The low rate of contamination by *L. monocytogenes* might merit future study.

Leaf age (Brandl and Amundson, 2008) and damage to the leaf (Brandl and Amundson, 2008; Erickson, 2010; Lapidotand, 2009) have been reported to affect pathogen contamination of leafy greens. These factors affect pathogen uptake through the roots and/or pathogen persistence on the leaves. However, this study used mature and intact leaves to focus on the comparison of contamination routes. The effects of leaf age and damage on the contamination of leaves with pathogens should be investigated in future research.

In conclusion, the studies so far indicate that the principal route of pathogen contamination of growing leaves in a hydroponic system was from the plant roots, rather than from direct contamination of the leaves. In addition, the contamination of the leaves from direct contact with the pathogens would mainly occur because of cross-contamination during the postharvest process. Thus, appropriate sanitary control of the hydroponic solution during cultivation and careful handling during the postharvest process will play key roles in the prevention of pathogen contamination of edible leaves.

12.4 How bacterial pathogens respond on fresh produce

12.4.1 Introduction

Fresh produce can be contaminated during growth from many sources, such as soil, water, wild animals, birds and insects, as described in the previous section. Following production, processes involving harvesting, washing, cutting, packaging and shipping can create additional contamination.

Although various sanitizers have been examined for their effectiveness on fresh produce, most treatments have been shown to have minimal effects, resulting in a less than 2 log cfu/g reduction of bacterial numbers. For example, chlorinated water, chlorine dioxide (ClO_2), hydrogen peroxide (H_2O_2), organic acid, calcinated calcium solution and ozonated water have been evaluated against pathogens on produce (Bari *et al.*, 1999; Han *et al.*, 2000; Kim *et al.*, 1999; Lin *et al.*, 2002). Acidic electrolyzed water, which is produced by the electrolysis of an aqueous sodium chloride solution in an anode cell, has also been reported to have a strong bactericidal effect on most pathogenic bacteria in lettuce (Koseki *et al.*, 2004; Park *et al.*, 2001), alfalfa seeds, sprouts (Kim *et al.*, 2003) and tomato (Bari *et al.*, 2003). These treatments, however, have been shown to have minimal effects, resulting in a less than 2 $\log_{10}$ cfu/g reduction of bacterial numbers. Furthermore, as bacteria remain in or on the contaminated produce after treatment, they will inevitably grow and/or be spread during subsequent distribution. Sanitizer treatment alone is not enough to keep the produce safe.

Thus, when fresh produce is contaminated with bacterial pathogens, as described in the previous section, we need to know quantitatively how the bacterial pathogens behave on the produce for microbial risk assessment. However, there have been limited reports on bacterial pathogen behaviour on fresh produce compared with the number of reports on bacterial pathogen growth on meat or meat products.

12.4.2 Overview of the studies published so far

Modelling the growth and survival of pathogenic and spoilage microorganisms is a basic tool in predicting food safety and the microbial deterioration of food products in the food chain (McMeekin *et al.*, 1993). Although numerous bacterial growth models have been published, few predictive models have been constructed with respect to fresh produce (Riva *et al.*, 2001; Rodriguez *et al.*, 2000; Viswanathan and Kaur, 2001). These models, however, provide predictions of bacterial growth during constant environmental conditions. A dynamic model has been developed by Baranyi and Roberts (Baranyi and Roberts, 1994) which can successfully provide predictions of bacterial growth in fluctuating temperature conditions (Baranyi *et al.*, 1995). Bovill *et al.* (2000, 2001) reported that in general, the Baranyi model provided accurate predictions of *L. monocytogenes* and *Salmonella* growth in a broth, as well as in meat and milk products. Shorten *et al.* (2004) applied the Baranyi model to *Erwinia carotovora* in vegetable juice under conditions of fluctuating temperature. Koseki and Isobe (2005a, 2005b) have investigated the bacterial pathogen growth on cut lettuce and have developed predictive models.

More recently, studies have revealed the pathogen behaviour on fresh produce. Carrasco *et al.* (2008) have modelled the growth of *L. monocytogenes* on iceberg lettuce at 5 °C and 13 °C with increments of 2.66 and 4.85 log cfu/g, respectively, after 14 days under modified atmosphere packaging (MAP) condition. Doering *et al.* (2009) indicated that on cut lettuce and whole leaf spinach that was packaged and stored at 4 °C, *E. coli* O157:H7 contamination could still be detected after typical handling practices, although populations decreased from initial levels in many cases by at least 1.5 log units. Although *E. coli* O157:H7 levels decreased on products handled and stored under recommended conditions, survivors persisted. Theofel and Harris (2009) illustrated that at 20 °C, pre-inoculation culture conditions had little impact on growth of *E. coli* O157:H7 on cut lettuce. However, survival at 5 °C was significantly better ($P<0.05$) for cultures grown at 15 °C or 37 °C in minimal medium and up to the late stationary phase. On the other hand, impact of pre-inoculation handling on survival on lettuce plants was less clear. Luo *et al.* (2010) indicated that storage at 5 °C allowed *E. coli* O157:H7 to survive, but limited its growth, whereas storage at 12 °C facilitated the proliferation of *E. coli* O157:H7. There was more than 2.0 log units increase in O157 populations on lettuce when held at 12 °C for three days. At 12 °C a significant decline in visual quality of lettuce was observable, although the quality of this lettuce was still fully acceptable when *E. coli* growth had reached a statistically significant level. McKellar and Delaquis (2011) and McKellar *et al.* (2012) developed a predictive model for growth/die-off of *E. coli* O157:H7 on lettuce and successfully simulated *E. coli* O157:H7 behaviour on lettuce under static and fluctuating temperature conditions. Sant'Ana *et al.* (2012a) examined *L. monocytogenes* and *Salmonella* growth on various vegetables. They indicated that *L. monocytogenes* was able to grow more in storage conditions (7 °C and 15 °C) and on vegetables than *Salmonella*. Growth of both microorganisms was inhibited in carrots, although a more pronounced effect has been observed against *L. monocytogenes*. Furthermore, Sant'Ana *et al.* (2012b) developed a growth model of *L. monocytogenes* and *Salmonella* on cut lettuce.

12.4.3 Case study

In this section, a case study on pathogen growth on cut lettuce is discussed as an example (Koseki and Isobe, 2005a, b). A series of five growth curves for each pathogen on lettuce was produced at temperatures of 5 °C, 10 °C, 15 °C, 20 °C and 25 °C. Representative data on the growth of *E. coli* O157:H7 and *Salmonella* on lettuce at different temperatures are shown in Fig. 12.1 with fitted growth curves produced using the Baranyi model; these provided a good statistical fit to the observed data. Square root analysis was carried out to determine the relationship between μ_{max} and the temperature of each bacterium on lettuce. The square root model is described as follows in general (Ratkowsky *et al.*, 1982):

$$\sqrt{\mu_{max}} = b(T - T_{min}) \qquad [12.1]$$

where b is a constant, T denotes temperature and T_{min} is the theoretical minimum temperature for growth. The relationship between $\sqrt{\mu_{max}}$ and temperature was linear with a high correlation coefficient of linearity (Fig. 12.2).

The real temperature history of lettuce collected from the farm to the retail store in early October in Japan is shown in Fig. 12.3. The transportation distance was about 400 km and the total time from farm to retail store was 72 hours. The temperature of the lettuce just after being harvested was 15–17 °C, and three hours later the lettuce was precooled by 5 °C in a vacuum refrigerator. The precooled lettuce was stored at 5 °C in a storehouse for around five hours until it was shipped. Trucks without any form of temperature control transported the lettuce. Lettuce temperature fluctuated in ranges between 3 °C and 15 °C during transportation. The lettuce was displayed on a showcase set at 7 °C after arrival in the retail store.

The viable counts of *L. monocytogenes* on lettuce under real temperature conditions are shown in Fig. 12.4 with growth curves predicted using the Baranyi-Ratkowsky model as shown below (Baranyi and Roberts, 1994).

$$\frac{dq}{dt} = \mu_{max} q(t) \qquad [12.2]$$

$$\frac{dN}{dt} = \frac{q(t)}{1+q(t)} \times \mu_{max} \times \left(1 - \frac{N}{N_{max}}\right) \times N \qquad [12.3]$$

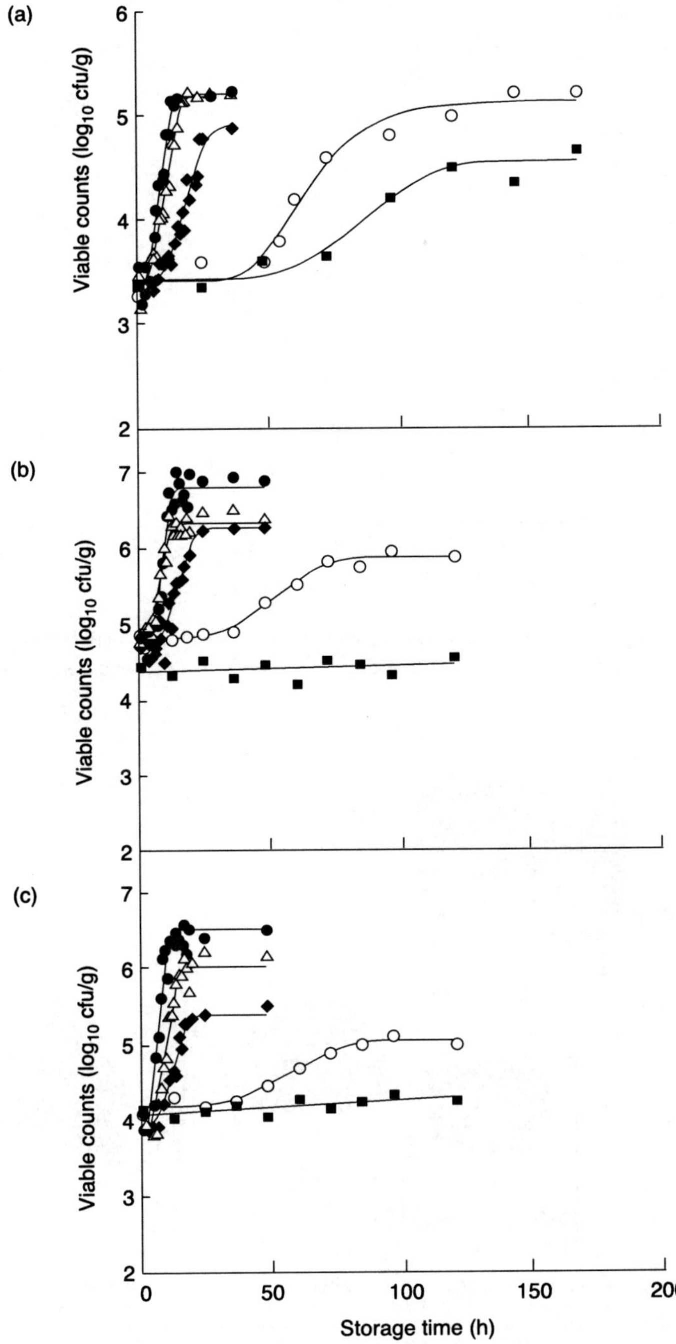

Fig. 12.1 The observed growth of *L. monocytogenes* (a), *E. coli* O157:H7 (b) and *Salmonella* spp. (c) on lettuce at constant temperature ranging from 5 °C to 25 °C. The curves were obtained using the Baranyi-Roberts model. ●, 25 °C; △, 20 °C; ♦, 15 °C; ○, 10 °C; ■, 5 °C.

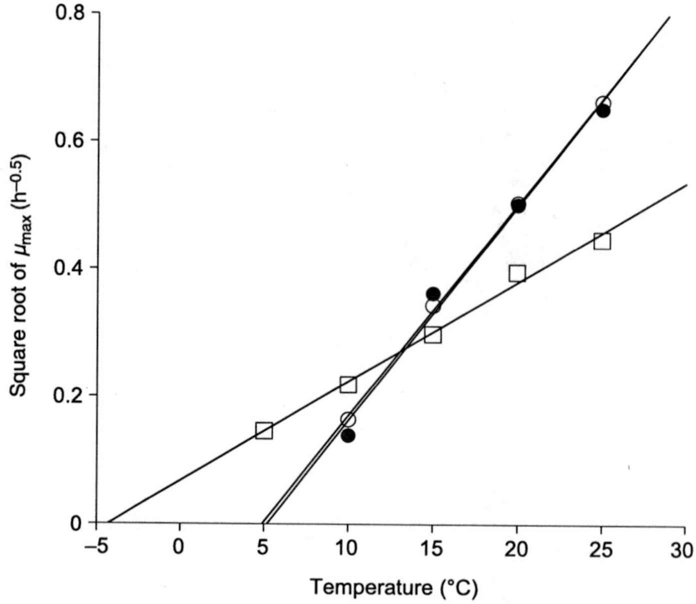

Fig. 12.2 The relationship between the square root of specific growth rate (μ_{max}) on lettuce and temperature in *L. monocytogenes* (□), *E. coli* O157:H7 (○) and *Salmonella* spp. (●).

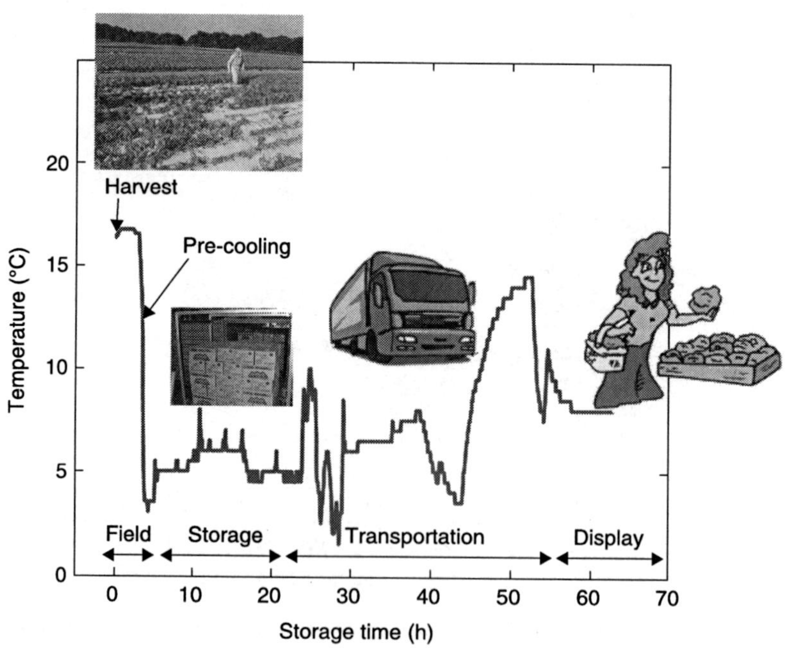

Fig. 12.3 A temperature history of lettuce from the field at harvesting to the retail store.

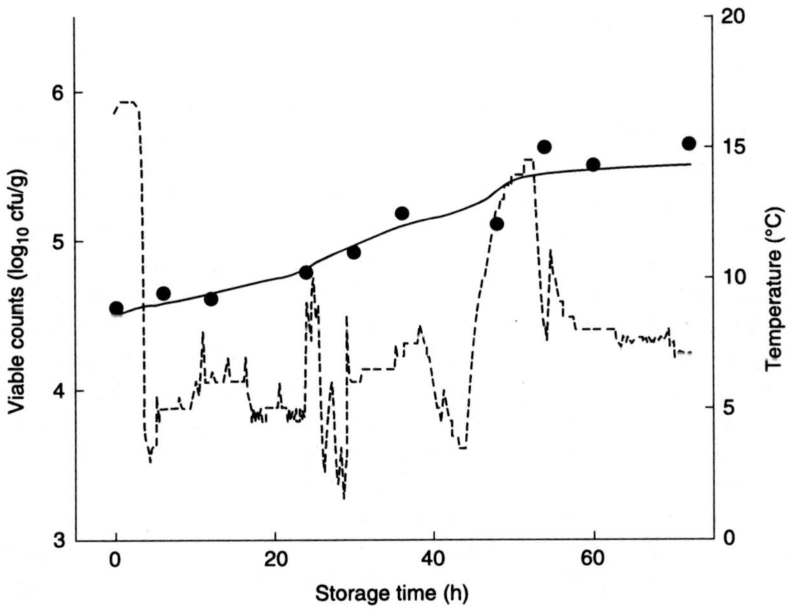

Fig. 12.4 The observed growth of *L. monocytogenes* (●) on lettuce under real temperature history (-----). The growth curves (——) were predicted using the Baranyi-Ratkowsky model.

where N denotes the bacterial cell concentration (cfu/g) at time t, q is a dimensionless quantity related to the physiological state of the cells, μ_{max} is the maximum specific growth rate (1/h), N_{max} represents the maximum population density of the bacteria (cfu/g). The model for μ_{max} was substituted into the above differential equation, and the temperature allowed to be dependent on time. The system was solved numerically by the fourth-order Runge-Kutta method as a means of obtaining predictions of bacterial concentration during time-dependent temperature fluctuations.

Overall predictions for each pathogen agreed well with observed viable counts. The results indicated that the Baranyi-Ratkowsky model is able to predict the growth of pathogens on lettuce using real temperature history during distribution from the farm to the retail store in most cases. Predicting the growth and behaviour of pathogenic bacteria in or on lettuce will help to reduce the microbial risks associated with the consumption of salad vegetables such as lettuce, and will provide valuable information to consumers concerning the shelf-life of products. As the prediction of pathogen growth during distribution will serve as proof of the importance of low-temperature management, it is useful to thoroughly investigate all aspects of temperature management for those concerned with the distribution of such products.

12.5 Future trends

12.5.1 Stochastic modelling

Predictive growth/death models have been applied to the quantitative microbial risk assessment (QMRA) of food products, in particular to exposure assessment, which describes the response of pathogenic bacterial populations in a product during distribution (Codex Alimentarius Commission, 1999). The variability and uncertainty of the estimated parameters of the predictive model are needed to specify risk assessment. Predictive models taking variability and uncertainty into account are indispensable in QMRA (FAO/WHO, 2008). However, as the various conventional predictive microbiology models are deterministic models based on kinetics, these deterministic growth models are problematic given the difficulty in separating model variability and uncertainty (Nauta, 2000). Nauta (2002) questioned their applicability to QMRA and especially their ability to describe variability and uncertainty in microbial growth. Therefore, a new type of predictive model is required based on probabilistic procedures.

Bayesian modelling approaches have recently been investigated to treat variability and uncertainty (Crépet *et al.*, 2009; Delignette-Muller *et al.*, 2006; Pouillot *et al.*, 2003; Powell *et al.*, 2006; Shorten *et al.*, 2004). In reports concerning the Bayesian modelling of bacterial growth in food products, the time series data are directly modelled and various parameters are estimated. These investigations are capable of estimating growth model parameters such as growth rate, minimum growth temperature, lag time, etc., with variability and uncertainty considered separately. Conventional predictive microbiology approaches that consist of primary and secondary models were applied to Bayesian Markov Chain Monte Carlo (MCMC) inference in these studies. These procedures involve a series of conventional approaches and their usefulness is widely accepted. However, these procedures cannot take into account the error derived from the process of data generation (Clark and Bjørnstad, 2004). An alternative predictive model based on a time series modelling technique, which is commonly applied in ecological modelling (Bjørnstad and Grenfell, 2001; Clark, 2007; Clark and Bjørnstad, 2004), might also be useful. Appropriate treatment of errors can be especially important for time series data, as errors of data generation process (variability) propagate forward with processing, and are different from the case of errors of data observation (uncertainty) (Clark, 2007). Among the most general and flexible of the alternative models is the Bayesian state-space model, which has been investigated in the field of ecology, separately from the conventional predictive microbiology approach (Durbin and Koopman, 2001).

Accordingly, future studies employing an alternative modelling procedure using the Bayesian state-space model will be convincing. The modelling procedure, which corresponds to the observed data and treats the data generation process as a latent variable, would be an alternative and new approach that takes into account the variability and uncertainty.

12.5.2 Key events dose-response framework

Dose-response models play an important role in characterizing the risk in quantitative microbial risk assessment (FAO/WHO, 2003). Currently available dose-response models of microbial risk assessment are derived from different source data as follows: human-feeding volunteer tests (FAO/WHO, 2002), extrapolation of the results of animal experiments to humans (Smith *et al.*, 2008; Williams *et al.*, 2007) and estimations from epidemiological studies of outbreaks of foodborne illness (Buchanan *et al.*, 1997). However, these approaches to dose-response modelling of low-dose exposures to pathogens rely on multiple assumptions and extrapolations, which reflect the current lack of knowledge regarding the fundamental mechanisms underlying the responses to a dose (Buchanan *et al.*, 2009).

Recently, a new concept, called the key events dose-response framework (KEDRF), has been introduced as an alternative approach for developing dose-response models (Buchanan *et al.*, 2009). In the KEDRF approach, five major and necessary steps, termed key events, are proposed as follows:

1. survival of pathogens in the upper gastrointestinal tract;
2. establishment in the intestine; attachment to and uptake into epithelial cells;
3. survival and escape from phagosomes in enterocytes and transfer of pathogen to phagocytes;
4. transfer of pathogens across the placenta;
5. pathogen growth leading to foetal morbidity and mortality.

At each step, the range of factors that may influence the response, including dose, pathogen factors and protective host mechanism, are systematically examined. The KEDRF can potentially help in understanding the mechanism by which these factors influence dose-response, that is how specific factors influence the probability of progression from one step to the next step towards the ultimate effect of concern, and how they may contribute to variability in the ultimate dose-response relationship. A deeper understanding of these factors and their relative importance can potentially improve the accuracy and reliability of the predictive dose-response models used in public health. Although there have been several studies on the first step of the KEDRF, in which survival of pathogens in the upper gastrointestinal tract (stomach) was examined (Barmpalia-Davis *et al.*, 2008a, b, 2009; Buchanan *et al.*, 2000; de Jonge *et al.*, 2003; Koseki *et al.*, 2011b; Minekus *et al.*, 1995; Takumi *et al.*, 2000), studies on the other steps of the KEDRF will be indispensable for more accurate and reasonable QMRA in the future.

12.6 Questions for discussion

1. Is fresh produce one of the critical vectors of foodborne illness?
2. What kind of bacterial pathogens are related to foodborne illness on consumption of fresh produce?

3. What kind of vegetables and/or fruits are related to foodborne illness?
4. How frequently do bacterial pathogens contaminate fresh produce?
5. How do bacterial pathogens contaminate fresh produce?
6. Is there any possibility of translocating the pathogenic bacteria from roots to aerial leaf?
7. Is it possible to completely eliminate contaminated bacterial pathogens from fresh produce through normal procedures?
8. How is the temperature dependency of microbial growth rate described mathematically?
9. Is it possible to predict the growth of pathogenic bacteria during distribution?
10. What kinds of studies are needed for more accurate risk assessment?

12.7 Acknowledgements

The studies described here were supported by research fellowships of the Japan Society for the Promotion of Science (JSPS), and a grant (Research project for ensuring food safety from farm to table, 7307) from the Ministry of Agriculture, Forestry, and Fisheries of Japan. Also, the authors wish to thank JA Tone-Numata and the SANYO Electric Co., Ltd for their kind cooperation with the measurement of lettuce temperature during distribution.

12.8 References

ABADIAS, M., USALL, J., ANGUERA, M., SOLSON, C., and VINAS, I. 2008. Microbiological quality of fresh, minimally-processed fruit and vegetables, and sprouts from retail establishments. *Int. J. Food Microbiol.* 123:121–9.

ARTHUR, L., JONES, S., FABRI, M., and ODUMERUZ, J. 2007. Microbial survey of selected Ontario-Grown fresh fruits and vegetables. *J. Food Prot.* 70:2864–7.

ARUSCAVAGE, D., MILLER, S. A., IVEY, M. L. L., LEE, K., and LEJEUNE, J. T. 2008. Survival and dissemination of *Escherichia coli* O157:H7 on physically and biologically damaged lettuce plants. *J. Food Prot.* 71:2384–8.

BARANYI, J., and ROBERTS, T. A. 1994. A dynamic approach to predicting bacterial-growth in food. *Int. J. Food Microbiol.* 23:277–94.

BARANYI, J., ROBINSON, T. P., KALOTI, A., and MACKEY, B. M. 1995. Predicting growth of *Brochothrix Thermosphacta* at changing temperature. *Int. J. Food Microbiol.* 27: 61–75.

BARI, M. L., KUSUNOKI, H., FURUKAWA, H., IKEDA, H., ISSHIKI, K., and UEMURA, T. 1999. Inhibition of growth of *Escherichia coli* O157: H7 in fresh radish (*Raphanus sativus* L.) sprout production by calcinated calcium. *J. Food Prot.* 62:128–32.

BARI, M. L., SABINA, Y., ISOBE, S., UEMURA, T., and ISSHIKI, K. 2003. Effectiveness of electrolyzed acidic water in killing *Escherichia coli* O157: H7, *Salmonella enteritidis*, and *Listeria monocytogenes* on the surfaces of tomatoes. *J. Food Prot.* 66:542–8.

BARMPALIA-DAVIS, I. M., GEORNARAS, I., KENDALL, P. A., and SOFOS, J. N. 2008a. Differences in survival among 13 *Listeria monocytogenes* strains in a dynamic model of the stomach and small intestine. *Appl. Environ. Microbiol.* 74:5563–7.

BARMPALIA-DAVIS, I. M., GEORNARAS, I., KENDALL, P. A., and SOFOS, J. N. 2008b. Survival of *Listeria monocytogenes* in a simulated dynamic gastrointestinal model during storage of inoculated bologna and salami slices in vacuum packages. *J. Food Prot.* 71:2014–23.

BARMPALIA-DAVIS, I. M., GEORNARAS, I., KENDALL, P. A., and SOFOS, J. N. 2009. Effect of fat content on survival of *Listeria monocytogenes* during simulated digestion of inoculated beef frankfurters stored at 7°C. *Food Microbiol.* 26:483–90.

BERNSTEIN, N. 2007. Evidence for internalization of *Escherichia coli* into the aerial parts of maize via the root system. *J. Food Prot* 70:471–5.

BERNSTEIN, N., SELA, S., and NEDER-LAVON, S. 2007. Assessment of contamination potential of lettuce by *Salmonella enterica* Serovar Newport added to the plant growing medium. *J. Food Prot* 70:1717–22.

BJØRNSTAD, O. N., and GRENFELL, B. T. 2001. Noisy clockwork: Time series analysis of population fluctuations in animals. *Science.* 293:638–43.

BOHAYCHUK, V. M., BRADBURY, R. W., DIMOCK, R., FEHR, M., GENSLER, G. E., *et al.* 2009. A microbiological survey of selected Alberta-grown fresh produce from farmers' markets in Alberta, Canada. *J. Food Prot.* 72:415–20.

BOVILL, R., BEW, J., COOK, N., D'AGOSTINO, M., WILKINSON, N., and BARANYI, J. 2000. Predictions of growth for *Listeria monocytogenes* and *Salmonella* during fluctuating temperature. *Int. J. Food Microbiol.* 59:157–65.

BOVILL, R. A., BEW, J., and BARANYI, J. 2001. Measurements and predictions of growth for *Listeria monocytogenes* and *Salmonella* during fluctuating temperature II. Rapidly changing temperatures. *Int. J. Food Microbiol.* 67:131–7.

BRANDL, M. T., and AMUNDSON, R. 2008. Leaf age as a risk factor in contamination of lettuce with *Escherichia coli* O157: H7 and *Salmonella enterica. Appl. Environ. Microbiol.* 74:2298–306.

BUCHANAN, R. L., DAMERT, W. G., WHITING, R. C., and VAN SCHOTHORST, M. 1997. An approach for using epidemiologic and microbial food survey data to develop a 'purposefully conservative' estimate of the dose-response relationship between *Listeria monocytogenes* levels and the incidence of foodborne listeriosis. *J. Food Prot.* 60:918–22.

BUCHANAN, R. L., HAVELAAR, A. H., SMITH, M. A., WHITING, R. C., and JULIEN, E. 2009. The key events dose-response framework: its potential for application to foodborne pathogenic microorganisms. *Crit. Rev. Food Sci. Nutr.* 49:718–28.

BUCHANAN, R. L., SMITH, J. L., and LONG, W. 2000. Microbial risk assessment: Dose-response relations and risk characterization. *Int. J. Food Microbiol.* 58:159–72.

CALIFORNIA FOOD EMERGENCY RESPONSE TEAM. 2007. *Investigation of an Escherichia coli O157:H7 outbreak associated with dole pre-packaged spinach.* In California Department of Health Services Food and Drug Branch, U.S. Food and Drug Administration San Francisco District. http://www.marlerclark.com/2006_Spinach_Report_Final_01.pdf/.

CARRASCO, E., PEREZ-RODRIGUEZ, F., VALERO, A., GARCIA-GIMENO, R. M., and ZURERA, G. 2008. Growth of *Listeria monocytogenes* on shredded, ready-to-eat iceberg lettuce. *Food Control* 19:487–94.

CDC. 2011. *CDC – Multistate outbreak of listeriosis linked to whole cantaloupes from Jensen Farms, Colorado – Listeriosis.* http://www.cdc.gov/listeria/outbreaks/cantaloupes-jensen-farms/index.html.

CLARK, J. S. 2007. *Models for Ecological Data: An Introduction.* Princeton University Press, Princeton.

CLARK, J. S., and BJØRNSTAD, O. N. 2004. Population time series: Process variability, observation errors, missing values, lags, and hidden states. *Ecology.* 85:3140–50.

CODEX ALIMENTARIUS COMMISSION. 1999. *Report of the Thirtieth Session of the Codex Committee on Food Hygiene (CCFH).* Alinorm 99/13 Appendix IV:58–64.

CRÉPET, A., STAHL, V., and CARLIN, F. 2009. Development of a hierarchical Bayesian model to estimate the growth parameters of *Listeria monocytogenes* in minimally processed fresh leafy salads. *Int. J. Food Microbiol.* 131:112–19.

DE JONGE, R., TAKUMI, K., RITMEESTER, W. S., and VAN LEUSDEN, F. M. 2003. The adaptive response of *Escherichia coli* O157 in an environment with changing pH. *J. Appl. Microbiol.* 94:555–60.

DELAQUIS, P., BACH, S., and DINU, L. D. 2007. Behavior of *Escherichia coli* O157: H7 in leafy vegetables. *J. Food Prot.* 70:1966–74.

DELIGNETTE-MULLER, M. L., CORNU, M., POUILLOT, R., and DENIS, J. B. 2006. Use of Bayesian modelling in risk assessment: Application to growth of *Listeria monocytogenes* and food flora in cold-smoked salmon. *Int. J. Food Microbiol.* 106:195–208.

DOERING, H. J., HARRISON, M. A., MORROW, R. A., HURST, W. C., and KERR, W. L. 2009. Use of the systems approach to determine the fate of *Escherichia coli* O157:H7 on fresh lettuce and spinach. *J. Food Prot.* 72:1560–8.

DOYLE, M. P., and ERICKSON, M. C. 2008. Summer meeting 2007 – the problems with fresh produce: an overview. *J. Appl. Microbiol.* 105:317–30.

DREUX, N., ALBAGNAC, C., CARLIN, F., MORRIS, C. E., and NGUYEN-THE, C. 2007. Fate of *Listeria* spp. on parsley leaves grown in laboratory and field cultures. *J. Appl. Microbiol.* 103:1821–7.

DURBIN, J., and KOOPMAN, S. J. 2001. *Time Series Analysis by State Space Models*. Oxford University Press, New York.

ERICKSON, M. C. 2010. Preharvest internalization of *Escherichia coli* O157:H7 into lettuce leaves, as affected by insect and physical damage. *J. Food Prot.* 73:1809–16.

ERICKSON, M. C., WEBB, C. C., DIAZ-PEREZ, J. C., PHATAK, S. C., SILVOY, J. J., *et al.* 2010. Surface and internalized *Escherichia coli* O157:H7 on field-grown spinach and lettuce treated with spray-contaminated irrigation water. *J. Food Prot.* 73:1023–9.

FAO/WHO. 2002. *Microbiological Risk Assessment Series 2: Risk Assessments of Salmonella in Eggs and Broiler Chickens*. World Health Organization, Food and Agricultural Organization of the United Nations, Geneva, Rome.

FAO/WHO. 2003. *Microbiological risk assessment series 3: Hazard characterization for pathogens in food and water*. World Health Organization, Food and Agricultural Organization of the United Nations, Geneva, Roma.

FAO/WHO. 2008. *Exposure assessment of microbiological hazards in food: guidelines*. Food and Agriculture Organization of the United Nations/World Health Organization, Rome.

FRANK, C., WERBER, D., CRAMER, J. P., ASKAR, M., FABER, M., *et al.* 2011. Epidemic profile of Shiga-toxin-producing *Escherichia coli* O104: H4 outbreak in Germany. *New Engl. J. Med.* 365:1771–80.

FRIESEMA, I., SIGMUNDSDOTTIR, G., VAN DER ZWALUW, K., HEUVELINK, A., SCHIMMER, B., *et al.* 2008. An international outbreak of Shiga toxin-producing *Escherichia coli* O157 infection due to lettuce, September–October 2007. *Euro Surveillance*. 13 (50): 1–5.

GIBBS, R., PINGAULT, N., MAZZUCCHELLI, T., O'REILLY, L., MACKENZIE, B., *et al.* 2009. An outbreak of *Salmonella enterica* Serotype Litchfield infection in Australia linked to consumption of contaminated papaya. *J. Food Prot.* 72:1094–8.

GREENE, S. K., DALY, E. R., TALBOT, E. A., DEMMA, L. J., HOLZBAUER, S., *et al.* 2008. Recurrent multistate outbreak of *Salmonella* Newport associated with tomatoes from contaminated fields, 2005. *Epidemiol. Infect.* 136:157–65.

HAN, Y., LINTON, R. H., NIELSEN, S. S., and NELSON, P. E. 2000. Inactivation of *Escherichia coli* O157: H7 on surface-uninjured and -injured green pepper (*Capsicum annuum* L.) by chlorine dioxide gases demonstrated by confocal laser scanning microscopy. *Food Microbiol.* 17:643–55.

HARAPAS, D., PREMIER, R., TOMKINS, B., FRANZ, P., and AJLOUNI, S. 2010. Persistence of *Escherichia coli* on injured vegetable plants. *Int. J. Food Microbiol.* 138:232–7.

HARRIS, L., FARBER, J., BEUCHAT, L., PARISH, M., SUSLOW, T., *et al.* 2003. Outbreaks associated with fresh produce: incidence, growth, and survival of pathogens in fresh and fresh-cut produce. *Comp. Rev. Food Sci. Food Safety* 2:78–141.

HEATON, J., and JONES, K. 2008. Microbial contamination of fruit and veg etables and the behaviour of enteropathogens in the phyllosphere: a review. *J. Appl. Microbiol.* 104:613–26.

HO, J. L., SHANDS, K. N., FRIEDLAND, G., ECKIND, P., and FRASER, D. W. 1986. An outbreak of type 4b *Listeria monocytogenes* infection involving patients from eight Boston hospitals. *Arch. Intern. Med.* 146:520–4.

ITOH, Y., SUGITA-KONISHI, Y., KASUGA, F., IWAKI, M., HARA-KUDO, Y., *et al.* 1998. Enterohemorrhagic *Escherichia coli* O157: H7 present in radish sprouts. *Appl. Environ. Microbiol.* 64:1532–5.

JOHANNESSEN, G. S., LONCAREVIC, S., and KRUSE, H. 2002. Bacteriological analysis of fresh produce in Norway. *Int. J. Food Microbiol.* 77:199–204.

JOHNSTON, L. M., JAYKUS, L. A., MOLL, D., ANCISO, J., MORA, B., and MOE, C. L. 2006. A field study of the microbiological quality of fresh produce of domestic and Mexican origin. *Int. J. Food Microbiol.* 112:83–95.

JOHNSTON, L. M., JAYKUS, L. A., MOLL, D., MARTINEZ, M. C., ANCISO, J., *et al.* 2005. A field study of the microbiological quality of fresh produce. *J. Food Prot.* 68:1840–7.

KIM, C., HUNG, Y. C., BRACKETT, R. E., and LIN, C. S. 2003. Efficacy of electrolyzed oxidizing water in inactivating *Salmonella* on alfalfa seeds and sprouts. *J. Food Prot.* 66:208–14.

KIM, J. G., YOUSEF, A. E., and CHISM, G. W. 1999. Use of ozone to inactivate microorganisms on lettuce. *J. Food Saf.* 19:17–34.

KOSEKI, S., and ISOBE, S. 2005a. Growth of *Listeria monocytogenes* on iceberg lettuce and solid media. *Int. J. Food Microbiol.* 101:217–25.

KOSEKI, S., and ISOBE, S. 2005b. Prediction of pathogen growth on iceberg lettuce under real temperature history during distribution from farm to table. *Int. J. Food Microbiol.* 104:239–48.

KOSEKI, S., MIZUNO, Y., KAWASAKI, S., and YAMAMOTO, K. 2011a. A survey of iceberg lettuce for the presence of *Salmonella, Escherichia coli* O157:H7, and *Listeria monocytogenes* in Japan. *J. Food Prot.* 74:1543–6.

KOSEKI, S., MIZUNO, Y., and SOTOME, I. 2011b. Modeling of pathogen survival during simulated gastric digestion. *Appl. Environ. Microbiol.* 77:1021–32.

KOSEKI, S., MIZUNO, Y., and YAMAMOTO, K. 2011c. Comparison of two possible routes of pathogen contamination of spinach leaves in a hydroponic cultivation system. *J. Food Prot.* 74:1536–42.

KOSEKI, S., YOSHIDA, K., KAMITANI, Y., ISOBE, S., and ITOH, K. 2004. Effect of mild heat pre-treatment with alkaline electrolyzed water on the efficacy of acidic electrolyzed water against *Escherichia coli* O157: H7 and *Salmonella* on lettuce. *Food Microbiol.* 21: 559–66.

LAPIDOTAND, A. 2009. Transfer of *Salmonella enterica* Serovar Typhimurium from contaminated irrigation water to parsley is dependent on curli and cellulose, the biofilm matrix components. *J. Food Prot.* 72:618–23.

LIANOU, A., and SOFOS, J. N. 2007. A review of the incidence and transmission of *Listeria monocytogenes* in ready-to-eat products in retail and food service environments. *J. Food Prot.* 70:2172–98.

LIN, C. M., MOON, S. S., DOYLE, M. P., and MCWATTERS, K. H. 2002. Inactivation of *Escherichia coli* O157: H7, *Salmonella enterica* serotype enteritidis, and *Listeria monocytogenes* on lettuce by hydrogen peroxide and lactic acid and by hydrogen peroxide with mild heat. *J. Food Prot.* 65:1215–20.

LONCAREVIC, S., JOHANNESSEN, G. S., and RORVIK, L. M. 2005. Bacteriological quality of organically grown leaf lettuce in Norway. *Lett. Appl. Microbiol.* 41:186–9.

LUO, Y., HE, Q., and MCEVOY, J. L. 2010. Effect of storage temperature and duration on the behavior of *Escherichia coli* O157:H7 on packaged fresh-cut salad containing romaine and iceberg lettuce. *J. Food Sci.* 75:M390–7.

MAHBUB, I. 2004. Persistence of enterohemorrhagic *Escherichia coli* O157:H7 in soil and on leaf lettuce and parsley grown in fields treated with contaminated manure composts or irrigation water. *J. Food Prot.* 67:1365–70.

MAKI, D. G. 2006. Don't eat the spinach — Controlling foodborne infectious disease. *New Engl. J. Med.* 355:1952–5.

MCKELLAR, R. C., and DELAQUIS, P. 2011. Development of a dynamic growth-death model for *Escherichia coli* O157:H7 in minimally processed leafy green vegetables. *Int. J. Food Microbiol.* 151:7–14.

MCKELLAR, R. C., LEBLANC, D. I., LU, J., and DELAQUIS, P. 2012. Simulation of *Escherichia coli* O157:H7 behavior in fresh-cut lettuce under dynamic temperature conditions during distribution from processing to retail. *Foodborne Pathog. Dis.* 9:239–44.

MCMAHON, M. A. S., and WILSON, I. G. 2001. The occurrence of enteric pathogens and *Aeromonas* species in organic vegetables. *Int. J. Food Microbiol.* 70:155–62.

MCMEEKIN, T. A., OLLEY, J., ROSS, T., and RATKOWSKY, D. A. 1993. *Predictive microbiology: Theory and Application*. Research Studies Press, Taunton, Somerset, England.

MINEKUS, M., MARTEAU, P., HAVENAAR, R., and HUISINTVELD, J. H. J. 1995. A multicompartmental dynamic computer-controlled model simulating the stomach and small-intestine. *ATLA-Altern. Lab. Anim.* 23:197–209.

MITRA, R., CUESTA-ALONSO, E., WAYADANDE, A., TALLEY, J., GILLILAND, S., and FLETCHER, J. 2009. Effect of route of introduction and host cultivar on the colonization, internalization, and movement of the human pathogen *Escherichia coli* O157:H7 in spinach. *J. Food Prot.* 72:1521–30.

MUKHERJEE, A., SPEH, D., DYCK, E., and DIEZ-GONZALEZ, F. 2004. Preharvest evaluation of coliforms, *Escherichia coli*, *Salmonella*, and *Escherichia coli* O157: H7 in organic and conventional produce grown by Minnesota farmers. *J. Food Prot.* 67:894–900.

MUKHERJEE, A., SPEH, D., JONES, A. T., BUESING, K. M., and DIEZ-GONZALEZ, F. 2006. Longitudinal microbiological survey of fresh produce grown by farmers in the upper midwest. *J. Food Prot.* 69:1928–36.

MUNNOCH, S. A., WARD, K., SHERIDAN, S., FITZSIMMONS, G. J., SHADBOLT, C. T., *et al.* 2009. A multi-state outbreak of *Salmonella* Saintpaul in Australia associated with cantaloupe consumption. *Epidemiol. Infect.* 137:367.

NAUTA, M. J. 2000. Separation of uncertainty and variability in quantitative microbial risk assessment models. *Int. J. Food Microbiol.* 57:9–18.

NAUTA, M. J. 2002. Modelling bacterial growth in quantitative microbiological risk assessment: is it possible? *Int. J. Food Microbiol.* 73:297–304.

PARK, C. M., HUNG, Y. C., DOYLE, M. P., EZEIKE, G. O. I., and KIM, C. 2001. Pathogen reduction and quality of lettuce treated with electrolyzed oxidizing and acidified chlorinated water. *J. Food Sci.* 66:1368–72.

POUILLOT, R., ALBERT, I., CORNU, M., and DENIS, J-B. 2003. Estimation of uncertainty and variability in bacterial growth using Bayesian inference. Application to Listeria monocytogenes. *Int. J. Food Microbiol.* 81:87–104.

POWELL, M., TAMPLIN, M. L., MARKS, B., and CAMPOS, D. T. 2006. Bayesian synthesis of a pathogen growth model: *Listeria monocytogenes* under competition. *Int. J. Food Microbiol.* 109:34–46.

RATKOWSKY, D. A., OLLEY, J., MCMEEKIN, T. A., and BALL, A. 1982. Relationship between temperature and growth rate of bacterial cultures. *J. Bacteriol.* 149:1–5.

RIVA, M., FRANZETTI, L., and GALLI, A. 2001. Microbiological quality and shelf life modeling of ready-to-eat cicorino. *J. Food Prot.* 64:228–34.

RODRIGUEZ, A. M. C., ALCALA, E. B., GIMENO, R. M. G., and COSANO, G. Z. 2000. Growth modelling of *Listeria monocytogenes* in packaged fresh green asparagus. *Food Microbiol.* 17:421–7.

SAGOO, S. K., LITTLE, C. L., and MITCHELL, R. T. 2001. The microbiological examination of ready-to-eat organic vegetables from retail establishments in the United Kingdom. *Lett. Appl. Microbiol.* 33:434–9.

SAGOO, S. K., LITTLE, C. L., and MITCHELL, R. T. 2003. Microbiological quality of open ready-to-eat salad vegetables: Effectiveness of food hygiene training of management. *J. Food Prot.* 66:1581–6.

SANT'ANA, A. S., BARBOSA, M. S., DESTRO, M. T., LANDGRAF, M., and FRANCO, B. D. G. M. 2012a. Growth potential of *Salmonella* spp. and *Listeria monocytogenes* in nine types of ready-to-eat vegetables stored at variable temperature conditions during shelf-life. *Int. J. Food Microbiol.* 157:52–8.

SANT'ANA, A. S., FRANCO, B. D. G. M., and SCHAFFNER, D. W. 2012b. Modeling the growth rate and lag time of different strains of *Salmonella enterica* and *Listeria monocytogenes* in ready-to-eat lettuce. *Food Microbiol.* 30:267–73.

SCHLECH, W. F., LAVIGNE, P. M., BORTOLUSSI, R. A., ALLEN, A. C., HALDANE, E. V., *et al.* 1983. Epidemic listeriosis — Evidence for transmission by food. *New Engl. J. Med.* 308:203–6.

SHORTEN, P. R., MEMBRÉ, J. M., PLEASANTS, A. B., KUBACZKA, M., and SOBOLEVA, T. K. 2004. Partitioning of the variance in the growth parameters of *Erwinia carotovora* on vegetable products. *Int. J. Food Microbiol.* 93:195–208.

SIVAPALASINGAM, S., FRIEDMAN, C. R., COHEN, L., and TAUXE, R. V. 2004. Fresh produce: A growing cause of outbreaks of foodborne illness in the United States, 1973 through 1997. *J. Food Prot.* 67:2342–53.

SMITH, M. A., TAKEUCHI, K., ANDERSON, G., WARE, G. O., MCCLURE, H. M., *et al.* 2008 Dose response for *Listeria monocytogenes* induced stillbirths in nonhuman primates. *Infect. Immun.* 76:726–31.

SOLOMON, E. B., PANG, H. J., and MATTHEWS, K. R. 2003. Persistence of *Escherichia coli* O157:H7 on lettuce plants following spray irrigation with contaminated water. *J. Food Prot.* 66:2198–202.

SOLOMON, E. B., YARON, S., and MATTHEWS, K. R. 2002. Transmission of *Escherichia coli* O157:H7 from contaminated manure and irrigation water to lettuce plant tissue and its subsequent internalization. *Appl. Environ. Microbiol.* 68:397–400.

STINE, S. W., SONG, I., CHOI, C. Y., and GERBA, C. P. 2005. Effect of relative humidity on preharvest survival of bacterial and viral pathogens on the surface of cantaloupe, lettuce, and bell peppers. *J. Food Prot.* 68:1352–8.

TAKUMI, K., DE JONGE, R., and HAVELAAR, A. 2000. Modelling inactivation of *Escherichia coli* by low pH: application to passage through the stomach of young and elderly people. *J. Appl. Microbiol.* 89:935–43.

THEOFEL, C. G., and HARRIS, L. J. 2009. Impact of preinoculation culture conditions on the behavior of *Escherichia coli* O157:H7 inoculated onto romaine lettuce (*Lactuca sativa*) plants and cut leaf surfaces. *J. Food Prot.* 72:1553–9.

U.S. DEPARTMENT OF AGRICULTURE, A. M. S. 2007, posting date. *Microbiological Data Program Progress Update and 2006 Data Summary*. http://www.ams.usda.gov/AMSv1.0/getfile?dDocName=STELPRDC5050633. Accessed 7 August 2012.

U.S. DEPARTMENT OF AGRICULTURE, A. M. S. 2008, posting date. *Microbiological Data Program Progress Update and 2007 Data Summary*. http://www.ams.usda.gov/AMSv1.0/getfile?dDocName=STELPRDC5067866. Accessed 7 August 2012.

U.S. DEPARTMENT OF AGRICULTURE, A. M. S. 2009, posting date. *Microbiological Data Program Progress Update and 2008 Data Summary*. http://www.ams.usda.gov/AMSv1.0/getfile?dDocName=STELPRDC5079908. Accessed 7 August 2012.

VISWANATHAN, P., and KAUR, R. 2001. Prevalence and growth of pathogens on salad vegetables, fruits and sprouts. *Int. J. Hyg. Environ. Health.* 203:205–13.

WACHTEL, M. R., WHITEHAND, L. C., and MANDRELL, R. E. 2002. Association of *Escherichia coli* O157: H7 with preharvest leaf lettuce upon exposure to contaminated irrigation water. *J. Food Prot.* 65:18–25.

WARRINER, K., IBRAHIM, F., DICKINSON, M., WRIGHT, C., and WAITES, W. M. 2003. Interaction of *Escherichia coli* with growing salad spinach plants. *J. Food Prot.* 66:1790–7.

WENDEL, A. M., HOANG JOHNSON, D., SHARAPOV, U., GRANT, J., ARCHER, J. R., *et al.* 2009. Multistate outbreak of *Escherichia coli* O157:H7 infection associated with consumption of packaged spinach, August–September 2006: The Wisconsin investigation. *Clin. Infect. Dis.* 48:1079–86.

WILLIAMS, D., IRVIN, E. A., CHMIELEWSKI, R. A., FRANK, J. F., and SMITH, M. A. 2007. Dose response of *Listeria monocytogenes* after oral exposure in pregnant guinea pigs. *J. Food Prot.* 70:1122–8.

Part III

Commercial solutions for fresh produce safety

13

Modified atmosphere packaging for fresh produce

S. Shayanfar, German Institute of Food Technologies (DIL), Germany

DOI: 10.1533/9781782420279.3.175

Abstract: One of the new trends in the world of foods is to satisfy the customer's expectations of having nutritious, fresh or fresh-like foods. Modified atmosphere packaging (MAP) can be regarded a revolutionary step that enables retailers to market fresh foods with extended shelf-life without any major processing, preservatives or chemical additives. This chapter discusses principles and key factors affecting MAP, regulatory and marketing issues, and innovations and novel trends in the application of MAP in fresh produce.

Key words: fresh produce, modified atmosphere packaging (MAP), shelf-life, packaging.

13.1 Introduction

Modified atmosphere packaging (MAP) is a term applied to a range of food packaging technologies that mainly rely on mixtures of different gases (e.g. O_2, CO_2, N_2, CO, O_3, etc.), depending on the product that needs to be packed. In the late 1940s, MAP was generated for the first time by reducing the oxygen concentration in apple packaging in order to slow down ripening. In other words, MAP induces a dormant phase in the living plant, similar to hibernation in animals. The most popular atmospheric gases for packaging fruits and vegetables are O_2, CO_2 and N_2. However, they are present in different concentrations from those in the atmosphere (78% nitrogen, 21% oxygen, 0.03% carbon dioxide and traces of other gases). MAP is also known as gas packaging, gas exchange packaging or protective atmosphere packaging. MAP can be broadly divided into two types, namely:

- passive modified atmosphere packaging, where the food products are hermetically sealed in a package and the ultimate composition of the atmosphere inside is the result of product respiration. Thus, by just sealing the food product

and not requiring any special gas additions or other requirements, MAP conditions can be generated;
- active modified atmosphere packaging, where after an initial application of vacuum to the package containing the food products, the package is flushed with a gas mixture at a very specific composition and concentration.

It is worth mentioning that active MAP should not be mistaken for controlled atmosphere packaging (CAP), in which the gas composition is actively monitored and maintained during the storage time.

13.1.1 Working principles of modified atmosphere packaging

Just like any other living being, fruits and vegetables respire, which ends up consuming their sugar reserves and producing CO_2 and H_2O. This process is primarily responsible for weight loss, texture softening, shrinkage and, ultimately, product deterioration. In addition, ethylene (C_2H_4) production, a result of plant respiration, promotes ripening and softening of tissues. Therefore, reducing the respiration rate in fresh fruits and vegetables can extend their shelf-life (Fig. 13.1).

The main working principle of MAP is that it mitigates the amount of oxygen accessible to the living cells and elevates carbon dioxide, thereby reducing the respiration rate.

13.2 Challenges of modified atmosphere packaging (MAP) storage

The challenges of 'fresh cuts' stored under MAP can be categorized into four main groups, which are of great importance.

13.2.1 Gas composition

Undoubtedly one of the most important challenges in designing an efficient MAP is the gas composition. Each of the three main gases (O_2, CO_2 and N_2) play different roles in MAP and thus optimizing their respective concentrations is of high importance. It is worth mentioning here that each cultivar of a fruit or vegetable can require different gaseous composition, which makes attaining the favorable gaseous mixture arduous to define. Carbon dioxide is the main element, which when dissolved into the water phase (forming carbonic acid), reduces the surrounding pH and thereby directly affects microbial growth. The bacteriostatic and fungistatic properties of this gas are manifested differently in different food products. However, because of the high solubility of CO_2 in water and considering the fact that the main compound in fresh cuts is water, it could nullify the application of 100% CO_2 as the ideal modified atmosphere in fresh cuts packages. The high amount of dissolved CO_2 in fresh cuts will result in off odors and a slight sour taste and could also damage the anthocyanin content (i.e. color).

$$CO_2 + H_2O \rightarrow H_2CO_3$$

Fig. 13.1 The effect of MAP on extending the shelf-life in fresh rocket/rucola stored at 5 °C. Packed under MAP (on the right) and the unpacked sample (on the left) (a) the first day of storage, (b) the fourth day of storage at 5 °C, (c) the seventh day of storage at 5 °C, (d) the tenth day of storage at 5 °C.

The other main gas in MAP is oxygen. The oxygen dose is normally kept as low as possible to inhibit the growth of aerobic spoilage bacteria. However, in fresh cuts (which are respiring products), care should be taken to set O_2 concentration no lower than a critical amount. Too low an oxygen content could lead to fermentation with the end result that ethanol, acetaldehyde and organic acids are produced. On the other hand, high levels of O_2 (80–90%) have been suggested to be promising for some special types of fruits and vegetables. A concentration of approximately 1–5% O_2 has been proven to be promising for most of the vegetables.

The other important gas in MAP is the inert gas, N_2. This tasteless, odorless and insoluble gas is used primarily as a filler to protect the package against the mechanical injuries.

13.2.2 Packaging material

There is a wide range of flexible packaging materials available in different forms for use as pouches, tray-seal packages, etc., which can be used for MAP. There are different parameters that characterize the packaging materials and these need to be taken into consideration before using them in MAP. In some cases, a single packaging material cannot meet all the desired expectations and therefore multilayered or co-extruded packaging films are adopted. Seal-ability, strength, and printability of the films are other parameters which are of importance. The main packaging machines used in MAP are either tray-sealer or form-fill-sealers (FFS). In the tray-sealer, the products are put in a rigid tray and wrapped over with a flexible film. In the latter, a tube of flexible packaging material is filled with the food product from above, sealed and cut from the bottom, forming pouches. In heavily loaded packages where the food product is densely compact, FFS is preferred over tray-sealer. This is because water vapor, a byproduct of respiration, may condense during cold storage on the trays. As there is neither a free circulation of air nor the food, there is the risk of microbial growth in tray-sealed packages, especially when the food is densely compacted.

However, the main characteristic of a packaging film, which dictates whether it is appropriate for MA packaging of a specific product is its permeability. As a result of different transpiration rates by various food products, different permeability characteristics are required. Permeability is expressed by different terms. These include the water vapor transmission rate (WVTR) and the oxygen transmission rate (OTR), which define the amount of water vapor or oxygen permeated through a specified surface of the packaging film, respectively. The packaging materials used for fruit and vegetables in particular have to be low in permeability. If the products were sealed in an airtight package, oxygen would be depleted (because of respiration) and fermentation could set in. On the contrary, if the packaging material is too porous, the optimized gaseous mixture will be depleted and no benefit can be expected. Therefore choosing the proper packaging material with the appropriate permeability and package design is a significant decision that must be made carefully.

13.2.3 Product

It should be noted that increasing the surface area of fresh cuts will cause significant increase in respiration compared with an intact or whole fruit. Thus, any cutting, peeling or dicing would increase the transpiration rate of the fruits and vegetables. The other parameters that affect respiration are the level of ripening, size and variety. Slicing or cutting of plant tissue will rupture the cell walls and facilitate the interaction of phenolic substrate to the polyphenol oxidase (PPO) enzyme, which results in tissue enzymatic browning. Although enzymatic browning does not cause any health risk, developing a strategy to prevent or at least mitigate this chemical reaction is of interest and value in maintaining the physical appearance of the final product. The sharpness of the knife and its material composition can also influence the extent of crushing, etc., that is quality of the final product. Blunt knives result in crushing of the texture and resultant tissue browning.

13.2.4 Processing and storage

Hygienic handling and maintaining HACCP principles and procedures in all processing steps of the fresh cut before packaging are key factors in maintaining quality of the final product. It is erroneously assumed that MAP is a method of quality improvement. MAP neither makes a product better, nor does it inactivate spoilage organisms. In reality, MAP only assists other hurdle technologies such as hygienic processing and cold storage to extend shelf-life in fresh products. Therefore, hygienic processing is critical for fresh cuts destined to be MA packaged and during the MAP procedures as well (Fig. 13.2).

The main extrinsic factor affecting MAP performance is temperature. Maintaining cold temperatures during the processing steps is critical as every 10 °C increase in temperature increases respiration rate by a factor of 3–4, which is defined as Q10 value. It is believed that the best storage temperature for MAP product is between 0 °C and 5 °C. As a result of the fresh cut's respiration and the accumulation of water vapor in the packaging film, there may be a foggy appearance in the packages. This can be alleviated by employing antifog assisted packaging films. Another extrinsic factor that can influence the shelf-life of MAP pouches is the relative humidity.

13.2.5 Significant factors affecting MAP efficiency

There are different intrinsic and extrinsic factors that affect microbial growth. The application of MAP exerts control on two different extrinsic factors, namely temperature and atmospheric composition. However, it should be noted that MAP cannot make a food safe if it is already contaminated. Thus precautions should be taken to prevent the contamination of the product during handling and processing. Standard sanitary operating procedures and Good Manufacturing Processes (GMP) should be enforced throughout the process.

A minimum level of 2–3% O_2 can hinder fermentation. However, *L. monocytogenes* and *Cl. botulinum* can grow under reduced O_2 levels and are not

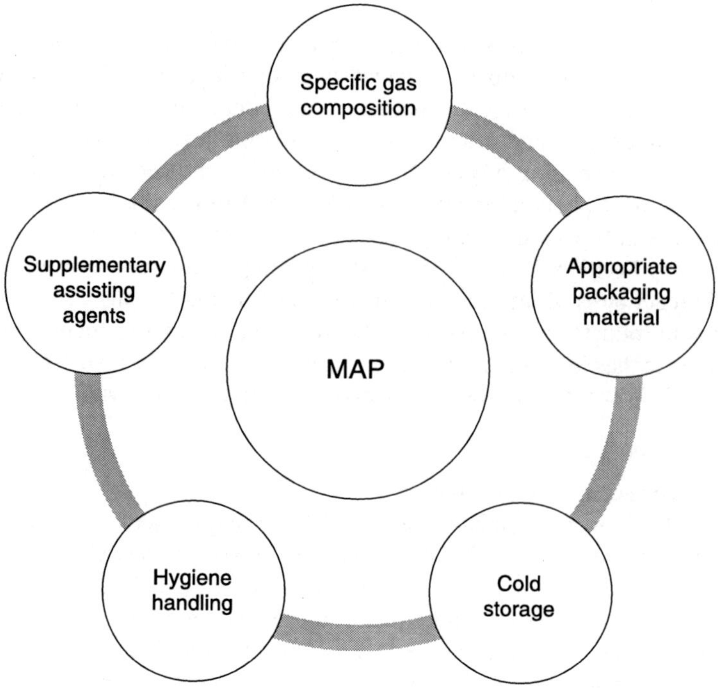

Fig. 13.2 Fundamental elements for implementation of MAP.

markedly inhibited by CO_2. The environment in which fruits and vegetables are grown can harbor pathogens including *Salmonella* species, toxigenic *E. coli* and viruses. Although these microorganisms may not be able to grow in MAP packs, especially if they are stored around 3 °C, they may be able to survive the storage conditions and proliferate when the temperature is more suitable and could ultimately cause food poisoning (Solomon *et al.*, 1990). These realities highlight the importance of hygienic processing, sanitary processing and rinsing and wiping prior to MAP to avoid any microbial growth.

Considering temperature is a key factor affecting MAP efficiency, countries with lower average temperatures can significantly profit from MAP in extending shelf-life. Additionally, in these countries the cold climate reduces the cost of cold storage, which increases the retailers' margin. The short transportation distances and the moderate temperature in such areas also provide the retailers with additional benefits that allow them to enlarge their distribution circles. Apart from this, as a result of different priorities in different countries, MAP has been applied to various food products. However, during the last decade, application of MAP in fresh cuts has been the strongest growth segment.

13.3 Regulatory aspects, economic aspects and market issues

13.3.1 Regulations in Europe and the USA

According to EU Directive No. 95/2/EC on 'Food additives other than colours and sweeteners', all the E-numbers of the gases applied in MAP need to be mentioned on the packaging label. Additionally according to the European regulation on 'Principles and requirements of food law' including traceability requirements (European Union regulation (EC) No 178/2002), the product should be traceable through all stages of production, processing and distribution. In the USA, the HACCP requirements are described in the FDA Food Code 2005 – Annex 6. All manufacturers producing products with reduced oxygen content need to set up critical control points (CCPs) regarding both the gas content and the seal integrity.

13.3.2 Regulations in Asia and Eurasia

Under the Australia New Zealand Food Standards Code, packaging must not cause food to be unsafe or tainted. However, there are no regulations regarding which foods can be stored in MAP. It is up to the manufacturer to protect the customer from illness, and provide storage instructions and date markings on how to store and handle the product safely.

13.3.3 Economic aspects and market issues

The fresh-cut produce market is a multi-billion dollar growth market bolstered by consumers' interest in fresh natural products, convenience and nutrition. Based on available data, the shares of fresh cut sales in Europe, the USA and Australia are 18%, 9% and 5%, respectively (Premier *et al.*, 2007). This trend is still increasing in different countries and, in Europe, the UK, France and Italy are considered to be the market leaders. The little or no need for chemical preservatives in MAP also provides for additional cost savings for the fresh-cut industry. On the other hand, as fresh-cut produce is a non-processed, living product, preserving freshness and minimizing quality degradation associated with temperature challenges during transit is crucial. This requires capital investigation in cold storage and cold chain transportation system. Furthermore, there are added costs associated with equipment, gases, packaging and control systems. There needs to be optimal control over dose and packaging and hence accuracy and authenticity of MAP filling and packaging is of high priority. The increased volume of the packages also adds to the distribution costs. However, the enlargement of the distribution territory coupled with the extended shelf-life that MAP provides could offset these additional costs. The ability to employ MAP can also open up the possibility to offer a larger variety of products. However, it is imperative that to completely evaluate the sustainability of MAP, a number of factors to be considered and a rigorous Life Cycle Assessment (LCA) needs to be performed.

The large product and economic losses that normally occur from the 'farm to fork' can be attributed to poor handling, distribution, storage, and customers' purchase behavior. Losses at almost every stage of the food chain may be reduced by using appropriate packaging. Packaging is an essential part of a long-term incremental development process to reduce losses, that will have to employ a blend of technologies and processes. As a result of modern urban habitation, retailers need to keep up with emerging issues such as how to deliver their goods with an improved level of quality at a fair price.

New packaging concepts are rapidly growing in the food packaging market. Packaging improves the product quality, freshness, increases the shelf-life of the product as well as providing convenience to the consumer and, ultimately, adds value to the product. It is mainly used to extend the shelf-life of fresh produce and perishable products. Under the influence of MAP, the product is held in a state similar to that of animal hibernation. MAP maintains this state for a long period of time, during which deterioration is effectively stopped or highly reduced. Additionally, the increasing fresh produce consumption in developed countries such as USA highlights the importance of packaging materials in order to distribute these perishable products (Table 13.1).

Table 13.1 Per capita (kg) consumption of raw fruits and vegetables in the USA, 1976–2009

Item	1976	2009	Growth 1976–2009
Fresh citrus fruits	28.5	20.7	−27.6%
Fresh non-citrus fruits	73.4	106.8	45.6%
All fresh fruits	101.9	127.5	25.1%
Fresh vegetables excluding potatoes	95.9	146.5	52.7%
Fresh potatoes	49.4	36.4	−26.2%
All fresh vegetables	145.3	182.9	25.9%
Fresh fruits and vegetables	247.2	310.3	25.5%

Source: Fruit and Tree Nut Situation and Outlook Report (USDA, 2009).

13.4 Novel trends in modified atmosphere packaging

13.4.1 Active packaging

Introduction of another 'assisting' agent or functionality to the typical MAP in order to control the internal environment of the package could be considered as 'active packaging'. Among the different functional agents applied in active packaging are O_2 scavengers, CO_2 emitters, desiccants, ethylene scavengers, antimicrobial and antioxidative emitters, etc., alone or in combination. These agents can be applied in different ways, namely as small sachets in the package or by co-extrusion in the packaging material.

One of the hot topics in this category is the application of antimicrobials in packaging. Undesirable microbial growth on fruit and vegetable surfaces can be actively controlled by the incorporation of natural antimicrobial substances into packaging material. The mode of action in these antimicrobial films is by releasing antimicrobials to the food. However, accurate dosing or control of this function is of high importance in order to avoid any potential safety risk to the consumers. Some of the major compounds proposed for incorporation in antimicrobial food packaging include sorbate, propionate and benzoate, nisin, pediocin, lysozyme, benomyl, etc. The application of ethanol vapor emitters as sachets has been evaluated to extend the shelf-life in apples. However, because of the high cost of the sachets, their application is still limited (Smith *et al.*, 1995). On the other hand, the limited application of antimicrobials by some religious (e.g. Halal foods) and ethnic groups should not be underestimated. When selecting an antimicrobial compound, its compatibility with the packaging material and its durability during the extrusion conditions should be taken into consideration. This is because some compounds could lose their activity during incorporation into the packaging materials. However, it is worth mentioning that many of the incorporated antimicrobials that have been reported in the literature have not yet been permitted for application in foods.

13.4.2 Smart packaging

The introduction of 'indicators' and 'sensors' in packages, which enables the end user to trace the current state and past conditions of a product, is regarded as smart packaging. The recorded information provides everyone involved in the farm to fork chain with necessary information to make an informed decision about the product quality. Under this type of packaging, different types of time-temperature indicators (TTIs), microbial growth biosensors, gas sensors, breakage indicators, ripeness indicators and radio frequency identification (RFID) tags can be included.

13.4.3 Edible coatings

Edible coverings, consisting of natural food ingredients, can control moisture and gas transfer between the product and the environment of the package, thereby improving or maintaining safety and product quality. It is believed that these edible coatings themselves can create modified atmosphere conditions within the product. However, when deciding on the different edible coatings, the compatibility of the coatings and the food in question should be carefully considered as some coatings are not completely tasteless. Also, these coatings are not uniform in layering and are not totally resistant against physical damage. Some examples of this group include antimicrobials, antibrowning agents and texture enhancers.

13.4.4 Hurdle technology

This field can be extended to application of a variety of different hurdles to bolster MAP in extending shelf-life. Application of organic acids such as ascorbic acid, which not only minimize enzymatic browning, but also act as a bactericide has been repeatedly proved to be successful.

13.5 Troubleshooting approaches

Because of the limited shelf-life of fresh cuts, any probable failure in practice is generally irrecoverable. Also, implementation of all hygienic regulations in processing may not be completely sufficient. Maintaining constant refrigeration storage temperature is a key factor in fresh cut success. Any fluctuations in the storage temperature will result in alteration of the situation in MAP, and will accelerate the respiration rate. Therefore, packaging, integrity of the seal and cold storage are of high importance.

Apart from the safety issues there are some visual aspects that may affect consumer perception about the safety of the fresh-cut packages. One of these visual off-sights is foggy appearance of the packaging materials, which prevents the consumer from seeing into the package clearly. Therefore plastic packages need to be clear and the content should be visible. New packaging materials can be co-extruded with antifog compounds that prevent large water drops from forming. This results in a more attractive package and a better view of the product.

13.6 Future trends

As MAP by itself cannot inactivate pre-existing microorganisms, the beneficial effects of different non-thermal technologies such as high pressure processing (HPP), electron beam irradiation, etc., which, have been proved to be promising in other fields, should be studied in combination with MAP. The application of these technologies can be studied not only for microbial inactivation but also for enzymatic inactivation such as polyphenol oxidase, the main enzyme responsible for enzymatic browning in fresh cuts.

Apart from combining MAP with other technologies, there are still significant limitations in packaging materials, which can be possibly overcome by combining nanotechnology into packaging material production.

The effect of natural antimicrobial agents customized for each individual food product and their mechanisms of action on target microorganisms under MAP conditions should also be studied in detail.

As one of the dominant quality characteristics in fresh cuts is their firmness, studying the effect of different texturizing agents on the retention of texture firmness need to be studied as well.

Another issue that has been less investigated is the effect and quality of cutting in fresh cuts. As mentioned earlier, one of the main reasons for enzymatic browning is the texture crush in tissue caused by dull cutters and knives. Therefore, cutting practices need to be improved or customized specifically for MAP products. Application of novel cutting instruments, such as ultrasonic cutting, can be successful in achieving precise and sharp cuts.

The other aspects that require further investigation are the effect of different relative humidity levels on the efficiency of MAP in fresh cuts and different edible coatings appropriate for each food product. Of course a great deal of work has already been done in these fields. However, considering each fruit and vegetable has its own specific requirements, highlights the importance of more research in this domain.

Although there is no thermal processing step within the scope of MAP, there is still some assumption regarding sorption of food flavors by polymeric packaging materials that can result in loss of flavor and taste. Therefore studying the flavor profile of the atmosphere inside the packages and on the surface of the packaging material can clarify if there is the incidence of any flavor sorption. However, the flavor sorption of some unwanted odors or flavors such as those responsible for grapefruit bitterness may actually be favorable. More research is needed to understand how to either enhance or retard flavor sorption within MAP conditions.

13.7 Questions for discussion

1. How might the fresh cut size affect its respiration rate?
2. In what products is CO suggested as a gas to be included in MAP?
3. What are the CCPs in a fresh produce processing unit?
4. What hurdles can be applied to MAP in order to improve its effectiveness?
5. What antibrowning agents can be considered?
6. How can the storage humidity affect the function of MAP?
7. How can the material used in knives applied for cutting fresh produce affect the visual appearance of the fresh cut?
8. What other processing technologies can be jointly applied with MAP in order to extend product shelf-life?
9. What are the common O_2-scavengers, and by which mechanism do they reduce oxygen in the packages?
10. In what fruits or vegetables is the application of ethylene scavengers not recommended? Why?

13.8 Sources of further information and advice

- *Innovations in food packaging*. J. Han, 2005. Academic Press.
- *Modified atmosphere packaging for fresh-cut fruits and vegetables*. A. L. Brody, H. Zhuang and J. H. Han, 2011. Wiley-Blackwell.

- *Principles and applications of modified atmosphere packaging of foods*. B. A. Blakistone, 1999. Aspen Publication.

13.9 References

PREMIER, R., J. JEAGER, and B. TOMKINS. 2007. Microbiological quality consideration for the ready-to-eat fresh produce industry. *Acta Hort.* 746: 25–32.

SMITH, J. P., J. HOSHINO, and Y. ABE. 1995. Interactive packaging involving sachet technology. In M. L. Rooney (ed.) *Active Food Packaging*, London, Blackie Academic and Professional, pp. 143–173.

SOLOMON H. M., D. A. KAUTTER, T. LILLY, and E. J. RHODEHAMEL. 1990. Outgrowth of *Clostridium botulinum* in shredded cabbage at room temperature under modified atmosphere. *J. Food Protect.* 53 (10): 831–3.

14

Biocontrol of *Listeria monocytogenes* on fresh produce

K. Jordan and A. Casey, Teagasc Food Research Centre, Ireland, A. Hoehl, BOKU – University of Natural Resources and Life Sciences, Austria and G. Bruggeman, Nutrition Sciences N. V., Belgium

DOI: 10.1533/9781782420279.3.187

Abstract: Outbreaks of listeriosis linked with fresh produce are rare. However, the recent food poisoning outbreaks associated with fresh produce – *Escherichia coli* O157:H7 linked to sprouts in Germany, but more particularly that linked to *Listeria monocytogenes* in Colorado – have demonstrated the need for vigilance with respect to *L. monocytogenes* in fresh produce. Despite improved hygiene management systems and increased regulation, *L. monocytogenes* remains a food safety risk. This review focuses on the potential opportunities presented by using bacteriocins and bacteriophages as biocontrol agents to reduce contamination of fresh produce with *L. monocytogenes*, and discusses the challenges faced in promoting more widespread use of these antimicrobial agents.

Key words: *Listeria monocytogenes*, fresh produce, biocontrol, bacteriophages, bacteriocin.

14.1 Introduction

The beneficial effect of fruits and vegetables for the prevention of chronic diseases is usually attributed to healthy plant components like vitamins, minerals, fibre and phytochemicals (Sánchez-Moreno *et al.*, 2012; Murphy *et al.*, 2011). Bioactive compounds like carotenoids, phenolics (e.g. flavonoids such as anthocyanins) or organosulfides have been repeatedly linked with a lower risk of cardiovascular disease, stroke, cancer or type 2 diabetes mellitus (Murphy *et al.*, 2011; Sánchez-Moreno *et al.*, 2012; Stanner *et al.*, 2004; Carter *et al.*, 2010; Hung *et al.*, 2004; Crowe *et al.*, 2011; Liu 2003; Stan *et al.*, 2008). Many of the health-related benefits of plant food have been proven, but the mechanisms of action are not fully understood, although it has been proposed that the combination of antioxidants and phytochemicals in the plant food counteract negative effects that are exerted by free radicals (Carter *et al.*, 2010; Stanner *et al.*, 2004).

Authorities and professional organizations worldwide have responded to this link between nutrition and chronic disease by issuing guidelines and launching campaigns to promote a healthy diet in general, and to strengthen the importance of the consumption of high amounts of fruits and vegetables. The World Health Organization (WHO) recommends a population dietary intake of at least 400 g fruits and vegetables per day (Anon, 2003a). Many countries promote campaigns of increased fruit and vegetable consumption under the umbrella term of a '5 a day program', and, although differing in detail and slogans, the key message of all these initiatives is to consume fruits and vegetables several times a day because of their health benefits (Anon, 2003b). In addition to healthier diets, convenience and pleasure are two other important consumer trends likely to affect the fresh produce market (Anon, 2009). Advanced modified atmosphere and antimicrobial packaging, UV treatment, ionizing irradiation, hurdle concepts, and an effective cold chain have resulted in improvements in shelf-life extension of fresh produce leading to increased volumes of international trade (Clemens 2004; Neff *et al.*, 2012; Anon, 2009).

On the other hand, fresh fruits and vegetables are increasingly responsible for many cases of foodborne illness. For example, between 1990 and 2003, there were at least 554 foodborne outbreaks associated with vegetables, and these outbreaks resulted in approximately 28 000 illnesses and several deaths (Dewaal *et al.*, 2006). The recent outbreaks associated with *Listeria monocytogenes* in cantaloupes and of *Escherichia coli* linked to bean sprouts produced on a German organic farm, highlight the need to develop better interventions to control the presence of foodborne pathogens in foods that are consumed raw.

The increasing consumption of fresh produce in combination with improved surveillance methods and/or systems, fast and international distribution channels, the increase in the amount of vulnerable people in the population, the accumulation of human pathogens in the environment as a consequence of intense animal production and changes in manure and sewage treatment, all contribute to raise concerns about the safety of fresh produce (Warriner, 2005). Consequently, this issue is on the agendas of major international organizations like the Codex Alimentarius Commission (CAC), Food and Agriculture Organization of the United Nations (FAO) and WHO, who are increasingly aware of this problem (Doyle and Erickson, 2008; Heaton and Jones, 2008; Critzer and Doyle, 2010; Berger *et al.*, 2010; Teplitski *et al.*, 2009; Anon, 2008).

The pathogenic bacteria most often involved in foodborne illness arising from fresh produce include *L. monocytogenes*, *Salmonella* spp. *E. coli* O157:H7, *Shigella* spp. and *Campylobacter* spp. Additionally, certain viruses are frequently associated with foodborne disease outbreaks (Warriner, 2005; Zhao, 2005; Anon, 2008; Anon, 2011). Leafy greens, berries, green onions, melons, sprouted seeds and tomatoes are product categories often linked with bacterial foodborne outbreaks (Anon, 2008; D'Lima and Suslow, 2009). Similarly, fresh produce categories often involved in viral outbreaks include salad crops (lettuce, green onions, other greens) and soft fruits like raspberries or strawberries (Bosch *et al.*, 2011; Rzeżutka and Cook, 2009; Croci *et al.*, 2008; Koopmans and Duixer, 2004).

Sources for contamination with pathogenic bacteria are varied and can be found along the whole production chain. At the preharvest level, irrigation water, run-off water from livestock operations, contaminated manure, manure compost, sewage sludge and soil can lead to contamination. Furthermore, pathogens can be transmitted on fresh produce via vectors such as animals and humans (Warriner, 2005; Zhao, 2005; Anon, 2008; Doyle and Erickson, 2008).

At the harvest and postharvest levels, pathogens can be transferred to fresh produce products via harvesters, washing/process water, humans, animals, process equipment and transport containers (Anon, 2008; De Roever, 1998; Warriner, 2005).

Although there are hygiene management systems in place to deal with foodborne pathogens like *L. monocytogenes*, persistence of *L. monocytogenes* in processing facilities can still be a challenge. New methods for control of *L. monocytogenes* are needed and biocontrol methods like using bacteriophages and bacteriocins offer opportunities to make advances in this area.

14.2 Outbreaks and control of *Listeria monocytogenes* on fresh produce

14.2.1 Occurrence of *Listeria monocytogenes* on fresh produce

L. monocytogenes is almost ubiquitous in the environment and is frequently isolated from environments such as soil and water (Fox *et al.*, 2009; Sauders *et al.*, 2012). Fresh produce can become contaminated from the soil or water or from manure used as fertilizer. Animals can carry the bacteria (without appearing ill) and excrete it in their faeces. Once *L. monocytogenes* contaminates food it can, in some cases, survive processing steps such as chlorine washes (Beuchat and Brackett, 1991), and therefore the food can pose a threat to public health. Modified atmosphere packaging, for example, does not inhibit growth completely (Francis and O'Beirne, 2006; Steinbruegge and Maxcy, 1988; Berrang *et al.*, 1989).

L. monocytogenes has been detected on a wide variety of fresh produce. However, sample sizes of surveys are often too small to be of statistical relevance for predictive modelling studies. This is especially the case as the probability of detection is low. There has also been little consistency in sample collection, laboratory test methods, data analysis methods, etc. Thus, results from these studies cannot be easily compared. Considering these limitations, Crépet *et al.* (2007) calculated the probabilities of fresh unprocessed and minimally processed vegetables being contaminated, based on data from 165 prevalence studies of *L. monocytogenes* in fresh vegetables (25 078 samples). Their results showed that the probabilities of contamination with concentrations higher than 10, 100 or 1000 viable *L. monocytogenes* organisms/g were 1.44, 0.63 and 0.17% respectively, indicating that there is approximately a 1.44% chance of fresh produce being contaminated with 10 cells/g. As the potential number of contaminants increased, the probability of observing those numbers decreased.

For *L. monocytogenes*, one of the key factors is the ability of the organism to grow on food. Many foods do not support the growth of *L. monocytogenes* and, for these, contamination at a low level (<100 cfu/g) is tolerated in some jurisdictions. Even at refrigerated temperatures, some foods can support growth of *L. monocytogenes* (Beuchat and Brackett, 1991; Castillejo Rodríguez *et al.*, 2000; Berrang *et al.*, 1989; Farber *et al.*, 1998). For these, detection of any number of organisms is a risk. The growth rate on these foods will depend on a number of factors, including temperature, strain, numbers, packaging atmosphere and intrinsic factors of the food such as pH. As each product will have different intrinsic factors, and some may have antibacterial properties, it is necessary to determine the ability of each food to support growth of *L. monocytogenes*. Challenge tests (such as at: http://ec.europa.eu/food/food/biosafety/salmonella/docs/shelflife_listeria_monocytogenes_en.pdf) are required for this and the factors to be considered include temperature, pH, atmosphere, inoculum size, strain used and intrinsic properties of the food.

14.2.2 Outbreaks of listeriosis associated with fresh produce

Although *L. monocytogenes* causes relatively mild gastroenteritis in healthy adults, it can be severe in susceptible individuals including pregnant women, neonates and immunocompromised individuals. The main source of *Listeria* infection is through consumption of contaminated food (Schlech *et al.*, 1983), primarily (although not exclusively) ready-to-eat and minimally processed foods. The infective dose has been estimated at about between 100 and 10^6 cells, depending on the immunological status of the host (Bhunia, 2008). Sukhaedo and Trinad (2009) also concluded that the rate of infection is affected by bacterial virulence and underlying host defences.

Although fresh produce is not regularly associated with listeriosis, the Colorado outbreak in 2011 (Anon, 2012) demonstrated the risks associated with fresh produce, and the need for continued vigilance. The outbreak resulted in 147 illnesses in 28 US states and in 33 deaths. Five different strains of *L. monocytogenes* were identified during the outbreak. It has been suggested that the grazing of sheep on the fields used by Jensen Farms to grow cantaloupes may have been the source of the Listeria, but this has not been confirmed. Water contamination and farming practices are also being investigated as possible sources.

Consequences of an outbreak

Listeriosis accounts for only 0.02% of gastroenteritis, but accounts for about 25% of deaths as a result of gastroenteritis (Ross *et al.*, 2000). Although the number of listeriosis cases is relatively low compared with other foodborne pathogens, the mortality rate is extremely high 20–30% (Scallan *et al.*, 2011). Various efforts have been made to estimate the cost of cases of listeriosis in the USA (Mead *et al.*, 1999; Scallan *et al.*, 2011). These costs include:

- *Human cost:* Once there are fatalities involved, the human costs are obvious. In addition to fatalities, the human costs include hospitalization, loss of earnings and potential ongoing medical expenses.
- *Litigation:* In the USA in particular, litigation in the case of a foodborne illness or death is becoming the norm. There are specialized law firms and lawyers promoting such litigation (http://www.pritzkerlaw.com/listeria/alfalfa-sprouts-listeria-lawyer.html). Such claims are made against the growers, processors, distributors, restaurants or other eating establishments that may be implicated in the case. Compensation can include claims for medical expenses, pain and suffering (includes physical pain, suffering, emotional distress and disability), loss of income, loss of potential earnings. In the case of death, funeral expenses, medical expenses, potential loss of earnings and others are included in the compensation claim.
- *Industry cost:* There is an onus on industry to protect the consumer from *L. monocytogenes* (Anon, 2005, 2007). This responsibility for consumer protection entails considerable costs on industry – implementation of hygiene controls, testing (processing environment and product samples). In addition to the financial costs, there are also the costs of reputational damage that can in some cases lead to closure.
- *Product recalls: L. monocytogenes* contamination of a product can be very low and not evenly distributed on the food. As some foods can support growth of *L. monocytogenes*, and they can grow at refrigeration temperatures, it is possible that contaminated food can be released onto the market. When such contamination is detected after product release, then a product recall, either voluntary or compulsory, is required. Such recalls impose considerable financial burden on the manufacturer, in addition to reputational damage.
- *Regulatory costs:* Regulatory authorities are a central part of protecting the consumer against *L. monocytogenes*. Each regulatory authority will have specific procedures in place to protect public health. This includes designing risk assessments, analyzing samples in outbreak and recall situations, in addition to responsibility for interacting with the public. There are considerable costs involved in putting these procedures in place, and keeping vigilance on their implementation.

14.2.3 Strategies already in place to control *Listeria*

As *L. monocytogenes* is ubiquitous in the environment, it is unrealistic to expect its complete absence in the processing environment. Therefore, its control in the processing environment is key to prevention of food contamination. There are strategies in place for the control of *L. monocytogenes*, these include a hazard analysis critical control point plan (HACCP plan), good manufacturing practice (GMP) and British Retail Consortium (BRC (or other)) certification. Such strategies are supported by microbiological criteria, performance criteria and performance objectives, etc. Despite these strategies, *L. monocytogenes* can contaminate a processing environment and become persistent, being regularly isolated despite apparent adequate attention to cleaning strategies. The mechanisms by which such persistence occurs have not been clarified, but may be due to opportunistic survival (Carpentier and Cerf, 2011), genetic adaptability (Fox *et al.*, 2011) or biofilm formation (Lemon *et al.*, 2007). In a processing environment, *L. monocytogenes* can adhere to surfaces and consequently form biofilm. This can constitute a protected mode of growth that allows it to survive in hostile environments.

In food processing, *Listeria* biofilms may be a source of recurrent contamination. Biofilms are difficult to eradicate because of their resistant phenotype. Conventional cleaning and disinfection regimes may even contribute to inefficient biofilm control and the dissemination of resistant strains around a facility. Consequently, in order to reduce the risk associated with *L. monocytogenes*, new control strategies are needed. Using biological methods to control *L. monocytogenes* (biocontrol) offers opportunities for application of novel technologies to impact positively on food safety and to minimize a public health risk.

14.3 Opportunities for biocontrol

In the 1980s the foodborne nature of listeriosis was realized. In the intervening years, the incidence of listeriosis has not been eliminated. There are still large outbreaks of listeriosis that are epidemiologically linked to food contamination. This, along with the current consumer demands for minimal processing and lack of chemical preservatives, presents the food processor with a dilemma. Food safety and hygiene strategies have not been successful at eliminating problems with *L. monocytogenes*. Yet, food processors are responsible for producing safe food. Biocontrol (using biological agents to control *L. monocytogenes*) offers opportunities to develop new strategies to solve this dilemma.

14.3.1 Bacteriocins

Bacteriocins are ribosomally synthesized peptides that are pore-forming agents, which act by disrupting the integrity of the target cell membrane leading to cell death. Therefore, they have the potential to act as a control mechanism to inhibit other bacteria, including pathogens. Lactic acid bacteria (LAB), in particular, are known to produce a variety of bacteriocins which possess characteristics that make them ideal for use in pathogen inactivation. These include their broad antimicrobial spectrum, their pH- and heat-tolerance, and the fact that they are non-toxic. In addition, they are inactivated by host proteases, resulting in little or no effect on the host gut microflora (Gálvez *et al.*, 2007). Bacteriocins may have several different modes of action, including inhibition of synthesis of the cell wall, inhibition of both DNAse and RNAse activity and, more commonly, through formation of pores in the membrane of the target cell (Cleveland *et al.*, 2001).

Lactic acid bacteria are ideal candidates for bacteriocin production in biocontrol as they also have a long history of safe use (Bourdichon *et al.*, 2012). They are also an integral part of the microflora in the gut of humans and animals. The best studied bacteriocin which exhibits anti-listerial activity is nisin; a class-I bacteriocin produced by the LAB, *Lactococcus lactis* (Rogers, 1928). Very little research has been undertaken to show the efficacy of nisin (or other bacteriocins) against *L. monocytogenes* on fresh produce. Most research has focused on higher risk pathogens such as *E. coli* O157:H7 that are more frequently associated with disease outbreaks. Some studies have focused on combination treatments, for example nisin in combination with acids and EDTA (Bari *et al.*, 2005).

Other bacteriocins, such as pediocin and enterocin, which are not currently used commercially, have been shown to have potential for application as food preservatives (Table 14.1). Pediocin is a class-IIa bacteriocin, which differs from nisin in that its host-range is considerably narrower; however, it is highly specific in its activity against *L. monocytogenes*. Enterocin AS–48 has had a similar effect on *L. monocytogenes* growth on vegetables (Cobo Molinos *et al.*, 2005). In that study, an immersion treatment with 25 μg/ml of the enterocin AS–48 resulted in a significant reduction of *L. monocytogenes* counts of up to 2.3 log cfu/ml when compared with an untreated control. Mundticin, a bacteriocin produced by *Enterococcus mundtii*, was found to have potential in biocontrol of modified atmosphere packaged and stored mungbean sprouts when used in a washing step or as a coating (Bennik *et al.*, 1999). BACTIBASE is a specific online database dedicated to bacteriocins (http://bactibase.pfba-lab-tun.org/main.php). Although the producer and target organisms are listed, there is little information relating to food substrates or efficacy in food systems.

Table 14.1 Studies on the efficacy of bacteriocins in reducing *Listeria monocytogenes* in fresh produce

Bacteriocin	Food studied	Result	Reference
LAB supernatant	Lettuce	Decreased viable pathogenic cells by 1.2–1.6 log units immediately after treatment	Allende *et al.* (2007)
LAB	Lettuce, cut apples	Growth of *L. monocytogenes* completely inhibited	Trias *et al.* (2008)
Enterocin AS–48	Alfalfa, soybean sprouts	Reduced viable counts by approx 2.7 log cfu/g upon application	Molinos *et al.* (2005)
Nisin	Cantaloupe melon	Reduced pathogen by 4 log cfu/cm^2 over seven-day period	Ukuku *et al.* (2005)
Pediocin and nisin	Cabbage, broccoli, mung beans	2.20- to 4.35-log reductions of *L. monocytogenes*	Bari *et al.* (2005)
Mundticin	Endives	Growth of *L. monocytogenes* inhibited on sterile vegetable-medium but not on fresh produce	Bennik *et al.* (1999)

LAB, lactic acid bacteria.

To date, insufficient data have been generated to get a complete picture of the potential use for many bacteriocins. The current regulatory situation dictates against the use of bacteriocins as biocontrol agents, as in many cases there are currently insufficient supporting data to assure the regulatory authorities. Nisin, discovered in the 1940s, is the only bacteriocin to have been granted a licence for use in food to date (Gálvez *et al.*, 2007). Considering that the efficacy of bacteriocins is dependent largely on surrounding environmental factors, there is much further research required in this area in order to ascertain the optimal environmental conditions under which each particular bacteriocin can be applied.

14.3.2 Bacteriophages

In 1919, Felix d'Herelle discovered bacteriophages (d'Herelle and Smith, 1922) and was the first to propose their use as biocontrol agents. However, the fact that early clinical trials involving phage therapy were deemed unreliable, coupled with the discovery of antibiotics, has meant that interest in studying phages as biocontrol agents seemed unnecessary. It has only been in recent times, with the observed increase in antibiotic resistance in pathogens, that the focus of research has shifted back to development of phage-based treatments.

Bacteriophages are logical candidates for biocontrol of *L. monocytogenes* in fresh produce. They are numerically the most abundant organisms on Earth, outnumbering bacteria in many environments by a factor of 10 (Brüssow *et al.*, 2004). They exhibit a high degree of specificity towards their target host bacterium, and, as a result, are safe for use in food processing, considering they will have no detrimental effect on the microflora of the eventual consumer, nor will they have an effect on any other desired bacteria in the food. In addition, they are composed entirely of proteins and nucleic acid; their eventual breakdown will be equally as harmless to the host. Moreover, bacteriophages also possess other desirable attributes, including a relative stability during storage, and the ability to self-perpetuate (Carlton *et al.*, 2005). Of particular importance in terms of suitability for biocontrol of *L. monocytogenes* is finding a virulent phage that is strictly lytic (Fig. 14.1), rather than a lysogenic phage which can be genetically unstable. Lytic phages are genetically stable, will always kill infected cells, and cannot therefore integrate their genomes into those of the bacterial chromosomes to form lysogens (Guenther *et al.*, 2009). In addition, it is preferable to choose phages that are incapable of transduction, and thus incapable of facilitating phage-mediated horizontal gene transfer (Brüssow *et al.*, 2004). It is also of critical importance that the full genome sequence of such phages is known, and that any phage applied to food does not encode any virulence factors or toxins which may be harmful (Hagens and Loessner, 2010).

The consensus among microbiologists is that bacteriophages do not have any known adverse effects on humans, animals or the environment. For this reason, many scientists and food safety experts predict that bacteriophages could become a useful tool in the reduction of dangerous pathogens in the food chain. However, there are concerns that limited safety data testing has been undertaken, although

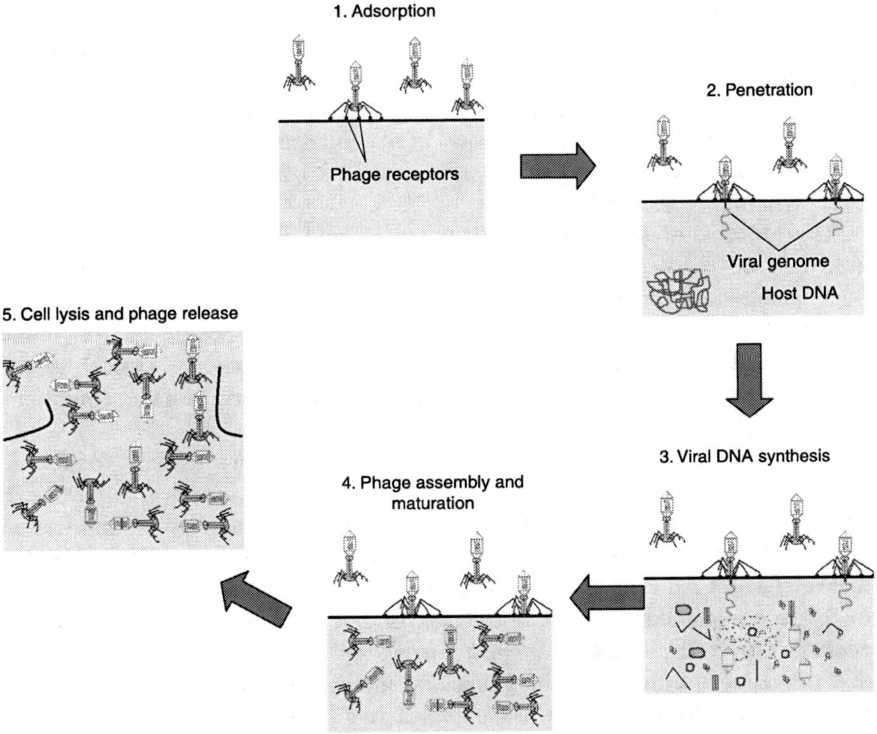

Fig. 14.1 Life cycle of a lytic bacteriophage. *Step 1:* bacteriophage initially adheres to the surface of the bacterial cell through attachment sites on the phage adsorbing to receptor sites on the host bacterium. *Step 2:* phage injects its genome into the bacterial cytoplasm. *Step 3:* phage then replicates its genome and uses the bacterium's metabolic machinery to synthesize phage enzymes and phage structural components. *Step 4:* The phage particles are then constructed and undergo maturation. *Step 5:* Phage-coded lysozyme breaks down the bacterial peptidoglycan causing osmotic lysis and release of the intact bacteriophages.

bacteriophages have been widely used for treatment of human diseases in the former Soviet Union (Chanishvili and Sharpe, 2010).

The renewed interest in use of bacteriophages as biocontrol agents has resulted in the development of several commercial products designed for this purpose. In 2006, the first such product, namely the LMP–102 phage preparation (now more commonly known as ListShield™), was granted GRAS (Generally Recognized As Safe) status by the FDA for use as a food additive in the control of *L. monocytogenes* in ready-to-eat foods (Lang, 2006). This preparation consists of six different phages, which have been shown to be an effective cocktail against 170 bacterial strains (Bren, 2007). Another phage preparation known as Listex™, was also granted GRAS status by the USDA's food safety and inspection service in 2006 for use on foods to control *L. monocytogenes* (Anon, 2006). This approval means that Listex™ no longer has to be listed as a food additive when used in the processing of food. In addition to this, a more recent alteration to the GRAS status of Listex™ identifies the phage product as a processing aid (Goodridge and Bisha, 2011). With regard to specificity for pathogenic *L. monocytogenes* serovars, the Listex™ P100 phage preparation has been demonstrated to be able to infect up to 95% of isolates belonging to the serovar groups 1/2a,b and 4b (Carlton *et al.*, 2005; Klumpp *et al.*, 2008).

The efficacy of the Listex™ P100 phage, and indeed all phage products used in food processing, depends on many extrinsic and intrinsic parameters. Such parameters include concentration of the phage applied, environmental conditions (such as the type of food, consistency, and specific matrix), ionic strength and pH of the bacterial surface (both of which are critical to the effectiveness at which phages bind to their ligands), and storage conditions (Guenther *et al.*, 2009).

A review of bacteriophage-based biocontrol strategies against foodborne pathogens (Goodridge and Bisha, 2011) summarized the applications of such strategies. More specifically, Leverentz *et al.* (2003, 2006) studied the application of lytic bacteriophages for biocontrol of *L. monocytogenes*. They showed that the phages reduced *L. monocytogenes* concentrations on honey dew melons by 2.0 to 4.6 logs compared with the control, but on fresh cut apples, the phage cocktail only reduced the *L. monocytogenes* by < 0.4 log units. It is possible that the reduced efficacy of the phages on the apple slices was due to the low pH (pH 4.4) on the cut surface of the apples. In combination with the bacteriocin nisin, a greater reduction in *L. monocytogenes* was seen (Leverentz *et al.*, 2003). A similar result was found by Dykes and Moorhead (2002), although in laboratory media, not in a practical application. In a further study with honey dew melons, Leverentz *et al.* (2006) showed that a cocktail of six phages resulted in a larger reduction in *L. monocytogenes* when the phages were applied at higher concentrations. This highlights the importance of phage dose during practical application.

A further study (Guenther *et al.*, 2009) demonstrated the efficacy of two broad host range phages (A511 and P100) for control of *L. monocytogenes* on lettuce leaves and cabbage. The vegetables were artificially contaminated with bacteria (10^3 cfu/g), followed by addition of phages at concentrations of (3×10^6 to 3×10^8 pfu/g), with storage at 6 °C for 6 days. The results indicated that the phages were able to reduce the concentrations of the *L. monocytogenes* strains by more than 2 logs in both the lettuce and cabbage.

These results demonstrate the effectiveness of using phages to control pathogens on fresh produce. In addition to the bacterial reductions, the studies are noteworthy because they show that phage treatments can control *L. monocytogenes* at refrigeration temperatures. However, further research on applications of bacteriophages, and combinations of bacteriophages and bacteriocins, is required. In particular, organic farming practices dictate that only natural antimicrobials be used in production, which makes phages an obvious choice as a biocontrol approach in such establishments.

Table 14.2 Reported applications of bacterial antagonists to improve the safety of fresh produce

Antagonists	Other hurdles	Target organisms	Tested food	Reference
Gluconobacter asaii	Bacteriophages	*L. monocytogenes*	Honeydew melon	Reynolds (2007)
Gluconobacter asaii		*L. monocytogenes*, *Salmonella* Poona	Apple plugs	Leverentz *et al.* (2006)
Native microflora (baby carrots)		*L. monocytogenes*, *Y. enterocolitica*, *Salmonella enterica*, *E. coli* O157:H7	Baby carrot homogenate, bell peppers	Liao (2007)
Native microflora		*L. monocytogenes*	Cantaloupe melon	Ukuku *et al.* (2005)
Lactobacillus casei LC34GF		*Staph. aureus*, *L. monocytogenes*, *A. hydrophila*, *E. coli*	Scarola salad leaves	Scolari and Vescovo (2004)
Pseudomonas fluorescens A3, Yeast D1		*S.* Chester, *E. coli*, *L. monocytogenes*	Green peppers	Liao and Fett (2001)
Pediococcus parvulus, *Enterococcus mundtii*		*L. monocytogenes*	Mungbean sprouts	Bennik *et al.* (1999)
Lactobacillus lactis, *Enterococcus faecium*		*L. monocytogenes*	Fresh-cut Caesar salad	Cai *et al.* (1997)
Native microflora		*L. monocytogenes*	Endive leaves	Carlin *et al.* (1996)
Lactobacillus casei, *Lactobacillus plantarum*, *Pediococcus* spp.		*Staph. aureus*, *L. monocytogenes*, *A. hydrophila*, *S.* Typhimurium	Vegetable salads, vegetable juice	Vescovo *et al.* (1996)

14.3.3 Antagonistic cultures

Many different cultures are used to inhibit pathogen growth on fresh produce (Table 14.2). In general, the efficacy of the antagonism has been studied, but not the mode of action. This lack of characterization of the compounds involved means that these antagonistic cultures are further from practical application than other, more characterized mechanisms of inhibition like bacteriocins and bacteriophages.

14.4 Conclusion

Despite strict regulations and improvements in hygiene management systems, *L. monocytogenes* remains a public health threat. Survival strategies and stress response mechanisms of the organism (that are not fully understood) enable it to persist in some processing environments where it can survive the normal cleaning regimes and potentially contaminate food. Although listeriosis outbreaks associated with fresh produce are rare, the recent 2011 outbreak associated with cantaloupe melons has demonstrated that vigilance is necessary. The application of novel technologies (including combination treatments) for biocontrol of *L. monocytogenes* on fresh produce has shown great potential. In addition to the anti-listeria potential, such biocontrol technologies do not damage the produce physically or organoleptically. The greatest concern about their use seems to be from a consumer perception. However, research to date has shown that there is no public health threat from use of fully characterized biocontrol technologies. As the threat of *L. monocytogenes* becomes greater, biocontrol technologies will become more attractive.

14.5 Acknowledgements

The authors acknowledge funding support from the EU-funded 7th Framework projects FOODSEG (project number 266061) and PROMISE (project number 265877).

14.6 References

ALLENDE, A., B. MARTÍNEZ, V. SELMA, M. I. GIL, J. E. SUÁREZ, *et al.* (2007). 'Growth and bacteriocin production by lactic acid bacteria in vegetable broth and their effectiveness at reducing Listeria monocytogenes in vitro and in fresh-cut lettuce.' *Food Microbiology* 24(7–8): 759–66.

ANONYMOUS (2003a). World Health Organisation/Food and Agriculture Organisation of the United Nations. Diet, nutrition and the prevention of chronic diseases Report of the joint WHO/FAO expert consultation – Global and regional food consumption patterns and trends. *WHO Technical report series, No. 916 (TRS 916)*. Available at: http://www.who.int/dietphysicalactivity/publications/trs916/en/gsfao_global.pdf [Accessed 2 December 2012].

ANONYMOUS (2003b). World Health Organization. Fruit and vegetable promotion initiative – A meeting report. Available at: http://www.who.int/dietphysicalactivity/publications/f&v_promotion_initiative_report.pdf [Accessed 2 October 2012].

ANONYMOUS (2005). Commission Regulation (EC) No 2073/2005 of 15 November 2005 on microbiological criteria for foodstuffs. Official Journal of the European Union. 2005; L 338: 1–26.

ANONYMOUS (2006). Food and Drug Administration. 'Submission of GRAS Notification of Bacteriophage P100 for Use in Food; Generally.' http://www.accessdata.fda.gov/scripts/fcn/gras_notices/701456A.PDF.

ANONYMOUS (2007). Commission Regulation (EC) No 1441/2007 of 5 December 2007 amending Regulation (EC) No 2073/2005 on microbiological criteria for foodstuffs. Official Journal of the European Union. 2007; L 322: 12–29.

ANONYMOUS (2008). Food and Agriculture Organisation of the United States/World Health Organisation. 'Microbiological hazards in fresh leafy vegetables and herbs.' Meeting Report Microbiological Risk Assessment Series 14.

ANONYMOUS (2010). International Markets Bureau, Consumer Trend Report. Available at: http://www.google.ie/url?sa=t&rct=j&q=&esrc=s&frm=1&source=web&cd=2&ved=0CDUQFjAB&url=http%3A%2F%2Fwww.gov.mb.ca%2Fagriculture%2Fstatistics%2Ffood%2Fconsumer_trends_convenience_en.pdf&ei=nB_wUbLKCKyh7AaKo4GwDA&usg=AFQjCNHZp-OCmz8mbbLTp3jQtS9TsSsrww&sig2=KLiVzLosw9ylA60kdxE77w. [Accessed 24 July 2013].

ANONYMOUS (2011). European Food Safety Authority. Scientific Opinion on an update on the present knowledge on the occurrence and control of foodborne viruses 1. *EFSA Journal*, 9(7), pp. 1–96.

ANONYMOUS (2012). Centers for disease control and prevention – Heart Disease Facts. Available at: http://www.cdc.gov/heartdisease/facts.htm [Accessed 2 October 2012].

BARI, M. L., D. O. UKUKU, T. KAWASAKI, Y. INATSU, K. ISSHIKI, *et al.* (2005). 'Combined Efficacy of Nisin and Pediocin with Sodium Lactate, Citric Acid, Phytic Acid, and Potassium Sorbate and EDTA in Reducing the *Listeria monocytogenes* Population of Inoculated Fresh-Cut Produce.' *Journal of Food Protection* 68(7): 1381–7.

BENNIK, M. H. J., W. VAN OVERBEEK, E. J. SMID, L. G. M. GORRIS (1999). 'Biopreservation in modified atmosphere stored mungbean sprouts: the use of vegetable-associated bacteriocinogenic lactic acid bacteria to control the growth of *Listeria monocytogenes*.' *Letters in Applied Microbiology* 28(3): 226–32.

BERGER, C. N., S. V. SODHA, R. K. SHAW, P. M. GRIFFIN, D. PINK, *et al.* (2010). 'Fresh fruit and vegetables as vehicles for the transmission of human pathogens.' *Environmental Microbiology* 12(9): 2385–97.

BERRANG, M. E., R. E. BRACKETT, L. R. BEUCHAT (1989). 'Growth of *Listeria monocytogenes* on fresh vegetables stored under controlled atmosphere.' *Journal of Food Protection* 52(10): 702–5.

BEUCHAT, L. R., R. E. BRACKETT (1991). 'Behavior of *Listeria monocytogenes* inoculated into raw tomatoes and processed tomato products.' *Applied and Environmental Microbiology* 57(5): 1367–71.

BHUNIA, A. K. (2008). *Foodborne microbial pathogens mechanisms and pathogenesis.* Springer, New York, USA

BOSCH, A., G. SÁNCHEZ, M. ABBASZADEGAN, A. CARDUCCI, S. GUIX, *et al.* (2011). 'Analytical Methods for Virus Detection in Water and Food.' *Food Analytical Methods* 4(1): 4–12.

BOURDICHON, F., S. CASAREGOLA, C. FARROKH, J. C. FRISVAD, M. L. GERDS, *et al.* (2012). 'Food fermentations: Microorganisms with technological beneficial use.' *International Journal of Food Microbiology* 154(3): 87–97.

BREN, L. (2007). 'Bacteria-eating virus approved as food additive.' *FDA Consumer* 41: 20–2.

BRÜSSOW, H., C. CANCHAYA, W.-D. HARDT (2004). 'Phages and the Evolution of Bacterial Pathogens: from Genomic Rearrangements to Lysogenic Conversion.' *Microbiology and Molecular Biology Reviews* 68(3): 560–602.

CAI, Y., L. K. NG, J. M. FARBER (1997). 'Isolation and characterization of nisin-producing Lactococcus lactis subsp. lactis from bean-sprouts.' *Journal of Applied Microbiology* 83(4): 499–507.

CARLIN, F., C. NGUYEN-THE, C. E. MORRIS (1996). 'Influence of Background Microflora on Listeria monocytogenes on Minimally Processed Fresh Broad-Leaved Endive (Cichorium endivia var. latifolia).' *Journal of Food Protection* 59(7): 698–703.

CARLTON, R. M., W. H. NOORDMAN, B. BISWAS, E. D. DE MEESTER, M. J. LOESSNER (2005). 'Bacteriophage P100 for control of *Listeria monocytogenes* in foods: Genome sequence, bioinformatic analyses, oral toxicity study, and application.' *Regulatory Toxicology and Pharmacology* 43(3): 301–12.

CARPENTIER, B., O. CERF (2011). 'Review – Persistence of *Listeria monocytogenes* in food industry equipment and premises.' *International Journal of Food Microbiology* 145(1): 1–8.

CARTER, P., L. J. GRAY, J. TROUGHTON, K. KHUNTI, M. J. DAVIES (2010). 'Fruit and vegetable intake and incidence of type 2 diabetes mellitus: systematic review and meta-analysis.' *BMJ (Clinical research ed.)* 341: c4229.

CASTILLEJO RODRÍGUEZ, A. M., E. BARCO ALCALÁ, R. M. GARCÍA GIMENO, G. ZURERA COSANO (2000). 'Growth modelling of *Listeria monocytogenes* in packaged fresh green asparagus.' *Food Microbiology* 17(4): 421–7.

CHANISHVILI, N., R. SHARPE (2010). '*A Literature Review of the Practical Application of Bacteriophage Research*'. Eliava Institute of Bacteriophage, Microbiology and Virology.

CLEMENS, R. (2004). 'The expanding U.S. market for fresh produce.' *Iowa Ag Review* 10(1): 8–9.

CLEVELAND, J., T. J. MONTVILLE, I. F. NES, M. L. CHIKINDAS (2001). 'Bacteriocins: safe, natural antimicrobials for food preservation.' *International Journal of Food Microbiology* 71(1): 1–20.

COBO MOLINOS, A., H. ABRIOUEL, N. BEN OMAR, E. VALDIVIA, R. LUCAS LÓPEZ, *et al.* (2005). 'Effect of Immersion Solutions Containing Enterocin AS–48 on *Listeria monocytogenes* in Vegetable Foods.' *Applied and Environmental Microbiology* 71(12): 7781–7.

CRÉPET, A., I. ALBERT, C. DERVIN, F. CARLIN (2007). *Estimation of Microbial Contamination of Food from Prevalence and Concentration Data: Application to Listeria monocytogenes in Fresh Vegetables*, American Society for Microbiology.

CRITZER, F. J., M. P. DOYLE (2010). 'Microbial ecology of foodborne pathogens associated with produce.' *Current Opinion in Biotechnology* 21(2): 125–30.

CROCI, L., E. DUBOIS, N. COOK, D. MEDICI, A. SCHULTZ, *et al.* (2008). 'Current Methods for Extraction and Concentration of Enteric Viruses from Fresh Fruit and Vegetables: Towards International Standards.' *Food Analytical Methods* 1(2): 73–84.

CROWE, F. L., A. W. RODDAM, T. J. KEY, P. N. APPLEBY, K. OVERVAD, *et al.* (2011). 'Fruit and vegetable intake and mortality from ischaemic heart disease: results from the European Prospective Investigation into Cancer and Nutrition (EPIC)-Heart study.' *European Heart Journal* 32(10): 1235–43.

D'HERELLE, F., G. H. SMITH (1922). *The Bacteriophage: Its Role in Immunity*, Williams and Wilkins Co. Waverly Press, Baltimore.

D'LIMA, C. B., T. V. SUSLOW (2009). 'Comparative Evaluation of Practical Functionality of Rapid Test Format Kits for Detection of Escherichia coli O157:H7 on Lettuce and Leafy Greens.' *Journal of Food Protection* 72(12): 2461–70.

DE ROEVER, C. (1998). 'Microbiological safety evaluations and recommendations on fresh produce.' *Food Control* 9(6): 321–47.

DEWAAL, C. S., G. HICKS, K. BARLOW, L. ALDERTON, L. VEGOSEN (2006). 'Foods associated with food-borne illness outbreaks from 1990 through 2003.' *Food Protection Trends* 26(7): 466–73.

DOYLE, M. P., M. C. ERICKSON (2008). 'Summer meeting 2007 – the problems with fresh produce: an overview.' *Journal of Applied Microbiology* 105(2): 317–30.

DYKES, G. A., S. M. MOORHEAD (2002). 'Combined antimicrobial effect of nisin and a listeriophage against *Listeria monocytogenes* in broth but not in buffer or on raw beef.' *International Journal of Food Microbiology* 73(1): 71–81.

FARBER, J. M., S. L. WANG, Y. CAI, S. ZHANG (1998). 'Changes in Populations of *Listeria monocytogenes* Inoculated on Packaged Fresh-Cut Vegetables.' *Journal of Food Protection* 61(2): 192–5.

FOX, E., T. O'MAHONY, M. CLANCY, R. DEMPSEY, M. O'BRIEN, *et al.* (2009). '*Listeria monocytogenes* in the Irish Dairy Farm Environment.' *Journal of Food Protection* 72(7): 1450–6.

FOX, E. M., N. LEONARD, K. JORDAN (2011). 'Molecular diversity of *Listeria monocytogenes* isolated from Irish dairy farms.' *Foodborne Pathogens and Disease* 8(5): 635–41.

FRANCIS, G. A., D. O'BEIRNE (2006). 'Isolation and Pulsed-Field Gel Electrophoresis Typing of *Listeria monocytogenes* from Modified Atmosphere Packaged Fresh-Cut Vegetables Collected in Ireland.' *Journal of Food Protection* 69(10): 2524–8.

GÁLVEZ, A., H. ABRIOUEL, R. L. LÓPEZ, N. B. OMAR (2007). 'Bacteriocin-based strategies for food biopreservation.' *International Journal of Food Microbiology* 120(1–2): 51–70.

GOODRIDGE, L. D., B. BISHA (2011). 'Phage-based biocontrol strategies to reduce foodborne pathogens in foods.' *Bacteriophage* 1(3): 130–7.

GUENTHER, S., D. HUWYLER, S. RICHARD, M. J. LOESSNER (2009). 'Virulent Bacteriophage for Efficient Biocontrol of *Listeria monocytogenes* in Ready-To-Eat Foods.' *Applied and Environmental Microbiology* 75(1): 93–100.

HAGENS, S., M. J. LOESSNER (2010). 'Bacteriophage for Biocontrol of Foodborne Pathogens: Calculations and Considerations.' *Current Pharmaceutical Biotechnology* 11(1): 58–68.

HEATON, J. C., K. JONES (2008). 'Microbial contamination of fruit and vegetables and the behaviour of enteropathogens in the phyllosphere: a review.' *Journal of Applied Microbiology* 104(3): 613–26.

HUNG, H.-C., K. J. JOSHIPURA, R. JIANG, F. B. HU, D. HUNTER, *et al.* (2004). 'Fruit and Vegetable Intake and Risk of Major Chronic Disease.' *Journal of the National Cancer Institute* 96(21): 1577–84.

KLUMPP, J., J. DORSCHT, R. LURZ, R. BIELMANN, M. WIELAND, *et al.* (2008). 'The Terminally Redundant, Nonpermuted Genome of Listeria Bacteriophage A511: a Model for the SPO1-Like Myoviruses of Gram-Positive Bacteria.' *Journal of Bacteriology* 190(17): 5753–65.

KOOPMANS, M., E. DUIXER (2004). 'Foodborne viruses: an emerging problem.' *International Journal of Food Microbiology* 90(1): 23–41.

LANG, L. H. (2006). 'FDA Approves Use of Bacteriophages to be Added to Meat and Poultry Products.' *Gastroenterology* 131(5): 1370.

LEMON, K. P., D. E. HIGGINS, R. KOLTER (2007). 'Flagellar Motility is Critical for *Listeria monocytogenes* Biofilm Formation.' *Journal of Bacteriology* 189(12): 4418–24.

LEVERENTZ, B., W. S. CONWAY, M. J. CAMP, W. J. JANISIEWICZ, T. ABULADZE, *et al.* (2003). 'Biocontrol of *Listeria monocytogenes* on Fresh-Cut Produce by Treatment with Lytic Bacteriophages and a Bacteriocin.' *Applied and Environmental Microbiology* 69(8): 4519–26.

LEVERENTZ, B., W. S. CONWAY, W. JANISIEWICZ, M. ABADIAS, C. P. KURTZMAN, *et al.* (2006). 'Biocontrol of the Food-Borne Pathogens *Listeria monocytogenes* and Salmonella enterica Serovar Poona on Fresh-Cut Apples with Naturally Occurring Bacterial and Yeast Antagonists.' *Applied and Environmental Microbiology* 72(2): 1135–40.

LIAO, C. H. (2007). 'Inhibition of Foodborne Pathogens by Native Microflora Recovered from Fresh Peeled Baby Carrot and Propagated in Cultures.' *Journal of Food Science* 72(4): M134–9.

LIAO, C. H., W. F. FETT (2001). 'Analysis of native microflora and selection of strains antagonistic to human pathogens on fresh produce.' *Journal of Food Protection* 64(8): 1110–15.

LIU, R. H. (2003). 'Health benefits of fruit and vegetables are from additive and synergistic combinations of phytochemicals.' *American Journal of Clinical Nutrition* 78(3): 517S–20S.

MEAD, P. S., L. SLUTSKER, V. DIETZ, L. F. MCCAIG, J. S. BRESEE, *et al.* (1999). 'Food-related illness and death in the United States.' *Emerging Infectious Diseases* 5(5): 607–25.

MOLINOS A. C., H. ABRIOUEL, N. BEN OMAR, E. VALDIVIA, R. L. LÓPEZ, *et al.* (2004). 'Effect of immersion solutions containing enterocin AS–48 on *Listeria monocytogenes* in vegetable foods.' *Applied and Environmental Microbiology* 71(12): 7781–7.

MURPHY, M. M., L. M. BARRAJ, D. HERMAN, X. BI, R. CHEATHAM, *et al.* (2011). 'Phytonutrient Intake by Adults in the United States in Relation to Fruit and Vegetable Consumption.' *Journal of the Academy of Nutrition and Dietetics* 112(2): 222–9.

NEFF, R. A., J. C. HARTLE, L. I. LAESTADIUS, K. DOLAN, A. C. ROSENTHAL, *et al.* (2012). 'A comparative study of allowable pesticide residue levels on produce in the United States.' *Global Health* 8(1): 2.

REYNOLDS, G. (2007). '*Bacteria effective against fresh cut fruit contamination.*' Retrieved 22 March 2007, from www.foodnavigator-usa.com/news/printNewsBis.asp?id=75142.

ROGERS, L. A. (1928). 'The Inhibiting Effect of Streptococcus Lactis on Lactobacillus Bulgaricus.' *Journal of Bacteriology* 16(5): 321–5.

ROSS, T., E. TODD and M. SMITH (2000). *Exposure assessment of Listeria monocytogenes in ready-to-eat foods*. Preliminary Report for Joint FAO/WHO Expert Consultation on Risk Assessment of Microbiological Hazards in Foods. Rome, Food and Agriculture Organization of the United Nations.

RZEŻUTKA, A., N. COOK (2009). Review of Currently Applied Methodologies used for Detection and Typing of Foodborne Viruses. In G. BARBOSA-CÁNOVAS *et al.*, eds. *Global Issues in Food Science and Technology*. San Diego, Academic Press: pp. 229–46.

SÁNCHEZ-MORENO, C., S. DE PASCUAL-TERESA, B. DE ANCOS, M. PILAR CANO (2012). Nutritional Quality of Fruits. *Handbook of Fruits and Fruit Processing*, Wiley-Blackwell: pp. 73–84.

SAUDERS, B. D., J. OVERDEVEST, E. FORTES, K. WINDHAM, Y. SCHUKKEN, *et al.* (2012). 'Diversity of Listeria species in urban and natural environments.' *Applied and Environmental Microbiology* 78(12): 4420–33.

SCALLAN, E., R. M. HOEKSTRA, F. J. ANGULO, R. V. TAUXE, M.-A. WIDDOWSON, *et al.* (2011). 'Foodborne Illness Acquired in the United States—Major Pathogens.' *Emerging Infectious Diseases* 17(1): 7–15.

SCHLECH, W. F., 3RD, P. M. LAVIGNE, R. A. BORTOLUSSI, A. C. ALLEN, E. V. HALDANE, *et al.* (1983). 'Epidemic listeriosis—evidence for transmission by food.' *New England Journal of Medicine* 308(4): 203–6.

SCOLARI, G., M. VESCOVO (2004). 'Microbial antagonism of Lactobacillus casei added to fresh vegetables.' *Italian Journal of Food Science* 16: 465–75.

STAN, S. D., S. KAR, G. D. STONER, S. V. SINGH (2008). 'Bioactive food components and cancer risk reduction.' *Journal of Cellular Biochemistry* 104(1): 339–56.

STANNER, S. A., J. HUGHES, C. N. KELLY, J. BUTTRISS (2004). 'A review of the epidemiological evidence for the "antioxidant hypothesis".' *Public Health Nutrition* 7(3): 407–22.

STEINBRUEGGE, E. G., R. B. MAXCY (1988). 'Nature and number of ground-beef microorganisms capable of growth at 25 degrees C but not at 32 degrees C.' *Journal of Food Protection* 51(3): 176–80.

SUKHAEDO, B., C. TRINAD (2009). 'Molecular mechanisms of bacterial infection via the gut.' *Current Topics in Microbiology and Immunology* 337: 173–95.

TEPLITSKI, M., J. D. BARAK, K. R. SCHNEIDER (2009). 'Human enteric pathogens in produce: un-answered ecological questions with direct implications for food safety.' *Current Opinion in Biotechnology* 20(2): 166–71.

TRIAS, R., E. BADOSA, E. MONTESINOS, L. BAÑERAS (2008). 'Bioprotective Leuconostoc strains against Listeria monocytogenes in fresh fruits and vegetables.' *International Journal of Food Microbiology* 127(1–2): 91–8.

UKUKU, D. O., M. L. BARI, S. KAWAMOTO, K. ISSHIKI (2005). 'Use of hydrogen peroxide in combination with nisin, sodium lactate and citric acid for reducing transfer of bacterial pathogens from whole melon surfaces to fresh-cut pieces.' *International Journal of Food Microbiology* 104(2): 225–33.

VESCOVO, M., S. TORRIANI, C. ORSI, F. MACCHIAROLO, G. SCOLARI (1996). 'Application of antimicrobial-producing lactic acid bacteria to control pathogens in ready-to-use vegetables.' *Journal of Applied Microbiology* 81(2): 113–19.

WARRINER, K. (2005). Pathogens in vegetables. *Woodhead Publishing in Food Science and Technology*. Cambridge, Woodhead Publishing Ltd: 3–43.

ZHAO, Y. (2005). Pathogens in fruit. In *Improving the safety of fresh fruit and vegetables*, W. JONGEN (ed.). Cambridge, Woodhead Publishing Ltd: pp. 44–88.

15

Commercial and novel solutions for fresh produce safety

A. Lee, Illinois Institute of Technology, USA

DOI: 10.1533/9781782420279.3.203

Abstract: Fresh produce has come under intense scrutiny as a result of several recent high profile outbreaks of foodborne illness involving both bacterial and viral agents. Pathogenic microorganisms on fresh produce are a particular concern because of minimal processing or absence of thermal treatments to the produce prior to human consumption. This chapter aims to address the development of sanitizers to minimize cross-contamination and their potential to inactivate pathogens in wash water. Similar applications of processing techniques or processing aids that may dislodge pathogens from the surfaces of produce or improve the efficacy of the sanitizer will be examined.

Key words: novel sanitizers, processing to minimize cross-contamination, inactivation of pathogens using sanitizers and processing.

15.1 Introduction

Fresh-cut produce, including leafy green vegetables, is a fast growing food category and an important component of a healthy diet. Fresh produce is a desirable food because the consumer perceives it as being healthy, tasty, convenient, and fresh, all of which are strong selling points to health-conscious and time-deprived consumers. For these reasons, the fresh-cut industry has experienced tremendous growth over the past 10 years by the increased consumption and sales of these products in supermarkets and restaurant menus, including quick serve establishments, around the USA and internationally. From 1994 to 2007, US supermarket salad sales increased from \$3.3 to \$15.5 billion (Cook, 2008). The continuing growth of the fresh produce market also means that the industry is facing new challenges such as the microbiological hazards associated with a product that is minimally processed without a kill step. These hazards, should they occur, potentially could be disastrous for the industry.

Until the mid-1990s, fresh produce was not considered to be a significant food matrix for the transmission of *Escherichia coli* O157:H7, but, since then, a number of significant outbreaks associated with *E. coli* O157:H7 have occurred on fresh-cut produce, most notably the 2006 outbreak associated with spinach. Other pathogens of concern include *Salmonella* spp., hepatitis A and noroviruses. In recent attribution data published by the US Centers for Disease Control and Prevention (US CDC), Painter *et al.* (2013) estimated the annual US foodborne illness, hospitalization and deaths attributed to 17 food commodities, and 46% of illnesses were attributed to produce, largely comprising norovirus. More illnesses were attributed to leafy greens (22%) than any other commodity examined in the study, and leafy greens were ranked second at 14% for hospitalization cases and fifth (6%) in cases leading to death. Such data further demonstrate the risk of fresh produce as a commodity and the need to have an effective prevention control that reduces or minimizes the risk of pathogens getting to consumers.

Pathogens can be introduced onto vegetables as a result of agricultural practices such as soil amendments and irrigation water used during growing of leafy green vegetables. Contamination of produce by pathogens may originate from animal wastes used as fertilizer. Washing of produce in the processing plant is designed to remove dirt, pesticide residues and microorganisms from the surface of the product, including exudates that may support microbial growth. The fresh-cut industry have widely used chlorine as a sanitizer to wash their products and to ensure the safety of their products. The efficacy of subsequent washing with chlorinated water to remove pathogens is questionable (Nichols *et al.*, 1971; Zhang and Farber, 1996) and many reports have indicated contradictory results in the efficacy of chlorine for inactivating pathogens and minimizing cross-contamination.

There has been a trend of replacing chlorine with alternative sanitizers because of concerns about its efficacy on the product and the environmental and health impacts associated with the formation of halogenated byproducts (Olmez and Kretzschmar, 2009). Commercial postharvest washing of fresh-cut leafy green vegetables is performed with water containing antimicrobial sanitizers in order to prevent cross-contamination of pathogenic microorganisms between contaminated leaves and uncontaminated leaves (FDA, 2008). A substantial body of research is available to suggest that although cross-contamination may be prevented by commercially available sanitizers, these are generally ineffective at reducing microbial counts from the surface of fresh-cut produce by more than 3-log (Beuchat *et al.*, 2004; Luo *et al.*, 2011; Beuchat and Ryu, 1997; Gil *et al.*, 2009; Delaquis *et al.*, 2002). Sanitizer effectiveness also tends to decrease in the presence of organic materials, such as those often found in the commercial wash tanks of fresh produce items (Poschetto *et al.*, 2007; Nou and Luo, 2010; Grove, unpublished, 2008). This has led to interest in the fresh produce industry for alternative sanitizers that are not affected dramatically by organic material and can achieve significant inactivation of pathogenic microorganisms that may be present.

Survival times for coliforms, bacterial pathogens and enteric viruses on most fresh produce are moisture- and temperature-dependent, and survival usually extends beyond the useful life of the product (Konowalchuk and Speirs, 1976). The presence of pathogens on produce is well documented (Garcia-Villanova Ruiz *et al.*, 1987; Kaneko *et al.*, 1999) with several reviews listing more than 30 kinds of vegetables from which pathogens have been isolated (Beuchat, 1996; Dreux *et al.*, 2007), along with a risk profile for microbial contamination of raw vegetables published in 2002 (ECSCF, 2002).

Human pathogens are generally not among the normal microbiota of fresh produce but rather they represent contamination of the primary production environment from human or animal sources. The microorganisms associated with fresh and fresh-cut produce are typically associated with primary production (Table 15.1) along with other microorganisms that may be acquired as a result of harvesting, packing and processing. These microorganisms are often associated with farm workers, harvesting, transportation equipment, and the production and harvesting environments. Many types of fresh produce support the growth of bacteria, including human pathogens, particularly at cut surfaces and therefore controlling microbial growth is critical for quality and safety. There are significant opportunities for cross-contamination to occur, particularly when water flumes are used during processing leading to the extensive spread of initial spot contamination. The microbial load on vegetables can be reduced to some degree (i.e. typically 1–2 logs) by washing and disinfection. However, this is generally restricted to microorganisms on the surface of the vegetable and not internalized microorganisms. Thus, care must be taken to ensure processes do not foster such uptake of microorganisms into the vegetable tissues. No chemical treatments can assure complete destruction of contaminating microorganisms on vegetable

Table 15.1 Some examples of microorganisms associated with produce-related outbreaks

Microorganism		Produce
Bacteria	*Clostridium botulinum*	Vegetable salad
	Escherichia coli O157:H7	Radish sprouts, apple juice, apple cider, iceberg lettuce, spinach
	Listeria monocytogenes	Leafy vegetables, lettuce
	Salmonella spp.	Coleslaw, onions, watermelon, cantaloupes, bean sprouts, alfalfa sprouts, root vegetables
	Shigella spp.	Iceberg lettuce, tossed salad
Parasites	*Cryptosporidium* spp.	Apple cider
	Cyclospora spp.	Raspberries
Viruses	Hepatitis A	Iceberg lettuce, green onions, raspberries, strawberries
	Noroviruses	Tossed salad, bell peppers

Source: compiled and adapted from Gil *et al.* (2009) and WHO (1998).

surfaces, and the primary purpose of antimicrobials or sanitizers added to wash or flume water is to prevent cross-contamination.

15.2 Sanitizers used in fresh-cut processing

As there are no practical technologies that provide a kill step in eliminating pathogens from fruits and vegetables without sacrificing product quality and shelf-life, potential sources of contamination from the environment to the consumer's table should be identified and specific measures or mitigation steps taken or implemented to minimize the risk associated from such products. The application of good hygienic practices during production, transport and processing, in combination with Good Agricultural Practices, Good Manufacturing Practices, Hazard Analysis Critical Control Point systems and use of validated systems with appropriately verified monitoring controls, will certainly minimize contamination of fruits and vegetables and significantly reduce the risk of illness associated with the consumption of such products.

The act of washing fruits and vegetables using potable water followed by a rinse step in potable water would assist in the removal of microorganisms from the surfaces of these products. Incorporation of an appropriate sanitizer could enhance the removal of microorganisms by 10- to 100-fold. However, efficacy of sanitizers or disinfectants in relation to the physical properties of fruits and vegetables requires further study, and effectiveness of sanitizers can be affected by factors such as pH, contact time, organic load, temperature and water properties, in addition to the physical properties of the fruit or vegetable being washed.

Legal use of various sanitizers or disinfectants differs from country to country. In the USA, wash water sanitizers or disinfectants used for fresh-cut produce are regulated by the US Food and Drug Administration (FDA) as a secondary direct food additive, unless the sanitizers are considered to be Generally Recognized As Safe (GRAS). If the product is a raw agricultural commodity that is washed in a food processing facility, such as a leafy green washing facility, both the US Environmental Protection Agency (EPA) and the FDA have regulatory jurisdiction and the sanitizer must be registered with the EPA as a pesticide. A list of approved solutions for use as sanitizers or disinfectants can be obtained from the Code of Federal Regulations 21 CFR, Sections 173.315 and 178.1010 and the FDA is recommending a 5-log reduction of pathogenic microorganisms on produce (FDA, 1995).

A number of new and novel sanitizers have emerged over the last decade with varying efficacy and effectiveness on various microbial pathogens. For the scope of this chapter, sanitizers that have microbial testing data or are commonly used by the industry will be addressed. This list is by no means exhaustive.

15.2.1 Chlorine

Chlorine and its hypochlorites have been used for many years as a sanitizer to wash, spray or flume waters in fresh produce industry and in application to food contact surfaces and processing equipment. Chlorine is commonly used at concentrations ranging from 50 to 200 ppm or mg/L with a typical contact time of 1–2 minutes. However, chlorine concentrations can rapidly decrease when chlorine comes into contact with organic matter, exposure to air, light and metals. Prolonged exposure to chlorine vapors during preparation and mixing of the sanitizer can cause irritation to the skin and respiratory tract, and pose potential occupational issues to workers in that environment.

Chlorine is a low-cost sanitizer, and its ability to kill pathogens in suspensions is established. This ability is largely the result of formation of hypochlorous acid as free chlorine (free active chlorine), which is the most effective form of chlorine (Zagory, 1999). Maintaining the appropriate amount of free chlorine in a wash flume setting can be technically challenging and the amount of free chlorine, like any other oxidant antimicrobial agents, is impacted by pH, temperature and the amount of organic matter in the system (Zagory, 1999; Gil *et al.*, 2009). Similarly, organic latex released into the wash flume from wounded tissues or cut surfaces as vegetable exudates can significantly impact chlorine effectiveness. This is particularly challenging at the front end of the wash flume, where large quantities of produce are continuously added and the surge of organic matter and exudates can rapidly deplete free chlorine, creating a situation where the demand for free chlorine exceeds availability and thereby increases the risk for cross-contamination or transfer of pathogens to uncontaminated product. The minimum amount of free chlorine that is effective in inactivating suspensions of *E. coli* O157:H7 to a non-detectable level has been shown to be around 0.5 mg/L or ppm, and increase the amount of free chlorine in the wash water required to inactivate *E. coli* O157:H7 to 10 mg/L or ppm or higher in the presence of product or organic matter before washing to prevent cross-contamination (Luo *et al.*, 2011). Repeated addition of chlorine to wash water that is high in organic load may be counterproductive as this can result in increased formation of toxic chlorine byproducts and generation of harmful chlorine off-gas (Suslow, 2001).

In order to minimize quick depletion of free chlorine by organic loads in the wash flume, free chlorine stabilization has been researched, like the use of cyanuric acid in maintaining free chlorine levels in outdoor swimming pools (Sommerfeld and Adamson, 1982) where formation of a relatively stable compound, chlorimide, is more resistant to degradation. Nou *et al.* (2011) explored the use of a proprietary formula named T-128 in stabilizing free chlorine in the presence of high organic load and evaluated its impacts on the quality and shelf-life of fresh-cut iceberg lettuce.

Nou *et al.* (2011) reported that T-128 significantly reduced the rate at which free chlorine was depleted in the presence of soil in the wash system and that T–128 also significantly reduced the survival of bacterial pathogens in the wash system with high organic loads, thereby potentially reducing the risk of cross-contamination when contaminated and uncontaminated produce were co-washed in the same system. The application of T-128 in the wash system did not have significant impact on the quality and shelf-life of the produce being washed. However, the authors noted that T-128 did not enhance the efficacy of chlorinated wash solutions for microbial reduction on contaminated produce.

15.2.2 Organic acid-based sanitizers

Organic acids are naturally present in fruits and vegetables, and accumulate in fermented products, which retards the growth of some microorganisms and prevents growth of others. Typically, microbial pathogens do not grow in pH environments below 4.0 so acidic pH environments or portions of the fruit or vegetable generally would not be areas of proliferation for microbial pathogens. Organic acids such as acetic, citric, succinic, malic, tartaric, benzoic and sorbic acids have been used in various applications to either control spoilage or pathogenic bacteria on foods such as the application of lemon juice to control *Campylobacter* species population in watermelon and papaya (Castillo and Escartin, 1994).

In an early study on organic acids, Shapiro and Holder (1960) used 1500 ppm of citric acid to treat salad vegetables. This did not affect the growth of bacteria in a 4-day storage trial at 10 °C, but when the vegetables were treated with 1500 ppm of tartaric acid, a 10-fold reduction in total counts was observed. Karapinar and Gonul (1992) used a solution containing 2% acetic acid or 40% vinegar to wash parsley that was artificially inoculated with 7-log of *Yersinia enterocolitica*. After the 15-minute wash, <1-log of *Y. enterocolitica* was detected. However, a 30-minute wash in 5% acetic acid resulted in no detectable bacteria.

One common example of organic acid-based sanitizer is peroxyacetic (peracetic) acid (PAA). PAA is a strong oxidizing agent that has been used extensively in processing equipment sanitization, and is FDA approved, up to 80 ppm, for applications on fruits and vegetables. A number of studies such as those of Simons (2001) and Gonzalez *et al.* (2004) have demonstrated PAA efficacy in pilot-scale or simulated wash conditions on minimally processed produce against *E. coli* O157:H7, *L. monocytogenes* and *Salmonella* spp. In these studies, up to 2-log reduction of inoculated pathogens on produce items such as carrot shreds was observed and PAA was not affected by the presence of organic matter compared with chlorine.

Gulati *et al.* (2001) and Baert *et al.* (2009) demonstrated efficacy of PAA on viruses inoculated onto various produce items. Gulati *et al.* (2001) showed 1- and 2-log reductions of FCV on strawberries and lettuce, respectively, when treated with 150 ppm PAA. No reduction was observed for a similar experimental set-up using 200 ppm chlorine. Allwood *et al.* (2004) showed comparable reductions of MS2 and FCV in 80 ppm PAA on lettuce.

15.2.3 Electrolyzed water

Electrolyzed water (EW) has been identified as a viable alternative disinfection tool for wash water (Ongeng *et al.*, 2006). Gil *et al.* (2009) listed the Ecodis technology in their review article, citing the use of anodic oxidation principles in the technology consisting of a highly efficient electrolysis cell equipped with coated permanent titanium electrodes. A direct low-voltage current that passes across the electrodes in a NaCl solution causes the formation of potent oxidizing agents derived from oxygen and free chlorine when chloride ions are present in the wash water. The oxygen and chloride radicals react with each other to form hypochlorous acid and hypochlorite ions, which may have residual activity after the product leaves the wash flume.

The anode end of the electrolysis process produces acidic EW with a low pH of around 2.5 or less, and is reported to have strong antibacterial activity against a number of pathogens. The effects of acidic EW were demonstrated by a number of researchers to decontaminate surfaces of lettuce, tomatoes, strawberries, cucumbers and spinach (Koseki *et al.*, 2001, 2004; Bari *et al.*, 2003; Guentzel *et al.*, 2008). Acidic EW was also effective for the inactivation of *E. coli* O157:H7, *Salmonella* spp., *L. monocytogenes* and *Bacillus cereus* on the surfaces of fresh produce as demonstrated by various researchers (Venkitanarayanan *et al.*, 1999; Kim *et al.*, 2000; Park *et al.*, 2001, 2008). However, the effectiveness of acidic EW can be impacted by organic matter and this will affect its ability to inactivate pathogens on surfaces of produce items (Park *et al.*, 2008).

In another study conducted by Kim *et al.* (2006), acidic EW was used to treat alfalfa and broccoli seeds. The acidic EW treatment at 55 °C for 10 minutes, produced from an acidic EW generator using deionized water and 10% NaCl solution resulted in 3.4- and 3.3-log reductions of *E. coli* O157:H7 from alfalfa and broccoli seeds, respectively, and did not significantly impact the germination rates of the seeds.

15.2.4 Chlorine dioxide

Chlorine dioxide (ClO_2) is another promising non-thermal technology that can be applied either as a gas or liquid for the reduction of pathogens in fresh fruits and vegetables. When applied as a gas, ClO_2 at 4 ppm with an exposure time of 30 minutes could achieve a 5-log reduction of *E. coli* O157:H7 and *Listeria monocytogenes* on strawberries (Han *et al.*, 2004), and a 4.4-log reduction of *Salmonella* spp. on strawberries was observed after exposure to 8 ppm ClO_2 gas for 120 minutes (Sy *et al.*, 2005). Similarly, the application of gaseous ClO_2 (4.1–8.0 ppm) to strawberries, blueberries and raspberries could have phytosanitary applications without impacting fruit quality, because it is effective at inactivating yeasts and molds that could extend the shelf-life of the product and provide opportunities to the industry to export to further markets (Sy *et al.*, 2005).

Studies on other produce have shown that treatment of uninjured green peppers with 3 ppm ClO_2 gas reduced the population of *L. monocytogenes* by more than 6-log after 30 minutes of exposure (Han *et al.*, 2001b). Additionally, Han *et al.* (2001a) found that treatments of uninjured green peppers with 0.6 ppm ClO_2 gas reduced the population of *E. coli* by 7.3-log after 30 minutes at 22 °C in a 90–95% relative humidity environment. Du *et al.* (2003) reported that treatment of apples with 3 ppm ClO_2 gas for 20 minutes resulted in 5.9-log reduction of *E. coli* after 30 minutes, and treatments conducted with 4 ppm ClO_2 gas for 30 minutes reduced *L. monocytogenes* on apple pulp skin by 6.5-log (Du *et al.*, 2003).

Chlorine dioxide, unlike traditional chlorine sanitizers, does not participate in chlorination reactions that result in harmful byproducts. Aqueous chlorine dioxide applied at 5 ppm was able to achieve greater than 5-log reduction of *L. monocytogenes* and *E. coli* O157:H7 on apples, lettuce and cantaloupe (Rodgers *et al.*, 2004). However, other studies have found that the same level of aqueous chlorine dioxide was only capable of reducing *L. monocytogenes* on lettuce by 1.7-log cfu/g (Zhang and Farber, 1996). It is possible that such variances of results could be a result of the method used to inoculate the produce (Rodgers *et al.*, 2004).

Fresh-cut leaves (25 g) of romaine or iceberg lettuce, inoculated by immersion in water containing 8-log of *E. coli* O157:H7 suspension for 5 minutes and dried in a salad spinner, were washed for 2 minutes in various concentrations (20–200 ppm) of aqueous chlorine dioxide. Populations of *E. coli* O157:H7 were reduced by up to 1.25-log on iceberg lettuce and approximately 1-log on romaine lettuce (Keskinen *et al.*, 2009)

15.3 Use of novel processing technologies

The use of novel processing technologies without thermal energy to either improve sanitizer effectiveness or a process to provide enhanced inactivation of microbial pathogens without compromising the visual and nutritive qualities of fresh produce, has gained traction over recent years. Application of such technologies has allowed new product development and maintenance of the freshness of the product. A number of technologies that could be applied to either fresh-cut produce, fruits and juices are available on the market, but, in the scope of this chapter, non-thermal technologies will be highlighted and again, the list is by no means exhaustive.

15.3.1 High powered ultrasound

Bacterial cells and viruses can attach to all surfaces of leafy green vegetables. If attached to freshly cut surfaces that are leaking nutrients or in pores or stomata, the ability of the antimicrobial sanitizer will be diminished (Takeuchi and Frank, 2000). Physical methods of removing bacterial cells and viruses are important, and shear forces generated during washing may assist in dislodging foreign particles into the sanitized water.

Ultrasound has been used for decades in various industries for cleaning purposes, and mounting evidence suggests that incorporating high powered ultrasound (HPU) into a commercial produce washing line can assist in removing bacterial cells and virus particles (virions) from produce surfaces or improving the access of the chemical sanitizers to the attached microorganisms.

Ultrasound creates a partial vacuum in the liquid that forms tiny bubbles that rapidly collapse under the vacuum. The bubbles collapse thousands of times each second, producing high-energy shockwaves that travel through the liquid. The shockwaves have the capacity to scour contamination and microorganisms from the surface, cracks and pore spaces, and internal cellular structure of agricultural products, which under conventional cleaning and sanitation processes would be difficult to remove. This mechanism accelerates physical and chemical reactions to enhance processes such as cleaning, plastic welding and sonochemistry. Ultrasound technology is easily adaptable to existing washing systems in industry, such as flumes and troughs, and can be used in combination with existing approved sanitizers. The combined technology of HPU with existing approved sanitizers can be developed in a number of ways for the agriculture industry to bring improved food safety, increased shelf-life and improved food quality to commercial products. The additional benefit of HPU technology in produce washing is the ease with which it may be installed into existing washing flumes and it may potentially have an impact on water reuse and conservation.

Maks *et al.* (2009) reported that effectiveness of chemical sanitizers to reduce the levels of pathogenic microflora was enhanced by 1–2-log by the addition of HPU into the washing of fresh-cut lettuce (Fig. 15.1). The authors demonstrated an additional 1-log reduction of total bacterial microflora on fresh-cut lettuce leaves with the combined use of HPU with peroxyacetic acid, compared with the sanitizer alone. Liu *et al.* (2009) demonstrated that treatments with 25 or 100 ppm sodium hypochlorite reduced infectious murine norovirus (MNV-1) titer by 1.7-log and 2.3-log, respectively, and the addition of HPU during treatment increased reduction to 2.7-log and 3.1-log, respectively. Similar reductions were recorded in treatments with 80 ppm PAA, with 2.5-log reduction achieved alone with PAA, and 3.7 log reduction when combined with HPU and PAA (Fig. 15.2). Seymour *et al.* (2002) reported that inactivation of *Salmonella* Typhimurium by a combined HPU and chlorinated wash water treatment was enhanced by 1-log; from 1.7-log reduction with chlorinated water alone, to 2.7-log reduction achieved with the

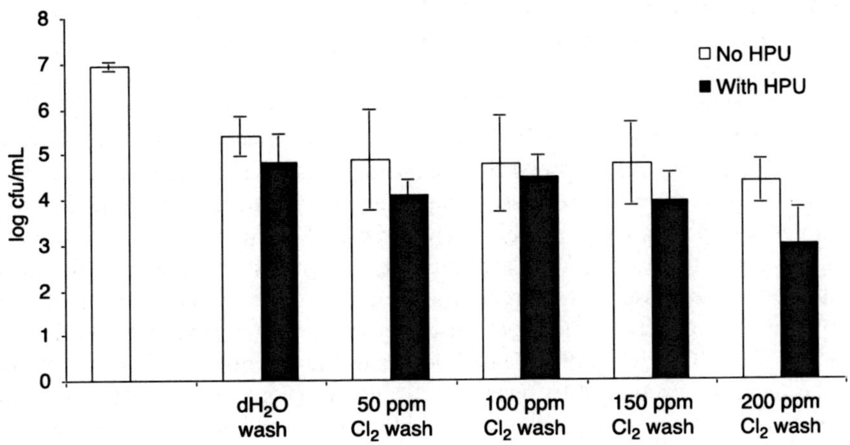

Fig. 15.1 *E. coli* O157:H7 TW14359 attached to romaine lettuce pieces treated with dH_2O or chlorine (50, 100, 150 and 200 ppm) with and without high powered ultrasound (HPU) at 10 °C (Maks *et al.*, 2009).

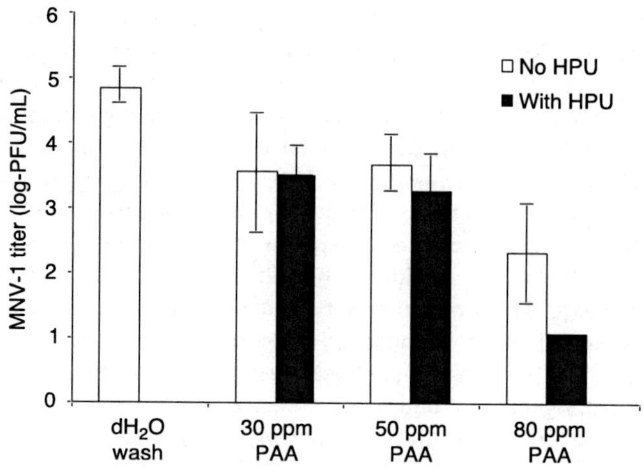

Fig. 15.2 Effect of peroxyacetic acid (PAA) on MNV-1 attached to romaine lettuce pieces treated with dH_2O or PAA (30, 50 and 80 ppm) with and without high powered ultrasound (HPU) at 10 °C (Liu *et al.*, 2009).

combined treatment. Zhou *et al.* (2009) demonstrated an additional 0.7–1.1 log reduction of *E. coli* O157:H7 from the surface of spinach leaves with the combined treatment over the use of water and various sanitizers alone.

15.3.2 High pressure processing

High pressure processing (HPP) has become a commercial reality with several fruit- and vegetable-based refrigerated food products currently on the international market, including a range of juices and fruit smoothies, jams, applesauce-fruit blends, guacamole and other avocado products, tomato-based salsas and fajita meal kits containing acidified sliced capsicum and onions, and heat-and-serve beef and chicken slices (pre-cooked) (Smelt, 1998; Stewart and Cole, 2001). A batch system is typically used where the product is placed into a final flexible consumer package before pressurization. The packages are loaded into a basket and placed into the pressure vessel where they are submerged in a liquid of low compressibility (typically potable water). Once loaded and closed, pressures ranging from 100 to 700 MPa are normally generated by pumping additional water into the vessel. The process is relatively energy efficient, requiring approximately the same energy to raise the pressure to 400 MPa as required to heat to 30 °C (Cheftel, 1992). Once the desired target pressure is obtained, no further energy is required to sustain that pressure (Farr, 1990). Unlike thermal processing, pressure is distributed instantaneously and uniformly throughout foods, ensuring a homogeneous treatment regardless of the size or shape of the product (Hoover *et al.*, 1989).

The required pressure treatment for microbiologically safe and stable products is dependent on the target microorganism to be inactivated. Bacterial vegetative cells, yeasts, molds and some viruses are sensitive to pressures between 200 and 700 MPa; bacterial spores can survive pressurization above 1000 MPa (Arroyo *et al.*, 1999; Cheftel, 1992; Sale *et al.*, 1970). Spoilage of food and/or food safety issues caused by the outgrowth of bacterial spores can be controlled via complementary means, such as refrigeration and acidification. Various common factors influence the pressure resistance of microorganisms, including the target microorganism and its physiological state, the intrinsic properties of the menstruum, the processing temperature, and the time and magnitude of pressure treatment (Hoover *et al.*, 1989). Considerable variation in susceptibility to high pressure has been observed among various microbial species and strains and on microorganisms in different substrates (Patterson *et al.*, 1995). Certain foods provide microorganisms protection from inactivation or injury from high pressure.

Neetoo *et al.* (2011) demonstrated the use of HPP in green onions artificially inoculated with *Salmonella* spp. and *E. coli* O157:H7. The degree of wetness (i.e. whether cut segments of green onions were dipped or sprayed with water or with no water added) in relation to microbial inactivation by HPP was evaluated. Green onions were treated at pressures ranging from 250 to 500 MPa for 2 minutes at 20 °C, which resulted in reduction of *Salmonella* and *E. coli* O157:H7 by 0.6 to >5-log. The amount of water or wetness of the green onions also dictated the extent of pressure inactivation, suggesting that certain fruit and vegetable products may be suitable for HPP treatment.

In another example, fresh whole and diced tomatoes inoculated with *Salmonella enterica* serovar Braenderup were subjected to HPP treatment at 350, 450 and 550 MPa for 120 seconds. Significant reductions of *S.* Braenderup were observed in diced tomatoes ranging from 0.46- to 3.67-log, while reduction in whole tomatoes ranged from 1.41- to 3.35-log (Maitland *et al.*, 2011).

HPP may be an effective postharvest strategy to reduce microbial pathogens from fresh produce but, because HPP is a batch process and requires the presence of a pressurizing fluid and certain level of water content to be effective, HPP is probably limited to juices, pre-packaged fruits, cut-fruits and certain vegetable products.

15.3.3 High-intensity pulsed light or ultraviolet

High-intensity pulsed light involves electrical ionization of a xenon lamp to emit a broadband white light with a spectrum resembling that of sunlight, containing wavelengths that include a large component (45%) of UV light; UV irradiation will inactivate viruses (Nuanualsuwan and Cliver, 2003). The intensity of pulsed white light required to inactivate viruses is estimated to be about 20 000 times the intensity of sunlight. Roberts and Hope (2003) investigated the potential of high-intensity broad-spectrum white light to inactivate viruses. Enveloped viruses (e.g. Sindbis and herpes simplex virus type 1) and non-enveloped viruses (e.g. encephalomyocarditis, poliovirus type 1, hepatitis A, bovine parvovirus and canine parvovirus) were diluted in phosphate-buffered saline and placed in small plastic sample dishes at a depth of 5 mm. A dose of 1.0 J/cm^2 was found sufficient to inactivate 4.8- to 7.2-log of all the listed viruses.

The efficacy of pulsed UV light for decontamination of minimally processed vegetables was investigated by Gomez-Lopez *et al.* (2005). A sensory evaluation was conducted with a semi-trained panel of four to six people who gave pulsed UV light-treated iceberg lettuce better scores than the control samples for off-odor, taste and leaf edge browning. This clearly indicates that pulsed UV light treatment helps in preserving the lettuce quality. In general, pulsed UV light treatment of food does not cause any adverse effect if applied in moderate amounts. This is required for microbial inactivation as strongly suggested by previous studies. However, modification and optimization of the treatment may be necessary for successful implementation of the process in some foods.

Gomez-Lopez *et al.* (2005) obtained 0.21- and 1.67-log reductions after treating minimally processed vegetables, such as spinach, celery, green paprika, soybean sprouts, radicchio, carrot, iceberg lettuce and white cabbage, with 7 J pulsed UV source for 45 s/side. Hoornstra *et al.* (2002) reported that carrots and paprika treated with two pulses of pulsed UV (0.30 J/cm^2) resulted in reduction of 1.6-log cfu/cm^2 and >2.6-log cfu/cm^2, respectively, in the total aerobic count on the surfaces. The authors reported no adverse effect on vegetable quality when stored at 7 °C or 20 °C for up to 7 days. Sharma and Demirci (2003) obtained more than 4-log reduction of *E. coli* O157:H7 when inoculated alfalfa seeds were treated with pulsed UV.

Bialka and Demirci (2007) treated blueberries with pulsed UV for inactivation of *E. coli* O157:H7 and *Salmonella* Typhimurium. They observed maximum reductions of 4.3- and 2.9-log cfu/g for *Salmonella* and *E. coli* O157:H7, respectively, with a 60-second treatment at 8 cm from the pulsed UV source. Pulsed UV did not change the sensory or color attributes of the blueberries. Similarly, *E. coli* O157:H7 and *Salmonella* Typhimurium populations were reduced 3.9-log cfu/g and 3.4-log cfu/g at 72 and 59.2 J/cm^2, respectively, in raspberries (Bialka and Demirci, 2008). When strawberries were treated with pulsed UV at 25.7 and 34.2 J/cm^2, maximum reductions of 2.1- and 2.8-log cfu/g, were obtained in populations of *E. coli* O157:H7 and *Salmonella* Typhimurium, respectively. The authors noted that there was no observable damage to the fruits caused by the pulsed UV treatment.

Because of the limited penetration capacity of UV, the efficacy of UV or light systems as a wash water disinfectant can be impacted by turbidity of the wash water. Application in wash water systems could be limited to filter disinfection or irradiation of recirculating water streams. As there is a required exposure time for UV to act on microorganisms, and because of the fast flowing rate of wash systems, it may be that not all microorganisms are inactivated when they pass through the UV source. However, UV or light systems could be incorporated onto conveyer belt systems or sorter systems prior to the product entry into the wash flume to reduce the microbial load on the product and enhance the efficiency of the disinfectant in the wash flume.

15.3.4 Irradiation

Although food irradiation has been shown to be an effective non-thermal means of preserving foods for the marketplace, its effectiveness against viruses is dependent on the size of the virus, the suspension medium, product characteristics and the exposure temperature (Patterson, 1993). Most viruses are far more resistant to irradiation than vegetative bacteria, parasites or fungi, which may be due in part to their smaller size and even smaller genome size (often single-stranded RNA) (Feng *et al.*, 2011). Additionally, as noted earlier, the treatment levels necessary to minimize organoleptic deterioration in irradiated foods have little effect on viruses, thus rendering irradiation impractical for the inactivation of viruses in commercial food products. The resistance of viruses to irradiation is comparable with that of bacterial endospores and extremely resistant vegetative types. For example, Monk *et al.* (1995) summarized pressure resistances in D_{10}-values (kGy) for important food microorganisms. Viruses had D_{10}-values ranging from 2 to 8 compared with varieties of *E. coli* with D_{10}-values of approximately 0.3. Radiation-resistant bacterial spores had D_{10}-values of 2–10, and the supremely resistant *Micrococcus radiodurans* (now *Deinococcus radiodurans*) showed D_{10}-values of 12–14.

The efficacy of using gamma irradiation in combination with a sanitizer such as chlorine was demonstrated by Foley *et al.* (2002) where chlorination in combination with 0.55 kGy of irradiation resulted in a 5.4-log reduction of *E. coli* O157:H7 on shredded iceberg lettuce. Irradiation applied at that level was also sufficient in inactivating yeast and molds without causing softening of lettuce, and had no impact on the sensory attributes.

After studying the safety of irradiating fresh iceberg lettuce and fresh spinach, and determining that these products retained nutrient value and are safe for consumers to consume, on 22 August 2008, the US FDA published the final rule that allowed the use of irradiation to ensure the safety of such products. It must be noted that irradiation is a complement to, not a replacement for, proper food handling practices and that it is an additional tool, not to take the place of washing, that may be used to reduce the level of microbial pathogens in iceberg lettuce and spinach.

Another example of non-thermal technology is non-thermal plasma (NTP), an emerging non-thermal sanitizing technology that has shown promise recently as an effective means for rapidly reducing pathogens on the surface of fresh produce and raw foods, and may be classified as a form of irradiation. Recent research at one of the application institutes showed that levels of *Salmonella* and *E. coli* O157:H7 were reduced on the surface of apples by 2.9–3.7-log cfu/mL and 3.4–3.6-log cfu/mL (Niemira and Sites, 2008), respectively, after 3 minutes of treatment, and almonds by as much as 1.34-log cfu/mL in 20 seconds (Niemira, 2010, unpublished data). NTP processes use high-voltage electricity to ionize gases near the electrodes of the device. Ionized gases, such as reactive oxygen species, are diffused toward the surface under the influence of electrical fields resulting in a rapid bactericidal effect. The antimicrobial efficacy of NTP is related to the specific design of the technology used, the power level used to generate the plasma, the gas composition and flow rate used in the plasma emitter, and the time and intensity of exposure.

15.4 Conclusion

Evaluation of sanitizer efficacy is affected by a number of factors such as organic loads, type of produce and other physiochemical properties. Sanitizers need to be carefully evaluated and the maintenance of effective sanitizer concentration in the wash flume is important to ensure wash water does not become a source of cross-contamination. Similarly, new technologies such as non-thermal processing technologies may not be suitable to all applications and will require careful consideration. The use of technologies to provide additional hurdles in addition to washing with sanitizer is welcome, and the use of new processing technologies to enhance washing could result in better sanitizer effectiveness and long-term sustainability. Intervention strategies that are developed to reduce and eliminate contamination of fresh produce must be flexible to serve both small and large establishments, and must be affordable, effective and efficient strategies. In all cases, validation studies in large-scale trials using microbial pathogens, where possible, will provide additional knowledge and confidence, and bridge the gap between laboratory-scale experiments conducted in well controlled environments to pilot scale or large scale where conditions are continuously changing.

15.5 References

ALLWOOD, P. B., Y. S. MALIK, C. W. HEDBERG and S. M. GOYAL. 2004. Effect of temperature and sanitizers on the survival of feline calicivirus, *Escherichia coli*, and F-specific coliphage MS2 on leafy salad vegetables. *J. Food Prot.* 67:1451–6.

ARROYO, G., P. D. SANZ, and G. PRESTAMO. 1999. Response to high pressure, low-temperature treatment in vegetables: determination of survival rates of microbial populations using flow cytometry and detection of peroxidase activity using confocal microscopy. *J. Appl. Microbiol.* 86:544–56.

BAERT, L., I. VANDEKINDEREN, F. DEVLIEGHERE, E. VAN COILLIE, J. DEBEVERE and M. UYTTENDAELE. 2009. The efficacy of sodium hypochlorite and peroxyacetic acid to reduce murine norovirus 1, B40–8, *L. monocytogenes* and *E. coli* O157:H7 on shredded iceberg lettuce and in residual wash water. *Int. J. Food Microbiol.* 131:83–94.

BARI, M. L., Y. SABINA, S. ISOBE, T. UEMERA, and K. ISSHIKI, K. 2003. Effectiveness of electrolyzed acidic water in killing *Escherichia coli* O157:H7, *Salmonella enteritidis* and *Listeria monocytogenes* on the surface of tomatoes. *J. Food. Prot.* 66:542–8.

BEUCHAT, L. R. (1996). *Listeria monocytogenes*: Incidence on vegetables. *Food Control* 7:223–8.

BEUCHAT, L. R. and J. H. RYU. 1997. Produce handling and processing practices. *Emerg Infect. Dis.* 3:459–65.

BEUCHAT, L. R., B. B. ADLER and M. M. LANG. 2004. Efficacy of chlorine and a peroxyacetic acid sanitizer in killing *Listeria monocytogenes* on iceberg and romaine lettuce using simulated commercial processing conditions. *J. Food Prot.* 67:1238–42.

BIALKA, K. L. and A. DEMIRCI. 2007. Decontamination of *Escherichia coli* O157:H7 and *Salmonella enterica* on blueberries using ozone and pulsed UV-light. *J. Food Sci.* 72:M391–6.

BIALKA, K. L. and A. DEMIRCI. 2008. Efficacy of pulsed UV-light for the decontamination of *Escherichia coli* O157:H7 and *Salmonella* spp. on raspberries and strawberries. *J. Food Sci.* 73:M201–7.

CASTILLO, A. and E. F. ESCARTIN. 1994. Survival of *Campylobacter jejuni* on sliced watermelon and papaya. *J. Food Prot.* 57:166–8.

CHEFTEL, J. C. 1992. Effects of high hydrostatic pressure on food constituents: an overview, pp. 195–209. In C. Balny, R. Hayashi, K. Heremans, and P. Masson (eds.), *High Pressure and Biotechnology*, INSERM/John Libbey Eurotext, Paris.

COOK, R. 2008. Trends in the marketing of fresh produce and fresh-cut products. Department of Agriculture and Resource Economics, University of California, Davis. http://agecon.ucdavis.edu/extension/presentations/files/cook/trends-in-the-marketing-of-fresh-produce-and-fresh-cut-products.pdf. Accessed 2 January 2013.

DELAQUIS, P., S. STEWART, S. CAZAUX and P. TOIVONEN. 2002. Survival and growth of *Listeria monocytogenes* and *Escherichia coli* O157:H7 in ready-to-eat iceberg lettuce washed in warm chlorinated water. *J. Food Prot.* 65:459–64.

DREUX, N., C. ALBAGNAC, M. FEDERIGHI, F. CARLIN, C. E. MORRIS and C. NGUYEN-THE. 2007. Viable but non-culturable *Listeria monocytogenes* on parsley leaves and absence of recovery to a culturable state. *J. Appl. Microbiol.* 103:1272–81.

DU, J., Y. HAN, and R. H. LINTON. 2003. Efficacy of chlorine dioxide gas in reducing *Escherichia coli* O157:H7 on apple surfaces. *Food Microbiol.* 20:538–91.

ECSCF (EUROPEAN COMMISSION SCIENTIFIC COMMITTEE ON FOODS). 2002. Risk profile on the microbiological contamination of fruits and vegetables eaten raw. http://ec.europa.eu/food/fs/sc/scf/out125_en.pdf. Accessed 28 September 2012.

FARR, D. 1990. High pressure technology in the food industry. *Trends Food Sci. Technol.* 1:14–16.

FENG, K., E. DIVERS, Y. MA and J. LI. 2011. Inactivation of a human norovirus surrogate, human norovirus virus-like particles, and vesicular stomatitis virus by gamma irradiation. *Appl. Environ. Microbiol.* 77:3507–17

FOLEY, D. M., A. DUFOUR, L. RODRIGUEZ, F. CAPORASO and A. PRAKASH. 2002. Reduction of *Escherichia coli* O157:H7 in shredded iceberg lettuce by chlorination and gamma irradiation. *Radiat. Phys. Chem.*, 63:391–6.

FOOD AND DRUG ADMINISTRATION (FDA). 1995. Secondary direct food additives permitted in food for human consumption. *Fed. Regist.* 60, 11899–900.

FOOD AND DRUG ADMINISTRATION (FDA). 2008. Guidance for industry: guide to minimize microbial food safety hazards of fresh-cut fruits and vegetables. http://www.fda.gov/food/guidancecomplianceregulatoryinformation/guidancedocuments/produceandplanproducts/ucm064458.htm. Accessed 4 October 2012.

GARCIA-VILLANOVA RUIZ, B., A. CUETO ESPINAR and M. J. BOLAÑOS CARMONA. 1987. A comparative study of strains of salmonella isolated from irrigation waters, vegetables and human infections. *Epidemiol. Infect.* 98:271–6.

GIL, M. I., M. V. SELMA, F. LOPEZ-GALVEZ, and A. ALLENDE. 2009. Fresh-cut product sanitation and wash water disinfection: problems and solutions. *Int. J. Food Microbiol.* 134:37–45.

GOMEZ-LOPEZ, V. M., F. DEVILEGHERE, V. BONDUELLE and J. DEBEVERE. 2005. Intense light pulses decontamination of minimally processed vegetables and their shelf-life. *Int. J. Food Microbial.* 103(1):79–89.

GONZALEZ, R. J., Y. LUO, S. RUIZ-CRUZ and J. L. MCEVOY. 2004. Efficacy of sanitizers to inactivate *Escherichia coli* O157:H7 on fresh-cut carrot shreds under simulated process water conditions. *J. Food Prot.* 67:2375–80.

GROVE, S. F., C. M. STEWART and A. LEE. 2008. Effect of sodium hypochlorite on murine norovirus, a surrogate for the human norovirus. P4–72, International Association for Food Protection Annual Meeting 2008, Columbus, OH.

GUENTZEL, J. L., K. L. LAM, M. A. CALLAN, S. A. EMMONS and V. L. DUNHAM. 2008. Reduction of bacteria on spinach, lettuce, and surfaces in food service areas using neutral electrolyzed oxidizing water. *Food Microbiol.* 25:36–41.

GULATI, B.R., P. B. ALLWOOD, C. W. HEDBERG and S. M. GOYAL. 2001. Efficacy of commonly used disinfectants for the inactivation of calicivirus on strawberry, lettuce, and a food contact surface. *J. Food Prot.* 64:1430–4.

HAN, Y., J. D. FLOROS, R. H. LINTON, S. S. NIELSEN and P. E. NELSON. 2001. Response surface modeling for the inactivation of *Escherichia coli* O157:H7 on green peppers (*Capsicum annuum* L.) by chlorine dioxide gas treatments. *J. Food Prot.* 64:1128–33.

HAN, Y., R. H. LINTON, S. S. NIELSEN and P. E. NELSON. 2001. Reduction of *Listeria monocytogenes* on green peppers (*Capsicum annuum L.*) by gaseous and aqueous chlorine dioxide and water washing and its growth at 7 degrees C. *J. Food Prot.* 64:1730–8.

HAN, Y., T. L. SELBY, K. K. SCHULTZE, P. E. NELSON and R. H. LINTON. 2004. Decontamination of strawberries using batch and continuous chlorine dioxide gas treatments. *J. Food Prot.* 67:2450–5.

HOORNSTRA, E., G. DE JONG, and S. NOTERMANS. 2002. Preservation of vegetables by light. In Society for Applied Microbiology (Ed.), *Frontiers in Microbial Fermentation and Preservation* (pp. 75–7). Wageningen, the Netherlands.

HOOVER, D. G., C. METRICK, A. M. PAPINEAU, D. F. FARKAS and D. KNORR. 1989. Biological effects of high hydrostatic pressure on food microorganisms. *Food Technol.* 43:99–107.

KANEKO, K. I., H. HAYASHIDANI, K. TAKAHASHI, Y. SHIRAKI, S. LIMAWONGPRANEE and M. OGAWA. 1999. Bacterial contamination in the environment of food factories processing ready-to-eat vegetables. *J. Food Prot.* 62:800–4.

KARAPINAR, M. and S. A. GONUL. 1992. Removal of *Yersinia enterocolitica* from fresh parsley by washing with acetic acid or vinegar. *Int. J. Food Microbiol.* 16:261–4.

KESKINEN, L. A., A. BURKE and B. A. ANNOUS. 2009. Efficacy of chlorine, acidic electrolyzed water and aqueous chlorine dioxide solutions to decontaminate *Escherichia coli* O157:H7 from lettuce leaves. *Int. J. Food Microbiol.* 132:134–40.

KIM, C., Y. C. HUNG and R. E. BRACKETT. 2000. Roles of oxidation-reduction potential in electrolyzed oxidizing and chemically modified water for the inactivation of food-related pathogens. *J. Food Prot.* 63:19–24.

KIM, H. J., H. FENG, M. M. KUSHAD and X. FAN. 2006. Effects of ultrasound, irradiation, and acidic electrolyzed water on germination of alfalfa and broccoli seeds and *Escherichia coli* O157:H7. *J. Food Sci.* 71:M168–73.

KONOWALCHUK, J. and J. I. SPEIRS. 1976. Virus inactivation by grapes and wines. *Appl. Environ. Microbiol.* 32:757–63.

KOSEKI, S., K. YOSHIDA, S. ISOBE and K. ITOH. 2001. Decontamination of lettuce using acidic electrolyzed water. *J. Food Prot.* 64:652–8.

KOSEKI, S., K. YOSHIDA, Y. KAMITANI and K. ITOH, K. 2004. Efficacy of acidic electrolyzed water for microbial decontamination of cucumbers and strawberries. *J. Food Prot.* 67:1247–51.

LIU, X., D. BATES, S. F. GROVE and A. LEE. 2009. Effect of antimicrobial sanitizers and high power ultrasound on murine norovirus on romaine lettuce. T8–01, International Association for Food Protection Annual Meeting 2009, Grapevine, TX.

LUO, Y., X. NOU, Y. YANG, I. ALEGRE, E. TURNER, *et al.* (2011). Determination of free chlorine concentrations needed to prevent *Escherichia coli* O157:H7 cross-contamination during fresh-cut produce washing. *J. Food Prot.* 74:352–8.

MAITLAND, J. E., R. R. BOYER, J. D. EIFERT, and R. C. WILLIAMS. 2011. High hydrostatic pressure processing reduces *Salmonella enterica* serovars in diced and whole tomatoes. *Int. J. Food Microbiol.* 149:113–17.

MAKS, N., S. GROVE, D. BATES, and A. LEE. 2009. Effect of sodium hypochlorite and high power ultrasound on *E. coli* O157:H7 in lettuce homogenate and on romaine lettuce. International Association for Food Protection Annual Meeting 2009, Grapevine, TX.

MONK, J. D., L. R. BEUCHAT and M. P. DOYLE. 1995. Irradiation inactivation of foodborne microorganisms. *J. Food Prot.* 58:197–208.

NEETOO, H., S. NEKOOZADEH, Z. JIANG and H. CHEN. 2011. Application of high hydrostatic pressure to decontaminate green onions from *Salmonella* and *Escherichia coli* O157:H7. *Food Microbiol.* 28:1275–83.

NIEMIRA, B. A. and J. SITES. 2008. Cold plasma inactivates *Salmonella* Stanley and *Escherichia coli* O157:H7 inoculated on golden delicious apples. *J. Food Prot.* 71:1357–65.

NOU, X. and Y. LUO. 2010. Whole-leaf wash improves chlorine efficacy for microbial reduction and prevents pathogen cross-contamination during fresh-cut lettuce processing. *J. Food Sci.* 75:M283–90.

NOU, X., Y. LUO, L. HOLLAR, Y. YANG, H. FENG, *et al.* 2011. Chlorine stabilizer T-128 enhances efficacy of chlorine against cross-contamination by *E. coli* O157:H7 and *Salmonella* in fresh-cut lettuce processing. *J. Food. Sci.* 76:M128–224.

NUANUALSUWAN, S. and D. O. CLIVER. 2003. Capsid functions of inactivated human picornaviruses and feline calicivirus. *Appl. Environ. Microbiol.* 69:350–7.

OLMEZ, H. and U. KRETZSCHMAR. 2009. Potential alternative disinfection methods for organic fresh-cut industry for minimizing water consumption and environment impact. *LWT-Food Sci. Technol.* 42:686–93.

ONGENG, D., F. DEVLIEGHERE, J. DEBEVERE, J. COOSEMANS and J. RYCKEBOER. 2006. The efficacy of electrolysed oxidising water for inactivating spoilage microorganisms in process water and on minimally processed vegetables. *Int. J. Food Microbiol.* 109:187–97.

PAINTER J. A., R. M. HOEKSTRA, T. AYERS, R. V. TAUXE, C. R. BRADEN, *et al.* 2013. Attribution of foodborne illnesses, hospitalizations, and deaths to food commodities by using outbreak data, United States, 1998–2008. *Emerg. Infect. Dis.* (Epub). Accessed on 28 January 2013. http://dx.doi.org/10.3201/eid1903.111866.

PARK, C. M., Y. C. HUNG, M. P. DOYLE, G. O. I. EZEIKE and C. KIM. 2001. Pathogen reduction and quality of lettuce treated with electrolyzed oxidizing and acidified chlorinated water. *J. Food Sci.* 66:1368–72.

PARK, E. J., E. ALEXANDER, G. A. TAYLOR, R. COSTA and D. H. KANG. 2008. Effects of organic matter on acidic electrolyzed water for reduction of foodborne pathogens on lettuce and spinach. *J. App. Microbiol.* 105:1802–9.

PATTERSON, M. F. 1993. Food irradiation and food safety. *Rev. Med. Microbiol.* 4:151–8.

PATTERSON, M. F., M. QUINN, R. SIMPSON and A. GILMOUR. 1995. Sensitivity of vegetative pathogens to high hydrostatic pressure treatment in phosphate-buffered saline and foods. *J. Food Prot.* 58:524–9.

POSCHETTO, L. F., A. IKE, T. PAPP, U. MOHN, R. BÖHM and R. E. MARSCHANG. 2007. Comparison of the sensitivities of noroviruses and feline calicivirus to chemical disinfection under field-like conditions. *Appl. Environ. Microbiol.* 73:5494–500.

ROBERTS, P. and A. HOPE. 2003. Virus inactivation by high intensity broad spectrum pulsed light. *J. Virol. Methods.* 110:61–5.

RODGERS, S. L., J. N. CASH, M. SIDDIQ and E. T. RYSER. 2004. A comparison of different chemical sanitizers for inactivating *Escherichia coli* O157:H7 and *Listeria monocytogenes* in solution and on apples, lettuce, strawberries, and cantaloupe. *J. Food Prot.* 67:721–31.

SALE, A. J., G. W. GOULD, and W. A. HAMILTON. 1970. Inactivation of bacterial spores by hydrostatic pressure. *J. Gen. Microbiol.* 60:323–34.

SEYMOUR, I. J., D. BURFOOT, R. L. SMITH, L. A. COX and A. LOCKWOOD. 2002. Ultrasound decontamination of minimally processed fruits and vegetables. *Int. J. Food Sci. Technol.* 37:547–57.

SHAPIRO, S. D. and I. A. HOLDER. 1960. Effect of antibiotic and chemical dips on the microflora of packaged salad mix. *Appl. Microbiol.* 8:341.

SHARMA, R. R. and A. DEMIRCI. 2003. Inactivation of *Escherichia coli* O157:H7 on inoculated alfalfa seeds with pulsed ultraviolet light and response surface modeling. *J. Food Sci.* 68:1448–53.

SIMONS, L. 2001. New washing treatments for minimally processed vegetables. Food Science Australia, Sydney, Australia.

SMELT, J. P. P. M. 1998. Recent advances in the microbiology of high pressure processing. *Trends Food Sci. Technol.* 9:152–8.

SOMMERFELD, M. R. and R. P. ADAMSON. 1982. Influence of stabilizer concentration on effectiveness of chlorine as an algicide. *Appl. Environ. Microbiol.* 43:497–9.

STEWART, C. M. and M. B. COLE. 2001. Preservation by the application of nonthermal processing, pp. 53–61. In C. J. Moir, C. Andrew-Kabilafkas, G. Arnold, B. M. Cox, A. D. Hocking, and I. Jenson. (eds.), *Spoilage of processed foods: causes and diagnosis.* Australian Institute of Food Science and Technology Incorporated, Waterloo DC, New South Wales, Australia.

SUSLOW, T. V. 2001. Water disinfection: a practical approach to calculating dose values for preharvest and postharvest applications. Davis, California, University of California. Available online at http://ucanr.org/freepubs/docs/7256.pdf. Accessed 6 February 2013.

SY, K. V., K. H. MCWATTERS and L. R. BEUCHAT. 2005. Efficacy of gaseous chlorine dioxide as a sanitizer for killing *Salmonella*, yeasts, and molds on blueberries, strawberries, and raspberries. *J. Food Prot.*, 68:1165–75.

TAKEUCHI, K. and J. F FRANK. 2000. Penetration of *Escherichia coli* O157:H7 into lettuce tissues as affected by inoculum size and temperature and the effect of chlorine treatment on cell viability. *J. Food Prot.* 63:434–40.

VENKITANARAYANAN, K. S., G. O. I. EZEIKE, Y. C. HUNG and M. P. DOYLE. 1999. Efficacy of electrolyzed oxidizing water for inactivation of *Escherichia coli* O157:H7, *Salmonella enteritidis* and *Listeria monocytogenes. Appl. Environ. Microbiol.* 65, 4276–9.

WORLD HEALTH ORGANIZATION (WHO). 1998. Surface decontamination of fruits and vegetables eaten raw: a review. (epub). http://www.who.int/foodsafety/publications/fs_management/en/surface_decon.pdf. Accessed 2 January 2013.

ZAGORY, D. 1999. Effects of post-processing handling and packaging on microbial populations. *Postharvest Biol. Technol.* 14:313–21.

ZHANG, S. and J. M. FARBER. 1996. The effects of various disinfectants against *Listeria monocytogenes* on fresh-cut vegetables. *Food Microbiol.* 13:311–21.

ZHOU, B., H. FENG and Y. LUO. 2009. Ultrasound enhanced sanitizer efficacy in reduction of *Escherichia coli* O157:H7 population on spinach leaves. *J. Food Sci.* 74: M308–13.

16

Ionizing irradiation for phytosanitary applications and fresh produce safety

S. Pillai and C. Bogran, Texas A&M University, USA and C. Blackburn, International Atomic Energy Agency, Austria

DOI: 10.1533/9781782420279.3.221

Abstract: The focus of this chapter is on the applications of electron beam (e-beam) and x-ray technologies to enhance food safety and quality and to meet international phytosanitary regulations that may limit world trade. The chapter provides background to the underlying technologies, current regulations governing the use of ionizing radiation and commercial acceptance of these technologies.

Key words: electron beam, x-ray, phytosanitary, fresh produce, pathogen, ionizing, irradiation.

16.1 Introduction

The use of ionizing irradiation technology for food preservation and sanitation is over 115 years old; however, its wider use really started in the 1950s following industrial and scientific advancements that made it possible to manufacture powerful sources of radiant energy. The technology has come a long way since its early days. Today, this technology is making significant improvements in the sterility of medical devices and products, the development of new materials, preventing the spread of detrimental insects and pathogens and enhancing the microbiological safety of foods. There are, however, significant differences in the modality of the underlying technologies as well as where the technology is gaining widespread use. Consumers in developed countries around the world are becoming more and more sophisticated in terms of their food needs and expectations. These demands extend the entire gamut from 'seasonal' fresh produce that is available year round, to pesticide-free foods, minimally processed foods and organically grown foods. Consumers in less well developed countries also require good-quality fresh foods but may also desire other qualities, for example foodstuffs that can be stored for prolonged periods at room temperatures. Irradiation technology, especially electron beam (e-beam) and x-ray technologies, can play a major role in making fresh fruits and vegetables available year-round, as well as preventing foodborne illnesses from microbial pathogen-contaminated fresh produce. The focus of this chapter is on the applications of e-beam and x-ray technologies to enhance food safety and quality and to meet international phytosanitary regulations that may limit world trade. The chapter also provides background to the underlying technologies, current regulations governing the use of ionizing radiation and commercial acceptance of these technologies.

16.2 Technology

Ionizing radiation, as the name implies, has enough energy to strip electrons away from atoms, hence 'ionization'. Ionizing radiation is able to cause ionization effects on DNA strands within organisms, which in turn leads to single strand and double strand breaks in DNA molecules. Although the organism attempts to repair these DNA breaks, the extent and number of such breaks are typically excessive for the organism to repair completely. Thus, the organism is lethally damaged, unable to replicate and will ultimately become non-viable. Ionizing radiation can be delivered using electrons or x-ray photons, as well as gamma ray photons. The methods for generating these types of ionizing radiation are quite different. The irradiation sources that have been internationally approved for food processing are gamma rays produced from radioisotopes cobalt-60 (1.17 and 1.33 MeV) or cesium-137 (0.662 MeV), machine generated electron beams (max. energy 10 MeV) and x-ray photons (max. energy 7.5 MeV in the USA, but 5 MeV in the rest of the world). Irradiation technologies that rely on gamma ray photons must necessarily involve the radioactive material cobalt-60 or cesium-137. The source is radioactive and cannot be switched off, the source strength decays with time as part of the natural radioactive decay process and the time of exposure needs to be adjusted to take this into account. Thus, facilities using gamma irradiation have inherently different designs to accommodate the normal operation, protection, shielding and safeguarding of the radioactive materials. In contrast, irradiation facilities that employ e-beam or x-ray are machine-based, using high energy linear accelerators (linacs) which generate the electrons or the x-ray. As the accelerators can be turned on and off, no radioactive residue is generated or needs to be protected.

16.2.1 Electron beam and x-ray technology

Electron beam and x-ray technologies do not involve radioactive isotope sources. Electron beam and x-rays are generated from regular electricity using linear accelerators. The single most important value of this technology is that it can be switched on and switched off (unlike gamma irradiators). It is because this technology relies on electricity that issues of safety and security in radioactive isotope transportation and disposal are non-existent. This may be one of the reasons why the number of e-beam irradiation facilities has grown to 1500 worldwide, and outnumbers gamma irradiation facilities by almost 10:1 (Hamm and Hamm, 2012). Most recent estimates suggest that e-beam systems account for approximately $80 billion of added value to commercial products. A variety of linear accelerator configurations have been developed since the 1950s. Different applications of the electron beam technology require different linear accelerator configurations. The ability to penetrate materials and products is dictated by the energy of the electrons that are generated by the accelerator. The higher the energy, the greater the penetration into the material. The penetrating potential depends not only on the thickness of the package product, but also the bulk density of the material. This makes it possible to calculate whether e-beam can be used for irradiating the product based on packaging-box dimensions and the product density (Miller, 2005). Table 16.1 provides a general description of the applications for accelerators of different energy profiles. It is important to note that e-beam energies greater than 10 MeV cannot be used for any food or food-related applications (IAEA, 1977; CODEX, 2003). The upper threshold is kept at 10 MeV to eliminate any possibility of inducing short-lived radioactivity in the products being irradiated with excessive energy.

The accelerator operating power dictates the irradiation processing rate, in other words, the product processing throughput is proportional to the operating power. Thus accelerators with a high power are required in commercial irradiation facilities that need to process large volumes of products with a quick turnaround time. Today, commercial irradiation facilities utilize accelerator power in the range 12–20 kW. With the financial viability of a commercial irradiation facility depending on the product volume that is processed, there is a demand for facilities to utilize accelerators with relatively higher power electron beams. Today accelerators in the range of 15–80 kW are commercially available and with the technology improvements that have occurred in accelerator designs, the conversion of e-beam to x-rays is now commercially feasible. X-rays are generated when electrons are accelerated into a high-density material such as gold or tantalum. Although much of the energy of the e-beam is lost as heat, a small percentage is converted into x-rays and the maximum energy of the incident e-beam limits the maximum energy of the x-rays produced. The primary advantage of x-rays compared with e-beam is that x-rays have a significantly greater penetration depth than e-beam. However, there are also technology drawbacks to x-ray irradiation, including lower processing speed, which is due to the lower dose rate compared with e-beam. In addition, because the conversion of e-beam into x-rays is not a very energy-efficient process,

Table 16.1 Commercial applications of accelerators of varying energy range

Application	Preferred electron beam energy
Surface curing of materials	80–300 kilo electron volts (keV)
Shrink film treatment	300–800 kilo electron volts (keV)
Wire and cable cross linking	0.4–3 million electron volts (MeV)
Food pasteurization and medical sterilization	3–10 million electron volts (MeV)

Source: adapted from Cleland (2012).

x-ray irradiation is slower and therefore the reduced thoroughput rate can make processing more expensive than e-beam irradiation. In the USA, X-ray accelerators cannot be greater than 7.5 MeV when used in food applications. However, in the rest of the world, the regulations restrict x-ray energies to no more than 5 MeV. In terms of x-ray processing power, commercial x-ray accelerators ranging between 18 kW and 800 kW are available. A major advantage of high power x-ray accelerators is that the x-ray processing speed can be significantly enhanced with higher power. Excellent descriptions of how e-beam and x-ray accelerators function and perform have been presented elsewhere (Miller, 2005; Hamm and Hamm, 2012). Presently, the capital investment required to build and develop an e-beam facility is estimated between \$6 and \$8 million. These costs are significantly lower than the capital investment required to build a gamma irradiation facility. Thus, from a financial standpoint as well, the future of e-beam technologies is promising.

There are many ways in which an e-beam irradiation facility can be designed and laid out. There are different layouts depending on whether the facility is designed as a multipurpose facility optimized to treat a broad range of products including food, whether it has both e-beam and x-ray capability, whether it has one or two e-beam accelerators, the product flow through, etc. At the National Center, the e-beam and x-ray irradiation equipment is housed in a 16 000 sq. feet dual modality (e-beam and x-ray) facility on campus that has two vertically mounted 10 MeV, 15 kW e-beam Varian linear accelerators and one horizontally mounted 5 MeV, 15 kW x-ray Varian linear accelerator. The facility utilizes a single conveyance system to move the product in and out of the process chamber (Fig. 16.1)

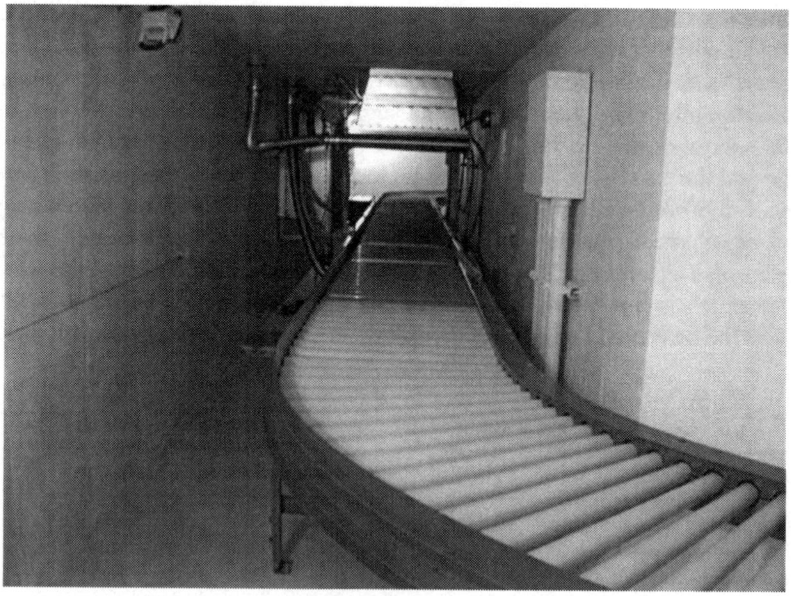

Fig. 16.1 View of the product conveyor system and the irradiation cell showing the e-beam 'horn' from the top and the x-ray 'horn' from the side.

16.3 Pathogen issues in fresh produce

Fresh produce is a food item that is vulnerable to contamination with pathogenic organisms. There can be an elevated risk of contamination where foods are grown intensively on a commercial scale, for example animal manure improves soil fertility but can also lead to the introduction of microorganisms that can cause food illness. Even crops grown to good agricultural or production practices can become contaminated, and this risk can be amplified by different gathering methods, storage under open field conditions, packing techniques, etc. This is particularly important for fresh produce that can be packed without a true 'pathogen-kill' step, and generally consumed fresh without any cooking or with just minimal heating. Fresh produce by virtue of its cultivation, handling and consumption practices is prone to pathogen contamination, and therefore a potential vehicle for widespread foodborne illness outbreaks. Monitoring or tracking of foodborne illnesses is difficult, and many foodborne diseases or outbreaks may go unreported. In the USA, fresh produce is now considered the number one cause of foodborne infections. Foodborne viruses are responsible for the majority of foodborne illnesses in the USA (Scallan *et al.*, 2011). Health care costs associated with foodborne viruses are currently estimated at $6 billion (Scharff, 2010). Table 16.2 is a compilation of survey data compiled by the US FDA regarding pathogen isolations on domestically grown and imported produce.

Very often, the only disinfectant step is chlorination. Although disinfection is helpful, it does not eliminate all pathogens, especially those that may be internalized within the fresh produce. A number of recent studies have demonstrated that microbial pathogens can become internalized either in the field or during cleaning and processing. Although the extent of enteric pathogen internalization is debatable, any technology that is aimed at eliminating pathogens should address the issue of internalized pathogens. Ionizing radiation technology is possibly the only known non-thermal technology that has the ability to inactivate pathogens that could be on the surface, within crevices, folds and internal to the product. A number of studies

Table 16.2 Compilation of a pathogen survey of imported and domestically grown fresh produce conducted by the US Food and Drug Administration

Fresh produce commodity	Domestically grown	Imported
Broccoli	0/36	Not performed
Cantaloupe	11/151 (7.3%)	5/164 (3%)
Celery	3/84 (3.6%)	0/120 (0%)
Cilantro	16/177 (9%)	1/85 (1.2%)
Lettuce	2/116 (1.7%)	1/142 (0.7%)
Parsley	2/84 (2.4%)	1/90 (1.1%)
Tomatoes	0/20 (0%)	0/198 (0%)
Scallion	3/180 (1.7%)	3/93 (3.2%)
Strawberries	1/143 (0.7%)	0/136 (0%)

Cilantro and scallion are also known as coriander and spring onion, respectively.

have shown that ionizing radiation can significantly reduce the levels of enteric pathogens on fresh produce without affecting the nutritional or sensory qualities of the produce item. A major collateral benefit of irradiation of fresh produce is the concomitant extension of shelf-life. This is because ionizing radiation does not discriminate between pathogenic and non-pathogenic microorganisms, and hence when fresh produce is irradiated, the entire microbial flora is reduced, which in turn results in extended shelf-life. A number of studies have demonstrated the extension in shelf-life when fruits and vegetables are irradiated. Enteric viruses are more resistant to irradiation compared with bacterial and protozoan pathogens. A recent study from our laboratory has shown that e-beam irradiation at doses that are currently approved by the US FDA for lettuce and spinach is able to inactivate viruses to certain levels (Espinosa *et al.*, 2012). Importantly, the study showed that the irradiation of lettuce and spinach at doses ≤4 kGy (current statutory dose limit in the USA) results in a reduction of infection risks. The study showed that if a typical US serving (~14 g) of lettuce was contaminated with 10 plaque forming units (pfu) of poliovirus, e-beam irradiation at 3 kGy will reduce the risk of infection from >2 in 10 persons to approximately 6 in 100 persons, a 14% reduction. Similarly, if a serving (~0.8 g) size of spinach is contaminated with 10 pfu of rotavirus, e-beam irradiation at 3 kGy will reduce infection risks from >3 in 10 persons to approximately 5 in 100 persons, a 25% reduction (Espinosa *et al.*, 2012). These reductions in infection risks translate to significant public health savings.

Table 16.3 shows the FDA approved dose limits and fresh produce commodities that can be treated with ionizing radiation. It should be emphasized that postharvest technologies such as e-beam irradiation were never designed to be used as clean-up technologies (Espinosa *et al.*, 2012). These technologies are meant to be used only as the final step of a comprehensive food safety program that starts with Good Agricultural Practices (GAP) in the field and Good Manufacturing Practices (GMP) in packing facilities. Unless the fresh produce commodity has manageable levels of contaminants to begin with, the use of e-beam or other such postharvest technologies cannot be expected to make significant reductions in the infection rates in the general public.

Table 16.3 US FDA approved irradiation dose limits for fresh produce and vegetables

Commodity	Intended application	Maximum dose limit (kGy)
Lettuce	Pathogen control	4.0
Spinach	Pathogen control	4.0
Seeds for sprouting	Pathogen control	8.0
Fresh fruits	Growth and maturation inhibition	1.0
Fresh fruits	Arthropod disinfestation	1.0

16.3.1 Phytosanitary treatment

The International Plant Protection Convention (IPPC) is an international agreement focused on preserving standardization of plant health practices around the world. Currently, there are 177 signatories of the IPCC. The primary objective of the IPPC is to prevent the introduction of regulated pests and pathogens to help protect cultivated and wild plants in different regions of the world. Different countries and regions face different threats from pests, and countries analyze risks to their plant resources and assess strategies to minimize these risks including the use of different technologies to satisfy their phytosanitary safety needs. Contracting parties to the IPPC strive to meet their shared goal of preventing the spread and introduction of plant pests while minimizing interference with the movement of goods and people. The IPPC provides countries with a framework to assess risks to their national plant resources and use science-based measures to safeguard their cultivated and wild plant systems. Countries employ a wide variety of technologies to prevent the spread of insects and pests that may be associated with fresh produce and agricultural commodities in international trade. These technologies include chemical treatments such as methyl bromide, heat treatments (hot air, hot water dip and steam), cold treatments, fumigation (phosphine gas, methyl bromide) and ionizing radiation. There is a growing trend in employing ionizing radiation as a phytosanitary treatment technology. One of the salient advantages of ionizing radiation is that it is a simple solution to an otherwise complex process and is effective against a wide variety of insects and growth stages. Ionizing radiation has minimal impact on commodity quality because it does not raise temperature during treatment, it does not leave residues, is environmentally friendly and is a sustainable alternative to methyl bromide, and may be the only practical treatment option for certain commodities in some circumstances.

There are international standards for the use of irradiation technologies to control pests in fresh produce. For example, the IPPC has established an International Standard for Phytosanitary Measures No. 18 (ISPM 18) which provides guidelines for the use of irradiation in phytosanitary treatments. Similarly, the American Society for Testing and Materials (ASTM International) has established ASTM F-1355-06 for standardized protocols for employing irradiation on fresh agricultural produce for phytosanitary purposes. The IPPC also has ISPM 28 on phytosanitary treatments for regulated pests and, as of 2012, this gives 14 internationally agreed irradiation phytosanitary treatments; 13 against specific pests and one 'generic treatment' for fruit flies.

In the USA, the USDA-APHIS has been encouraging the use of methyl bromide alternatives including ionizing radiation for phytosanitary treatments. In 2006, it established 400 Gy as the generic dose for all insects and 150 Gy for all fruit flies. Table 16.4 shows the commodities that can be imported into the USA using ionizing radiation. A current trend is the approval by USDA-APHIS of ionizing radiation as the only viable treatment for new commodities. An example is the stipulation that only ionizing radiation can be used for the import of carambola, sweet lime, guava, manzano pepper, sweet orange and tangelo from Mexico. The radiation dose that is used to inactivate microbial pathogens (e.g. those responsible

Table 16.4 USDA-APHIS approved minimum irradiation doses for phytosanitary treatment purposes

Commodity	Country of import	Approved dose (Gy)
Eggplant, okra, pepper	Ghana	400
Mango, pomegranate	India	400
Rambutan	Malaysia	400
Guava	Mexico	400
Mango	Pakistan	400
Dragon fruit, litchi, longan, mango, mangosteen, pineapple, rambutan	Thailand	400
Dragon fruit, rambutan	Vietnam	400
Banana, breadfruit, cowpea (pod), curry leaf, dragon fruit, guava, jackfruit, mangosteen, melon, moringa pods (drumstick), star fruit, abiu, atemoya, breadfruit, carambola, *Cucurbita* spp., sapodilla, and sweet potato	Hawaii islands	400
Grape, stone fruit, pears, persimmons	South Africa	400

for foodborne infections) is generally much higher than the doses required for phytosanitary treatment. This is partly because the endpoints are different. When used as a phytosanitary treatment the purpose of irradiation is not to kill the target organism, but is instead to prevent it from being able to multiply or develop and so preventing egg hatch or inducing infertility may suffice (Table 16.5).

Table 16.5 USDA-APHIS mandated minimum doses required for specific insect pests

Target insect family	Required minimum dose (Gy)
All fruit flies of the family Tephritidae	150 (generic dose)
All insects except adults and pupae of the order Lepidoptera	400 (generic dose)
Rhagloletic pomonella	60
Anastrepha ludens, Anastrepha oblique, Anastrepha suspense	70
Conotrachelus nenuphar	92
Anastrepha serpentine, Bactrocera jarvisi, Bactrocera tryoni, Ceratitis capitata, Copitarsia declora	100
Bactrocera cucurbita, Aspidiotus destructor, Bactrocera dorsalis, Cylas formicarius, Escepes postfasciatus, Omphisa anastomosalis, Pseudaulacapsis pentagona	150
Cydia pomonella, Grapholita molesta	200
Cryptophlebia ombrodelta, Cryptophlebia illepida	250
Brevipalpus chilensis, Sternochetus mangiferae	300

16.4 Regulatory aspects and consumer acceptance

16.4.1 Regulatory aspects

With the recent USDA approval of irradiated mangos from Mexico and India, and guavas from Mexico, the volume of imports of these fruits has grown significantly. To date, 10 countries have signed the USDA's Framework Equivalency Agreement which develops the framework and work plans for exporting and importing fruits from countries into the USA. These work plans cover specific topic areas such as field pest risk mitigations, inspection procedures, irradiation treatment and dosimetry systems, documentation, treatment verification procedures, post treatment quarantine and handling and product labeling. The work plans provide the operational procedures for the cross-border movement of irradiated products to and from the USA. A more recent regulatory change by the USDA-APHIS is the permitting of imported agricultural products to be treated at irradiation facilities in the USA after importation but before distribution. This new approval has significant ramifications in terms of access to technology and pricing. All irradiated fruits and vegetables have to have the 'radura' symbol (treated with irradiation) either on the package or on the fruit themselves

16.4.2 Consumer acceptance of irradiated foods

It has been incorrectly assumed that US consumers will not purchase irradiated foods. Since 2000 when irradiated foods became available quite extensively, there has not been a consumer backlash against these products in grocery stores. On the contrary, the popularity of these products actually increased between 2000 and 2002. In 2002 there were at least 2000 stores that carried irradiated foods in the USA. In 2002, the main supplier of irradiated foods (SureBeam, Inc.) filed for bankruptcy and the interest among store owners declined. Peer-reviewed studies have conclusively shown that consumers not only are accepting of irradiated foods when information is accurately presented, but also are willing to purchase these products at a higher price (Ibarra *et al.*, 2010; Nayga *et al.*, 2006). Today, irradiated spices, irradiated fresh ground beef, irradiated frozen ground beef, irradiated mangoes, irradiated guavas, irradiated papaya, irradiated dragon fruit, irradiated litchi, and irradiated rambutan are available in grocery stores and from online stores across many parts of the USA. These products have the 'radura' symbol placed on the 'country of origin' or product locator unit (PLU) stickers. It is also important to note that no irradiated product has ever been removed from the shelves because of consumer resistance. Thus, the arguments 'consumers will not buy', or there are 'consumer acceptance issues' are anecdotal references unsupported by empirical evidence.

16.5 Challenges facing food irradiation

There are, however, multiple challenges or hurdles that stand in the way of greater adoption of this technology globally. These include lack of understanding among decision makers, lack of profitable business models, lack of packaging standards, lack of current survey-based data on acceptance, and lack of irradiation capacity and trained personnel. There is a surprising lack of understanding among decision makers and many in the food industry of what this technology is, or is not, what does it take to adopt and how does one adopt this technology in a profit-sustainable manner. There is no doubt that the food industry knows the potential of this technology. However, most seem unaware of how to exploit this technology in their business context. What are missing are business models that help with the adoption of this technology for different sizes or types of producers, distributors and retailers.

There are no business models for the ground beef industry, the seafood industry, the fresh produce industry or even the ingredient industry that include an irradiation step. Today, most of the food irradiation is done using so-called service facilities to which companies send their products for treatment. This approach may work for perhaps half a dozen companies. However, this need not be the only model for adopting food irradiation. If this technology is to gain widespread use in the food industry, sustainable business models must be developed and tested that clearly identify the required capital investment, expected profit margins and potential threats. Such models have to be developed and validated based on multiple scenarios.

There is also a lack of irradiation capacity and trained personnel, not just in the USA but worldwide. If the FDA and the USDA approved the use of irradiation technology for all different types of foods, there is not enough capacity of irradiators in the USA to serve all the irradiation needs. The relative scarcity of this technology has also resulted in a small pool of trained individuals who are involved in different roles such as those who are qualified to operate and service the instruments and qualified to serve as dosimetrists. This lack of installed capacity and trained labor force is a serious issue in the USA and around the world. The education of regulators in e-beam and x-ray technology is also important because many countries currently do not have the regulations that allow importation of accelerators. Thus, if this technology is expected to increase, then rules and regulations must be in place to allow importation of such equipment.

16.5.1 Cost-benefit aspects

Another key issue that is a barrier to the adoption of this technology is the cost, and who is willing to bear the cost for this technology? Whether it is gamma or e-beam/x-ray irradiation, there are significant costs involved. These costs can be broken down into different classes; up-front costs, operating costs or the costs associated with transportation, storage and disposal of radioactive material. However, the break-even costs per unit of commodity treated are not large. Currently, costs for irradiation processing range from about \$0.10 per kg for fruits to approximately \$0.20 per kg for frozen ground beef. Business models that incorporate the new metrics must be developed for the different types of food industries.

Conventional wisdom has been to use irradiation technology for processing large volumes of product quickly. This argues for using this technology to process large pallets or crates. However, this bulk processing is contradictory to what the reality is today in the marketplace. Today, foods, especially fresh fruits and produce, are sold in either single or small serving sizes with the product often cleaned, cored, sliced and packaged as a snack or convenience food. If this is the reality today, why are irradiation technology providers still advocating high product throughput as the key price point? Why should a processor or distributor or retailer invest in a facility that is designed primarily for bulk processing? Are the current packaging materials used in such single serving/small serving packages compatible with the federal approved packaging material? It would appear that the smaller packaging design would be suitable for e-beam processing and more experimental data to support this are required so that this can be taken forward by industry.

Enhanced consumer survey instruments need to be designed to capture the type of information that is really needed for the decision makers to understand the value of this technology. Outreach and educational materials that no longer focus on this technology as just a 'pathogen-kill' technology must be developed. There is a strong unfulfilled demand for these tools and methods. The outreach materials to different audiences such as corporate decision makers, corporate legal departments, corporate risk reduction officers, investment bankers, etc., must be carefully thought through and developed with input from target audience representatives.

16.6 Conclusion and future trends

There is a growing interest in processing of fruits and vegetables by ionizing radiation both to enhance their quality and to ensure that shipments are free from insect pests and meet quarantine requirements. Ensuring high quality is of considerable competitive value in today's globalized food production, distribution and supply chains. Irradiation technology should not be used as a clean-up technology of last resort, instead we envision its commercial future potential to add value and enhance food quality. Quality enhancement is achieved by minimizing the need for fumigants, pesticides, artificial ripening agents and other chemical substances. Also, by employing irradiation technologies, product shelf-life can be extended, making fresh commodities keep better and avoiding losses. In addition, irradiation also enables international trade in fresh commodities by providing a means to meet quarantine standards. These factors allow consumers to access fresh fruits and vegetables from all around the world at a fair price. Irradiation of fruits and vegetables could become a cornerstone technology for any country in order to protect and secure its food supplies and provide access to international markets.

16.7 Acknowledgments

This work was supported in part by funds from Hatch project H8708 from the Texas A&M System's AgriLife Research. This work was supported in-kind by the IAEA Coordinated Research Project D6.20.08, 'Development of generic irradiation doses for quarantine treatment'.

16.8 References

CLELAND, M.R. 2012. Electron beam materials irradiators In: *Industrial accelerators and their applications* (eds. R.W. Hamm, and M.E. Hamm) World Scientific Co. Pte. Ltd. Singapore.

CODEX, 2003. General Standard for irradiated foods. CODEX STAN 106–1983 REV 1–2003.

ESPINOSA, A.C., P. JESUDHASAN, R. ARREDONDO, M. CEPEDA, M. MAZARI-HIRIART, *et al.* 2012. Quantifying the reduction in potential health risks by determining the sensitivity of poliovirus type 1 chat strain and rotavirus SA-11 to electron beam irradiation of iceberg lettuce and spinach. *Appl. Environ. Microbiol.* 78: 988–993.

HAMM, R.W., and M.E. HAMM, 2012. *Industrial accelerators and their applications*. World Scientific Co. Pte. Ltd. Singapore.

IAEA, 1977; Report of a Joint FAO/IAEA/WHO Advisory Group on International Acceptance of Irradiated Foods, Wageningen, Netherlands, 28 Nov–1 Dec. 1977. International Atomic Energy Agency, Vienna, 1979 (STI/PUB/530).

IBARRA, A.A., A.S. VARGAS, and R. NAYGA. 2010. Water quality concerns and acceptance of irradiated food: a pilot study on Mexican consumers. *J. Sci. Food and Ag.* 90: 234.

MILLER, R.B., 2005. *Electronic irradiation of foods: an introduction to the technology*. Springer.

NAYGA JR., R.M., R. WOODWARD, and W. AIEW. 2006. Willingness to Pay for Reduced Risk of Foodborne Illness: A Nonhypothetical Field Experiment. *Can. J. Agrl. Econ.* 54: 461–75.

SCALLAN, E., R.M. HOEKSTRA, F.J. ANGULO, R.V. TAUXE, M.A. WIDDOWSON, *et al.* 2011. 'Foodborne Illness Acquired in the United States-Major Pathogens.' *Emerg. Inf. Dis.* 17: 7–15.

SCHARFF, R.L. 2010. Health related costs from foodborne illnesses in the United States. Produce Safety project. George Washington University. http://producesafetyproject.org.

17

Edible coatings for fresh and minimally processed fruits and vegetables

L. Ciolacu and A. I. Nicolau, Dunarea de Jos University of Galati, Romania and J. Hoorfar, Technical University of Denmark, Denmark

DOI: 10.1533/9781782420279.3.233

Abstract: A solution for increasing the shelf-life of fresh and minimally processed fruits and vegetables is coating with edible films. A presentation of the materials that can be used as edible coatings and their properties is related to the fruits and vegetables on which they were tested. The combination of coating materials with antimicrobials is shown as a possibility to enhance the safety of ready-to-eat fresh produce. Technological, regulatory and economical aspects regarding usage of edible coatings are reviewed along with consumer perception, challenges for ecology and new research needs.

Key words: edible coatings, antimicrobials, packaging, nanotechnology.

17.1 Introduction: development of edible coatings

The development of edible coatings began as a solution to rapid dehydration, enzymatic browning and texture decay of fruits and vegetables, which all have a major impact on shelf-life. As well as being a partial barrier to moisture, gases (oxygen, carbon dioxide, volatile substances) and microorganisms and improving mechanical resistance of fruit and vegetables, edible coatings have also been utilized as carriers for a variety of food additives. Furthermore, they can be used to contribute to the production of volatile aromas (Olivas and Barbosa-Cánovas, 2005). In other words, a technology that started as an alternative to plastic bags or tray packaging has become the 'packaging of the future' (Fundazioa, 2006).

'Edible film' is a term that has only been used specifically for food applications over the last 50 years. There is usually no differentiation between the terms 'film' and 'coating'. Both refer to a thin layer of a material that covers the surface of a food (Pavlath and Orts, 2009). Despite this relatively recent application of the term, edible coatings have been used to preserve the quality of fruit and vegetables for many years. For example, as far back as the twelfth century, there are reports that wax coatings were used in China to preserve oranges and lemons (Hardenburg, 1967). According to Soliva-Fortuny, Rojas-Graü and Martin-Beloso (2011), in the USA, the first patent on the use of edible coatings belonged to A.F. Hoffman who, in 1916, invented a method of covering whole fruits with molten wax. In 1972, this continued with D.S. Bryan who patented an edible coating made of carrageenan for the conservation of grapefruit halves.

Currently, edible films and coatings are used in various applications: mainly to decrease the transmission rate of water vapor thereby preventing texture decay (Chien *et al.*, 2007; Dhanapal *et al.*, 2012; Garcia and Barret, 2002; Olivas *et al.*, 2003); to preserve the texture of fruits and vegetables (Kader, 1986; Knee, 1980; Maftoonazad *et al.*, 2007; Olivas *et al.*, 2007); to reduce respiration and hence increase the shelf-life of a commodity (Kays and Paull, 2004); to control browning (Bai *et al.*, 2003; Baldwin *et al.*, 1996; Olivas *et al.*, 2007; Perez-Gago and Krochta, 2001); and to preserve flavor (Baldwin *et al.*, 1999; Fellman *et al.*, 2003; Ke *et al.*, 1994; Olivas *et al.*, 2003, 2007).

The most important parameters to consider when developing edible coatings are how the coating can be applied to the product and the capacity of the coating to adhere to the surface. Food products are usually coated by dipping or spraying, which forms a thin film on the food surface that acts as a semi-permeable membrane (Dhanapal *et al.*, 2012). Currently, dipping is the most common method for applying coatings to fruits and vegetables (Vargas *et al.*, 2008), but other methods such as brushing, spraying, solvent casting and extrusion are also used depending on the food product and the characteristics of the forming film.

17.2 Types of edible coatings

Edible coatings can be composed of a variety of substances of different types. These can be grouped into carbohydrates, proteins and lipids, and have different characteristics (Table 17.1).

In addition to the edible coatings presented in Table 17.1, there are also composite films. These comprise a blend of polysaccharides, proteins and/or lipids to improve the permeability or mechanical properties of the coating depending on its specific applications. These heterogeneous films can be applied using various methods, either in the form of an emulsion, suspension or dispersion of the non-miscible constituents, or in successive layers (multilayer coating or films), or in the form of a solution in a common solvent. The method of application affects the barrier properties of the films obtained (Khin *et al.*, 2005).

17.3 Antimicrobial properties of edible films

As fresh fruits and vegetables have been linked to many foodborne bacterial outbreaks (EFSA, 2012), new methods of packing used today also include antimicrobial

Table 17.1 Substances utilized as edible coatings

Substance name			Properties	Used for
Polysaccharides	Cellulose and derivatives	Cellulose Methyl-cellulose Hydroxyl-propyl-methyl-cellulose Hydroxyl-propyl-cellulose Carboxy-methyl-cellulose	Excellent oxygen barrier properties Maintain aroma Good oil barriers Good mechanical properties Not effective moisture barriers because of their hydrophilic nature	Avocado, apple, apricot, chestnuts fruitberries, guava, melon, litchi, longan fruit, papaya pieces, peach, pear, plum, walnuts Bell pepper, carrot, cucumber, green pepper, shiitake mushroom, tomato
	Chitin and derivatives	Chitin Chitosan, N-acetyl-chitosan, mono-acetyl-chitin di-butyryl-chitin, chitosan acetate, chitosan butyrate		
	Starch	Corn, potato or rice starch		
	Gums	Alginate Guar Arabic Xanthan		
	Pectin and derivatives	Pectin Pectate Amidated pectin		
Lipids	Waxes	Paraffin Carnauba Candelilla Beeswax	Limited oxygen barrier properties Good water vapor barrier properties Relatively inflexible	Avocado, apple, banana, grape, lime, lemon, mango, melon, orange, passion fruit, peach Peppers, cucumber, sweet potato, tomato
	Resins	Shellac		

(Continued)

Table 17.1 Continued

Substance name			Properties	Used for
	Oils	Mineral Vegetal		
Proteins	Vegetal	Zein Gluten Soy	Excellent oxygen and oil barriers Provide strength and structural integrity Maintain aroma Not effective moisture barriers	Apple, pear, nuts, strawberry, kiwi Baby carrot, tomato, bell pepper, celery, potato, zucchini
	Animal	Gelatin Whey Casein		

compounds for the prevention of spoilage and/or pathogenic bacteria. In one study, antimicrobial coating with acid-chitosan solutions provided an alternative method for decontaminating tomatoes that had been inoculated with a four-strain cocktail of *Salmonella* (serovars Montevideo, Newport, Saintpaul and Typhimurium) (Jin and Gurtler, 2012). In another study, after 8 days at 10 °C, a population of *L. monocytogenes* was 10-fold lower on coated sweet corn than on non-coated sweet corn, indicating that the zein (corn protein) coating acts as a barrier to the bacterium (Carlin *et al.*, 2001). Lee *et al.* (2012) reported an antimicrobial edible film developed from defatted mustard which could be used to control the growth of *L. monocytogenes* when applied to foods as wraps or coatings. Furthermore, some edible coatings, or different antifungal compounds that may be incorporated in the edible films, proved to be effective for preserving different fruits and vegetables during refrigerated storage (Mehyar *et al.*, 2011; Valencia-Chamorro *et al.*, 2009, 2010).

Antimicrobial substances can be usefully incorporated into edible coatings. These antimicrobials may be divided into different categories (Franssen and Krochta, 2003):

- organic acids (acetic, benzoic, lactic, propionic and sorbic acid);
- fatty acid esters (glyceryl monolaurate);
- polypeptides (lysozyme, peroxidase, lactoferrin and nisin);
- plant essential oils (cinnamon, oregano and lemongrass);
- nitrites;
- sulfites.

Of these categories, the main antimicrobial agents used in coatings for fresh-cut fruits are organic acids and plant essential oils. Use of the above substances in edible coatings does help to preserve fresh foods for longer. It has been shown that some volatile essential oils can be used to prepare apple-based antimicrobial edible films with good physical properties, and that the films are effective against major foodborne pathogens (Du *et al.*, 2009). However, use is currently limited in fresh-cut fruits. One reason for this is that a major disadvantage of using essential oils in coatings (among otherwise positive results) is their strong flavor, which can negatively affect the original taste of the food (Rojas-Graü *et al.*, 2011). For example a residual aromatic herbal taste was detected in fresh-cut apples coated with an apple puree-alginate film containing a low concentration of oregano oil, which had been added as an antimicrobial film (Rojas-Graü *et al.*, 2007).

17.4 Challenges for ecology

Increased consumer demand for higher quality food in combination with the environmental need to reduce disposable packaging waste has led to more interest in research into edible films (Del-Valle *et al.*, 2005). Food packaging specialists are more and more concerned about the preservation and protection of all types of foods and their raw materials, particularly from oxidative and microbial spoilage in order to extend their shelf-life. Awareness of environmental pollution and the need for a safe, eco-friendly atmosphere has led to a shift away from renewable agricultural feedstock and marine food processing industry wastes towards the use of biodegradable materials (Tharanathan, 2003).

Edibility, biodegradability and increased food safety are the main benefits of active edible films. Their environmentally friendly aspects make them a preferable alternative to packaging systems that use synthetic non-biodegradable materials (Geraldine *et al.*, 2008). The main problems with biopolymers are their higher costs and less optimal physical and processing properties when compared with synthetic plastics. In addition, there have not been sufficient incentives for downstream processors to incorporate biodegradable materials into their products (Siracusa *et al.*, 2008). Polymer cross-linking and graft copolymerization of natural polymers with synthetic monomers are other alternatives of value in biodegradable packaging films. It would be impossible, and perhaps not even wholly necessary, for these to completely replace synthetic plastics. However, more research is needed into other possible applications (Tharanathan, 2003).

17.5 Consumer perceptions

In order to maximize the potential of a new technology, it is important to consider consumer concerns and acceptance. Two main aspects that consumers use to make judgments about the quality of fresh-cut foods are appearance and freshness at the time of purchase (Barry-Ryan *et al.*, 2007). Some edible coatings are invisible to the eye but consumers must still be informed of their existence. Regarding the safety and health impact of edible coatings, the regulations command that edible polymer films must be declared appropriately to the consumer, no matter how small the amount used.

Consumers are concerned about the types of products that are coated, safety of the coating material, sensory qualities of the resulting products, end benefits of edible film coatings and the cost of the food products packaged with edible film coatings (Wan *et al.*, 2007). In 2003, a study showed that, compared with Caucasians, Asians were more aware and Hispanics less aware of the use of edible coatings. A 7% increase in purchase intent was observed after the advantages of edible coatings had been described to consumers (Sonti, 2003).

Research in the field of sensory analysis of fruits and vegetables coated with edible films is still in its early stages. One example of consumer opinion regarding edible coating compared samples of fresh strawberries after 1 week of storage at 2 °C and strawberries with chitosan coatings. The samples had similar sensory descriptors. It was shown that the coatings did not change consumer acceptability of flavor, sweetness or firmness of samples (Han *et al.*, 2005).

17.6 Regulatory and economic aspects

17.6.1 Regulatory aspects

Edible films are considered to be either food ingredients, food additives, food contact substances or food packaging materials. They have to meet the relevant regulations for substances in one of these categories. In Europe, most components of edible coatings are presented as general food additives, although some are mentioned specifically for coating applications (e.g. acacia and karaya gums, beeswax). In both instances, components of edible coatings have to meet the requirements of EU Regulation 1331/2008, which lays down specific purity criteria for food additives and quantities allowed for use. For those that are harmless enough to have no specific quantity restriction, it is permitted to add as much as is needed to achieve the desired result, but not more. This is known as the 'quantum satis' principle (Rojas-Graü *et al.*, 2011).

In the USA, the FDA regulations have established that edible coatings must be 'generally recognized as safe' (GRAS) in keeping with the conditions of intended use and with the quantity of additive applied in accordance with good manufacturing practices. Furthermore, the US-FDA requires that all products containing film-coating-substances are safe for their intended use before they are permitted for sale in the USA. It is the responsibility of the manufacturer of the coatings to ensure that food contact materials comply with the specifications and limitations in all applicable authorizations.

A very important aspect that also must be regulated when using edible coatings is the presence of allergens. Pre-packaged fruits and vegetables that have a protective edible coating or wax must include the source of any allergen on their labels. In particular, for coatings derived from milk, wheat, soy or peanut proteins, their source is declared as they could cause allergic reactions.

17.6.2 Economic aspects

Despite advantages claimed for edible coatings, they are not yet used at industrial level on a large scale. Costs are considered to be the main obstacle for large scale application, although the use of edible films is still increasing. In 1967, their commercial use was mainly limited to wax layers on fruits. Between 1986 and 1996 the number of companies offering these types of product grew from about 10 to 600. The application of edible films to preserve the quality of foods has now expanded to a greater variety of foods. The industry has a total annual revenue of more than $100 million (Pavlath and Orts, 2009). Table 17.2 shows some of the specific coatings that have been developed and are already available with commercial names.

Very encouraging for the use of edible coatings is the report 'Advances in Food Antimicrobial Coatings' made by the market research outfit Frost & Sullivan, which indicates that food safety concerns spur a rise in antimicrobial coatings (Spinner, 2013).

17.7 Production and market issues

The development of new edible coatings with improved functionality and performance for fresh and minimally processed fruits is one of the main challenges

Table 17.2 Commercially available edible coatings

Name	Company	Composition	Fruits/vegetables
Fresh Seal™	BASF Corporation, NJ, USA	Cellulose derivatives and emulsifiers	Avocado, cantaloupe, mangoes, papaya
Natural Shine™	Pace International, LLC, USA	Carnauba wax	Citrus
NatureSeal™ AgriCoat	Mantrose-Haeuser Co., Inc., USA	Cellulose derivatives without sucrose fatty acid esters	Most fruits and vegetables
PacRite®	Pace International, LLC, USA	Water-based, carnauba-shellac emulsions, shellac and resin water emulsion, water-based mineral oil fatty acid emulsions	Citrus, stone fruits, vegetables, potatoes
PrimaFresh®	Pace International, LLC, USA	Mineral-oil base	Cherries, stone fruits, mangoes, melons, apples
Schield-Brite products	Pace International LLC, USA	Shellac/carnauba-based, natural wax vegetable oil/wax and xanthan gum	Citrus, pears, stone fruits
Semperfresh™	Mantrose Ltd, UK	Sucrose esters of fatty acids, sodium carboxymethyl cellulose	Most fruits and vegetables, nectarines
TAL Pro-Long (Pro-Long)	Courtaulds Group, UK	Sucrose polyesters of fatty acids and the sodium salt of carboximethylcellulose	Pears, apples, bananas, plums

for keeping fruit fresh in the postharvest industry. In the past few years, research efforts have focused on the design of new eco-friendly coatings based on biodegradable polymers. These not only reduce the requirements of packaging but also lead to the conversion of byproducts of the food industry into value-added film-forming components (Vargas *et al.*, 2008). The success of edible coatings for fruits and/or vegetables depends mainly on selecting films or coatings appropriate for a specific product.

As well as their advantages, the edible coatings used have their limitations. For example, the efficacy of films and coatings is dramatically affected by water availability so the moisture content of the foodstuff and the relative humidity in the storage environment should be taken into account in order to develop effective films and coatings with antioxidant activity (Bonilla *et al.*, 2012).

Furthermore, wax and sucrose fatty acid ester (SFAE) mixtures, the most widely used edible coatings for fruits and vegetables, are not equally effective on all produce. Moreover, consumers tend to be wary of waxy coatings because of their undesirable taste (Park, 1999). The production of edible films from fruit purees is an interesting way of reducing the huge postharvest losses of tropical fruits. However, their main limitation is caused by the relatively poor mechanical and barrier properties of biopolymers when compared with synthetic polymers (Azeredo, 2012).

17.8 Further developments

Currently, there are many methods that can be used to improve quality and increase safety of fresh food products. However, the use of edible coatings stands out among other techniques in the struggle for maintaining the freshness of fresh vegetables and cut fruits. Many studies concerning food coatings have dealt with the extraction and characterization of low-cost coating materials, the optimization of methods for coating application, and the effects and implications of applications on different food matrices. However, in most cases, the commercial implementation of edible coatings still requires further research and consideration of some critical issues such as consumer acceptance and regulatory aspects (Rojas-Graü *et al.*, 2010).

One major advantage of using edible coatings is that several active compounds can be incorporated into the polymer matrix and consumed with the food. In this way, nanotechnology can be used to incorporate active ingredients into edible coatings for fresh and minimally processed fruits and vegetables. However, current applications of nanosystems to fresh products are still scarce. More research about the effect of nanotechnologies on quality and nutritional aspects on such fruits and vegetables should be conducted.

Further studies on edible film coatings should focus on understanding the sensory attributes of integrated coated products, on proper declaration of coating usage with a focus on marketing the natural ingredients added, and finally should target marketing strategies to advertise the direct consumer benefits of the resulting edible film coating products (Wan *et al.*, 2007).

17.9 Questions for discussion

1. What materials can be used to obtain edible coatings?
2. Is there a 'perfect edible coating material'?
3. What are the properties that make edible coatings so attractive?
4. Why should edible substances used for food coating be declared on the food label?
5. Can nanotechnology help edible coating and, if so, how?
6. Can we classify the nano-type of packaging as 'active' or 'passive'?
7. Do we really need edible films? Please explain your answer.
8. Are edible coatings visible or invisible?
9. Are edible coatings 'the packaging of the future'?
10. Do you consider food coating with edible films an economically viable technology?

17.10 Sources of further information and advice

EMBUSCADO, M.E. and HUBER, K.C. (eds.) (2009). *Edible Films and Coatings for Food Applications*. Springer Science and Business Media, DOI 10. 1007/978-0-387-92824-11.

HAN, J.H. (2005). New technologies in food packaging: Overview. In *Innovations in Food Packaging*, J. H. Han, ed., Academic Press, London, 3–11.

OLIVAS, G.I. and BARBOSA-CÁNOVAS, G. (2009). Edible Films and Coatings for Fruits and Vegetables. In *Edible films and coatings for food applications*, Milda E. Embuscado and Kerry C. Huber (Eds.), Springer NY.

SOZER, N., and KOKINI, J.L. (2012). Chapter 6 – The Applications of Nanotechnology. In *Chemical Analysis of Food: Techniques and Applications*, pp. 145–176. Academic Press, Boston.

17.11 References

ANON. (2012). What You DON'T See Is Also What You Eat (retrieved on 09.10.2012 from www.rense.com)

AZEREDO, H.M.C. (2012). Edible Films and Coatings from Tropical Fruits (retrieved on 10.10.2012 from www.worldfoodscience.org/cms/?pid=1005081)

BAI, J., HAGENMAIER, R.D., and BALDWIN, E.A. (2003). Coating selection for 'Delicious' and other apples. *Postharvest Biology and Technology* 28, 381–90.

BALDWIN, E.A., BURNS, J.K., KAZOKAS, W., BRECHT, J.K., HAGENMAIER, R.D., *et al.* (1999). Effect of two edible coatings with different permeability characteristics on mango (Mangifera indica L.) ripening during storage. *Postharvest Biology and Technology* 17, 215–26.

BALDWIN, E.A., NISPEROS, M.O., CHEN, X., and HAGENMAIER, R.D. (1996). Improving storage life of cut apple and potato with edible coating. *Postharvest Biology and Technology* 9, 151–63.

BARRY-RYAN, C., MARTIN-DIANA, A., RICO, D., and BARAT, J. (2007). Extending and measuring the quality of fresh-cut fruit and vegetables: a review. *Trends in Food Science & Technology* 18, 373–86.

BONILLA, J., ATARÉS, L., VARGAS, M., and CHIRALT, A. (2012). Edible films and coatings to prevent the detrimental effect of oxygen on food quality: Possibilities and limitations. *Journal of Food Engineering* 110, 208–13.

CARLIN, F., GONTARD, N., REICH, M., and NGUYEN-THE, C. (2001). Utilization of Zein Coating and Sorbic Acid to Reduce Listeria monocytogenes Growth on Cooked Sweet Corn. *Journal of Food Science* 66, 1385–9.

CHIEN, P.-J., SHEU, F., and YANG, F.-H. (2007). Effects of edible chitosan coating on quality and shelf life of sliced mango fruit. *Journal of Food Engineering* 78, 225–9.

DEL-VALLE, V., HERNÁNDEZ-MUÑOZ, P., GUARDA, A., and GALOTTO, M.J. (2005). Development of a cactus-mucilage edible coating (Opuntia ficus indica) and its application to extend strawberry (Fragaria ananassa) shelf-life. *Food Chemistry* 91, 751–6.

DHANAPAL, A., SASIKALA, P., RAJAMANI, L., KAVITHA, V., YAZHINI, G., and BANU, M.S. (2012). Edible films from Polysaccharides. *Food Science and Quality Management* 3.

DU, W.X., OLSEN, C.W., AVENA-BUSTILLOS, R.J., MCHUGH, T.H., LEVIN, C.E., and FRIEDMAN, M. (2009). Effects of allspice, cinnamon, and clove bud essential oils in edible apple films on physical properties and antimicrobial activities. *Journal of Food Science* 74, M372–8.

EFSA. (2012). The European Union Summary Report on Trends and Sources of Zoonoses, Zoonotic Agents and Food-borne Outbreaks in 2010. *EFSA Journal* 10, 2597.

FELLMAN, J.K., RUDELL, D.R., MATTINSON, D.S., and MATTHEIS, J.P. (2003). Relationship of harvest maturity to flavor regeneration after CA storage of 'Delicious' apples. *Postharvest Biology and Technology* 27, 39–51.

FRANSSEN, L.R., and KROCHTA, J.M. (2003). Edible coatings containing natural antimicrobials for processed foods. In Roller, S. (Ed.) *Natural Antimicrobials for Minimal Processing of Foods*. CRC Press, Boca Raton. 250–62.

FUNDAZIOA, E. (2006). Edible coatings: the packaging of the future? *Science Daily* (retrieved 7.10.2012 from http://www.sciencedaily.com/releases/2006/09/060905084809.htm).

GARCIA, E., and BARRET, D.M. (2002). Preservative treatments for fresh cut fruits and vegetables. In Lamikanra, O. (Ed.) *Fresh-Cut Fruits and Vegetables*. CRC Press, Boca Raton, FL.

GERALDINE, R.M., SOARES, N.D.F.F., BOTREL, D.A., and DE ALMEIDA GONÇALVES, L. (2008). Characterization and effect of edible coatings on minimally processed garlic quality. *Carbohydrate Polymers* 72, 403–9.

HAN, C., LEDERER, C., MCDANIEL, M., and ZHAO, Y. (2005). Sensory Evaluation of Fresh Strawberries (Fragaria ananassa) Coated with Chitosan-based Edible Coatings. *Journal of Food Science* 70, S172–8.

HARDENBURG, R.E. (1967). Wax and Related Coatings for Horticultural Products In A Bibliography. *Agricultural Research Service Bulletin* 51–55, United States Department of Agriculture, Washington, DC.

JIN, T., and GURTLER, J.B. (2012). Inactivation of salmonella on tomato stem scars by edible chitosan and organic Acid coatings. *Journal of Food Protection* 75, 1368–72.

KADER, A.A. (1986). Biochemical and physiological basis for effects of controlled and modified atmospheres on fruits and vegetables. *Food Technology* 40, 99–104.

KAYS, S.J., and PAULL, R.E. (2004). Stress in harvested products. In Kays S.J and Paull R.E. *Postharvest Biology*. Exon Press Athens, GA.

KE, D., YAHIA, E., MATEOS, M., and KADER, A. (1994). Ethanolic fementation of Barlett pears as influenced by ripening stage and atmospheric composition. *Journal of the American Society for Horticultural Science* 32, 593–600.

KHIN, M.M., ZHOU, W., and PERERA, C. (2005). Development in the Combined Treatment of Coating and Osmotic Dehydration of Food – A Review. *International Journal of Food Engineering* 1.

KNEE, M. (1980). Physiological responses of apple fruits to oxygen concentrations. *Annals of Applied Biology* 96, 243–53.

LEE, H.B., NOH, B.S., and MIN, S.C. (2012). Listeria monocytogenes inhibition by defatted mustard meal-based edible films. *International Journal of Food Microbiology* 153, 99–105.

MAFTOONAZAD, N., RAMASWAMY, H.S., MOALEMIYAN, M., and KUSHALAPPA, A.C. (2007). Effect of pectin-based edible emulsion coating on changes in quality of avocado exposed to Lasiodiplodia theobromae infection. *Carbohydrate Polymers* 68, 341–9.

MEHYAR, G.F., AL-QADIRI, H.M., ABU-BLAN, H.A., and SWANSON, B.G. (2011). Antifungal effectiveness of potassium sorbate incorporated in edible coatings against spoilage molds of apples, cucumbers, and tomatoes during refrigerated storage. *Journal of Food Science* 76, M210–7.

OLIVAS, G.I., and BARBOSA-CÁNOVAS, G.V. (2005). Edible coatings for fresh-cut fruits. *Critical Reviews in Food Science and Nutrition* 45, 657–70.

OLIVAS, G.I., MATTINSON, D.S., and BARBOSA-CÁNOVAS, G.V. (2007). Alginate coatings for preservation of minimally processed 'Gala' apples. *Postharvest Biology and Technology* 45, 89–96.

OLIVAS, G.I., RODRIGUEZ, J.J., and BARBOSA-CANOVAS, G.V. (2003). Edible coatings composed of methylcellulose, stearic acid, and additives to preserve quality of pear wedges. *Journal of Food Processing and Preservation* 27, 299–320.

PARK, H.J.H.J. (1999). Development of advanced edible coatings for fruits. *Trends in Food Science and Technology* 10, 254–60.

PAVLATH, A.E., and ORTS, W. (2009). Edible Films and Coatings: Why, What, and How? In: Milda E. Embuscado and Kerry C. Huber (Ed.). *Edible films and coatings for food applications*. Springer NY.

PEREZ-GAGO, M.B., and KROCHTA, J.M. (2001). Lipid Particle Size Effect on Water Vapor Permeability and Mechanical Properties of Whey Protein/Beeswax Emulsion Films. *Journal of Agricultural and Food Chemistry* 49, 996–1002.

ROJAS-GRAÜ, M.A., RAYBAUDI-MASSILIA, R.M., SOLIVA-FORTUNY, R.C., AVENA-BUSTILLOS, R.J., MCHUGH, T.H., and MARTÍN-BELLOSO, O. (2007). Apple puree-alginate edible coating as carrier of antimicrobial agents to prolong shelf-life of fresh-cut apples. *Postharvest Biology and Technology* 45, 254–64.

ROJAS-GRAÜ, M.A., SOLIVA-FORTUNY, R., and MARTÍN-BELLOSO, O. (2010). Edible coatings: past, present and future. *Stewart Postharvest Review* 6, 1–5.

ROJAS-GRAÜ, M.A., SOLIVA-FORTUNY, R., and MARTIN-BELOSO, O. (2011). Use of edible coatings for fresh-cut fruits and vegetables. In: Martin-Beloso, Olga and Soliva-Fortuny, Robert (Ed.). *Advances in fresh-cut fruits and vegetables processing*. CRC Press.

SIRACUSA, V., ROCCULI, P., ROMANI, S., and ROSA, M.D. (2008). Biodegradable polymers for food packaging: a review. *Trends in Food Science & Technology* 19, 634–43.

SOLIVA-FORTUNY, R., ROJAS-GRAÜ, M.A., MARTIN-BELOSO, M. (2011). Polysaccharides coatings. In *Edible Coatings and Films to Improve Food Quality*, Second Edition, Baldwin, E.L.A., Hagenmayer, R.D., Bai, J. (Eds.), CRC Press, 120.

SONTI, S. (2003). Consumer Perception and Application of Edible Coatings on Fresh-Cut Fruits and Vegetables – Mater thesis *Osmania University College of Technology*.

SPINNER, J. (2013). Food safety concerns spurs rise in antimicrobial coatings, *Food Production Daily* (retrieved 2.05.2013 from http://www.foodproductiondaily.com/Quality-Safety/Food-safety-concerns-spurs-rise-in-antimicrobial-coatings).

THARANATHAN, R.N. (2003). Biodegradable films and composite coatings: past, present and future. *Trends in Food Science & Technology* 14, 71–8.

VALENCIA-CHAMORRO, S.A., PEREZ-GAGO, M.B., DEL RIO, M.A., and PALOU, L. (2009). Curative and preventive activity of hydroxypropyl methylcellulose-lipid edible composite coatings containing antifungal food additives to control citrus postharvest green and blue molds. *Journal of Agricultural and Food Chemistry* 57, 2770–7.

VALENCIA-CHAMORRO, S.A., PEREZ-GAGO, M.B., DEL RIO, M.A., and PALOU, L. (2010). Effect of antifungal hydroxypropyl methylcellulose-lipid edible composite coatings on Penicillium decay development and postharvest quality of cold-stored 'Ortanique' mandarins. *Journal of Food Science* 75, S418–26.

VARGAS, M., PASTOR, C., CHIRALT, A., MCCLEMENTS, D.J., and GONZALEZ-MARTINEZ, C. (2008). Recent advances in edible coatings for fresh and minimally processed fruits. *Critical Reviews in Food Science and Nutrition* 48, 496–511.

WAN, V. C.-H., LEE, C.M., and LEE, S.-Y. (2007). Understanding consumer attitudes on edible films and coatings: focus group findings. *Journal of Sensory Studies* 22, 353–66.

Part IV

Laboratory testing for fresh produce safety

18

Pathogen testing in fresh produce: Earthbound Farm

W. Daniels, Earthbound Farm, USA

DOI: 10.1533/9781782420279.4.247

Abstract: Fresh and fresh-cut produce had been linked to numerous foodborne illnesses leading up to the summer of 2006, when the largest outbreak associated with fresh spinach took shape. Earthbound Farm packaged the spinach that led to the outbreak and recall that sickened over 200 and killed three. Earthbound emerged with a new, multi-hurdle approach to food safety in fresh-cut produce. Many of the enhanced controls focused on testing. In this chapter, the reader will learn about the recall, as well as how the multi-hurdle program was launched and the nuances of managing a new concept in an adolescent industry.

Key words: test-and-hold, pathogen, enterohemorrhagic *E. coli* (EHEC), food safety in fresh produce, spinach.

18.1 Introduction

Fresh and fresh-cut produce had been in the news several times by 2006. Between 1997 and 2006, produce had been implicated in recalls and outbreaks about 19 times, with little consistency. The pathogen varied, the produce varied and the cause was never revealed. All the while, Earthbound Farm had been growing and packing organic produce (mainly fresh-cut salads) since 1984 and took pride in operating at industry-best standards. They were active participants in food safety initiatives, including several technical working groups. Their facility was regularly subjected to second and third party audits as well as regulatory inspections. In every instance, they were found to exceed expectations. The fall of 2006 changed all that.

The Earthbound team was notified by California Department of Public Health (then known as California Department of Health Services) that they may be implicated in the outbreak of *E. coli* O157:H7 in spinach that had sickened consumers, was paralyzing the fresh spinach market and concerning the federal government. Although several brands were being named by case patients, a brand that Earthbound Farm co-packed was being mentioned often. On Friday, 15 September 2006, with just that early anecdotal evidence and careful deliberation, the choice was clear: the Earthbound Farm Incident Management Team initiated a recall of all brands of spinach it had produced. By Friday afternoon, the team had advised the Food & Drug Administration (FDA) of its decision to recall. Inspectors had arrived to start the investigation and the Earthbound Farm Recall Team and Incident Management Team were fully engaged.

18.2 The investigation

The ensuing weeks and months were packed with intensive investigation by FDA, CDPH and Earthbound Farm teams. Everyone was determined to find the answer to what led to this outbreak. And yet, despite the rigor and comprehensiveness of the investigations, no clear cause of the contamination was identified. No contamination in the processing facility, no malfunctioning equipment, no records out of line, no smoking gun.

But they got closer than ever before. The investigation eventually pointed to one ranch from which spinach had been harvested on that fateful day. Bacteria matching the outbreak strain were found in the vicinity of the field where the spinach had been grown and harvested. Nothing was found on the field, but about a mile away, matches were found in wild pig intestines, cow feces and samples from standing water. But the vector to the field was never identified (http://www.cdph.ca.gov/pubsforms/Documents/fdb%20eru%20Spnch%20EC%20Dole032007wph.PDF).

This was enough to catalyze Earthbound Farm to unprecedented action. Knowing that contamination likely occurred in the field, and that the system in place at the time was not able to mitigate the contamination, Earthbound Farm set out to re-envision food safety for fresh and fresh-cut produce. They had been operating at the industry's best standards and yet this tragic incident happened; it was clear that a complete overhaul was necessary to prevent this from happening again.

18.3 A multi-hurdle approach to food safety

The leadership at Earthbound Farm was clear with their instruction to Will Daniels, then director of food safety, and the team responsible for food safety: don't let this happen again. And to make this a reality in a process without a true kill step, they knew they had their work cut out for them. So the team tore apart their ideas about what was possible in their food safety program and starting building a true multi-hurdle approach to food safety. They looked to industries other than fresh-cut salads that had faced these challenges to see what might be applicable in their process. Seed, for example, is a risk to the sprout industry as a vector for contamination. As baby greens are not that far removed from a sprout, seeds were considered a potential risk. The beef industry had instituted a test-and-hold program for beef trim and they were seeing improvements. All ideas were on the table and nothing was considered a bad idea until it was proven to be inappropriate.

Beyond the manufacturing facility, the team also identified as many potential controls as possible in the field. They increased the number and frequency of inspections in the field, instituting unannounced inspections in addition to regularly scheduled visits. They placed more stringent controls on suppliers and instituted requirements for pathogen testing of all inputs (water, fertilizer, compost, etc.). And they put together a blue ribbon panel of the nation's top food safety scientists to help them build the best food safety program possible.

At the core of Earthbound Farm's new food safety program was the Test-and-Hold firewall. Despite widespread skepticism from others in the industry, Earthbound Farm persevered, believing that test-and-hold had the potential to revolutionize food safety reliability in a process with no kill step. Pathogen-specific test-and-hold was being implemented by the meat industry with success. Test methods were now available that would allow rapid analysis for this highly perishable product. The testing method was polymerase chain reaction (PCR) testing, which afforded a 12–16 hour turnaround, fast enough not to adversely affect the freshness of the product or field planting schedules.

In the field, the testing focused on ensuring that all the inputs going into the field were free from pathogens. Seed, soil amendments, water and even transplants were required to be tested for the presence of *E. coli* O157:H7, non-O157 enterohemorrhagic *E. coli* (EHEC) and *Salmonella*. Inputs that cleared testing were allowed to be used, inputs that got a positive test were rejected. Water was tested throughout the growing cycle at a frequency relevant to its source (more frequently for groundwater; less frequently for deep well water).

The most significant undertaking from a testing standpoint that Earthbound Farm implemented was tissue testing. Product would be tested for the presence of pathogens at the 'raw product' stage (product that has been harvested and delivered to the manufacturing facility) in addition to testing performed at the 'finished goods' stage (product that has been washed and packed into its consumer package). Some product would also be tested in the field before harvest but with limited benefit. The tests performed in the raw and finished goods stages were significantly more rigorous than those that could be performed in the field, because once in Earthbound's facility, the product was under control. However, the team believed that in cases where the lettuce was going to be both used in a fresh-cut salad (processed) and field-packed as a whole commodity (e.g. romaine hearts from the same field could wind up in a retail 3-pack and also chopped in a salad), it would be prudent to have cross-referenced field testing. If field testing did not occur and a positive was found on raw material for further processing, any product from that field that was not tested would be implicated in that positive test.

Most of the focus early on was put into raw product testing, which is where pathogens were most likely to be caught and where the product was in Earthbound's control. If the outbreak investigation was correct in the assumption that the implicated ranch did, in fact, supply the contaminated spinach to the manufacturing facility on that day in August 2006, Earthbound Farm only used one pallet of product from that ranch on that day (approximately 500 pounds). To put this into context, the facility that made the contaminated product was processing about 200 000 pounds of spinach daily and of that, just 500 pounds had led to a major outbreak. The test-and-hold sampling program took this into consideration as well as the International Commission for the Microbial Specifications of Food's (ICMSF) Micro-organisms in Foods 7, Microbiological Testing in Food Safety Management, which categorized risk for food products. Fresh-cut salad fell into Case 15, the riskiest category, which said that once the product was in its finished package, there was the ability for microorganisms to not only survive the process but also potentially grow through the shelf-life of the product. At the time, the sampling program was designed assuming uniform contamination and that approximately 10–15% of an acre of spinach could lead to foodborne illness. The confidence that this level of contamination could be detected 99.99% of the time was very high (see Fig. 18.1).

To get this level of confidence, field lots were broken down into production units comprising four pallets of harvested product (approximately 2000 pounds). From that production unit, 60 grabs of product would be randomly taken to create one, 150 gram sample. This sample was then submitted to the on-site third party lab (IEH Laboratories) for PCR analysis, looking for *E. coli* O157:H7, non-O157 *enterohemorrhagic E. coli* (EHEC), *Salmonella* and *Shigella*. Because of the

Fig. 18.1 Raw product sampling.

short time the testing required, this new procedure had little impact on the production flow of the company's processing environment.

Not only would this hurdle act as a firewall to prevent contamination from entering into the manufacturing process, it also allowed for near real-time investigation of contamination events. Because of the rapid nature of the testing, when a positive occurs in the raw material phase, inspectors are in the field and lot from which the product came the very next day. These inspections are meant to try and identify the cause of the contamination. Typically nothing is found, but there have been a handful of events over hundreds of positives where the potential source of the contamination was found. In addition, when a positive occurs, there is a great deal of information that is put into a database around that positive. Not only are the basics such as product type, grower and location recorded, but Earthbound Farm is also recording things like soil type, temperature, humidity, animal activity, inputs used and more. Even when these data points do not seem valuable today, tomorrow they could be the answer to many contamination events (Fig. 18.2).

18.4 Testing is not the only answer

Although Earthbound Farm believes in the power of testing, it is important to emphasize that testing alone cannot substitute for a comprehensive food safety program. It is important to note that these testing protocols are simply another hurdle in the multi-hurdle approach. The testing is designed to catch gross contamination events and provide verification of upstream efforts. At the time of this writing, they have tested raw material for almost 6 years and have over 700 positives. They do not believe that all of these events are gross contamination events, so the sampling program is able to catch low-level contamination and remove that risk from the product stream. In addition, over time, as new interventions are implemented in the field, the company believes they will see a reduction in the number of positives found on the raw material.

The second firewall implemented around tissue testing is the finished goods test-and-hold. To achieve this, the production unit is defined as a 2-hour window of production on each packing line. For each production unit, a total of 60 grabs through the 2 hours are collected, equaling a 150 gram sample. This sample is submitted to the lab and PCR tested for *E. coli* O157:H7, non-O157 *enterohemorrhagic E. coli* (EHEC), *Salmonella* and *Shigella*. Production units that get a negative test are released, positives are destroyed. At this phase, wash water samples are collected before and after the 2-hour production unit and libraried for analysis in the event of a positive. In the event of a positive, the water samples are submitted for testing of the same pathogens. If the test comes back negative, the positive is considered to be isolated within the 2-hour unit. If it were positive, the flanking units would be considered positive regardless of the finished goods test. The flanking water is tested and the scope expands until negative water results are obtained. Earthbound Farm maintains a rolling 6-hour hold in finished

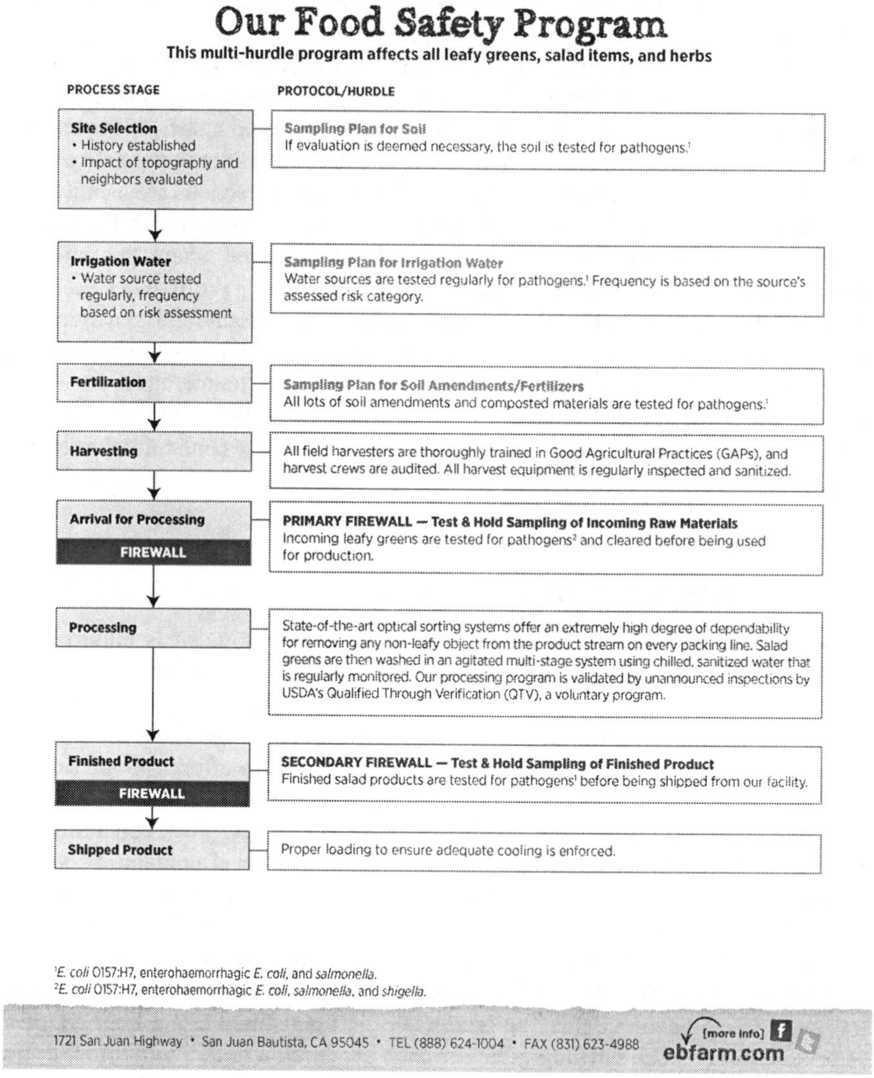

Fig. 18.2 Earthbound Farm's multi-hurdle food safety program, post event.

goods at all times. Holding the unit being tested, the test unit before it and the test unit after are also held. This helps to ensure that there is control of the product in the event of a positive and prevents the need to withdraw or recall product. One would also conclude that in the event of a finished goods positive, the plant may be the vector for contamination. Earthbound considers this every time they get a positive and performs a thorough investigation of the process records in addition to performing environmental sampling to see if the contamination can be found.

In the time that the program has been in place, they have never found the plant to be the cause of the contamination.

18.5 Examining the data

At the time of this writing, Earthbound Farm has seen dozens of finished goods positives over the course of five and half years. With both raw and finished goods testing, there is a seasonality to the positives. The peak of activity is July through October. However, there have been positives found in product from all growing regions (Central California, Southern California, Arizona, Nevada, Northern Mexico) and at all times of the year with all ingredients. If you average the positives annually, the company sees a positive about every 3 days on the raw product side, about every 5 weeks on the finished goods side (Fig. 18.3).

It is clear that contamination of leafy greens is a sporadic event. Over the course of this testing program, never has an entire field lot, nor an entire production day tested positive. This supports the strategy to sample a field lot in smaller increments than the entire planting. The conventional wisdom is that because a field lot has all the same inputs, that contamination would be uniform across the entire lot. And that might be true if the contamination were caused by those things. However, the testing that Earthbound Farm has conducted over the course of several years suggests that this is not the case. The same can be said for the finished goods testing and implicating the entire production day. Conventional wisdom here suggests that if contamination occurs, you need to take out the entire day, from sanitation to sanitation. The positive events in finished goods are again sporadic and not spread through an entire day, suggesting that breaking the manufacturing day down into smaller units is a good strategy (Fig. 18.4).

The potential vectors of contamination are many. As an example, a ranch that has been considered high risk because of deer and wild boar intrusion has been protected with fences 3 feet deep and 6 feet high to prevent boar and deer intrusion, respectively. However, even with this extreme mitigation effort, contaminated product is still detected from this ranch, suggesting that there are other vectors at work like perhaps birds, small animals, wind, insects, etc.

18.6 Lessons learned

For a food safety testing program to be sound, the testing must be comprehensive. One cannot sample part of a field or part of a production day to ensure that all the product is safe. Every pallet of raw material must be tested from every field. Every production line must be tested throughout the day. Without this exacting, comprehensive rigor, the impact of a positive would simply make the test-and-hold program unaffordable. If, for example, one test were conducted in finished goods over the course of a manufacturing day and it came back positive, one must assume that everything that was not tested is positive and therefore the entire

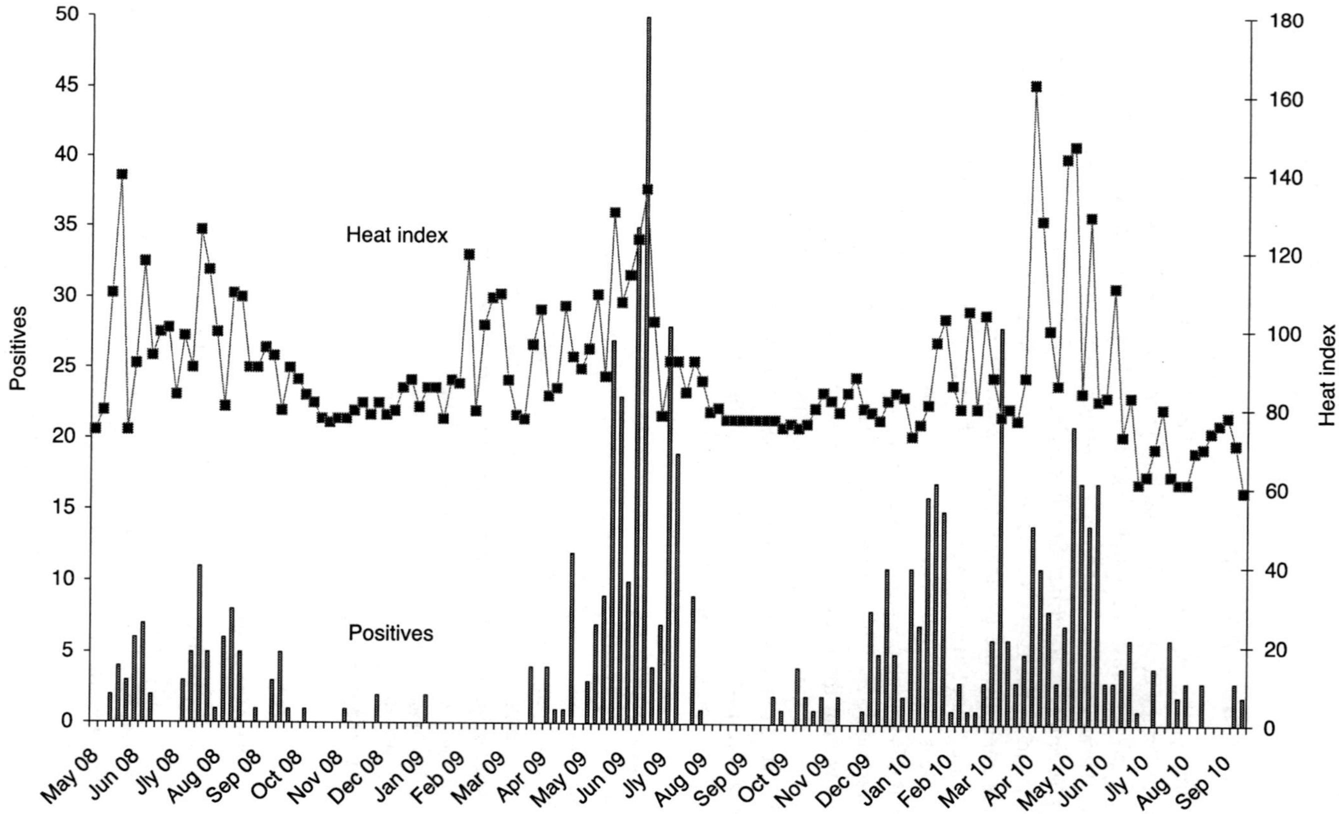

Fig. 18.3 Raw positives versus heat index.

Fig. 18.4 Depiction of raw product sampling in a field showing sporadic nature of contamination when laid out in field.

day's worth of product must be destroyed. Same for the field: if one test is pulled for a 40 acre planting and it comes back positive, the entire field lot must be destroyed because one must conclude the untested product is positive. The sample size in the testing program is, in the event of a positive, tolerable to the grower or to the manufacturing facility. Throwing away 2000 pounds of product is much different from 70 000 pounds in a 10-acre planting. Likewise on the finished goods side, 2000 pounds of finished goods is a much more tolerable loss than 300 000 pounds produced in a day.

This program is not designed to be a one size fits all approach. It was specifically designed for Earthbound Farm's system of production and the final use of the finished goods (packaging, distribution, ready-to-eat, etc.). The risks are different with every model and a thorough risk assessment must be conducted to understand what controls must be implemented to prevent contaminated product from getting to the consumer. However, given that the fresh-cut process does not have a kill step, that the distribution model is nationwide in a container that allows bacterial growth, testing appears to be a realistic strategy to reduce the risk of foodborne illness.

In addition to the testing performed on the leafy green salads manufactured by Earthbound Farm, they are testing and holding tomatoes, green onions, melons,

Fig. 18.5 Other products that go through a test-and-hold program for pathogens.

sliced apples and mini-peeled carrots. Tomatoes, green onions and melons should not be surprising as they have all been associated with recalls, outbreaks and foodborne illness. Sliced apples and mini-peeled carrots are a bit more surprising but Earthbound Farm tests these products because of the target consumer, children, being more susceptible to foodborne illness (Fig. 18.5).

Before a pathogen testing program is initiated, it must be clear what will happen in the event of a positive test. What is the impact to the product, the grower, the organization, etc.? Does the positive implicate other finished products that may be in distribution? Does the positive implicate other areas of the plant as product moves through the manufacturing process? What raw materials are implicated and what is the impact of that analysis? Running through mock drills will help to identify potential holes in the program. It is extremely important to be sure your program is well thought out, has several control steps to prevent violations to procedures and is audited regularly to ensure the program is being followed. The last thing one wants is to get a positive and then discover that there has been a breach in procedure leading to the need to withdraw or recall product. No matter what level of sophistication or automation your testing program has, there are humans involved in the process that may cause errors, so regular monitoring is crucial.

18.7 Conclusion

In summary, it is important to reiterate that Earthbound Farm does not treat their testing program as the answer to food safety; it is merely an important component of their multi-hurdle approach to food safety. It is meant to catch gross contamination events, provide verification of upstream efforts and provide valuable data to an issue that is yet to be fully understood. As industry, academia and government work to reduce the threat of foodborne illness in fresh and fresh-cut produce, these data may prove to be helpful in getting to the answers faster and possibly cracking the code to answer the question of how our baby greens are getting contaminated. At the very least, the test-and-hold program is eliminating contaminated product from the process and not challenging the system to eliminate it through its many controls, therefore reducing the risk to consumers.

18.8 Questions for discussion

1. Why is a multi-hurdle approach considered a good strategy for fresh-cut salad manufacturing?
2. What are two advantages of rapid analysis testing (PCR)?
3. Does Earthbound Farm consider contamination events to be sporadic or uniform in nature?
4. What are the benefits of smaller sample sizes in a test-and-hold program?
5. According to Earthbound Farm, what is the raw material test-and-hold program designed to do?
6. What must one assume in the event of a finished goods positive?
7. What is the purpose of testing wash water in the event of a finished goods positive?
8. Was the cause of the contamination in 2006 ever determined?
9. What is the benefit of a comprehensive testing program?
10. Is Earthbound Farm's test-and-hold program designed to fit any system? Why or why not?

19

Capacity building of legislative fresh produce testing in China

Z. Ye, S. Chen, F. Wang and T. Chen, Chinese Academy of Agricultural Sciences, China and L. Xiao, National Certification and Accreditation Administration, China

DOI: 10.1533/9781782420279.4.258

Abstract: Agro-products are vital to Chinese international trade. Between 2002 and 2010, Chinese export of agro-products saw an annual increase of 11.7%. In Chinese cuisine most vegetables are cooked, therefore microbial contamination is not a major issue. Instead, chemical contaminants and residues are the major concern in China. The inspection and testing system plays a key role in policy-making, technology transfer and technical consultation services to ensure quality and safety of fresh produce. After years of effort to build an inspection and testing system, remarkable achievements have been made. However, many aspects of the system need adjustment to meet new challenges.

Key words: fresh produce, safety, testing, inspection, capacity.

19.1 Introduction

China's legislative testing system for fresh produce is dedicated to providing scientific and impartial monitoring, identification and evaluation services for agro-products, agricultural inputs and production environments. This testing system was established on the basis of advanced analytical equipment and is guaranteed by reliable laboratory facilities in accordance with national regulations and standards (Xiao, 2006). This system is one of the major technical supports to the national network of agro-product quality and safety and it also plays a crucial role in governmental efforts to ensure sound management of the quality and safety of agro-products. It provides vital information for policy-making, technical services and consultation. Strengthening and building the capacity of the inspection and testing system for agro-product quality and safety is of critical significance for China in order to ensure food safety, to improve strategic structural adjustment in the agricultural sector and to upgrade the competitiveness of Chinese agro-products in the global market.

19.2 General situation of the Chinese legislative testing system for agro-product quality and safety

Since the reforms and opening-up policy adopted in China in the early 1980s, agricultural production and the rural economy of the country have experienced a rapid and continued development, which has given rise to a historical breakthrough from deficiency to surplus in terms of the supply of major agricultural products. As a result of the massive increase in quantity of stable food commodities, the issue of agro-product quality and safety has gradually emerged as a major concern. A new era with dual emphasis on quantity and safety has emerged. As the most direct and important measure in supervising and managing the quality and safety of fresh produce, legislative testing has received much attention from the Chinese government in the way of initiating capacity building efforts and projects (Qian *et al.*, 2007).

In China, the Ministry of Agriculture (MOA) is the sole government sector responsible for management of agro-product quality and safety (Lin *et al.*, 2004). Since the 1980s, MOA has taken active measures to accelerate the establishment of an inspection and testing system for agro-product quality and safety and to strengthen relevant management works, in order to meet the urgent need for agriculture development, as well as to fulfill the country's overall objective to build a comprehensive monitoring, supervision and testing system for agricultural commodities. Initially, efforts were focused on establishing inspection and testing centers for monitoring the quality of agricultural inputs such as chemical fertilizer, pesticide, feed, agricultural machinery and seed. In the 1990s, MOA initiated plans and carried out projects to set up national and ministerial inspection and testing bodies for agro-product quality and safety in accordance with the shift of agricultural production targets to high yields, good quality and high efficiency (Yu *et al.*, 2011). These testing centers were mainly established at national and provincial agricultural research, education and extension organizations with good hardware and software facilities, such as available laboratory equipment and appropriate technical staff. Through authorized certification and accreditation, MOA has set up testing centers for agro-product quality and safety both at national and ministerial levels in four batches. In addition, agricultural departments at provincial, municipal and county levels also established corresponding testing and inspection bodies. Up until the period of 2006–2010, construction of the national inspection and testing system for agro-product quality and safety was among the key construction projects in the framework of building modern agriculture in the country and was included in the *Eleventh Five-Year Plan for Economic and Social Development* (Ministry of Agriculture, 2006). A National Construction Plan for Inspection and Testing Systems for Agro-product Quality and Safety was launched with a grand investment of over 5 billion RMB. By 2009, more than 860 agricultural inspection and testing centers had been established, including 275 ministerial, 105 provincial and 489 municipal centers, in addition to 1356 county testing stations. The laboratories of these centers/ stations took up a total of 1.3 million square meters (Xiao *et al.*, 2007). Testing centers at various levels have been working on a more reasonable layout, more advanced inspection and testing equipment, and better supporting services (Zhen, 2010). Testing stations at county level have gradually developed their capabilities, and their testing approaches included both qualitative and quantitative methods, with their testing scope expanded to all kinds of fresh produce. For testing centers at ministerial and provincial level, testing capability has upgraded greatly; sophisticated instruments at each testing center have increased over 40 sets and testing items have increased to 300 on average. Limit of Quantification (LOQ) has increased 20 times from 0.1 mg/kg to 0.005 mg/kg for the contaminant detection. With regards to testing capacity, more than 3000 samples of various fresh produces have increased on a yearly basis.

The current system has formed in such a shape that the testing center at ministerial level takes the lead in work arrangements and planning. Testing centers at provincial and municipal levels function as the main working body, testing stations at county level work as a fundamental network and rapid testing laboratories in production bases and wholesale markets at township level function as auxiliary forces (Zhang *et al.*, 2007). The entire system plays a critical role in upgrading agro-product quality and safety, and ensuring safe consumption of fresh produce in China.

19.3 Challenges in building capability of legislative testing

As agriculture in China has evolved to a new phase of development and China has been accepted as a formal member of the World Trade Organization (WTO), it is clear that there are significant deficiencies in the existing inspection and testing system for fresh produce when compared with developed countries. These include loopholes in the system and a lack of controlling approaches, as well as its particular inability to ensure safety of consumers, control hazardous substances and elevate international competitiveness of fresh produce, as well as management level, which have resulted in rigorous challenges.

19.3.1 Increasing public expectations for food quality and safety

Requirements for food quality and safety increase with rapid socio-economic development. Currently, China has entered the intermediate stage of its industrialization process and Engel's Coefficients (the proportion of income spent on food) for people in urban and rural areas were 35.7% and 41.1%, respectively, in 2010, which means people have focused increasingly on the quality and safety aspects of food and health impact of food. It is therefore

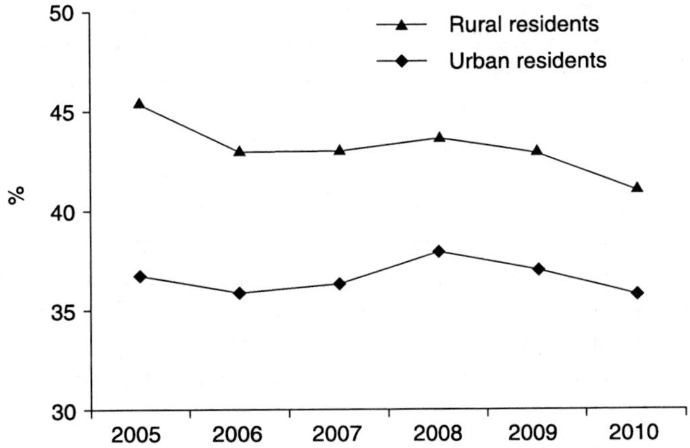

Fig. 19.1 Engel's Coefficients of residents in urban and rural areas.

necessary to put agro-product security and safety in an equally important position and maintain a balanced development among quantity, safety and economic profit so as to meet the ascending requirement of consumers. Figure 19.1 demonstrates the decreasing trend of Engel's Coefficients for urban and rural residents in China.

19.3.2 Growing contaminant residues

With the rapid socio-economic development and agricultural science and technology growth, more types and larger quantities of toxic and hazardous substances are being used and their application scope has expanded (Guo *et al.*, 2011). In recent years, the wide application of chemical fertilizers, pesticides, veterinary drugs, feed additives and hormones has facilitated the fast growth of agricultural production, but it has also posed risks to fresh produce quality and safety. Pollution resulting from industrial wastes and urban rubbish poses a great threat to agricultural producing environments; food incidents originating from over-dosing of pesticides, veterinary drugs and other contaminants have greatly jeopardized the health condition of consumers and have become a key public concern. Contaminants are in every corner of the environment, and it is unnecessary and impossible to apply a zero tolerance rule toward contaminants. The key is to set a safe limit under which the risk can be acceptable to both people and environment, which raises a harsh question about agro-product quality and safety management.

19.3.3 Demanding technical barriers to trade

With the rapid development of globalization and China's entry into the WTO, fresh produce in China faces competition from peer products both at home and overseas. In order to protect domestic agriculture from competition, many countries have put in place Technical Barriers to Trade (TBT) and hurdles based on the Agreement on Sanitary and Phyto-sanitary (SPS). Labor-intensive fresh produce in China such as vegetables, fruits, aquatic products and meat products usually enjoy discounts in price, but, because of technical barriers to trade, export of Chinese fresh produce has experienced setbacks (Gao *et al.*, 2005). In 2010, a direct decrease in fresh produce export registered at US$3.74 billion. Insufficiencies in the inspection and testing system have led to problems in the export sector. For example, many tests of items for export cannot be undertaken in China, which results in difficulties in quality and safety conditions of fresh produce and direct economic losses at overseas customs. Similarly, potential risks can be introduced to China without reliable testing methods when importing international agro-products (Duan, 2009).

19.3.4 Rapid development in inspection and testing technology

In order to ensure the quality and safety of fresh produce, a strategy of whole chain testing, including pre-production testing, in-production testing and postharvest testing, is universally applied as an international practice. China is currently addressing the quality and safety of agro-products from the origin of production through to application of standardized farming procedures. Aided by advances in inspection and testing technology, capability of testing has been greatly elevated in terms of sensitivity, accuracy and selectivity, as well as speed, which has contributed to the promotion of quality and safety level of agro-products (Cui *et al.*, 2010). However, the capability of the testing system in China is still far behind the demands for whole chain control of fresh produce quality and safety in terms of hardware condition, capacity of testing and personnel proficiency.

19.4 Regulations and policies for legislative testing

Strengthening the capacity building of legislative testing for agro-products in China is an essential way of ensuring consumer safety, and also significantly marks the level of national control over agro-product quality and safety. After years of development, much progress has been made with regards to legal system establishment and management.

19.4.1 Formation of legal framework

At present, China has approved and issued 14 laws and 18 relevant regulations concerning inspection and testing for fresh produce quality and safety (Sun, 2008). The Law of Agro-Products Quality and Safety is the core of the legal system, which has laid a favorable foundation for set-up and operation of inspection and testing entities. For example, according to the Law, agricultural

Table 19.1 Laws and regulations relevant to legislative testing for agro-product quality and safety

Levels	Names	Year of issuance
Laws	Law of Agro-Products Quality and Safety	2006
	Law on Animal Husbandry	2005
	Law on Promotion of Agricultural Mechanization	2004
	Law on Seed	2004
	Law on Quality of Products	2000
Acts	Act of Pesticide Management	1997
	Act of Veterinary Drug Management	1987
	Act of Feed and Feed Additives Management	2011
	Act of Commodities Inspection	2005
	Act of Quarantine for Animals and Plants	1996
	Special Act of Strengthening Safety Management Upon Food and Other Products	2007
Regulations	Regulation on Safe Agro-Product Management	2003
	Regulation on Aptitude Recognition of Laboratories and Examining Bodies	2005
	General Requirement on Testing and Calibration of Laboratory Capability	2008
	Criterion on Laboratory Aptitude Recognition	2007
	Regulation on Management of Inspection and Testing Centers at Ministerial Level	2007
	Basic Requirements of Inspection and Testing Centers at Ministerial Level	2007
	Criterion of Examination on Inspection and Testing Centers at Ministerial Level	2007

departments in governments over county level should authorize entities with eligible conditions and capability to undertake monitoring and testing of fresh produce in either the production or the markets, and wholesale markets should establish or entrust testing entities to carry out random checks on fresh produce for sale. Other laws and regulations on animal husbandry, seeds, pesticides, veterinary drugs, feed and plant quarantine also articulate specific requirements and requests for corresponding aspects of agro-product quality and safety (Ye *et al.*, 2000a). These legal provisions have significant impacts on the establishment, operation and development of inspection and testing bodies, and provide a legal basis for these bodies to fulfill their responsibilities. Table 19.1 lists a series of Chinese laws and regulations to address agro-product quality and safety.

19.4.2 Gradual optimization of management policies

In order to address the emerging challenges in the new era of agricultural development, the Chinese government has issued a portfolio of policies to regulate legislative testing at both national and ministerial level (Table 19.1). MOA has enacted regulations for the development of safe agro-food and green food, as well as an overall upgrade scheme for testing bodies of agro-product quality and safety. In doing so, the inspection and testing system has been given special focus to aid its development. To guarantee the capability of testing laboratories, management policies have also been put in place to ensure that all eligible entities pass the necessary certification and accreditation procedures (Chen, 2011).

19.5 The role of legislative testing in agricultural economy development

Agriculture and agricultural economy in China have entered a new phase, and it is inevitable that China will facilitate its structural adjustment in the agricultural sector, pursue a balanced development and stimulate the vitality of producers and business entities. Legislative testing plays a key role in the whole reform process.

19.5.1 Strengthening legislative testing: an important route to facilitate structural adjustment in the agricultural sector

In recent years, the Chinese government has placed special focus on agricultural production, farmers' income and rural development, and a package of incentive policies to benefit and improve agriculture has been publicized. Capacity building of science and technology innovation and transfer has strengthened, and investment in agriculture has increased. The structure of the agricultural sector has been optimized, and a rough balance of supply and demand for major fresh produce has been reached and a surplus can be expected in affluent years. Comprehensive productivity of agriculture has been remarkably improved. As the balance between supply and demand has changed, structural surplus has occurred and agricultural development has experienced dual restrictions from resources and the market. Some deep-seated problems in agricultural development have been exposed, such as poor quality, lack of famous brands, weak regional competitiveness, low level of industrialization and so on. It is imperative that structural adjustment is accelerated in the agricultural sector, as it has been facilitating the transformation of the agricultural economy from an extensive mode to being quality-and-benefit oriented.

19.5.2 Strengthening legislative testing to steer development of the agricultural sector in a healthy direction

The status of fresh produce quality and safety is directly connected to the development of agriculture. Numerous fresh produce incidents, such as melamine-tainted infant formula and clenbuterol-contaminated pork products, seriously jeopardized the corresponding companies and the related industries (Zhang *et al.*, 2009). Thanks to modern mass media, even a small case could cause public panic, but an effective way of responding to such incidents and preventing them from happening lies in strengthening legislative testing. The case of 'the incident of pesticide residue in cowpeas in Hainan', resulted in a direct economic loss of billions of RMB, and the provincial government and agricultural department put in place a series of measures to control the quality and safety of fresh produce by undertaking 100% testing over products going out of the island and training 3000 personnel to carry out inspection and testing in the production (Guo, 2007). Legislative testing also played a critical role in the response to the case of the melamine-contaminated infant formula and series of clenbuterol incidents.

19.5.3 Strengthening legislative testing to direct producers and market players

China's reform and expansion in rural areas made hundreds of millions of farmers the main body of market economy and laid a favorable foundation for the rise of enterprises at township level. This rise explored a route for deployment of excessive labor force in rural areas and became one of the most vivid sectors in the national economy (Sheng, 2006). In 2010, the number of enterprises at township level was registered at 1 107 967, and the number of people within the sector has reached 27.78 million. Industrial-added value made by enterprises with over 5 million RMB Yuan in turnover registered at 1.81 trillion RMB Yuan, which was an increase of 15.8% over the last year. Additionally, the industrial-added value made by enterprises at township level with a certain scale and above accounted for 30.9% of this. China has attached great importance to developing specialized farmer cooperatives as a significant approach to promoting industrialization and modernization of agriculture. Based on the system of contracted responsibilities on the household basis with remuneration linked to output, in connecting individual agricultural producers and gathering limited production materials and funds in rural areas, advancements in agricultural science and technology innovation are made possible. At the same time, land productivity, resource utility and labor efficiency have been upgraded, which facilitates the tight connection between agricultural production and the market. Currently, there are roughly 552 000 specialized farmer cooperatives with legal registration and more than 43 million agricultural households involved, which accounts for 17.2% of the agricultural households in China. However, despite favorable development by agricultural producers and market players, most agro-products are produced in poorly organized and small-scale premises with poor conditions and infrastructure, and products for sale in the market are mostly in raw material and bulk. Given the status, government departments carried out legislative testing along the whole chain, namely, preharvest, on-farm and postharvest phases. The preharvest phase focuses mainly on monitoring production environment, and for the on-farm phase emphasis is given to agricultural inputs, including pesticides, chemical fertilizers, seeds, feeds, veterinary drugs and so on. In the postharvest phase, focus is attached to monitoring various contaminants, such as residues of pesticide and veterinary drug, heavy metals, etc.

19.6 The role played by legislative testing in production and trade

Along with the adjustment of agricultural structure and improvement of living conditions of consumers, yield and amount in circulation of fresh produce has increased on a year-to-year basis. The international market exerts ever-rigorous requirements on quality and safety of fresh produce, which means legislative testing is now even more important.

19.6.1 Legislative testing: an effective way to normalize safe production of fresh produce

In terms of agricultural production and consumption of fresh produce, China is a big nation. Statistics show that the gross grain output in 2011 registered at 5.71 billion tons, with a 24.73 million ton increase (or an increase of 4.5%) over that in 2010, which indicates that the gross grain output achieved an eighth consecutive year of growth (Fig. 19.2). The outputs of vegetables and fruits reached 677 million tons and 142 million tons respectively. The total yields of meat, poultry, egg and dairy products registered at 79.5 million tons, 27.8 million tons and 38.2 million tons respectively. The output of aquatic products registered over 56 million tons. Consumption of fresh produce has grown rapidly with population expansion, increasing personal income and further urbanization. The per capita consumption of most agro-products in urban and rural areas has surpassed the international average level. For example, the per capita consumption of vegetables

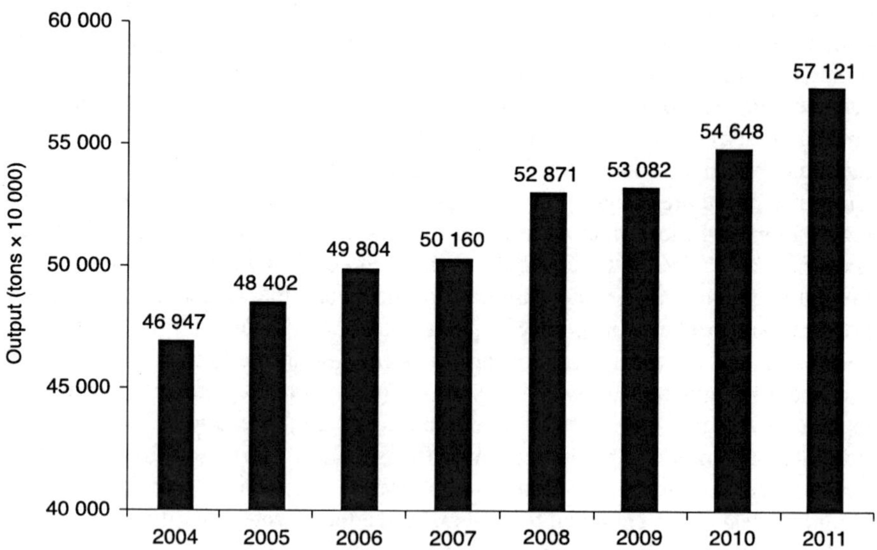

Fig. 19.2 Trend of grain output for China.

in 2010 registered at 370 kilos, far higher than the international average level of 105 kilos. However, it has presented a challenge for small-scale agro-products producers in China to deal with food security and safety simultaneously. In order to acquire information and control the risk of fresh produce quality and safety, MOA has been launching routine monitoring endeavors and supervisory testing efforts since 2001. In so doing, the proportion of qualified fresh produce has been kept, on average, over 95%.

19.6.2 Legislative testing: a firewall to guarantee quality and safety of fresh produce in the circulation sector

Currently, over 70% of fresh produce in China is supplied by agricultural wholesale markets, which have become the main channel for fresh produce circulation and connect a large number of small-scale agricultural producers with consumers. Further building of the agricultural wholesale markets has ensured increases in the number of markets, number of booths, area of markets, trade aggregate and volume of business. Up to the end of 2010, there were 4093 agricultural wholesale markets with an area of over 20,000 m^2, among which 1672 wholesale markets had a business volume of over 100 million RMB. These 1672 wholesale markets accounted for 40.9% of all markets in China and covered almost every urban area and zone with concentrated fresh produce resources. But most wholesale markets in China have faults such as small scale, weak competitiveness, and poor organization and management. The majority of fresh produce has been confronted with common difficulties in tracing origin and tracking destination, which has resulted in quality and safety risks (Qian *et al.*, 2008). In order to have reliable market access control over agro-products, the nation has enacted the Law of Agro-products on Quality and Safety to address this issue by setting compulsory requirements of on-site legislative testing.

19.6.3 Legislative testing: a warning light to break through the technical barriers to trade

Agro-product trade has a vital role in Chinese international trade. Between 2002 and 2010, Chinese export volume of agro-products saw an average increase of 11.7% every year, and in 2010 the export volume of agro-products registered at US$49 billion with a 25% year-on-year growth, and accounted for 3.1% of the entire export volume (Fig. 19.3). According to statistics from WTO, in 2008 China became the fifth largest country in terms of fresh produce export and the largest exporter of aquatic products, which made up 12% of the global export of aquatic products. China has also secured its leading position in products exports, including garlic, peanuts, apple concentrate, mushrooms and honey. As competition in the international fresh produce market becomes even more fierce, each country has attached great importance to innovation and created more technical barriers to trade, such as food inspection standard, trade standard and GMO, as well as labor and environmental standards (Ying *et al.*, 2005). In

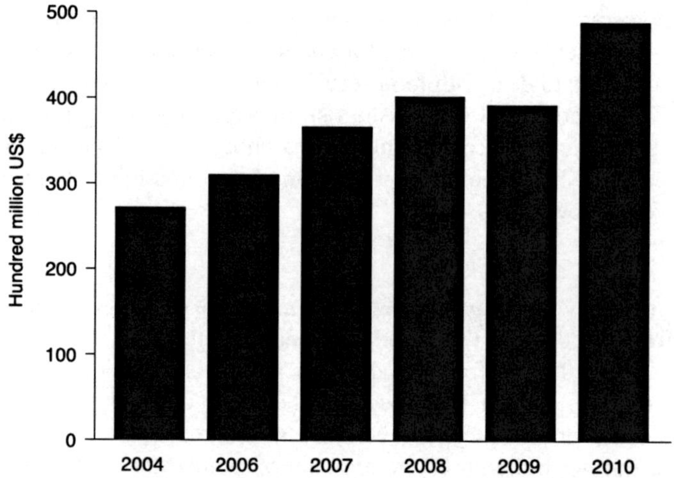

Fig. 19.3 Export volume of agro-products.

response to the changing situation, the Chinese government has strengthened management upon inspection and quarantine of import and export businesses by optimizing the layout of laboratories and upgrading instruments and equipments, as well as ensuring personnel training. As a result, a group of laboratories with international recognition have been established (Yang *et al.*, 2009), but innovation has made instruments and equipments even more capable of telling the 'true story', and reasonable resource allocation in various areas has provided the industry with better precaution ability.

19.7 Achievements in capacity building of legislative testing for agro-product quality and safety

Inspection and testing is a critical component and important technical support in legislative management on agro-product quality and safety (Jin *et al.*, 2003). China has plotted a specific scheme in order to enhance capacity building of legislative testing for agro-product quality and safety in its Twelfth Five-Year Plan Period to meet the emerging risks and adapt to the volatile situation.

19.7.1 Accelerating building of inspection and testing systems

In China, MOA has taken the lead in plotting an inspection and testing system across the country since the 1990s. Up to now, a network consisting of 49 inspection and testing centers at ministerial level mainly undertaking risk assessment research has been formed, organizing monitoring activities and national standard setting, 30 inspection and testing centers at provincial level mainly undertaking monitoring and local standard setting, and 936 inspection and

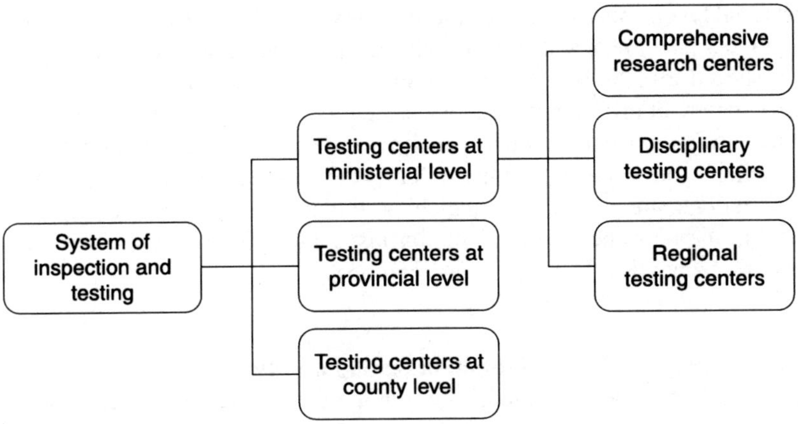

Fig. 19.4 Diagram of system of inspection and testing centers in agricultural sector.

testing stations at county level mainly undertaking daily inspection works and technical consultations. This serves the country in legislative testing service for agro-products (Fig. 19.4).

19.7.2 Establishing risk assessment laboratories

Risk assessment provides scientific evidence for management of agro-product quality and safety (Wang *et al.*, 2008). By the joint efforts of the National Development and Reform Commission and MOA, China plans to promote capacity building efforts in risk monitoring and precaution of agro-product quality and safety through two strategic steps. In so doing, the risk of quality and safety of major fresh produce can be monitored on a regular basis and early warnings made as required, which will allow for scientific evaluation of and responsive solution to unacceptable risks.

19.7.3 Improved testing equipment

The inspection and testing system is designed to meet the requirements in quality and safety monitoring and control of agro-products in pre-production, in-production and post-production stages. In accordance with the different needs of inspection and testing centers at different levels, corresponding configurations of instruments and equipments have been deployed across the country (Xiao, 2008). State-of-the-art analytical apparatus has been equipped to meet the requirements of risk assessment research, analysis, data collection and processing for comprehensive research centers at ministerial level. Equipment and instruments have been configured to adapt to quantitative and qualitative analytical needs of pesticides, veterinary drugs, bio-toxins, pathogenic microorganisms, heavy metals and other food contaminants for disciplinary inspection and testing centers at ministerial level, and sample pretreatment apparatus, virus detectors, rapid testing equipments, sampling and transmitting equipments, as well as data processing units, should also be put in place in such centers. For regional inspection and testing centers at ministerial level, according to various products, equipments for quality and functional components analysis and analysis of agricultural inputs and environment have been installed. For provincial testing centers, equipment configurations are tailored by matching the local characteristics of product, agricultural inputs and environment. For inspection and testing stations at county level, basic analytical instruments are equipped.

19.7.4 Strengthening evaluation of testing ability

In order to further develop capacity building of the agricultural sector, MOA has enacted specific regulations to organize annual proficiency testing programs every year for the inspection and testing laboratories at different levels (Cui *et al.*, 2006). Since 2003, such proficiency testing programs have been put forward over pesticide residues in vegetables, veterinary drug residues in animal products, drug residues in aquatic products, recombined milk in dairy products and heavy metals in soil and fertilizers (Ye *et al.*, 2000b). Over 200 inspection and testing centers and other relevant testing laboratories have participated in the programs, and proficiency testing programs have been constantly optimized in terms of procedure, design and coverage by timely taking problems, new situations and emerging requirements into consideration, which has greatly improved the credibility of these inspection and testing centers.

19.7.5 Focusing on personnel training

Having realized the importance of eligible personnel and considered the actual needs in management of agro-product quality and safety, MOA has issued regulations on personnel training in a bid to build and reserve qualified comprehensive human resources in administration, legislative, testing and technology extension (Ni *et al.*, 2005). Training programs have been organized and assigned to relevant personnel groups in testing entities at different levels, and training materials have been carefully designed and compiled in a bid to raise legal awareness, management abilities and testing skills of trainees. Evaluations of personnel in key positions relevant to agro-product quality and safety could also be made permanent. In 2012, experts from 100 testing centers at ministerial level shared their expertise with staff from 100 testing stations at county level, covering areas such as testing technologies and laboratory management in order to raise credibility of testing bodies at lower levels (Zhuang *et al.*, 2003).

19.7.6 Strengthening legislative testing of agro-product quality and safety

In recent years, according to the unified decisions of Chinese government, MOA has attached great importance not only to fresh produce security, but also to fresh produce quality and safety. Notable momentum in elevating the safety level of fresh produce has been gained in terms of legal establishment, management, testing and monitoring, as well as standardized production. Special campaigns have been launched to deal with such problems as chemical residues and illegal addition of hazardous substances, as well as to further safeguard agro-product quality and safety on significant occasions like the Beijing Olympics and Shanghai Expo (Dai *et al.*, 2004). Annual routine monitoring schemes on pesticide residues, veterinary drug residues, feed and feed additives have also been put in place. A comprehensive network covering 138 cities, 101 varieties of fresh produce and 86 items for testing has been established across the nation. To satisfy the dynamic needs of agro-product quality and safety, MOA will be dedicated to further promoting its monitoring network building by enlarging the monitoring range and enriching the testing items. In order to establish effective mechanisms to control the risks in agro-product quality and safety, MOA is also focusing on data mining and evaluation of monitoring and testing results.

19.8 Future trends

An inspection and testing system for agro-product quality and safety plays a key role in providing the government with support in policy-making, technology transfer and technical consultation services to ensure the quality and safety of fresh produce. After years of efforts to build an inspection and testing system, remarkable achievements have been made; however, many aspects of the system need to be perfected to meet challenges in the new era (Qian *et al.*, 2010). In the National Twelfth Five-Year Social and Economic Development Plan Period, the system is expected to cover producing site environment, agricultural inputs and fresh produce based on the principles of reasoning layout, suitability, inclusiveness and effective managing performance. Inspection and testing centers at ministerial, provincial and county levels will be further increased in both quantity and scale. Web-based risk monitoring and precaution networks will be further perfected to guarantee information flow is responsive and accurate. Our vision is to establish an inspection and testing system consisting of testing bodies at different levels, covering the whole production process of major fresh produce for domestic use and export, and being capable of ensuring the quality and safety of agro-products in China.

19.9 References

CHEN X. 2011. Multi-ways to strengthen management of agro-product quality and safety. *Communication of Rural Works*. 12: 6–9.

CUI Y, LIU P, WANG Y. 2006. Status quo and suggestions for laboratory proficiency testing. *Agricultural Quality and Standard*. 5: 35–8.

CUI Y, LIU P, WANG Y, WANG W, ZHOU Y. 2010. Analysis and suggestion on development of Chinese inspection and testing centers at ministerial level. *Quality and Safety of Agro-Products*. 6: 35–8.

DAI T, NI Y, LU P, LI W, ZHOU H. 2004. Current advances of testing technology for pesticide residue. *Journal of Pesticide*. 9: 389–93.

DUAN L. 2009. Discussion on how to elevate testing level of testing bodies[J]. *Grain Issues Research*. 4: 39–40.

GAO Q, CUI K. 2005. Discussion on the effects of green technical barriers to trade on Chinese vegetable export. *Issues in Agricultural Economy*. 7: 41–4.

GUO L, QIAN Y, QI Y, XING F. 2011. Research on the advances of technical barriers to trade of major trade partners. *Standard Science*. 12: 88–91.

GUO T. 2007. Discussion on the major issues affecting quality and safety of animal products and the countermeasures. *Chinese Journal of Animal Quarantine*. 3: 21–2.

JIN F, QIAN Y. 2003. Status quo and countermeasures of system of inspection and testing for agro-product quality and safety in China. *Agricultural Quality and Standard*. 2: 26–9.

LIN D, QU Y. 2004. Comparative research on management systems of US and China. *Journal of Wuhan University of Technology (Information & Management Engineering)*. 3: 154–8.

MINISTRY OF AGRICULTURE. 2006. Decisions of Ministry of Agriculture on strengthening the capacity building of management of agro-product quality and safety. Communiqué of Ministry of Agriculture of P. R. China. 12.

NI Y, LI Z. 2005. Status quo of pesticide contamination in food and countermeasures for a better supervision and management. *Chinese Journal of Health Inspection*. 1: 58–60.

QIAN Y, WANG F. 2008. Analysis on the problems and factors in Chinese agro-product quality and safety. *Journal of Agricultural Economy*. 2: 78–9.

QIAN Y, WANG F, CHEN S. 2007. Analysis on the enforcement of the law on quality and safety of agro-products. *Agricultural Quality and Standard*. 6: 4–6.

QIAN Y, WANG F, GAO S, GAO G. 2010. Discussion on the management for agro-product quality and safety in the new era. *Quality and Safety of Agro-Products*. 1: 13–15.

SHENG G. 2006. Discussion on the establishment of inspection and testing system for agro-product quality and safety. *Agricultural Quality and Standard*. 5: 41–3.

SUN Z. 2008. Management section, pp. 417–428. In Yao Hong, Liu Mingchang, Yanbaorong (ed.), *30 years in reform of agriculture and rural areas*. China Agriculture Press, Beijing.

WANG F, CHEN S, QIAN Y. 2008. Research on the establishment and features of risk analysis mechanisms in overseas. *World Agriculture*. 9: 44–7.

XIAO L. 2006. Requirements on aptitude management for laboratories and testing bodies. *Journal of Real Estate*. 11: 68–9.

XIAO L. 2008. Research on the inspection and testing system for agro-product quality and safety. Beijing: Chinese Academy of Agricultural Sciences.

XIAO L, YE Z. 2007. Introduction of animal quarantine laboratory in Canadian Food Inspection Agency. *Chinese Journal of Animal Quarantine*. 3: 51–2.

YANG Y, GAO J, YU S, XUE L, WANG Q. 2009. Research on the aptitude recognition of laboratories in agro-product testing bodies. *Shaanxi Journal of Agricultural Sciences*. 3: 168–9.

YE J, HE Y, TAO CH. 2000b. Introduction of national standards on pesticide residue in China. *Pesticide Science and Administration*. 4: 20–3.

YE J, HUANG X, ZONG F, LIU S. 2000a. Status quo and outlook of establishment of monitoring system for pesticide quality. *Pesticide Science and Administration*. 1: 36–9.

YING X, XU X, ZHU Z, CHEN M, MOU R, SHI Y. 2005. Analysis on preferences of vegetable importing countries upon testing items of pesticide residue. *Chinese Vegetable*. 1: 4–7.

YU Q, ZHANG T. 2011. Outlook of domestic testing bodies in China. *China's Foreign Trade* (English Edition). 24: 473.

ZHANG S, GE M. 2007. Proficiency testing is key to ensure capability of testing laboratory. *Advanced Measurement and Laboratory Management*. 5: 59–60.

ZHANG X, JIN F, QIAN Y, WANG J. 2009. Research on advancement in melamine determination in food and feed. *Science and Technology of Food Industry*. 7: 337–42.

ZHEN R. 2010. Status quo and developing trend of testing laboratories in China. *Journal of Chinese Standardization*. 9: 35–8.

ZHUANG Y, YE Y. 2003. Comparative research on establishment and management of national key laboratories in US and China. *Journal of Science and Technology Management*. 12: 21–4.

20

Bottlenecks and limitations in testing for pathogens in fresh produce

J. Hoorfar, Technical University of Denmark, Denmark, P. Feng, US Food and Drug Administration, USA, G. Duffy, Teagasc Food Research Centre, Ireland, B. Malorny, Bundesinstitut für Risikobewertung (BfR), Germany and R. Binet, US Food and Drug Administration, USA

DOI: 10.1533/9781782420279.4.274

Abstract: Microbial content of produce can be impacted by sanitary conditions of the production region, use of composted manure, type and quality of irrigation water and many other factors, which can introduce risk factors. Furthermore, the types of produce, uneven distribution of contamination and heterogeneity of samples all pose challenges to obtaining representative samples. Other variables to consider when testing for pathogens in produce include method sensitivity, specificity, applicability, etc. These statistical and analytical uncertainties can have a significant impact as the data are used to infer the health risk of larger product lots, which are broadly sold and are often ready-to-eat.

Key words: sampling, sample preparation, cultivation, PCR, sequencing.

20.1 Introduction

Fresh produce is a diverse group of foods that can potentially be contaminated with different human pathogenic microorganisms (Olaimat and Holley, 2012). Of these, the greatest concerns are *Salmonella* (in tomato, seeds sprouts, spices), *Shigella* (in green onion), verotoxigenic *Escherichia coli* (in sprouts and leafy greens like lettuce and spinach), *Listeria monocytogenes* (in melons and fresh cut salad) and norovirus (in berries) (see Chapters 21 and 22).

This diversity of pathogens makes it difficult to propose general guidelines for microbiological analysis, and, furthermore, methods may vary depending on the commodity being tested, as well as production and postharvest processing conditions. In addition, there may be differences between methods performed in government laboratories, those used in field testing, producer quality control testing, as well as methods that are used in outbreak traceback investigations. For instance, many countries are performing whole genome sequencing (WGS) of foodborne pathogens, which has generated useful information in some *Salmonella* outbreak investigations. However, the need for a pure culture isolate for WGS remains a rate limiting step in generating data quickly, and also investigators are using various technologies, so the procedures are not standardized and the data may not always be comparable.

In public health laboratories, produce analyses are often performed to detect the occurrence of specific pathogens or to test for indicators to assess sanitary quality. At food companies that have established internal inspection and process controls, it would be less effective to perform end product testing because of the sporadic and very low occurrence of pathogens in products. However, testing for indicators such as *E. coli* or coliphages may be useful (Rambo and Pillai, 2011), especially to monitor insanitary processing conditions. Nevertheless, many fresh produce manufacturers are actively testing incoming raw materials for the presence of pathogens to remove contaminated produce from their processing line.

When analyzing for pathogens in fresh produce, it is relevant to know the production region or country to assess the sanitary and health condition of the production region, watering condition, use of compost manure and other production conditions that can be risk factors (see Chapter 12). For zoonotic pathogens such as *Salmonella*, *Campylobacter* and parasites that are found widely around the world in animals and humans, it is often the production conditions, rather than production region, which is relevant for testing schemes. The same is valid for ubiquitous microorganisms such as *L. monocytogenes, Bacillus* sp., *Clostridium* sp., *S. aureus* and norovirus.

Curiously, in 2011, an outbreak of illness caused by *Yersinia enterocolitica* in Norway was linked to ready-to-eat salad mix, which was an unusual vehicle for this pathogens (MacDonald *et al.*, 2012). During traceback investigations, the suspected radicchio rosso salad samples were cold-enriched for as long as 21 days and tested both by culture and PCR methods but were negative, so the case remains unsolved. The lengthy enrichment period often needed in *Yersinia* testing is most certainly a major bottleneck, especially in outbreak investigations when data are needed fast.

Hygienic standards vary from country to country; in developed countries in Europe or North America, there are usually higher standards, thus, there should be closer monitoring to make a distinction between domestic and imported fresh produce. This is especially important when considering testing products for human-specific pathogens such as *Shigella*, *Salmonella* typhi/parathyphi, enterotoxigenic *E. coli* (ETEC) and hepatitis A, where contaminations are usually rooted in hygienic breach by production staff.

Depending on the objectives, there will be differences between testing schemes for qualitative and quantitative methods. In the case of product safety and market release, qualitative (presence/absence) data are sufficient. However, quantitative (enumeration) assays can generate data that are useful for risk assessment or epidemiological knowledge on human exposure. For example, dependent on country, low presence of *L. monocytogenes* can be acceptable in products that are held under conditions that do not support the growth of *L. monocytogenes*, so in this case, both qualitative and quantitative data are relevant. However, for environmental pathogens such as *Clostridia*, *Bacilli* or *S. aureus* and for indicator microorganisms like *E. coli* and *Enterobacteriaceae*, quantitative data on microbial load are most relevant and of importance in assessing insanitary conditions. The situation is more complex for VTEC, viruses or parasites, as the strains detected in the produce samples may have major differences in virulence potential.

When selecting a testing method, it is important to know what the results are to be used for. In addition, before implementing methods in the laboratory, one should consider the sampling plan, how to prepare the samples, the product tested, tolerable and permitted limits, if any, of the pathogen or microbial indicator in the product, and not-the-least, the action to be taken if the pathogen content exceeds the permitted limits.

There are a large number of well-established and internationally recognized standards for microbial testing of fresh produce (www.ISO.org, www.CENorm.be, http://www.fda.gov/Food/FoodScienceResearch/LaboratoryMethods/ucm2006949.htm). Similarly, several standards deal with the enumeration aspect of pathogenic microorganisms or hygienic indicators. However, all these standards and guidelines are based on reference culture methods that can take several days to complete, are labor-intensive and are low-throughput (Hoorfar *et al.*, 2011). In addition, each reference method has some application limitations that must be taken into consideration before setting up a testing scheme. Finally, microbiological methods (both culture and advanced), despite great promise, suffer from some technological bottlenecks, in particular with regard to the complexity of produce samples, so that sample preparation steps must be addressed and adopted to specific sample types. The present chapter discusses the testing methods for some major pathogens of concern to point out the bottlenecks and viable solutions, if any, in testing for pathogens in produce. For virus testing, the reader is referred to Chapters 21 and 22.

20.2 Logistics in sample preparation

In EU legislation, all microbiological tests are referred to as open and horizontal (across different food types) reference standards that are developed and maintained by ISO (International Organization for Standardization) and CEN (European Committee for Standardization). From a legislative view, however, the EU is open to the application of alternative methods other than ISO or CEN approved culture methods (EU legislation 2073/2005), provided that these have been validated according to ISO 16140.

When sampling, the integrity and identity of the sample must be preserved, and the sample must be in the same microbiological condition when it reaches the laboratory for testing as when it was originally taken. Although these measures can be achieved, variability in the sample itself presents major challenges. Most matrices that are not in liquid form are not homogeneous, so they can have uneven distribution of contaminating microorganisms. Thus, in produce sampling, it is difficult to obtain a sample that represents the entire lot. Furthermore, a major limitation of laboratory methods is that only a small amount of a sample is tested, typically 10–100 g, so it is often necessary to test many samples in order to increase confidence in the test results. Even so, no number or amount of samples tested from a batch or a lot can ensure the absence of pathogens unless the entire batch or lot is tested, which is not possible.

The two main types of sampling plans are attribute and variable samplings (Ross *et al.*, 2011). Attribute sampling plans test against a single criterion or attribute, such as the presence of *Salmonella* in 25 g, or for instance the proportion of sample units that contain >100 cfu of *L. monocytogenes*/g. In other words, the samples either do or do not have the attribute being assessed, or simply pass or fail. In contrast, 'variable sampling' uses the full range of numerical data describing microbial loads. Variable sampling is more appropriate to food producers who regularly perform in-house microbiological testing of their products to assess quality, but such information is rarely made available to regulatory authorities or a food purchaser. Hence, in terms of food safety, attributes sampling plans are most often used as these test foods for a criterion related to the safety of product.

For handling samples, it is important to note that, once collected, the sample should be refrigerated but not frozen as freezing will substantially reduce the level of microorganisms present and thus will not provide a realistic indication of the contamination levels in the initial product. Unfortunately, there are quite a few published studies where microbial levels in sample data were derived from frozen samples, so these should be interpreted with caution.

Sample preparation methods can vary depending on the sample type and also microbial load. For example, levels of normal flora present can affect test results, so samples known to have heavy microbial loads may need to be prepared differently and using more selective enrichment media to suppress growth of background flora. Also, preparation methods such as rinsing, soaking, blending, etc., may be applicable in some but not all produce, depending on the types of produce and whether the pathogen has firmly attached or perhaps internalized into the sample.

Foods are notorious for containing ingredients that can interfere with assays. An increasing number of laboratories now use PCR as the only screening method for produce samples, where a preliminary enrichment step is used to resuscitate sub-lethally injured pathogens or to increase the target pathogen levels prior to the PCR screening step. However, PCR inhibitory substances are known to exist in produce, so it is important to include internal amplification controls to ensure that a negative result obtained was not caused by assay inhibition. PCR testing is further challenged by the fact that the presence of a gene sequence is not indicative of gene expression or the viability of the bacteria. Furthermore, bacteria can enter into a viable-but-non-culturable (VBNC) state where they are detectable by PCR and, thus, indicative of potential health risk, but may not actually become a threat as the pathogen may not grow or propagate to become infectious. The VBNC phenomenon has been reported for some bacteria but there are some uncertainties that it occurs in others. Also, as to whether VBNC cells are infectious continues to be investigated.

Finally, it is often desirable to test a larger number of samples in order to increase the confidence level of the results, but cost and time factors often limit the number of samples that can be tested. One way to circumvent this problem is by using sample pooling or compositing strategies. For example, Jarvis (2007) proposed pooling the incubated enrichment buffers rather than the actual food samples. However, one should bear in mind that the dilution effect resulting from pooling a number of samples together can also reduce the detection limit of an assay. Regardless of whether reference culture methods, alternative rapid methods or pooling and compositing strategies are used, when setting up a testing scheme it is important to focus on the objective of the testing in order to determine a proper sampling plan. The various factors that need to be considered and are associated with sampling and testing, are referred to as 'Microbiological Criteria' (Commission Regulation (EC) No 2073/2005) and include factors such as the method of analysis to be used, sampling strategies, permitted level in the sample, if any, the number of sample units that can exceed the permitted microbe levels and, finally, descriptions of the action plan if the permitted level is exceeded or pathogens are detected.

20.3 Logistics in field testing

Sampling of fresh produce to determine the presence of foodborne microbial pathogens presents many challenges. The term 'fresh produce' can cover a very wide variety of products ranging from root vegetables, sprouted seeds, leafy greens to soft fruits. In general, produce are grown in areas covering many acres of land, and in some regions, the growing area can extend to hundreds or thousands of acres. Contamination of fresh produce can arise from many sources, including irrigation waters, manure or compost applied to soil as a fertilizer, field workers, wild life or indeed farmed animals, which have access to the area in which the produce is grown. The potential size of the growing area and the multiple sources of contamination pose difficult questions in terms of what, where and how to sample, as well as the frequency at which sampling should be carried out, in order to obtain some indication of the microbial quality of fresh produce.

Field testing generally involves labor- and resource-intensive procedures, and unless performed in a scientific and risk-based manner, the results may not be useful. On the contrary, it may even hinder the management and production of safe produce if the testing methodology or sampling plan are not suitably robust, leading to inaccurate conclusions regarding the microbiological status of the produce or growing area.

A field testing program should have a clear objective and purpose, a statistically appropriate sample plan, a relatively good certainty of detecting a pathogen if present, and a clear action plan if a pathogen is recovered or acceptance criteria are exceeded.

For analysis from the growing areas, justification is needed for the selection of the sample type (produce or environmental samples). Regulatory requirements may require testing of the product, but often environmental sampling can provide more useful information for management of produce safety. For example, irrigation water may be tested for generic *E. coli* to give an indication of whether fecal contamination of the water source has occurred. This is especially useful for static irrigation ponds, but more complex to interpret for flowing irrigation sources. Testing of fertilizer or compost for thermotolerant coliforms may be used as an indicator that the compositing or waste treatment process has been adequate and will not pose a contamination risk for produce. If there is evidence that animals have intruded into an area of the field, testing animal droppings, soil and products in the area is warranted to determine if the product may have been contaminated with animal feces and that a pathogen may be present.

Whether it is the product or an environmental sample, microbial contamination is heterogeneous, hence selecting an appropriate number of samples to test, and an understanding of the level of confidence in the result that those samples represent, is one of the biggest challenges in microbiological testing of any food, including fresh produce. There needs to be awareness among risk managers that unless very large numbers of samples are taken, there may be poor confidence in the result and low probability to detect a contamination event. The International Commission on Microbiological Specifications for Foods (ICMSF, volume 7) examined the statistical confidence of test results on the basis of numbers of samples tested, which illustrated the limitations of sampling. If a lot is heavily contaminated (98%), then taking three samples would give a 95% probability of finding the contamination, but if there is a low level of contamination (2%), 100 samples would have to be tested to give an 87% probability of detection. Pooling samples and compositing them may or may not improve the chances of detecting contamination, depending on the test. Research focus is now on ways to effectively sample large amounts of product. Some examples of this include on-site washing processes to recover and concentrate pathogens from large batches of produce and on ways to concentrate large volumes of irrigation water *in situ* in the field. Developments in biosensor technology may also offer real opportunities in the future, for rapid *in situ* testing of produce and environmental samples for presence of pathogens or levels of microorganisms prior to harvest (Kramer and Lim, 2004; McGrath *et al.*, 2012; Terry *et al.*, 2005).

The accuracy of a test result is also critically dependent on the use of proper sampling techniques. Aseptic sampling procedures, such as the use of sterile containers, sanitized gloves and handling utensils and careful handling procedures are essential to minimize the potential for cross-contamination from other sources, including from the individual collecting the sample and from other samples already collected, which may result in false positive tests. The exact field location from which samples were taken needs to be clearly and accurately identified and be readily traceable should contamination be detected. Depending on the field size, this will probably require detailed mapping and coding of sampling areas prior to sampling so that each specific sampling site can be identified. Representative samples of the produce must be taken with each plant or fruit having an equal chance to be selected, and the samples taken should be parts of the crop that normally constitute the part of the produce that will be sold or is likely to be consumed. Produce that is diseased or under-sized should not be taken and samples should be taken in such a way as to be reasonably representative of typical harvesting practice.

Unless specified otherwise by the test method, fresh produce, water and environmental samples should be chilled (0–4 °C) as soon after collecting as practical, and kept cold but without freezing, until tested. A time/temperature recorder, or other device to verify proper temperature control, is recommended if the samples are shipped or held for more than a few hours before testing. Even at low temperatures, microorganisms in the sample may grow or die if held for too long before testing, thereby potentially causing erroneous results. Samples should be tested as soon after collection as practical, but this should be within 24 or 48 hours, depending on the test, with <30 hours (1 day) being highly recommended, especially for environmental samples.

Lastly, it must also be recognized that a microbial testing program by itself will not guarantee or enhance product safety and thus, Good Agricultural Practices (GAP), Good Manufacturing Practices (GMP) and Hazard Analysis Critical Control Points (HACCP) programs, etc., should also be in place to minimize the risk of product contamination from the various sources (see Chapters 1 and 4). However, properly designed and implemented sampling plans and methodologies can provide important information to enable identification and traceback of possible contamination sources, including in the growing environment, and in specific product lots.

20.4 Logistics in product testing

20.4.1 *Salmonella*

Salmonella-contaminated vegetable and fruits are known to be the causative agent for large outbreaks (Gill *et al.*, 2003; Lynch *et al.*, 2009; Mody *et al.*, 2011; Nygård *et al.*, 2008; Rimhanen-Finne *et al.*, 2011). Potential routes for preharvest contamination of fresh produce with *Salmonella* are exposure to irrigation water, soil, dust, manure, human or animal feces, contaminated transport containers or processing equipment.

Notably, sprouts and sprouted seeds have been identified as a special problem because of the potential for the pathogen to multiply during production (National Advisory Committee on Microbiological Criteria for Foods, 1999). The pathogen can easily grow during the sprouting process from initial low levels up to 6 log10 cfu/g (Gabriel, 2005). Because of the potential health risk, sprout seed is often decontaminated before germination by soaking the seeds in different chemical solutions such as calcium hypochloride (Gabriel, 2005; National

Advisory Committee on Microbiological Criteria for Foods, 1999) or with gaseous acetic acid treatment (Nei *et al.*, 2011) to reduce microbial proliferation. However, procedures or washing operations are often not able to eliminate completely the microbial load including *Salmonella* from seeds (Montville and Schaffner, 2004). Furthermore, by chemical, heat or irradiation of the seeds, *Salmonella* cells could be sub-lethally injured leading to the inability to multiply sufficiently during the enrichment steps used in the testing protocols, resulting in false negative data.

For alfalfa seed, it has been shown that the probability of detecting *Salmonella* was much higher in sprouted samples and in soaking and rinse water than in seeds (Rimhanen-Finne *et al.*, 2011). Therefore, testing for *Salmonella* should be performed on sprout samples rather than decontaminated seed. Because of an uneven distribution of the contamination in seeds, it is recommended to test large-size samples to aim for a high sensitivity. Sampling should be carried out where the highest probability is expected to find *Salmonella*. Producers should, therefore, implement an efficient internal quality control and sampling plan depending on the risk of finding *Salmonella* in a product. Automated packing machines need to be monitored regularly for the absence of *Salmonella* contamination.

Beside sprouts, *Salmonella* has been isolated from several raw vegetables and fruits (Beuchat, 2002). Outbreaks were associated with certain produce such as tomatoes, pepper, rucola, raw baby corn, cantaloupe or water melon (Lynch *et al.*, 2009). Prior to implementing specific levels of process control or testing schemes, one should consider the microbiological hazards associated with the particular vegetable or fruits and on the conditions of use (eaten raw or cooked). For example, watermelon and cantaloupes provide sufficient nutrients for rapid growth of *Salmonella* (Golden *et al.*, 1993), whereas fruits with $pH \leq 4.0$ do not generally promote growth of *Salmonella*. If fresh produce is cultivated next to livestock production or wild animal areas, sampling should include monitoring for possible cross-contamination.

Salmonella is one of the pathogens with zero-tolerance limit and, consequently, testing methods have to be highly sensitive. For testing for *Salmonella* in fresh produce, the ISO 6579 method is usually recommended but alternative methods may be used when they have been validated against the ISO method (International Commission on Microbiological Specifications for Foods (ICMSF), 2011). Because of the perishable nature of produce, rapid analytical assays such as PCR are more frequently applied to obtain the test result within 24–36 hours. However, as many PCR-based methods (and even some cultural methods) have not been validated, fresh produce farmers and food business operators should confirm that those methods can reliably detect low numbers of *Salmonella* in their products. It is especially important that the enrichment media and procedure used prior to the analytical assay be clearly elaborated and validated, as prolonged recovery time and well buffered enrichment media might be important considerations, because of possible stress damage of *Salmonella* cells by disinfection procedures or general low pH values derived from acid sample matrices.

Although surface contamination of fresh produce is regarded as the main route for human infections, *Salmonella* may be able to internalize during cultivation, harvesting or processing (Hirneisen *et al.*, 2012; Lynch *et al.*, 2009). Thus, sanitizing or washing will not eliminate the *Salmonella* cells inside the plant products, and testing methods might be not sensitive enough to detect the internalized pathogen. Despite the growing knowledge on pathogen penetration of produce, many questions remain unanswered with respect to the risk of internalized pathogens. It is also unclear if those samples can be efficiently detected by standard analytical test systems.

20.4.2 Verotoxigenic *Escherichia coli* (VTEC)

Enterohemorrhagic *Escherichia coli* (EHEC) is often regarded as the pathogenic subset of the much larger verotoxigenic *E. coli* (VTEC) group that is also known as shigatoxigenic *E. coli* (STEC). The prototypic EHEC strain is serotype O157:H7, a pathogen of primary safety concern worldwide and it has caused numerous fresh produce outbreaks with lettuce, spinach and alfalfa sprouts (see Chapter 25).

Testing for O157:H7 in produce is challenging because of the presence of assay inhibitors, interference from the high levels of normal flora often found in produce (Valentin-Bon *et al.*, 2008) and the infrequent and low levels of O157:H7 present. As a result, produce rinses or products have to be soaked in selective broths to enrich for the target and suppress background flora prior to testing. Enriched produce samples are most often screened by multiplex PCR or qPCR assays that target key virulence factor genes such as *stx* and *eae* that encode for Shiga toxin (Stx) and the intimin binding protein, respectively. Assays can also include targets like the O157 antigen operon genes (wzx, wzy, *rfb*E, etc.) or single nucleotide polymorphisms (SNP) such as the +93 SNP in the *uidA* gene that encodes for ß-D-glucuronidase (Feng *et al.*, 2012), which is conserved in O157:H7 and therefore, is a unique marker for this serotype (Feng *et al.*, 1998). Samples found to be positive for O157:H7 markers by PCR are plated onto selective/differential media to isolate and identify the organism. Although these confirmation steps can be labor-intensive, this has been facilitated by unique traits of O157:H7 strains, such as the absence of sorbitol fermentation and ß-D-glucuronidase activity, as well as the O157 and H7 antigen markers that are easily identified by serology.

More recently, other VTEC serotypes have emerged as important foodborne pathogens worldwide and several have been implicated in produce outbreaks in the USA. Some of these include: O111, O121 and O145 in lettuce and O26 in clover sprouts. Testing for VTEC in produce is more complex as the VTEC group is large and comprises diverse strains and serotypes. Hence, it is essential to ensure that the selective enrichment used is not inhibitory to the VTEC strain.

As VTEC is characterized solely by the production of vero or shigatoxins (Vtx or Stx), screening produce enrichments for the presence of VTEC is fairly straightforward (Feng *et al.*, 2012) and most PCR and qPCR assays that target *stx* genes will suffice. However, there are estimated to be ~300 VTEC serotypes and not all have been implicated in illness, so to determine which strains are of health concern is more challenging. The production of Vtx alone does not appear to be sufficient to cause severe illness, as a binding or attachment factor is also required for pathogenesis. Intimin is a common attachment factor among pathogenic VTEC, hence many multiplex PCR assays also include *eae* as a target.

Implication of selected VTEC serotypes in infections has also prompted the inclusion of follow-up PCR, where the samples that are positive for virulence gene are further tested for serotype-specific genes. However, a drawback of this strategy is that produce enrichment comprises mixed bacterial populations so that the targets detected may be derived from different cells. As a result, it is essential to isolate the bacteria to confirm that all the virulence and O antigen genes are from the same bacteria. To isolate VTEC from produce is more complex and challenging than for O157:H7 because of the large diversity of VTEC strains and serotypes. Hence, it often entails tedious screening of many colonies or pools of colonies by virulence gene-specific PCR until a pure culture is obtained. If the VTEC isolate has *eae* and is one of the major serotypes of concern, the product is considered a serious health risk. Even if the serotype is unknown, a VTEC strain that has *eae* should be of public health concern.

However, intimin is not the only known attachment factor among pathogenic VTEC, as *eae* negative strains like those in serotype O113:H21 and O91:H21 have also caused severe illness worldwide and these carry other putative binding factors such as the STEC agglutinating adhesin. Also, the O104:H4 strain that caused the large 2011 outbreak in Germany was *eae* negative, but its enteroaggregative properties coupled with Stx production caused severe illness. Such *eae* negative VTEC strains like O113:H21 have been found in fresh produce in the USA (Feng *et al.*, 2011). Thus, VTEC isolated from produce and found to be *eae* negative should be further tested for other putative virulence genes to determine if they are of health concern.

In summary, methods to test for O157:H7 in produce are fairly straightforward because of its unique markers, so it is sometimes possible to obtain timely results. To test for the other 'known' pathogenic VTEC serotypes takes longer but is feasible using combinations of *stx, eae* and serotype-specific gene targets. Although the isolation of these serotypes can be laborious, it can be facilitated by the use of serotype-specific antibodies. These procedures, however, pose serious challenges, as the lengthy and labor-intensive isolation procedures coupled with the battery of confirmatory tests that may or may not identify health risk, preclude the possibility of obtaining timely data for a short shelf-life product as fresh produce.

20.4.3 *Shigella*

As a result of its low infectious dose and the potential for rapid dissemination of *Shigella* spp., the hallmark of a good method for the recovery and identification of these pathogens from fresh produce has to be its sensitivity. Conventional methods to test for *Shigella* in foods can take 7–10 days, and it has a low probability of success because the pathogen may be present in a stressed physiological state, and be outnumbered by the native microbial flora (Binet and Lampel, 2013). The ancestral strain of *Shigella* has lost many metabolic pathways during the evolutionary path to become an intracellular pathogen adapted to humans and, as a consequence, it is a challenge to devise media specifically selective and/or differential for *Shigella*. In particular, *Citrobacter* spp., *Enterobacter* spp., *Klebsiella* spp. and *Acinetobacter* spp. have been shown to out-compete *Shigella* spp. in traditional enrichment cultures (Uyttendaele *et al.*, 2001). During the last 20 years, antibiotic resistance and multiresistance in *Shigella* spp. have increased greatly, causing a lot of concerns to the medical community in treating *Shigella* outbreaks.

Poor personal hygiene practices of food handlers are probably the major factor for food contamination by *Shigella*, hence routine inspections and testing of particular produce types are difficult to justify for *Shigella*. As a consequence, most shigellosis outbreaks are identified initially in the clinical laboratory and not at food manufacturing facilities. In testing, the analyst may use the antibiotic resistance profile established for the isolated pathogen and incorporate the relevant antibiotics into the enrichment medium or in post-enrichment isolations agar plates to help in identifying the source of the foodborne outbreak.

The US FDA's Bacteriological Analytical Manual (BAM) method selects for the pathogen by enrichment in a low carbohydrate medium (*Shigella* broth) and incubation in an anaerobic atmosphere at an elevated temperature in the presence of the antibiotic novobiocin (FDA, 2001; Mehlman *et al.*, 1985). High throughput monitoring of *Shigella* contamination in fresh produce is, however, performed preferably in typical incubators to avoid the use of anaerobic jars and water baths (FDA, 2005a, b). However, as microflora bacteria can grow to higher population density under aerobic conditions, *Shigella* may be under-represented among the indigenous microflora and missed on the post-enrichment isolation plates (Zhang and Lampel, 2010). To improve the sensitivity of detection, various methods based on genetic rather than phenotypic attributes have been developed for *Shigella* (Levin, 2009).

The PCR assays to detect *Shigella* rely mainly on the amplification of DNA fragments from the 220-kbp invasive plasmid shared by *Shigella* and enteroinvasive *Escherichia coli* (EIEC). This virulence plasmid can be spontaneously lost during growth outside the host, thus many PCR protocols use primers directed against the invasion plasmid antigen gene *ipaH* that is present in multiple copies in *Shigella* and EIEC bacterial genomes (Sethabutr *et al.*, 1993). Although PCR has advantages of sensitivity and specificity over conventional culture techniques, molecular detection does not generally prove that an organism is viable and able to cause disease. This is particularly important considering that *Shigella* spp. are not natural inhabitants of the environment and they are so adapted to the intracellular lifestyle that they lack typical surface appendages (i.e. flagella, fimbriae and curli) that are involved in bacterial attachment and in the formation of biofilms to protect them from damaging environmental factors (Binet and Lampel, 2013). Consequently, once excreted from the host, *Shigella* are very fragile to environmental conditions and die off rapidly, especially when exposed to direct sunlight or dried. Nevertheless, the presence of surface microsites on produce such as textured skin (Islam *et al.*, 1993; Warren *et al.*, 2007) or a bruise (Reller *et al.*, 2006) can extend the survivability of the pathogen on fresh produce by offering some degree of protection from desiccation. Typical methods for *Shigella* detection and isolation recommend a product rinse to collect the microbial flora from the surfaces of presumably contaminated fresh produce prior to enrichment, and to limit the introduction of substances that would be inhibitory to growth and/or PCR detection.

20.4.4 *Campylobacter*

Campylobacter jejuni is a leading cause of acute gastroenteritis worldwide, and most cases are identified as sporadic events rather than as recognized outbreaks. Several case-control studies of sporadic cases of human campylobacteriosis have identified eating fresh, unfrozen chicken as the main risk factor (Wingstrand *et al.*, 2006). However, *Campylobacter* can also be found in produce, perhaps contaminated from chicken run-off, so risk of infection from eating fresh produce also exists.

There are few published investigations on the occurrence of *Campylobacter* in fresh produce. A Dutch study (Verhoeff-Bakkenes *et al.*, 2011) reported 0.25% (13 samples) *Campylobacter* positive results in more than 5000 samples of leafy greens, fruit and fresh-cut salads investigated. Among these, 10 out of the 13 positive samples were vegetables. The samples were taken randomly from various vegetable and fruit retail sources, mainly supermarkets and fruit and vegetable shops. The detection method used was based on the ISO10272: 1995 (ISO, 1995), which consisted of a selective enrichment of 25 g of the homogenized sample in CCD-broth (charcoal cefoperazone deoxycholate broth) incubated under microaerobic conditions (5% O_2, 10% CO_2 and 85% N_2) for 24 hours at 42 °C. *Campylobacter* spp. were isolated by streaking a loopful of enrichment onto CCD-agar and incubated under microaerobic conditions for 48 hours at 42 °C. Typical colonies were confirmed using wet mount and latex agglutination test. The theoretical detection limit for this method was reported to be 1 cfu/25 g.

In 2009–2011, the Danish food control authorities, using the culture method based on the ISO10272 standard, investigated about 100 lots of Danish products, including 30 lots of sprouts, 33 lots of leafy greens, 36 lots of spices and one lot of sugar snap peas. Only one salad and one parsley sample were found to be contaminated with *Campylobacter* (Anon, 2011).

In the USA, a *Campylobacter* outbreak was associated with consumption of raw peas (Gardner *et al.*, 2011). Epidemiological surveillance and traceback identified 63 cases of laboratory-confirmed infection and matched case-control study showed that only raw peas, consumed by 30/45 (67%) case-patients and by 15/90 (17%) control participants, were associated with illness (adjusted odds ratio: 8.2; $P<0.001$). An additional 69 patients (26 laboratory-confirmed) who reported eating raw peas within 10 days of the illness onset were identified through enhanced surveillance. In total, five cases were hospitalized, and Guillain-Barré syndrome developed in one case; but there were no fatalities. The implicated pea farm was located near a stopover and breeding site for migrating Sandhill cranes (*Grus canadensis*). To trace the source of the contamination, clinical and environmental specimens were tested for *Campylobacter*, and isolates were subtyped by pulsed-field gel electrophoresis (PFGE). Samples of food, water, and processing surfaces were obtained and also, crane fecal samples were collected in Cary-Blair transport media, placed in coolers, and shipped within 24 hours. Environmental samples were sent to the lab, where they were tested with the *Campylobacter* isolation methods described in the US FDA Bacteriological Analytical Manual. Crane fecal samples that were presumptive positive for *Campylobacter* were also sent to the CDC and tested with three different isolation methods: direct plating onto selective blood agar plates and incubated microaerobically for 48 hours at 42 °C; filtered onto non-selective brain heart infusion (BHI) agar containing 5% (vol/vol) rabbit blood, and incubated at 37 °C for up to 7 days in microaerobic conditions; and enriched in two different broths (Preston enrichment broth (PEB) and Bolton enrichment broth (BEB)), where PEB was incubated at 37 °C for 48 hours in microaerobic conditions, and BEB was incubated for 4 hours at 37 °C, followed by further incubation at 42 °C for 44 hours. After enrichment, all broths were inoculated onto *Campylobacter* blood-free agar (modified charcoal cefoperazone deoxycholate agar) at 48 hours and incubated at 37 °C for 48 hours in microaerobic conditions. Of the 36 environmental samples collected, 16 were positive for *C. jejuni*, including four crane-fecal samples and two pea samples (Gardner *et al.*, 2011).

Kärenlampi and Hänninen (2004) used seven human and two chicken *C. jejuni* strains, with known Penner heat-stable serotypes and PFGE genotypes to investigate their survival on fresh-cut iceberg lettuce. In addition, the survival of four selected *C. jejuni* strains was assessed on cantaloupe pieces, cucumber slices, grated carrot and strawberries. Fresh produce was inoculated with 10^5–10^7 cfu/g of *C. jejuni* and survivors were enumerated using standard methods after sample storage at 7 °C and 21°C for 24, 48 and 72 hours. At 7 °C, the mean death rates per day were 0.44 on cantaloupe, 0.41 on cucumber slices, 0.43 on grated carrot, 0.59 on iceberg lettuce and 1.02 on strawberries. The corresponding death rates at 21 °C for the same products above were 1.52, 1.55, 2.61, 1.39 and 8.74, respectively. The death rate of *C. jejuni* on strawberries (pH 3.4) at 21 °C was significantly higher ($P<0.05$) than on other produce, but the study suggested that after contamination of fresh produce, especially strawberries, *C. jejuni* may survive sufficiently long to pose a risk to the consumer.

Detection of *Campylobacter* in foods, in general, can be carried out according to the ISO reference culture methods (ISO TS 10272, Fig. 20.1), or enumerated according to ISO TS 10272-Part 1, or semi-enumerated according to ISO TS 10272-Part 3. Alternatively, the Nordic method, NMKL no. 119 can be used. However, traditional culture-based detection of *Campylobacter* bacteria, including enrichment, isolation and confirmation, is a time-consuming procedure requiring 5–6 working days (Josefsen *et al.*, 2010). Furthermore, *Campylobacter* cells may enter a viable but non-culturable (VBNC) state in which they have the potential to cause human infection but are not detected by the culture method.

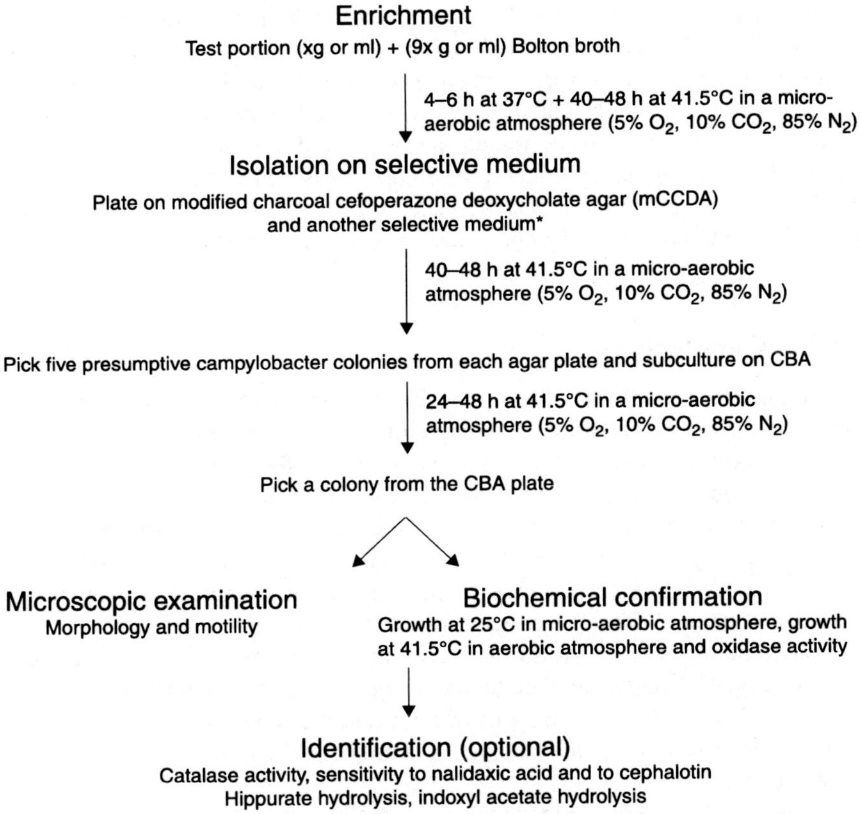

Fig. 20.1 Campylobacter culture methods. CAB is Columbia Blood Agar

A large number of qualitative rapid kits are available and may be used for detection of *Campylobacter* in foods. However, most commercial kits or in-house developed methods have focused on testing poultry samples. Hence, there is a need to further develop or evaluate existing methods or kits for testing fresh produce samples.

Introduction of qPCR for *C. jejuni* and *C. coli* has enabled faster, more sensitive and less labor-intensive quantitative detection of these pathogens. However, these methods are used mainly in testing poultry, which is recognized as an important source of human *Campylobacter* infections. Furthermore, as process control strategies mostly focus on reducing the number of bacterial contaminants, the usefulness of these qPCR methods for risk assessment could be limited, as the qPCR will detect all of the *Campylobacter* bacteria present in a sample, including the dead cells (Josefsen *et al.*, 2010). An example of qPCR detection of non-viable cells is the study of Chai *et al.* (2007), who used PCR to survey fresh produce for the presence of *Campylobacter*. They found a surprisingly high number of positive samples (49.5%), but only 28 of 153 PCR positive samples were culture confirmed to contain *Campylobacter* spp. giving a prevalence rate of 9.1% (Cl: 6.11–12.83). As PCR detects not only viable but also non-viable *Campylobacter*, Verhoeff-Bakkenes *et al.* (2011) have argued that their findings suggest the number of cultureable *C. jejuni* in a product to be a good measure for the infection risk of a product, as non-culturable *Campylobacter* were shown not to be infectious *in vitro*. Thus, PCR as such may not be a suitable option for this purpose.

20.5 Conclusion

It is important that any use of testing is founded on scientific assessment of each particular sampling situation in combination with a broader analysis of the practical and economic impacts. Food quality and safety are typically monitored by end product testing of randomly sampled lots. Sample-based monitoring programs, however, may lead to statistical uncertainty, as the contaminant is often heterogeneously distributed in a lot and it is not possible to test the entire lot to ensure that it is free of presence of pathogenic microorganisms. The risk of releasing a batch of food that is actually contaminated will be higher if the level of contamination is low, as it is less likely that the contamination will be detected by sampling. The statistical uncertainty can be reduced by increasing the number of samples being tested, but this will also increase the economic burden associated with added testing costs and reduce the amount of food available for sale and distribution.

In addition to statistical uncertainties in sampling, the user should also be aware of the analytical sensitivity of the method, which is a measure of the effectiveness of the method in detecting the microorganisms in a contaminated sample. The rule of thumb is that no method is 100% effective. Hence, even though the contaminated units in a lot have been sampled and tested, the level of contaminant in the sample may be so low that the test is unable to detect them. Hence, low analytical sensitivity will also increase the uncertainty of sample results.

Both the statistical uncertainty and analytical sensitivity are important factors to consider when determining the purpose of a given monitoring or control program; limitations of both factors can affect the outcome of test results, which can have severe impact as the data are used to infer the health risk for the remainder of the batch and/or the producer.

20.6 Questions for discussion

1. What is the most important bottleneck in fresh produce testing, if you were to mention just one?
2. What is the main difference between producer testing and regulatory authority testing programs?

3. In which situations do you recommend a sample composite (pooling) strategy?
4. Why culture cultivation of *Shigella* is difficult?
5. What are the main bottlenecks of PCR testing of fresh produce?
6. Is Whole Genome Sequencing (WGS) going to replace current molecular typing methods? and why?

20.7 Acknowledgements

The work of J. Hoorfar was supported by a Network Grant (www.FreshProduce.dk) from the Danish Ministry of Science, Innovation and Higher Education. The assistance of Dr Iben-Bang Berthelsen is much appreciated.

The authors responsible for each section were:

- 20.2 Logistics in sample preparation: Jeffrey Hoorfar
- 20.3 Logistics in field testing: Geraldine Duffy
- 20.4 Logistics in product testing
  - 20.4.1 *Salmonella*: Burkhard Malorny
  - 20.4.2 Verotoxigenic *Escherichia coli* (VTEC): Peter Feng
  - 20.4.3 *Shigella*: Rachel Binet
  - 20.4.4 *Campylobacter*: Jeffrey Hoorfar

20.8 References

ANON. (2011) Kampagner og projekter – slutrapport – Kontrol af patogener I importeret og danske spiseklart grønsager. J.nr. 2008–20–64–00910. (Campagns and projects: Final report on the control of imported and Danish, ready-to-eat, fresh produce). Copenhagen, Denmark.

BEUCHAT L.R. (2002) Ecological factors influencing survival and growth of human pathogens on raw fruits and vegetables. *Microbes Infect* 4:413–23.

BINET R. and LAMPEL K.A. (2013) *Shigella* species. In *Food Microbiology: Fundamentals and Frontiers*. Doyle M.P., and Buchanan, R.L. (eds). ASM Press, Washington DC.

CHAI L.C., ROBIN T., RAGAVAN U.M., GUNSALAM J.W., BAKAR F.A., *et al.* (2007) Thermophilic *Campylobacter* spp. in salad vegetables in Malaysia. *Int J Food Microbiol* 117: 106–11.

FENG P.C.H., COUNCELL T., KEYS C. and MONDAY S.R. (2011) Virulence characterization of Shigatoxigenic *Escherichia coli* serotypes isolated from wholesale produce. *Appl Environ Microbiol* 77:343–5.

FENG P., LAMPEL K., KARCH H. and WHITTAM T. (1998) Sequential genotypic and phenotypic changes in the emergence of *Escherichia coli* O157:H7. *J Infect Dis* 177:1750–53.

FENG P., WEAGANT S.D. and JINNEMAN K. (2012) Chapter 4A. Diarrheagenic *Escherichia coli*. FDA Online Bacteriological Analytical Manual http://www.fda.gov/Food/FoodScienceResearch/LaboratoryMethods/ucm070080.htm.

FOOD AND DRUG ADMINISTRATION (2001) Chapter 6. Shigella. eBAM http://www.fda.gov/Food/FoodScienceResearch/LaboratoryMethods/ucm070789.htm.

FOOD AND DRUG ADMINISTRATION (2005a) FDA Survey of Domestic Fresh Produce – Domestic Produce Assignment (DFP # 05–20). 8–2-2005a. 9–27–0012a. Ref Type: Internet Communication. http://www.fda.gov/Food/GuidanceRegulation/GuidanceDocumentsRegulatoryInformation/ProducePlantProducts/ucm118297.htm

FOOD AND DRUG ADMINISTRATION (2005b) FDA Survey of Imported Fresh Produce – Import Produce Assignment (DFP # 05–16). 3–18–2005b. Ref Type: Internet Communication. http://www.fda.gov/Food/GuidanceRegulation/GuidanceDocuments RegulatoryInformation/ProducePlantProducts/ucm118884.htm.

GABRIEL A.A. (2005) Microbial quality of chlorine soaked mung bean seeds and sprouts. *Food Sci Technol Res* 11:95–100.

GARDNER T.J., FITZGERALD C., XAVIER C., KLEIN R., PRUCKLER J., *et al.* (2011) Outbreak of campylobacteriosis associated with consumption of raw peas. *Clin Infect Dis* 53:26–32.

GILL C.J., KEENE W.E., MOHLE-BOETANI J.C., FARRAR J.A., WALLER P.L., *et al.* (2003) Alfalfa seed decontamination in a *Salmonella* outbreak. *Emerg Infect Dis* 9:474–9.

GOLDEN D.A., RHODEHAMEL E.J. and KAUTTER D.A. (1993) Growth of *Salmonella* spp. in cantaloupe, watermelone, and honeydew melons. *J Food Prot* 56:194–6.

HIRNEISEN K.A., SHARMA M. and KNIEL K.E. (2012) Human enteric pathogen internalization by root uptake into food crops. *Foodborne Pathog Dis* 9:396–405.

HOORFAR J., CAHILL S., CLARKE R., BARKER G.C., FAZIL A., *et al.* (2011). The public health, industrial, and global significance of rapid microbiological food testing. In: *Rapid Detection, Characterization and Enumeration of Food-Borne Pathogens*. Hoorfar J. (ed.). ASM Press, Washington, DC pp. 1–12.

ICMSF (INTERNATIONAL COMMISSION ON MICROBIOLOGICAL SPECIFICATIONS FOR FOODS) (2002) Microorganisms in Foods, 7, Microbiological Testing In Food Safety Management, Springer, New York, USA.

ICMSF (INTERNATIONAL COMMISSION ON MICROBIOLOGICAL SPECIFICATIONS FOR FOODS) (2011) Vegetable and vegetables products. In: *Microorganisms in Foods 8* (K.M.J. Swanson, ed.), pp. 147–176. Springer, New York.

ISLAM M.S., HASAN M.K., and KHAN S.I. (1993) Growth and survival of *Shigella flexneri* in common Bangladeshi foods under various conditions of time and temperature. *Appl Environ Microbiol* 59:652–54.

ISO (1995) Microbiology of food and animal feeding stuffs – Horizontal method for detection and enumeration of *Campylobacter* spp. International Organisation for Standardisation (ISO), Geneva, Switzerland.

JARVIS B. (2007) On the compositing of samples for qualitative microbiological testing. *Lett Appl Microbiol* 45:592–8.

JOSEFSEN M.H., LÖFSTRÖM C., HANSEN T.B., CHRISTENSEN L.S., OLSEN J.E. and HOORFAR J. (2010) Rapid quantification of viable *Campylobacter* bacteria on chicken carcasses, using real-time PCR and propidium monoazide treatment, as a tool for quantitative risk assessment. *Appl Environ Microbiol* 76:5097–104.

KRAMER M.F. and LIM D.V. (2004) A rapid and automated fiber optic-based biosensor assay for the detection of *Salmonella* in spent irrigation water used in the sprouting of sprout seeds. *J Food Prot* 67:46–52.

KÄRENLAMPI R. and HÄNNINEN M.L. (2004) Survival of *Campylobacter jejuni* on various fresh produce. *Int J Food Microbiol* 97:187–95.

LEVIN R.E. (2009) Molecular methods for detecting and discriminating *Shigella* associated with foods and human clinical infections – A Review. *Food Biotechnol* 23:214–28.

LYNCH M.F., TAUXE R.V. and HEDBERG C.W. (2009) The growing burden of foodborne outbreaks due to contaminated fresh produce: risks and opportunities. *Epidemiol Infect* 137:307–15.

MACDONALD E., HEIER B.T., NYGÅRD K., STALHEIM T., CUDJOE K.S., *et al.* (2012) *Yersinia enterocolitica* outbreak associated with ready-to-eat salad mix, Norway, 2011. *Emerg Infect Dis* 18:1496–9.

MCGRATH T.F., ELLIOTT C.T. and FODEY T.L. (2012) Biosensors for the analysis of microbiological and chemical contaminants in food. *Anal Bioanal Chem* 403:75–92.

MEHLMAN I.J., ROMERO A. and WENTZ, B.A. (1985) Improved enrichment for recovery of *Shigella sonnei* from foods. *J Assoc Off Anal Chem* 68: 552–5.

MODY R.K., GREENE S.A., GAUL L., SEVER A., PICHETTE S., *et al.* (2011) National outbreak of *Salmonella* serotype Saintpaul infections: importance of Texas restaurant investigations in implicating jalapeno peppers. *PLoS. One* 6, e16579.

MONTVILLE R. and SCHAFFNER D.W. (2004) Analysis of published sprout seed sanitization studies shows treatments are highly variable. *J Food Prot* 67:758–65.

NATIONAL ADVISORY COMMITTEE ON MICROBIOLOGICAL CRITERIA FOR FOODS (1999) Microbiological safety evaluations and recommendations on sprouted seeds. *Int J Food Microbiol* 52:123–53.

NEI D., LATIFUL B.M., ENOMOTO K., INATSU Y. and KAWAMOTO S. (2011) Disinfection of radish and alfalfa seeds inoculated with *Escherichia coli* O157:H7 and *Salmonella* by a gaseous acetic acid treatment. *Foodborne Pathog Dis* 8:1089–94.

NYGÅRD K., LASSEN J., VOLD L., ANDERSSON Y., FISHER I., *et al.* (2008) Outbreak of *Salmonella* Thompson infections linked to imported rucola lettuce. *Foodborne Pathog Dis* 5:165–73.

OLAIMAT A.N. and HOLLEY R.A. (2012) Factors influencing the microbial safety of fresh produce: A review. *Food Microbiol*. Doi:10.1016/j.fm.2012.04.016.

RAMBO C.H. and PILLAI S. (2011) Pathogen testing in fresh produce and irrigation water. In: *Rapid detection, characterization and enumeration of foodborne pathogens.* Hoorfar J. (ed.) ASM Press, Washington, DC.

RELLER M.E., NELSON J.M., MOLBAK K., ACKMAN D.M., SCHOONMAKER-BOPP D.J., *et al.* (2006) A large, multiple-restaurant outbreak of infection with *Shigella flexneri* serotype 2a traced to tomatoes. *Clin Infect Dis* 42:163–9.

RIMHANEN-FINNE R., NISKANEN T., LIENEMANN T., JOHANSSON T., SJOMAN M., *et al.* (2011) A nationwide outbreak of Salmonella Bovismorbificans associated with sprouted alfalfa seeds in Finland, 2009. *Zoonoses Public Health* 58:589–96.

ROSS T., FRATAMICO P.M., JAYKUS L. and ZWIETERING M.H. (2011) Statistics of sampling for microbiological testing of foodborne pathogens. In: *Rapid detection, characterization and enumeration of foodborne pathogens*. Hoorfar J. (ed.). ASM Press, Washington, DC.

SETHABUTR O., VENKATESAN M., MURPHY G.S., EAMPOKALAP B., HOGE C.W. and ECHEVERRIA P. (1993) Detection of *Shigellae* and enteroinvasive *Escherichia coli* by amplification of the invasion plasmid antigen H DNA sequence in patients with dysentery. *J Infect Dis* 167:458–61.

TERRY L.A., WHITE S.F. and TIGWELL L.J. (2005) The application of biosensors to fresh produce and the wider food industry. *J Agric Food Chem* 5:1309–16.

UYTTENDAELE M., BAGAMBOULA C.F., DE S.E., VAN W.S. and DEBEVERE J. (2001) Evaluation of culture media for enrichment and isolation of *Shigella sonnei* and *S. flexneri*. *Int J Food Microbiol* 70:255–65.

VALENTIN-BON I., JACOBSON A., MONDAY S. and FENG P. (2008) Microbiological quality of bagged cut spinach and lettuce mixes. *Appl Environ Microbiol* 74:1240–2.

VERHOEFF-BAKKENES L., JANSEN H.A., INT'VELD P.H., BEUMER R.R., ZWITERING M.H. and VAN LEUSDEN F.M. (2011). Consumption of raw vegetables and fruits: A risk factor for *Campylobacter* infections. *Int J Food Microbiol* 144:406–12.

WARREN B.R., YUK H.G. and SCHNEIDER K.R. (2007) Survival of *Shigella sonnei* on smooth tomato surfaces, in potato salad and in raw ground beef. *Int J Food Microbiol* 116:400–4.

WINGSTRAND A., NEIMANN J., ENGBERG J., NIELSEN E.M., GERNER-SMIDT P., *et al.* (2006) Fresh chicken as main risk factor for campylobacteriosis in Denmark. *Emerg Infect Dis* 12:280–5.

ZHANG G. and LAMPEL K.A. (2010) Comparison of chromogenic Biolog Rainbow agar *Shigella/Aeromonas* with xylose lysine desoxycholate agar for isolation and detection of *Shigella* spp. from foods. *J Food Prot* 73:1458–65.

21

New developments in safety testing of soft fruits

R. Morales-Rayas and M. W. Griffiths, University of Guelph, Canada and A. C. Shultz, Technical University of Denmark, Denmark

DOI: 10.1533/9781782420279.4.292

Abstract: The chapter begins by establishing the definition of soft fruit, then discusses microbial and chemical hazards that might be found in soft fruit. Methods developed for virus detection in soft fruit are reviewed and, finally, the factors that mainly affect virus detection are detailed as these pathogens are currently linked to outbreaks caused by soft fruit consumption.

Key words: berries, foodborne viruses, parasites, sample preparation, molecular detection.

21.1 Introduction

Fruits and derived products are known carriers of foodborne pathogens. In recent years, an increase in their consumption has been observed as fresh fruits have increased in popularity as part of the human diet. As a consequence, an increase in the number of outbreaks related to fresh fruit consumption has been observed. Soft fruits have been frequently linked to outbreaks, especially to emerging pathogens that have been recently identified in this type of fruit. Identification of the causative agents in an outbreak has been facilitated by recent developments in foodborne pathogen detection. New methodologies for identification of pathogens have helped not only to identify emerging pathogens, but also to support improvement of surveillance and agricultural practice systems.

Traditionally, pathogens have been separated from food matrices by homogenization of samples in a diluent. Analysis of the resulting suspension requires selective enrichment and subsequent biochemical identification. Traditional testing of soft fruits has therefore been slow and, in some cases, non-causative agents have been identified as a result of a lack of proper testing methods.

Some standard methods for microbiological testing are still based on conventional microbiological methods. However, the inclusion of molecular methods to microbiology has extensively supported the identification of some non-culturable pathogens such as viruses. As many viruses cannot be propagated *in vitro*, their routine identification could not be included in routine testing.

As part of the laboratory testing for fresh produce safety, this chapter will provide information about new pathogen detection methods developed for soft fruits. Some of considerations for the development of new methodologies include fruit surface factors and pathogen surface factors that allow pathogen survival, extrinsic factors such as where the soft fruit is produced and stored and physical characteristics of the target pathogen. Furthermore, sample preparation is a key step to remove inhibitors present in soft food matrices.

21.2 Soft fruit

Soft fruit is a generic term used for small, fleshy fruit. Although most of the fruits included in this term are berries, botanically, the term berry not only comprises blueberries, gooseberries, blackcurrant and grapes but also avocados, tomatoes and kiwis (Stuppy and Kesseler, 2008). For the purposes of this chapter, the term 'soft fruit' will refer to small edible fruit with multiple seeds. Therefore, berries such as blueberry, strawberry, raspberry, gooseberry, elderberry and currants will be included.

Recent statistics about the major crops of soft fruit worldwide include strawberry, blueberry, raspberry, currants and gooseberry in ascending order of metric tonnes produced (FAOSTAT, 2010). The main producers of soft fruit from 2006 to 2010 were the USA for strawberry and blueberry and the Russian Federation for raspberry and currant production (Fig. 21.1). The Russian Federation was also the major producer of gooseberries during this period of time (FAO, http://faostat.fao.org/site/339/default.aspx). The remaining soft fruit producers include countries such as Turkey, Spain, Poland, Serbia, Canada, Ukraine, Spain and Mexico (Fig. 21.1).

Soft fruits are offered to consumers directly in farms or as U-pick, pre-packed fresh in local or international markets or processed in the form of juice, puree, jams or frozen (Strik, 2007). In general, data observed during 5 years show increased production of soft fruit worldwide. This might be attributable to the popularity that berries have gained because of their content of compounds associated with health benefits. In addition to their high nutritional value because of their vitamin and mineral contents, soft fruits are also highly valued for their antioxidant compounds, flavonoid and phenolic contents (Parish *et al.*, 2012).

Production of soft fruit has also been increased by exportations from new countries. Production of soft fruit in different countries has also expanded their availability in all seasons in most regions (Parish *et al.*, 2012). With the myriad of production and safety systems performed worldwide, it is not surprising that hygienic practices during soft fruit production are quite diverse. The safety of

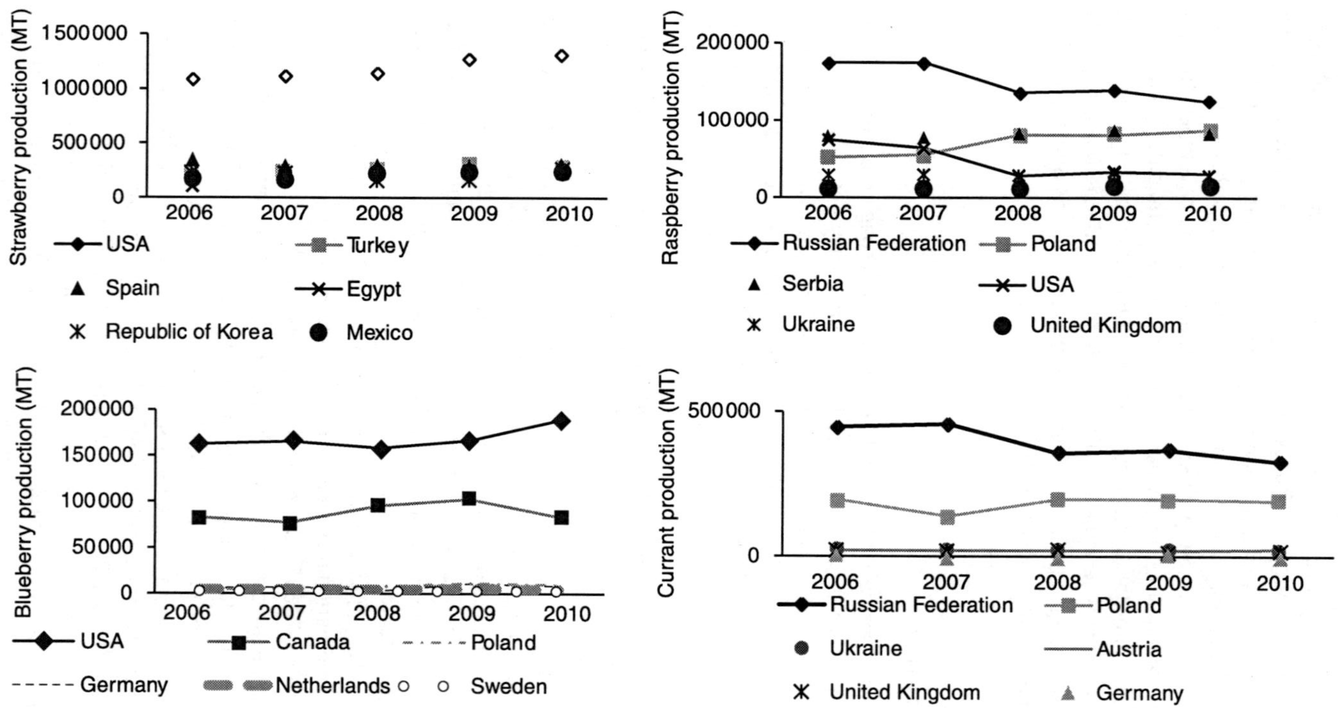

Fig. 21.1 Production of major soft fruit crops by main growers in 2010. Data are presented from 2006 to 2010 but countries included are in order of production in 2010 (derived from United Nations Food and Agriculture Organization, FAO, FAOSTAT, accessed 18 December 2011, at http://faostat.fao.org/site/339/default.aspx).

food, and in this case of soft fruit, involves potential contamination from chemical, physical and microbiological sources (Fig. 21.2). Potential microbiological risks are the main safety issue with fresh fruit. Fresh produce is most often contaminated as a result of poor practice in primary production and/or misuse of natural and environmental resources (Brassard *et al.*, 2012; Newell *et al.*, 2010), for example the irrigation of produce with polluted water (including possible contamination through roots owing to drop irrigation (Urbanucci *et al.*, 2009)), contact with human faeces or faecal soiled materials and poor hygiene practice by food handlers during harvest (León-Félix *et al.*, 2010). Furthermore, contamination may arise by inappropriate practices during processing or at the point of sale/consumption (Boxman *et al.*, 2011; Daniels *et al.*, 2000; Schmid *et al.*, 2007). Finally, there may be cross-contamination from polluted wash water, working instruments or surfaces, which had been previously contaminated by infected food handlers or contaminated food items (D'Souza *et al.*, 2006; Boxman *et al.*, 2009; Dreyfuss, 2009).

Chemical and physical contamination of food may also become a risk to consumers. Physical contamination may not represent a high risk in soft fruit unless the fruits are processed for juices, jam or frozen. Physical contamination of processed soft fruit occurs when any material such as metal fragments, gravel, jewellery and glass particles, which is not normally in processed products, can

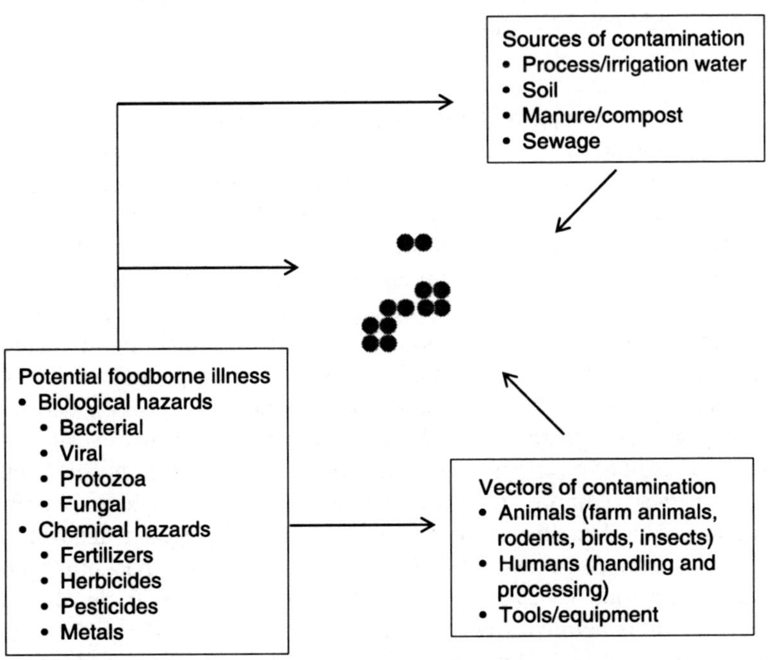

Fig. 21.2 Potential contamination factors for soft fruit during preharvesting, harvesting and processing (adapted from Bower *et al.*, 2003).

produce illness or injury to consumers (Daeschel and Pathima, 2007). Chemical contamination might occur during production or processing of soft fruit (Fig. 21.2). Use of pesticides in berry production can be carried to either fresh or processed fruit, representing a high risk to consumers. It is therefore crucial to follow integral agricultural practices in the field to ensure that chemical residues are not passed through the soft fruit processing to reach consumers (Parish *et al.*, 2012). Chemical contaminants can also be introduced during processing when compounds recognized as safe are not used in agreement with regulatory guidelines. Compounds recognized as safe might include antioxidants, preservatives, sulfiting agents and colorants, among others (Daeschel and Pathima, 2007).

21.3 Microbial pathogens of safety concern in soft fruits

In general, fruit and processed fruit products are believed to be safe from pathogenic bacteria because of their low pH (Zhao, 2005). However, contamination of soft fruit can occur during different points at the preharvest, harvest and postharvest (Fig. 21.2) stages. The main factors that contribute to a high microbial load in a soft fruit can be divided into the source of contamination, the vector for contamination and the potential microorganisms that can be found in the different fruits. Outbreaks linked to soft fruit have shown that pathogens are able to survive high acid content products derived from soft fruit, showing that soft fruits are vulnerable to microbial pathogens.

21.3.1 Bacterial and parasitic pathogens in soft fruits

Preliminary analysis of the trend of pathogens present in food has shown a decrease in infections caused by *Campylobacter*, *Listeria*, *E. coli* O157, *Shigella* and *Yersinia* in 2010, compared with 1996–1998 (CDC, 2011). Contrary to infections where *Salmonella* and *Vibrio* were the causative agents, the reduction of illnesses caused by these five pathogens is attributed to improved detection and surveillance systems, regulatory banning of contaminated ground beef with *E. coli* O157 and increased awareness of the risk of consumption of undercooked beef in food service organizations and at home. However, pathogens that are not included in these surveillance systems such as norovirus (NoV), *Clostridium perfringens* and *Toxoplasma* are still the main microorganisms causing human illness. As a result of developments in methods for viral testing in soft fruits and recent strategies to improve the production, quality and safety of soft fruit production, recent data on the presence of bacteria and parasites in soft fruit are scarce. Some reports of outbreaks linked to bacterial and parasitic pathogen presence in soft fruit were reported more than 10 years ago (Table 21.1). From 1973 to 1997, nine outbreaks were linked to berries including strawberry (four), raspberry (four) and blackberry (one). Seven of those outbreaks were associated with *Cyclospora* and one with *Staphylococcus aureus* (Sivapalasingam *et al.*, 2004).

Table 21.1 Outbreaks linked to bacterial and parasitic contamination of soft fruits

Pathogen	Type of berry	Year	Location	Notes
Bacteria				
Staphylococcus aureus	Strawberry	1985	New York, USA	14 persons were ill during this outbreak reported by the Foodborne Outbreak Surveillance System
Salmonella	Strawberry	2001	NR	1/143 samples collected for analysis from imported products to USA from five different countries was positive
Parasite				
Cyclospora cayetanensis	Raspberries likely	1995	Florida, USA	Raspberries from Guatemala were involved in 87 cases, but causative agent was not isolated from the berries
C. cayetanensis	Raspberries	1996	20 USA states and two Canadian provinces	Contamination was attributed to contaminated water mixed with insecticides and fungicides that were sprayed onto the fruit
C. cayetanensis	Raspberries	1997	Multiple states in USA and Ontario, Canada	The source of contamination was not identified
C. cayetanensis	Raspberries	1998	Ontario, Canada	The source of contamination was not identified
C. cayetanensis	Blackberries	1999	Ontario, Canada	The source of contamination was not identified
C. cayetanensis	Raspberries	2000*	Philadelphia, USA	The source of contamination was identified after interviewing guests at a wedding reception and DNA identification by PCR

Source: adapted from FDA, *Ho *et al.* (2002).

More recent studies have shown that *Salmonella* and *Cyclospora cayetanensis* are able to survive in juices made from soft fruit (Bower *et al.*, 2003). *Listeria monocytogenes* and *Salmonella* have been isolated from frozen blueberries and fresh strawberries respectively (Zhao, 2005). In addition, the attachment of bacteria or parasites to soft fruit surfaces is also another factor that could lead to an increment of the microbial load. Kniel *et al.* (2002) demonstrated that oocysts of *Toxoplasma gondii* can attach to berry epidermis and persist during 8 weeks. *T. gongii* is a coccidial parasite related to *Cyclospora* and it can be used as a *Cyclospora* surrogate. Raw soft fruit is usually harvested by hand and packed in containers that will eventually reach consumers without being washed. For example, strawberries that will be frozen, are destemmed in the fields, most of the time using thumbnail or metal devices. The extra human handling of the soft fruit therefore represents an opportunity for a pathogen to attach to the food surface and survive. Although no recent data are available about outbreaks linked to contaminated soft fruit with bacterial or parasitic pathogens, the hygienic practices used for their production could eventually lead to high bacterial or parasitic loads.

21.3.2 Viruses in soft fruits

Soft fruits that are eaten raw or lightly heat-treated are frequently causes of outbreaks of viral illnesses. As examples, frozen raspberries have caused NoV outbreaks in Finland, Denmark, Sweden and France (Little and Gillespie, 2008), and, in 2012, Germany experienced their biggest foodborne NoV outbreak ever caused by frozen strawberries implicated in outbreaks that involved 500 institutions with 11 000 cases (Anonymous, 2012). Even though several viruses such as enterovirus, human rotavirus, adenovirus, sapovirus, astrovirus, coronaviruses or aichi virus appear capable of causing foodborne illness or emerging as foodborne pathogens, all recent data show that hepatitis A virus (HAV) and NoV continue to be the most common viral pathogens causing foodborne illness (Yadav *et al.*, 2010). Statistical data have shown that HAV is a higher health risk than other pathogens such as *Salmonella* Typhi, *Shigella sonnei* or *Staphylococcus aureus* because of the number of deaths caused by HAV outbreaks (Todd *et al.*, 2009). Besides the risk that soft fruits can be contaminated with viruses during growth, the role that food workers play in the spread of illness has been assessed in HAV and NoV outbreaks in which over 50% of the cases were hospitalized. Bidawid *et al.* determined that 9% of HAV and nearly 18% of a NoV surrogate, feline calicivirus (FCV), could be transferred from artificially contaminated finger pads to lettuce by contacts during food handling. When studying the opposite route of transmission, they found that 14% of FCV could be transferred from lettuce to bare hands (Bidawid *et al.*, 2004, 2000a). This ease of reversible viral transfer between workers and produce was later supported in a field study (León-Félix *et al.*, 2010) evaluating the impact of harvesting and packinghouse operations on the contamination of green bell pepper with NoV. When testing the workers' hands for NoV before and after 3 hours of work, they were able to show a change in the rates of positivity of 0% to 14% for pickers, 53% to 33% for classifiers and 45% to 25% for packers. Although these data are for fresh produce, similar viral transmission is expected to occur during soft fruit production because of the similar practices of picking.

The recent year's significant progress in the method development for virus detection in foods reflects the growing acknowledgement of the risk of food acting as a vector for virus transmission. With the introduction of TaqMan real-time

RT-PCR (RT-qPCR), it has been possible to simultaneously confirm the amplified product (Mattison *et al.*, 2009b; Lees and CEN/WG6/TAG4, 2010) as well as to estimate the amount of detectable virus genomes (Le Guyader *et al.*, 2009). However, the lack of reliable and harmonized sampling plans and standardization in the detection methods limits their suitability for routine analysis of viruses in soft fruit. Therefore, data on the prevalence and titres of viruses in these matrices are rare. In a collection of data on outbreaks that occurred in USA during 1990–2005 with known aetiology, fresh produce accounted for 13% (713/5416) of all foodborne related outbreaks (DeWaal and Bhuiya, 2007; Doyle and Erickson, 2008). Of these, NoV was the most common pathogen accounting for 40% of outbreaks associated with fresh produce (i.e. fruits and vegetables). Specifically, NoV in lettuce and salad greens alone accounted for nearly 25% of the produce-associated outbreaks (Doyle and Erickson, 2008). Data from the European Rapid Alert System for Food and Feed (RASFF), outbreak investigations, and a few monitoring studies verify this importance of fresh produce as a high-risk food product for NoV. The application of recently developed methods in outbreak investigations has begun to provide an insight into the magnitude of contamination of soft fruits that has resulted in disease after consumption. Several outbreak reports of NoV-contaminated raspberries describe large numbers (up to 11 000) of people being infected (Falkenhorst *et al.*, 2005; Sarvikivi *et al.*, 2011a; Anonymous, 2012). Table 21.2 illustrates published outbreaks where NoV contaminated soft fruits were epidemiologically linked to the patients and available for viral testing. Only a few screening studies have been conducted of viruses in soft fruits. In a French study of soft red fruits, 6.7% of 150 samples tested positive for NoV genogroup I (GI), GII or both (Baert *et al.*, 2011). In a Canadian study of irrigated, field-grown strawberries, 26% (16/60) of samples were positive for NoV GI, and a few samples also tested positive for swine hepatitis E virus genogroup 3 and human rotavirus (Brassard *et al.*, 2012). In a minor screening study in Belgium, five samples from each of two lots of raspberries and 10 samples from each of two lots of strawberries were tested for NoV. Infrequently, samples from all four lots were found to be positive for NoV GI, GII or both (Stals *et al.*, 2011). The quantified levels of NoV found in these studies are summarized in Table 21.3. No illnesses could be connected to these batches of soft fruit, despite some being contaminated with NoV in levels similar to or exceeding the estimated infectivity dose of 18–1000 viral particles (Teunis, 2008) and what is commonly found for oysters associated with disease outbreaks (10^2–10^4 NoV genomic copies/g digestive tissue) (Dore *et al.*, 2010; Le Guyader *et al.*, 2008). According to the authors, suggested reasons for this could be under-reporting of sporadic gastroenteritis and that the amount of NoV genomic copies detected by RT-qPCR (which does not differentiate between infective and non-infective viruses) may overestimate the amount of infectious viruses (Baert *et al.*, 2011).

Finally, although rather similar analytical principles are used among the laboratories when analyzing soft red fruits, there are important factors which differ such as variation in the number of samples (29 to 150) and most likely also batches tested, which may affect the prevalence of positive findings. Moreover,

Table 21.2 Outbreaks caused by NoV and HAV contaminated soft fruits where foods were available for analysis

Country	Year	Product	Origin	Outbreaks	Cases	Patient sample	Food sample	Reference
Canada	1997	Raspberries	Bosnia	2	100	GI*	GI	Sarvikivi *et al.* (2011b)
Finland	1998	Raspberries	East Europe	1	108	GII	Neg.	Ponka *et al.* (1999)
Sweden	2001	Raspberries	Slovakia	1	30	GI	GI, GIIb	Le Guyader *et al.* (2004)
New Zealand	2002	Blueberries	New Zealand	1	39	HAV	HAV	Calder *et al.* (2003)
Denmark	2005	Raspberries	Poland	6	1100	GII.4, II.b, II.7	GII.4	Falkenhorst *et al.* (2005)
Germany	2005	Blackberries	Unknown	1	241	GI	Neg.	Fell *et al.* (2007)
France	2005	Strawberries	Serbia?	1	75	GI.5	Neg.	Cotterelle *et al.* (2005)
Sweden	2006	Raspberries	China	4	43	NoV*	Neg.	Hjertqvist *et al.* (2006)
Finland	2009	Raspberries	Poland/Serbia	13	900	GI.4, GII, II.b, II.4, II.7	GII	Sarvikivi *et al.* (2011a)
Denmark	2009	Raspberries	Serbia	1	6	GII.8	GI, GII	Anonymous (2011)
Finland	2009	Raspberries	Poland	3	200	GI.4	GI.4	Maunula *et al.* (2009)
Denmark	2010	Raspberries	China	1	10	NoV*	GII	Anonymous (2011)
Denmark	2010–11	Raspberries†	Serbia	7	224	GI.b / I.6	GI, GI.6 GII	Anonymous (2011)
Denmark	2011	Raspberries	China	1	8	GI.4	GII	Anonymous (2011)

Source: Schultz *et al.* (2012) PhD thesis.

*NoV genogroup or type is not described.
†Multiple batches of different package date from same producer.

Table 21.3 Amount of detected NoV in monitoring studies of soft fruits

	NoV GC/g min–max*		
Sample source	GI	GII	Reference
Raspberries/ strawberries (BE)	2.0×10^{1}–1.3×10^{3}†	1.0×10^{2}–6.3×10^{2}†	Baert *et al.* (2011); Stals *et al.* (2011)
Raspberries (FR)	2.5×10^{2}–1.0×10^{5}†	1.0×10^{2}–6.3×10^{5}†	Baert *et al.* (2011)
Strawberries (CA)	5.0×10^{1}–3.0×10^{3}	Neg.	Brassard *et al.* (2012)

* Samples were considered valid for analysis if the recoveries of the process control were higher than 0.1% or if NoV was detected in the sample.
† The GC/g-values (detectable genomic copies/g) are reverted from the log values originally given in the references.
Abbreviations: BE, Belgium; FR, France; CA, Canada.

the difference in the initial and analyzed sample sizes (10–25 g or 0.025–1.25 respectively) is likely to affect the detection limit of the target virus as well as the risk of co-extracted RT-PCR inhibitors. Thus, interpretation of the NoV prevalence and quantitative findings in relation to general occurrence and comparison of results between laboratories should be handled with care. To assist source attribution studies, a recent EFSA (European Food Safety Authority) report (EFSA BIOHAZ, 2011) recommended systematic strain typing in combination with routine harmonized surveillance of viral outbreaks, and of virus occurrence in food commodities. Successful typing of the NoV findings was achieved in only a few of the outbreak studies listed in Table 21.2 (Maunula *et al.*, 2009; Ponka *et al.*, 1999; Le Guyader *et al.*, 2004). The failure to obtain NoV sequences from these samples could be because the strains initially detected by RT-qPCR were not recognized by the applied primers used for conventional RT-PCR or below the assay detection limit (Baert *et al.*, 2011). Other common issues that can challenge successful sequencing of virus strains detected in soft fruits could be insufficiently pure RNA extract or simultaneous RT-PCR amplification of multiple strains (Sarvikivi *et al.*, 2011a).

21.4 Methods for evaluation of microbial safety in soft fruit

Procedures to evaluate microbial load in soft fruit aim to detect and quantify pathogens that represent a risk to human health. The final objective of the evaluation is, therefore, to determine the number of microorganisms for which regulatory systems have adopted a zero tolerance level. For emerging pathogens for which no safety regulations are well established, such as NoV or sapovirus, the objective is set to detect the infectious dose causing infection in consumers. Consequently, the detection limit of methods for evaluation of microbial load in soft fruit requires the amount of colony forming units (cfu), plaque forming units (pfu) or oocysts (infectious stage of some parasites). In order to quantify any microbial pathogen, the objective of the methodology is to concentrate and separate the target from the food matrix. The particular case of soft fruit represents a challenge as fruits are fleshy and easily create big particulates, which could entrap the microbial target for detection. Furthermore, some soft fruit surfaces contain crevices that could obstruct the separation of the microbial target from the soft fruit surface (Fig. 21.3). In addition to physical entrapment in a rough surface of a berry, smooth surfaces can interact with microbial particles by electrostatic interactions. Methods developed for quantification and detection of foodborne pathogens must take into account the rupture of these interactions in order to improve the microbial recovery, and, hence, promote a sensitive assessment of soft fruits.

Conventional methods for detection of pathogens in food are based on cell resuscitation in the case of bacteria or on identification by indirect methods. In the case of pathogens for which there are no *in vitro* propagation methods, the use of cultivation-independent methods represents the only alternative for evaluation of soft fruit safety. Therefore, new developments in testing of soft fruit have been based on the molecular identification of pathogen. As has been stated in previous sections, recent outbreaks linked to soft fruit consumption are related to the presence of viruses. Therefore, most of the methods reviewed in this section will be focused mainly on these microorganisms.

21.4.1 Sampling

The major challenges for conventional detection of pathogens in soft fruits as well as for molecular identification are: the uneven distribution of presumably low

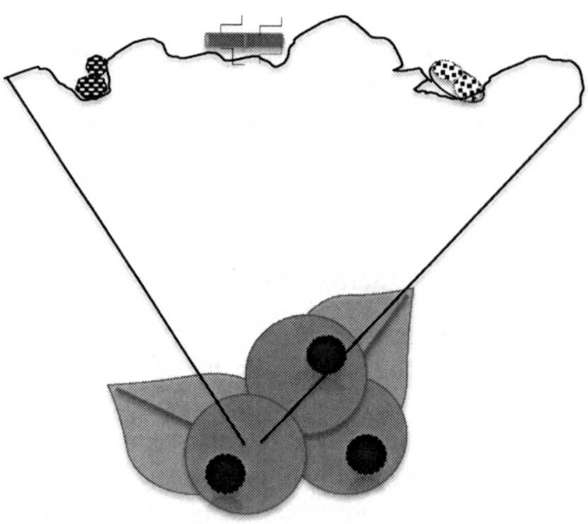

Fig. 21.3 Representation of the interaction between soft fruit surfaces and bacterial, viral and parasitic pathogens. Interaction on a smoother surface, such as blueberry, might be similar because of neutral, hydrophilic and hydrophobic forces (adapted from Wang *et al.*, 2009 and Morales-Rayas and Jean, 2011).

viral loads, the limited reproducibility of certain pathogens or nucleic acid extraction efficiencies, the presence of inhibiting substances that jeopardize molecular detection and the genetic variability of pathogens such as NoV (Le Guyader *et al.*, 2006).

As for other types of foods, an appropriate sampling strategy is important to undertake effective pathogen monitoring. This is particularly important for methods applied within a legislative context and used for risk management and control (Bosch *et al.*, 2011). A plan for sampling should be established on a risk-based approach and the sample or subsamples must represent the original matrix. In addition, the sampling process must not alter the condition of the sample and thus not affect the subsequent analysis (Bosch *et al.*, 2011). To develop a sampling plan for pathogen analyses, it is important to consider the characteristics of the matrix to be analyzed (nature, food category, composition and amount) as the choice of sample size may rely on the capability of the available method to remove inhibitors along the sample processing (Rodriguez-Lazaro *et al.*, 2011). Currently, there is no official standard for sampling for virus detection in foods (Rodriguez-Lazaro *et al.*, 2011) and the limited amount of available harmonized data makes it difficult to enable effective application of virological methods for monitoring virus contamination in food. The low viral infectious dose requires sampling and testing of relatively large volumes of food. In general, published protocols for viral analyses of soft fruits use an initial sample size of 10–100 g (Table 21.4). Although viruses have been recovered from relatively small amounts of soft fruit (Stals *et al.*, 2011), the concentration of virus particles often varies within a consignment, between sub-batches, and even between single pieces of food because of uneven distribution of contamination (Falkenhorst *et al.*, 2005).

21.4.2 Parasitic and bacterial testing in soft fruit

Conventional methods for bacteria in food are generally used for the analysis of pathogens in soft fruit. Standard methods in different countries are based on the methods recommended by the International Organization of Standardization (ISO). Methods for bacteria such as *Salmonella*, *Listeria*, *Campylobacter* or *E. coli* have been published basing the detection on enrichment of cells in selective media. Some of the new methods for detection of bacteria in soft fruit base the pathogen identification on combining culture methods followed by a molecular identification of the target. Although the advantage of this strategy is the speed in assessment of microbial safety, the downside is the identification of the total microbial load. This is because selective media and culture conditions are used for a specific target, making undetectable other infectious pathogens that could be present in low numbers in the soft fruit analyzed. The identification of parasites is based on indirect quantification of the number of oocysts present in a food matrix. Immunoassays are most commonly used for the separation of oocysts; however, the bottleneck in the identification of pathogens is still the sample preparation (Ortega and Sterling, 2006). A recent study compared the use of different wash solutions for identification of *C. cayetanensis* in raspberries (Shields, 2012). A

Table 21.4 Selected methods used for virus detection in soft fruits

Food sample (s)	Sample mass (g)	Viral elution	Concentration and clarification	RNA extraction	References
Blueberries	5	Distilled water, glycine buffer	Receptor-binding capture and magnetic sequestration	TIANamp viral RNA extraction	Pan *et al.* (2012)
Blueberries Raspberries Strawberries	15	TGBE buffer (pectinase), alkaline TGBE buffer, pectinase	Ultra-filtration, PEG, chloroform-butanol	QIAamp (or NucliSens miniMAG), RNeasy Mini kit	Butot *et al.* (2007)
Raspberries	10	Glycine, chloroform-butanol, CatFloc	PEG	Prot. K, phenol-chloroform CTAB	Le Guyader *et al.* (2004)
	60	Alkaline	Ultracentrifugation	QIAamp	Rzezutka *et al.* (2005)
	25	Alkaline TGBE buffer, pectinase, alkaline glycine buffer	PEG NaCl, chloroform-butanol, PEG, recirculating magnetic capture with cationic beads	QIAamp, viral RNA Mini Kit, NucliSens miniMAG, TrizolTM	Dubois *et al.* (2007)
	50	Alkaline TGBE buffer, pectinase	PEG, chloroform-butanol	RNEasy kit	Baert *et al.* (2008)
Raspberries Strawberries	60–90	Alkaline sodium bicarbonate+soya protein	Ultracentrifugation	QIAamp	Rzezutka *et al.* (2006)
Strawberry	One piece of food	PBS	Immunomagnetic separation, positively charged filter filtration	95 °C for 5 min	Bidawid *et al.* (2000b)
Strawberry rinses	25	PBS, immunomagnetic separation, monoclonal antibody	PBS	TRIzol reagent	Shan *et al.* (2005)
Strawberries	20	Beef extract, immunomagnetic separation, polyclonal antibody	Chloroform, PEG, sodium phosphate buffer	95 °C for 5 min	Park *et al.* (2008)
	25	Glycine buffer, cationic beads	Pathatrix™	QIAamp, viral RNA Mini Kit	Mattison *et al.* (2010, 2009a)

Source: modified from Schultz *et al.* (2009).

CatFloc, polydiallyldimethylammonium chloride; CTAB, cethyl trimethyl ammonium bromide; GuSCN, guanidinium thiocyanate; HAV, hepatitis A virus; NoV, norovirus; PBS, phosphate buffered saline; PCRU, RT-PCR units; PEG, polyethylene glycol; pfu, plaque forming units; Prot. K, proteinase K; $TCID_{50}$, 50% tissue culture infective dose.

high recovery of oocysts was obtained (80.2 ± 11.3%) using 0.1% of a solution containing a synthetic detergent and surfactant and a food additive with emulsifying and dispersing properties, sodium dodecylbenzenesulfonate ($C_{12}H_{25}C_6H_4SO_3Na$) and tetrasodium pyrophosphate ($Na_4P_2O_7$) respectively. The recovery using this solution represented about twofold of the recovery obtained from raspberries using 1 M glycine pH 5.5 in a standard method reported by Cook *et al.* (2006a). Although the recovery was improved, the detection method for oocysts was microscopy in both cases, which is the conventional detection method for oocysts in water (Ortega and Sterling, 2006; Cook *et al.*, 2006a, b).

21.4.3 Viral testing in soft fruit

Evaluation of the safety of soft fruits has recently focused on detection of viruses as a result of the frequent soft-fruit-borne outbreaks of NoV and HAV (Table 21.2). As viruses do not replicate in food and the most important foodborne viruses (such as NoV) cannot be propagated *in vitro*, conventional identification for clinical samples based on microscopy or immunoassays, which requires a large amount of viral particles, cannot be used. Therefore, the development of molecular methods for identification of viruses in different food matrices has greatly helped in identification of pathogens that were not conventionally identified some years ago. However, compared with bacterial pathogens that can be cultured, the demands of high assay sensitivity and the lack of ability to differentiate between PCR-detected infectious and non-infectious particles are still major drawbacks in the detection methods for viruses in foods. New developments in the safety testing of soft fruit therefore aim to improve the sample preparation methods to overcome the low number challenge. Whichever method is used, it is crucial to ensure that issues associated with low virus numbers and the complexity of the matrix do not result in false negative or false positive interpretations of results. Therefore, effective quality controls must be included as part of the detection method. Such controls were included to a varying degree in the methods used for the screening and outbreak studies listed in Tables 21.2 and 21.3 and are also described by D'Agostino *et al.* (2011). After the initial sampling, most methods for the detection of viruses in soft-fruits include the following three steps: elution and concentration of virus from the food matrix; extraction of nucleic acid; and detection, confirmation and quantification of target viral RNA. The most critical part of sample processing before detection of viral RNA is the elution and concentration of virus particles from the food matrix. Each step must be designed to achieve reduction of sample volume while simultaneously recovering the viruses and eliminating matrix-associated inhibitors. Various strategies for each step have been described, and an overview of selected methods used for soft fruit is summarized in Table 21.4.

In Fig. 21.4, a flow diagram of a procedure for virus detection in raspberries is shown as an example of virus detection in soft fruit. Promising new developments for the concentration of virus from eluates focus on virus capture using cationic charged beads or filters, or anion exchange filtration (Morales-Rayas *et al.*, 2010;

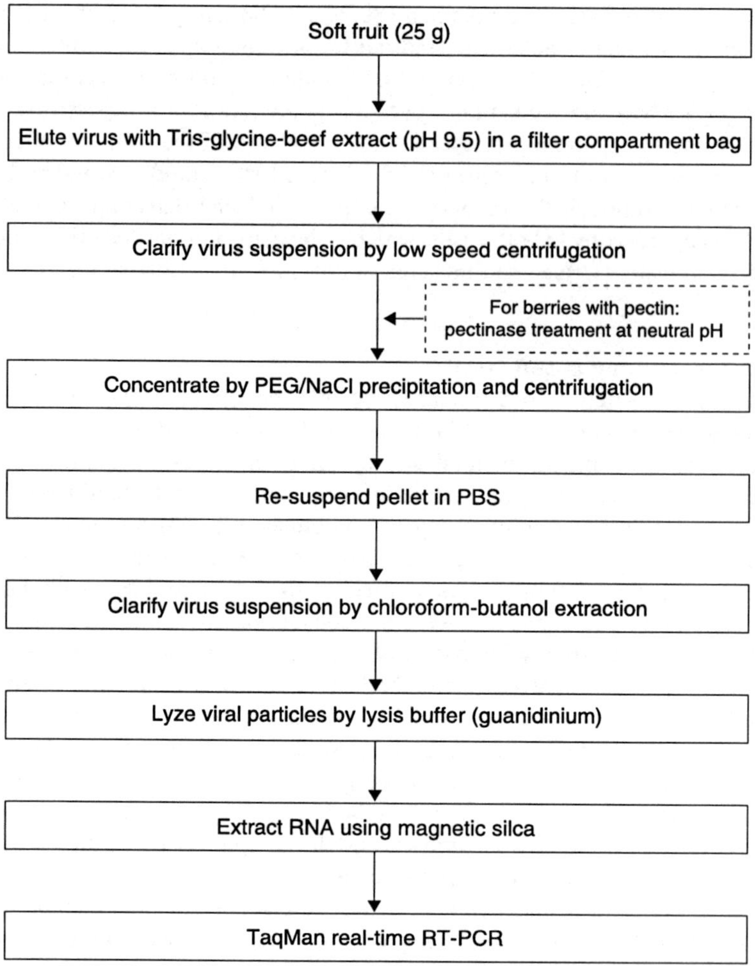

Fig. 21.4 Procedure for norovirus detection in raspberries (Blaise-Boisseau *et al.*, 2010).

Papafragkou *et al.*, 2008) and carbohydrate-conjugated magnetic beads in a recirculating affinity magnetic separation system (Papafragkou *et al.*, 2008; Tian *et al.*, 2011). For detection of NoV and HAV RNA, a large variety of published RT-qPCR assays are available in the scientific literature. For NoV, some of these can be found in the references in Tables 21.2–4, and for HAV, assays developed by, for example, Costafreda *et al.* (2006) and Jothikumar *et al.* (2005) have been successfully used for analysis of foods. Examples of the use of quality controls can also be found in some of these studies (Baert *et al.*, 2011; Costafreda *et al.*, 2006). To quantify NoV detected by RT-qPCR, the value of a positive signal must be converted into RNA copies by interpolation to a standard curve. Based on the primers and probe used in the detection assay, a standard curve can be generated from different types of nucleic acids: purified synthetic RNA or DNA oligonucleotides spanning the complete PCR amplicon, plasmid DNA constructs or *in vitro* transcribed RNA (Costafreda *et al.*, 2006; Le Guyader *et al.*, 2009). Because of lack of validation and harmonization of the type and use of standard curves, exact comparison between viral loads detected in different studies may not be possible.

A qualitative and quantitative standard (including a suite of proper controls) for NoV and HAV detection in foods has recently been developed by the technical working group within the European Committee on Standardisation, CEN/TC275/WG6/TAG4 and accepted by EU member states. This standard is expected as a technical specification in 2013, with the potential of being incorporated into EU legislation as a reference method for virus control (Lees, 2010). However, as the method does not distinguish between detection of infective and non-infective viruses, better understanding on how to interpret a positive result in terms of health risk must be further discussed (Lees and CEN/WG6/TAG4, 2010; Le Guyader *et al.*, 2009).

21.5 Conclusion and future trends

Despite the lack of evidence and limited knowledge of the exact hygiene and growing conditions for soft fruit products, it is realistic to believe that irrigation water is among the main routes of virus transmission (Newell *et al.*, 2010; Kurdziel *et al.*, 2001). It is therefore necessary to develop ways to control and monitor for pathogen introduction during production (EFSA BIOHAZ, 2011). Clear legislation on the use, handling and treatment of water applied in industrial fresh produce production should be efficient to reduce the risk of pathogens from faecal contamination (EFSA BIOHAZ, 2011). Moreover, it has been shown that handling of fresh produce by workers during harvesting and packing is an important risk factor for not least NoV contamination of fresh produce (León-Félix *et al.*, 2010). Implementation of improved hygienic measures is therefore needed in agricultural operations to avoid risk of viral contamination (Rodriguez-Lazaro *et al.*, 2011). Farm and restaurant owners are responsible for the safety of the food they harvest or prepare, and for providing sufficient training to their employees who handle foods (Hicks *et al.*, 1996). Decontamination processes are also important, as they affect pathogen viability, survival, persistence and cross-contamination between food items (Hirneisen *et al.*, 2010). However, for fragile foods, for example raspberries, which lose their visual and sensory qualities when exposed to washing, heating or high pressure processing, there may not be many options for decontamination. To implement control measures such as detection of viruses in the food production, the EU requires the inclusion of a standardized and quality controlled reference method. To meet this requirement, a protocol for viral RNA extraction and detection has been developed within the CEN/TC275/WG6/TAG4 (virus detection in foods). Still, it is essential to gain more information about the practical application (e.g. sampling plans and result interpretation) of these methods for the assessment of product safety (Lees and CEN/WG6/TAG4, 2010). Current molecular assays to detect viruses can supplement prevention efforts as part of the HACCP program of producers of soft fruit (Baert *et al.*, 2011; Brassard *et al.*, 2012) and provide knowledge of the extent of viral distribution (Mattison *et al.*, 2010). However, the available molecular detection methods must be accommodated by harmonized sampling plans and procedures to interpret results before they can be used for routine monitoring and control (Rodriguez-Lazaro *et al.*, 2011).

As a result of the development of molecular detection methods, the presence of human pathogens not quantified by conventional methods is now evidenced in specific food matrices. In this sense, soft fruits thought to be a conventionally 'safe' food, have turned out to be one of the main channels for introduction of viral pathogens to the food supply. Therefore, new developments on testing of soft fruit are focused towards sample preparation methods—fast, easy to use and able to remove inhibitors to molecular identification. The viral load in soft fruits, as with other complex matrices, still represents a challenge because of inhibitors that may be introduced downstream to molecular identification. It is therefore evident that until the sample preparation bottleneck is passed, availability of molecular methods such as real-time PCR, microarrays, immunoassays or next generation sequencing to identify/quantify a pathogen is still restrained. Furthermore, the lack of differentiation between infectious and non-infectious viral particles, and possible bacterial and parasitic pathogens, remains a major issue in standardization of soft fruit testing. As future remarks, the detection step in pathogen identification should evolve towards identification of bacterial, viral and parasitic targets in one step. In addition, a continuous record of the presence of each pathogen in soft fruit as well as development of in-field applications for identification should be expanded to satisfy the demands of the dynamic food chain.

21.6 Questions for discussion

1. What are the potential contamination factors in soft fruit during preharvest?
2. What are the potential contamination factors in soft fruit during postharvest?
3. What are the potential contamination factors for soft fruit during processing?
4. What microbial contaminants in soft fruits are of main concern for human risk?
5. What viruses are more commonly found in soft fruits?
6. What methods can be applied for detection of bacteria in soft fruits?
7. What methods can be applied for detection of parasites in soft fruits?
8. What detection methods can be applied for screening of viruses in soft fruit?
9. Why are soft fruits and other foods currently not routinely tested for viruses?

21.7 References

ANONYMOUS, 2011. Annual Report on Zoonoses in Denmark 2010. National Food Institute, Technical University of Denmark. Available at http://www.food.dtu.dk/Publikationer/Fødevaresikkerhed/Annual_report_on_zoonosis.aspx. Accessed 25 January 2013.

ANONYMOUS, 2012. BfR opinion No. 038/2012. Tenacity (resistance) of noroviruses in strawberry compote. Bundesinstitut für Risikobewertung (BfR). Available at http://www.bfr.bund.de/en/press_information/2012/32/outbreak_of_acute_vomiting_with_diarrhoea_in_children_and_youths_over__cause_identified-131816.html. Accessed 28 January 2013.

BAERT, L., MATTISON, K., LOISY-HAMON, F., HARLOW, J., MARTYRES, A., *et al.*, 2011. Review: Norovirus prevalence in Belgian, Canadian and French fresh produce: A threat to human health? *Int. J. Food Microbiol.* 151:261–9.

BAERT, L., UYTTENDAELE, M., and DEBEVERE, J., 2008. Evaluation of viral extraction methods on a broad range of Ready-To-Eat foods with conventional and real-time RT-PCR for Norovirus GII detection. *Int. J. Food Microbiol.* 123:101–8.

BIDAWID, S., FARBER, J. M., and SATTAR, S. A., 2000a. Contamination of foods by food handlers: experiments on hepatitis A virus transfer to food and its interruption. *Appl. Environ. Microbiol.* 66:2759–63.

BIDAWID, S., FARBER, J. M., and SATTAR, S. A., 2000b. Rapid concentration and detection of hepatitis A virus from lettuce and strawberries. *J. Virol. Methods* 88:175–85.

BIDAWID, S., MALIK, N., ADEGBUNRIN, O., SATTAR, S. A., and FARBER, J. M., 2004. Norovirus cross-contamination during food handling and interruption of virus transfer by hand antisepsis: experiments with feline calicivirus as a surrogate. *J. Food Protect.* 67:103–9.

BLAISE-BOISSEAU, S., HENNECHART-COLLETTE, C., GUILLIER, L., and PERELLE, S., 2010. Duplex real-time qRT-PCR for the detection of hepatitis A virus in water and raspberries using the MS2 bacteriophage as a process control. *J. Virol. Methods* 166:48–53.

BOSCH, A., SÁNCHEZ, G., ABBASZADEGAN, M., CARDUCCI, A., GUIX, S., *et al.*, 2011. Analytical Methods for Virus Detection in Water and Food. *Food Anal. Methods* 4:4–12.

BOWER, C.K., S. STAN, M. DAESCHEL and Y. ZHAO., 2003. Promoting the safety of northwest fresh and processed berries. Oregon State University. Extension Service. 1–35.

BOXMAN, I. L., DIJKMAN, R., TE LOEKE, N. A., HAGELE, G., TILBURG, J. J., *et al.*, 2009. Environmental swabs as a tool in norovirus outbreak investigation, including outbreaks on cruise ships. *J. Food Protect.* 72:111–19.

BOXMAN, I. L., VERHOEF, L., DIJKMAN, R., HAGELE, G., TE LOEKE, N. A., and KOOPMANS, M., 2011. Year-Round Prevalence of Norovirus in the Environment of Catering Companies without a Recently Reported Outbreak of Gastroenteritis. *Appl. Environ. Microbiol.* 77:2968–74.

BRASSARD, J., GAGNE, M. J., GENEREUX, M., and COTE, C., 2012. Detection of Human Foodborne and Zoonotic Viruses on Irrigated, Field-Grown Strawberries. *Appl. Environ. Microbiol.* 78:3763–6.

BUTOT, S., PUTALLAZ, T., and SANCHEZ, G., 2007. Procedure for rapid concentration and detection of enteric viruses from berries and vegetables. *Appl. Environ. Microbiol.* 73:186–92.

CALDER, L., SIMMONS, G., THORNLEY, C., TAYLOR, P., PRITCHARD, K., *et al.*, 2003. An outbreak of hepatitis A associated with consumption of raw blueberries. *Epidemiol. Infect.* 131:745–51.

CDC, 2011. Center for Disease Control, Trends in Foodborne Illness, 1996–2010. www.cdc.gov/foodsafety.

COOK, N., PATON, C. A., WILKINSON, N., NICHOLS, R. A. B., BARKER, K. and SMITH, H. V., 2006a. Towards standards methods for detection of *Cryptosporidium parvum* on lettuce and raspberries. Part 1. Development and optimization methods. *Int. J. Food Microbiol.* 109:215–21.

COOK, N., PATON, C. A., WILKINSON, N., NICHOLS, R. A. B., BARKER, K. and SMITH, H. V., 2006b. Towards standards methods for detection of *Cryptosporidium parvum* on lettuce and raspberries. Part 2. Validation. *Int. J. Food Microbiol.* 109:222–8.

COSTAFREDA, M. I., BOSCH, A., and PINTO, R. M., 2006. Development, evaluation, and standardization of a real-time TaqMan reverse transcription-PCR assay for quantification of hepatitis A virus in clinical and shellfish samples. *Appl. Environ. Microbiol.* 72: 3846–55.

COTTERELLE, B., DROUGARD, C., ROLLAND, J., BECAMEL, M., BOUDON, M., *et al.*, 2005. Outbreak of norovirus infection associated with the consumption of frozen raspberries, France, March 2005. *Euro Surveill.* 10:E050428.

D'AGOSTINO, M., COOK, N., RODRIGUEZ-LAZARO, D. and RUTJES, S., 2011. Nucleic Acid Amplification-Based Methods for Detection of Enteric Viruses: Definition of Controls and Interpretation of Results. *Food Environ. Virol.* 2:55–60.

D'SOUZA, D. H., SAIR, A., WILLIAMS, K., PAPAFRAGKOU, E., JEAN, J., *et al.*, 2006. Persistence of caliciviruses on environmental surfaces and their transfer to food. *Int. J. Food Microbiol.* 108:84–91.

DAESCHEL, M. A. and PATHIMA, U., 2007. Microbial safety concerns of berry fruit. In Zhang, Y. (ed.), *Berry fruit: value-added products for health promotion*, pp. 229–59. CRC Press, Boca Raton, FL.

DANIELS, N. A., BERGMIRE-SWEAT, D. A., SCHWAB, K. J., HENDRICKS, K. A., REDDY, S., *et al.*, 2000. A foodborne outbreak of gastroenteritis associated with Norwalk-like viruses: first molecular traceback to deli sandwiches contaminated during preparation. *J. Infect. Dis.* 181:1467–70.

DEWAAL, C. S. and BHUIYA, F., 2007. Outbreaks by the numbers: fruits and vegetables 1990–2005. Presented at the International Association for Food Protection Poster Presentation P3–03, 8–11 July, Orlando, Florida. 2007. 7-4-2012.

DORE, B., KEAVENEY, S., FLANNERY, J., and RAJKO-NENOW, P., 2010. Management of health risks associated with oysters harvested from a norovirus contaminated area, Ireland, February-March 2010. *Euro Surveill.* 15:ii/19567.

DOYLE, M. P. and ERICKSON, M. C., 2008. Summer meeting 2007 – the problems with fresh produce: an overview. *J. Appl. Microbiol.* 105:317–30.

DREYFUSS, M. S., 2009. Is norovirus a foodborne or pandemic pathogen? An analysis of the transmission of norovirus-associated gastroenteritis and the roles of food and food handlers. *Foodborne Pathog. Dis.* 6:1219–28.

DUBOIS, E., HENNECHART, C., MERLE, G., BURGER, C., HMILA, N., *et al.*, 2007. Detection and quantification by real-time RT-PCR of hepatitis A virus from inoculated tap waters, salad vegetables, and soft fruits: characterization of the method performances. *Int. J. Food Microbiol.* 117:141–9.

EFSA BIOHAZ, 2011. Scientific Opinion on an update on the present knowledge on the occurrence and control of foodborne viruses. *EFSA Journal* 9(7):2190 [96 pp.]. European Food Safety Authority (EFSA) Panel on Biological Hazards (BIOHAZ). Available at http://www.efsa.europa.eu/en/efsajournal/pub/2190.htm.

FALKENHORST, G., KRUSELL, L., LISBY, M., MADSEN, S. B., BOTTIGER, B., and MOLBAK, K., 2005. Imported frozen raspberries cause a series of norovirus outbreaks in Denmark, 2005. *Euro Surveill.* 10:E050922.

FAOSTAT, 2010. United Nations Food and Agriculture Organization, FAO, FAOSTAT, accessed 18 December 2011, at http://faostat.fao.org/site/339/default.aspx.

FELL, G., BOYENS, M., and BAUMGARTE, S., 2007. [Frozen berries as a risk factor for outbreaks of norovirus gastroenteritis. Results of an outbreak investigation in the summer of 2005 in Hamburg]. *Bundesgesundheitsblatt. Gesundheitsforschung. Gesundheitsschutz.* 50:230–6.

HICKS, N. J., BEYNON, J. H., BINGHAM, P., SOLTANPOOR, N., and GREEN, J., 1996. An outbreak of viral gastroenteritis following a wedding reception. *Commun. Dis. Rep. CDR Rev.* 6:R136–9.

HIRNEISEN, K. A., BLACK, E. P., CASCARINO, J. L., FINO, V. R., HOOVER, D. G., and KNIEL, K. E., 2010. Viral inactivation in foods: a review of traditional and novel food-processing technologies. *Compre. Rev. Food Sci. Food Safety* 9:3–20.

HJERTQVIST, M., JOHANSSON, A., SVENSSON, N., ABOM, P. E., MAGNUSSON, C., *et al.*, 2006. Four outbreaks of norovirus gastroenteritis after consuming raspberries, Sweden, June-August 2006. *Euro Surveill.* 11:E060907.

HO, A. Y., LOPEZ, A. S., EBERHART, M. G., LEVENSON, R., FINKEL, B. S., *et al.* 2002. Outbreak of cyclosporiasis associated with imported raspberries, Philadelphia, Pennsylvania, 2000. *Emerg. Infectious Diseases.* 8:783–8.

JOTHIKUMAR, N., CROMEANS, T. L., SOBSEY, M. D., and ROBERTSON B. H., 2005. Development and Evaluation of a Broadly Reactive TaqMan Assay for Rapid Detection of Hepatitis A Virus. *Appl. Environ. Microbiol.* 71:3359–63.

KNIEL, K. E., LINDSAY, D. S., SUMNER, S. S., HACKNEY, C. R., PIERSON, M. D. and DUBEY, J. P., 2002. Examination of attachment and survival of *Toxoplasma gondii* oocysts on raspberries and blueberries. *J. Parasitol.* 88:790–3.

KURDZIEL, A. S., WILKINSON, N., LANGTON, S., and COOK, N., 2001. Survival of poliovirus on soft fruit and salad vegetables. *J. Food Protect.* 64:706–9.

LE GUYADER, F. S., BON, F., DEMEDICI, D., PARNAUDEAU, S., BERTONE, A., *et al.*, 2006. Detection of multiple noroviruses associated with an international gastroenteritis outbreak linked to oyster consumption. *J. Clin. Microbiol.* 44:3878–82.

LE GUYADER, F. S., LE SAUX, J. C., MBERT-BALAY, K., KROL, J., SERAIS, O., *et al.*, 2008. Aichi virus, norovirus, astrovirus, enterovirus, and rotavirus involved in clinical cases from a French oyster-related gastroenteritis outbreak. *J. Clin. Microbiol.* 46:4011–17.

LE GUYADER, F. S., MITTELHOLZER, C., HAUGARREAU, L., HEDLUND, K. O., ALSTERLUND, R., *et al.*, 2004. Detection of noroviruses in raspberries associated with a gastroenteritis outbreak. *Int. J. Food Microbiol.* 97:179–86.

LE GUYADER, F. S., PARNAUDEAU, S., SCHAEFFER, J., BOSCH, A., LOISY, F., *et al.*, 2009. Detection and quantification of noroviruses in shellfish. *Appl. Environ. Microbiol.* 75:618–24.

LEES, D. and CEN/WG6/TAG4, 2010. International Standardisation of a Method for Detection of Human Pathogenic Viruses in Molluscan Shellfish. *Food Environ. Virol.* 2:146–55.

LEÓN-FÉLIX, J., MARTÍNEZ-BUSTILLOS, R., BÁEZ-SAÑUDO, M., PERAZA-GARAY, F., and CHAIDEZ, C., 2010. Norovirus Contamination of Bell Pepper from Handling During Harvesting and Packing. *Food Environ. Virol.* 2:211–17.

LITTLE, C. L. and GILLESPIE, I. A., 2008. Prepared salads and public health. *J. Appl. Microbiol.* 105:1729–43.

MATTISON, K., BRASSARD, J., GAGNE, M. J., WARD, P., HOUDE, A., *et al.*, 2009a. The feline calicivirus as a sample process control for the detection of food and waterborne RNA viruses. *Int. J. Food Microbiol.* 132:73–7.

MATTISON, K., GRUDESKI, E., AUK, B., CHAREST, H., DREWS, S. J., *et al.*, 2009b. Multicenter comparison of two norovirus ORF2-based genotyping protocols. *J. Clin. Microbiol.* 47:3927–32.

MATTISON, K., HARLOW, J., MORTON, V., COOK, A., POLLARI, F., *et al.*, 2010. Enteric viruses in ready-to-eat packaged leafy greens. *Emerg. Infect. Dis.* 16:1815–17.

MAUNULA, L., ROIVAINEN, M., KERANEN, M., MAKELA, S., SODERBERG, K., *et al.*, 2009. Detection of human norovirus from frozen raspberries in a cluster of gastroenteritis outbreaks. *Euro Surveill.* 14.

MORALES-RAYAS, R. and JEAN, J., 2011. Persistence of viruses in food. *Curr. Topics Virol.* 9:1–17.

MORALES-RAYAS, R., WOLFFS, P. F. G., and GRIFFITHS, M. W., 2010. Simultaneous separation and detection of hepatitis A virus and norovirus in produce. *Int. J. Food Microbiol.* 139:48–55.

NEWELL, D. G., KOOPMANS, M., VERHOEF, L., DUIZER, E., IDARA-KANE, A., *et al.*, 2010. Food-borne diseases – the challenges of 20 years ago still persist while new ones continue to emerge. *Int. J. Food Microbiol.* 139 Suppl 1:S3–15.

ORTEGA, Y. R. and STERLING, C. R., 2006. Outbreaks associated with *Cyclospora* and *Cryptosporidium*. In *Microbiology of fruits and vegetables*, pp. 267–80. CRC Press, Boca Raton, FL.

PAN, L., ZHANG, Q., LI, X., and TIAN, P., 2012. Detection of human norovirus in cherry tomatoes, blueberries and vegetable salad by using a receptor-binding capture and magnetic sequestration (RBCMS) method. *Food Microbiol.* 30:420–6.

PAPAFRAGKOU, E., PLANTE, M., MATTISON, K., BIDAWID, S., KARTHIKEYAN, K., *et al.*, 2008. Rapid and sensitive detection of hepatitis A virus in representative food matrices. *J. Virol. Methods.* 147:177–87.

PARISH, M., DANYLUK, M. and NARCISO, J. A., 2012. Safety of fruit, nut, and berry products, pp. 109–125. In O. A. Oyarzabal, and S. Baskert (eds.), *Microbial Food Safety: an introduction*, Food Science text series. Springer Science.

PARK, Y., CHO, Y. H., JEE, Y., and KO, G., 2008. Immunomagnetic separation combined with real-time reverse transcriptase PCR assays for detection of norovirus in contaminated food. *Appl. Environ. Microbiol.* 74:4226–30.

PONKA, A., MAUNULA, L., VON BONSDORFF, C. H., and LYYTIKAINEN, O., 1999. An outbreak of calicivirus associated with consumption of frozen raspberries. *Epidemiol. Infect.* 123:469–74.

RODRIGUEZ-LAZARO, D., COOK, N., RUGGERI, F. M., SELLWOOD, J., NASSER, A., *et al.*, 2011. Virus hazards from food, water and other contaminated environments. *FEMS Microbiol. Rev.* 2011:1–29.

RZEZUTKA, A., ALOTAIBI, M., D'AGOSTINO, M., and COOK, N., 2005. A centrifugation-based method for extraction of norovirus from raspberries. *J. Food Protect.* 68:1923–5.

RZEZUTKA, A., D'AGOSTINO, M., and COOK, N., 2006. An ultracentrifugation-based approach to the detection of hepatitis A virus in soft fruits. *Int. J. Food Microbiol.* 108:315–20.

SARVIKIVI, E., ROIVAINEN, M., MAUNULA, L., NISKANEN, T., KORHONEN, T., *et al.*, 2011a. Multiple norovirus outbreaks linked to imported frozen raspberries. *Epidemiol. Infect.* 140:260–7.

SARVIKIVI, E., ROIVAINEN, M., MAUNULA, L., NISKANEN, T., KORHONEN, T., *et al.*, 2011b. Multiple norovirus outbreaks linked to imported frozen raspberries. *Epidemiol. Infect.* 140:260–7.

SCHMID, D., STUGER, H. P., LEDERER, I., PICHLER, A. M., KAINZ-ARNFELSER, G., *et al.*, 2007. A foodborne norovirus outbreak due to manually prepared salad, Austria 2006. *Infection.* 35:232–9.

SCHULTZ, A. C., 2012. Detection and inactivation of noroviruses in raspberries: Ph.D. Thesis. pp. 31–50. Faculty of Health and Medical Science, University of Copenhagen and The National Food Institute, Technical University of Denmark. SL Grafik, Frederiksberg C, Denmark.

SCHULTZ, A. C., NØRRUNG, B., and VINJÉ, J., 2009. Noroviruses. In Liu, D. (ed.), *Molecular Detection of Foodborne Pathogens*. CRC Press, Taylor & Francis Group, New York. 75–98.

SHAN, X. C., WOLFFS, P., and GRIFFITHS, M. W., 2005. Rapid and quantitative detection of hepatitis A virus from green onion and strawberry rinses by use of real-time reverse transcription-PCR. *Appl. Environ. Microbiol.* 71:5624–6.

SHIELDS, J. M., LEE, M. M. and MURPHY, H. R., 2012. Use of common laboratory glassware detergent improves recovery of *Cryptosporidium parvum* and *Cyclospora cayetanensis* from lettuce, herbs and raspberries. *Int. J. Food Microbiol.* 153:123–8.

SIVAPALASINGAM, S., FRIEDMAN, C. R., COHEN, L. and TAUXE, R. V., 2004. Fresh produce: a growing cause of outbreaks of foodborne illness in the United States, 1973 through 1997. *J. Food Protect.* 67:2342–53.

STALS, A., BAERT, L., JASSON, V., VAN COLLIE, E., and UYTTENDAELE, M., 2011. Screening of fruit products for norovirus and the difficulty of interpreting positive PCR results. *J. Food Protect.* 74:425–31.

STRIK, B. C., 2007. Berry crops: worldwide area and production systems. In Zhang, Y. (ed.), *Berry fruit: value-added products for health promotion*, pp. 3–72. CRC Press, Boca Raton, FL.

STUPPY, W. and KESSELER, R., 2008. The truth about berries. In A. Papadakis (ed.), *Fruit, edible, enedible, incredible*, pp. 60–62. Papadakis Publisher, London, UK.

TEUNIS, P. F. M., MOE, C. L., LIU, P. E., MILLER, S., LINDESMITH, L., *et al.*, 2008. Norwalk virus: How infectious is it? *J. Med. Virol.* 80:1468–76.

TIAN, P., YANG, D., and MANDRELL, R., 2011. A simple method to recover Norovirus from fresh produce with large sample size by using histo-blood group antigen-conjugated to magnetic beads in a recirculating affinity magnetic separation system (RCAMS). *Int. J. Food Microbiol.* 147:223–7.

TODD, E. C. D., GREIG, J. D., BARTLESON, C. A. and MICHELS, B. S., 2009. Outbreaks where food workers have been implicated in the spread of foodborne disease. Part 6. Transmission and survival of pathogens in the food processing and preparation environment. *J. Food Protect.* 72:202–19.

URBANUCCI, A., MYRMEL, M., BERG, I., VON BONSDORFF, C. H., and MAUNULA, L., 2009. Potential internalisation of caliciviruses in lettuce. *Int. J. Food Microbiol.* 135:175–8.

WANG, H., FENG, H., LIANG, W., LUO, Y., and MALYARCHUK, V., 2009. Effect of surface roughness on retention and removal of *Escherichia coli* O157:H7 on surfaces of selected fruits. *J. Food Sci.*, 74, E8–E15.

YADAV, R., DWIVEDI, S., KUMAR, S. and CHAUDHURY, A., 2010. Trends and perspectives of biosensors for food and environmental virology. *Food Environ. Virol.* 2:53–63.

ZHAO, Y., 2005. Pathogens in fruit. In Jongen, W. (ed), *Improving the safety of fresh fruit and vegetables*, pp. 44–84. CRC Press, Boca Raton, FL.

22

Cases of public emetic events caused by foodborne viruses and potential issues for fresh produce

S. M. Markland, D. H. D'Souza and K. E. Kniel,
University of Delaware, USA

DOI: 10.1533/9781782420279.4.314

Abstract: Gastrointestinal illnesses, specifically those caused by foodborne viruses, may strike at any time. Contaminated food can be consumed hours or even days before onset of symptoms. Human norovirus is a leading cause of foodborne illness worldwide, and may be spread by food, water, person-to-person contact or even aerosolized vomitus droplets. Because of this airborne transmission and sudden onset of symptoms, public emetic events lead to other subsequent infections. Such events are common in places where large numbers of people are housed in close proximity. Several real-life situations are described in this chapter, and while fresh produce may not be named specifically, concerns for norovirus contamination exist as described in this chapter.

Key words: norovirus, outbreaks, risk management, tracing, ready-to-eat foods, catering.

22.1 Introduction

An outbreak of gastrointestinal disease may affect hundreds of individuals, or only a handful. Frustration mounts when the initial assumption that the outbreak was foodborne turns out to be false and recalls of other common risk factors also fall short. What happens when specific pathogenic bacteria cannot be isolated from clinical or environmental samples?

Gastrointestinal illnesses, specifically those caused by foodborne viruses, may strike at any time. The food originally contaminated with the virus may have been consumed hours or even days before the onset of symptoms, which is often unexpected. Human norovirus, with a highly infectious single-stranded RNA genome, is a leading cause of foodborne illness worldwide. It is estimated that only 10–18 virus particles are necessary to induce illness involving symptoms such as diarrhea, vomiting, nausea, fever and fatigue (Teunis *et al.*, 2008). Human norovirus may be spread by food, water, person-to-person contact or even by aerosolized vomitus droplets (Fig. 22.1). It is because of this airborne transmission and sudden onset of symptoms that public emetic events lead to other subsequent infections. Such events are common in places where large numbers of people are housed in close proximity such as in nursing homes, college dormitories, military bases, hospitals and cruise ships. Emetic events caused by human norovirus and followed by outbreaks have also taken place in restaurants, at parties and at major sporting events. A couple of example situations are described here.

Our first example involves an especially curious human norovirus outbreak involving football players and staff during a collegiate football game (Becker *et al.*, 2000). According to the epidemiological investigation, on Friday, 18 September 1998, the members and staff of a North Carolina football team were served a luncheon in their locker room prior to departing on a flight to Florida for a 7 p.m. game the next evening. During the football game, several players on the North Carolina football team began to vomit and have diarrhea. The report of the investigation indicates that players on both sides had a difficult time avoiding contact with vomitus and feces. By the following evening, football players from the Florida team began experiencing gastrointestinal symptoms. Interestingly, the only contact Florida players had with North Carolina players was on the field. Keep

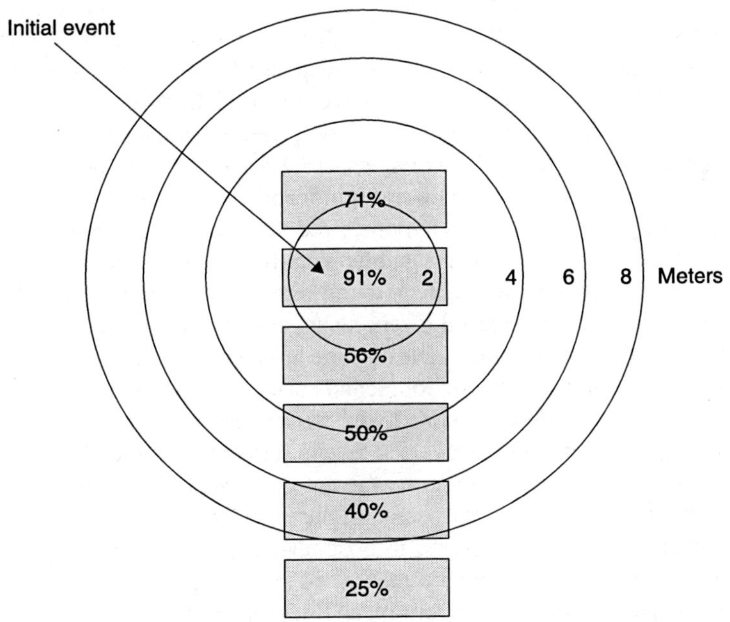

Fig. 22.1 Human norovirus transmission. During a public emetic event, for example at a restaurant, viral transmission via vomitus droplets shows higher attack rates at closer tables. Image adapted from Marks *et al.* (2000).

in mind that football is a contact sport. Laboratory analysis of stool samples from both North Carolina and Florida players identified human norovirus as the etiologic agent. The genetic sequences of virus obtained from players from North Carolina and Florida were compared and found to be identical. It was determined that both fecal-oral transmission and aerosol transmission of vomitus were the most likely routes of transmission from one team to another given the high degree of physical contact and use of hands that occurs during a game of football. Here, it is evident that the emetic events that occurred in a public place (the sideline of the North Carolina football team) led to subsequent illnesses of the Florida football players by direct person-to-person contact occurring during the game. It was hypothesized that the initial infection was a result of contaminated turkey sandwiches consumed by the North Carolina football team during the lunch that was served the day before the game. The sandwiches were thought to have been contaminated by food handlers who did not experience gastrointestinal illness at any time before or after the luncheon, but were found to have high titers of antibodies toward the virus.

A second common means of exchanging norovirus caused by one or more public emetic events are cruise ships. According to the Centers for Disease Control and Prevention (CDC) which monitors cruise ships closely through the Vessel Sanitation Program (www.cdc.gov/nceh/vsp), human norovirus outbreaks are often connected with cruise ships because of close living quarters and subsequent increases in amount of group contact. Additionally, high passenger turnover can expose current passengers and crew to the virus. Over 90% of recent gastrointestinal illness outbreaks investigated by the CDC on cruise ships with a confirmed etiology were caused by norovirus (Wikswo *et al.*, 2011). An outbreak that occurred on a cruise ship in January of 2009 prompted response by the CDC (Wikswo *et al.*, 2011). The outbreak was suspected to have occurred fairly early in the 15-day cruise on Cruise Ship X. According to a completed questionnaire handed out to every cabin, 236 of the 1842 passengers (15.4%) fit the case definition of the outbreak. Curiously, ill passengers were significantly more likely to have witnessed vomiting during boarding. It was hypothesized that a passenger brought the virus onto the vessel where they had a public vomiting incident on Deck 5 during embarkation. This is certainly a tribute to the aerosolization and impact of a single norovirus attack. The authors of this report suggest that passengers who witnessed this initial vomiting incident may have become infected via ingestion of aerosolized vomitus or from contact with environmental surfaces around the area. It was also suggested that this incident explained why passengers residing on Deck 5 of the ship or those who visited a café on Deck 5 were statistically more likely to become ill than other passengers. Although it was not confirmed, it was most likely this initial emetic event that led to the spread of the virus on this particular cruise ship.

Similar outbreaks have occurred in hospitals, vacation resorts, daycares, and camps where large groups of people are housed in close quarters. For example, the largest norovirus outbreak in the Dominican Republic occurred at a single vacation resort, where over 800 people were affected in a 15-day period (Domenech-Sanchez *et al.*, 2011). The outbreak reportedly began on 29 July 2007 when seven people began to experience symptoms including diarrhea and vomiting. Food and water at the resort, specifically high-risk foods including salads and seafood, were tested but were found to be negative for the pathogen. The resort removed such high-risk foods from their menus and began to hypochlorinate their recreational and potable water supplies. Despite their efforts, several new cases of illness were reported by visitors at the resort each day. By 23 August 2007 over 100 people at the resort reportedly were experiencing gastrointestinal symptoms. This led to a high number of public emetic events all of which occurred in the lobby and in restaurants. Despite attempts to disinfect resort facilities including the use of disposable paper towels and public toilet cleaning and disinfection after each use, new cases continued to arise. Finally, the resort had to cancel any further entries into the resort. It was not until after this that the number of cases began to drastically decline. It was determined by investigators that one ill person brought this illness into the resort, and it was then spread via a public emetic event.

22.2 Challenges in containing virus spread

These examples are testament to the biological robustness of human norovirus and the cascade of events that may ensue following an emetic event. Through direct contact with the ill and through contact with aerosolized virus particles via vomitus or feces, many people and even places can become contaminated in an extremely short period of time. As described, basic control and prevention measures are not enough. But how can one stop a public emetic event from occurring? That seems likely to be an impossible task. As a result of an initial emetic event, secondary attacks are possible, and the rate of these secondary attacks is largely dependent on the initial response to the initial event compounded by the situation. Secondary attacks are those illnesses that occur as a result to those persons who fell ill or surfaces that became contaminated as a result of the first emetic event. For example, a person witnessing the emetic event falls ill (this is the primary attack) via ingestion or inhalation of aerosolized virus particles. This person then passes the illness to someone they live with (this is a secondary attack). It is easy to see how so many people over a short period of time can be exposed to any particular pathogen caused by a single emetic event.

22.3 Significant factors affecting outbreaks

22.3.1 Pathogen significance

Human norovirus is a member of the *Caliciviridae* family of viruses (Fig. 22.2), which are single-stranded RNA non-enveloped viruses that cause acute gastroenteritis in humans (Saeed and Naji, 2007). Norovirus was originally identified as Norwalk-like virus, also referred to as small round structured viruses (SRSVs). Currently, noroviruses are classified into at least five genogroups based on amino acid identity of their viral structural proteins (VP1); strains that infect humans are currently represented as G1, GII and GIV (CDC, 2011). Noroviruses

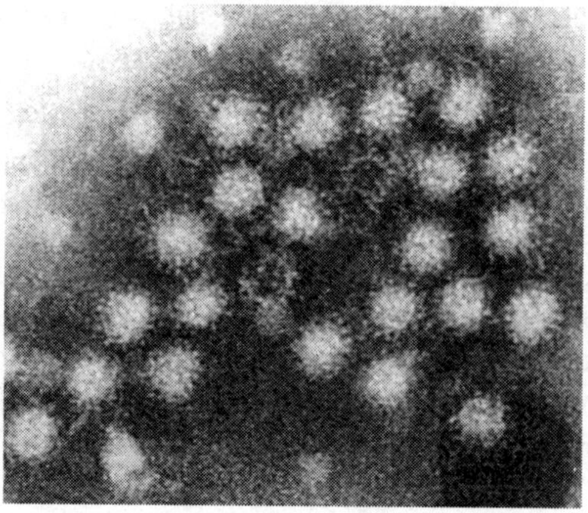

Fig. 22.2 Human norovirus particles visualized by electron microscopy.

are not routinely believed to be zoonotic, but strains that have infected pigs have been found in GIV and are very similar to those that infect humans. Human norovirus strains have been found in feces of cattle and swine (Mattison *et al.*, 2007). Norovirus remains an enigma in many ways and is an important pathogen of human concern, responsible for the majority (50%) of gastroenteritis outbreaks worldwide and for 90% of all non-bacterial gastroenteritis outbreaks (CDC, 2011).

22.3.2 Geographical and regional significance, climate and general consumer perceptions

Another challenge in the prevention of illnesses caused by norovirus is the identification of those persons at risk. Outbreaks can occur in a wide variety of locations and can affect people of all ages (CDC, 2011). There are also several routes by which the virus can be transmitted including exposure to contaminated food resulting in person-to-person spread and secondary attacks. These factors make it difficult to identify the at-risk population for norovirus infection; however, in a study conducted by the CDC from 1994 to 2006, 35.4% of the laboratory confirmed cases of norovirus were from long-term care facilities, 31.1% were from restaurants, parties, and events, 20.5% were from vacation settings including cruise ships and resorts, and 13.0% were from schools and communities (CDC, 2011). Studies such as these can help identify areas with higher risk for transmission.

22.4 Regulatory and economic aspects

22.4.1 Regulatory aspects

It is important to determine whether there is a food product that contributed to the initial illness. If an emetic event occurs in a restaurant or other facility serving food, it is important for people to understand that it is more than likely that the food at that restaurant or event was not the cause of the emetic event. The incubation period for norovirus is typically between 24 and 48 hours, but illness can occur within 12 hours after exposure to the contaminated product (Saeed and Naji, 2007). This is why it is important for physicians to obtain clinical specimens, especially if an outbreak is suspected or physicians begin to see an increasing number of patients experiencing gastrointestinal symptoms. More laboratories and public state health facilities are able to test for norovirus or at the very least eliminate the chance of bacterial origin. Testing will also help determine if the suspect cases are linked and if a single source is responsible.

Containment and proper cleanup of the initial emetic event is crucial in order to prevent subsequent illnesses. Noroviruses are extremely contagious and as few as 10 virus particles can cause illness in an individual (Saeed and Naji, 2007). According to the CDC, 1 g of feces may contain 5 billion infectious doses (CDC, 2011). Vomitus should be immediately covered in order to prevent aerosolization. The area should immediately be disinfected with sodium hypochlorite (chlorine bleach) or with another product registered with the EPA as effective against norovirus (CDC, 2011). All other environmental surfaces that are considered high contact and therefore high risk, should be disinfected regularly, including bathrooms, door handles and kitchen surfaces. In addition, frequent hand washing is crucial to prevent the spread of the virus. Federal and state food codes along with industry-specific Standard Sanitation Operating Procedures (SSOPs) are implemented to ensure the safety of issues described here concerning restaurants and other food-service establishments.

These issues are part of the inspection of cruise ships covered by the Vessel Sanitation Program (VSP) maintained by the CDC. The VSP started in 1970 in response to foodborne illness outbreaks and is a surveillance system for cruise ships carrying ≥ 13 passengers. The VSP is voluntary but nearly all cruise lines join. The objective of the VSP is to protect passenger and crew health by minimizing the risk of gastrointestinal illness aboard cruise ships.

22.4.2 Economic aspects

Costs of emetic events from norovirus are vast. Their impact is difficult to estimate because so many people are affected and many illnesses go unrecognized. Costs include lost wages from work, healthcare, epidemiological and environmental investigations, discarded food and more. In 2012, the CDC estimated that each case of foodborne illness cost approximately $1626 with an estimated yearly cost at $77.7 billion in the USA.

22.5 Production and market issues

Early recognition of disease symptoms and identifying the source by rapid detection methods are important components in the prevention of foodborne viral outbreaks. Rapid, improved diagnostic methods with increased sensitivity for the detection of foodborne viral contamination are needed. Although progress has been made in this field of diagnostics, challenges exist that need to be addressed in the future. Proper hygienic practices, adequate washing of ready-to-eat (RTE) foods (in particular fresh produce), and proper handling and cooking of foods are recommended to prevent outbreaks. Proper personal hygiene practices include the use of barrier protection (no bare-hand contact with foods and use of gloves during handling and preparation) as well as appropriate hand and surface decontamination, that serve as initial lines of defense for preventing viral contamination of RTE foods (D'Souza *et al.*, 2006). Various surface and hand sanitizers are being researched in order to optimize prevention and control of human norovirus (Macinga *et al.*, 2008; D'Souza and Su, 2010). Cross-contamination issues caused by improper washing of utensils, cutting boards, improper wearing of gloves or inadequate hand-washing by food handlers can lead to cases of outbreaks. Traditional chlorine washes (bleach) have proven to be successful in decontamination of surfaces and inanimate objects. Proper cooking practices may be appropriate only for specific foods that require a terminal heating step before consumption. For ready-to-eat items, such as raw fruits and vegetables, washing produce with clean water with or without the addition of chemical disinfectants may help decrease virus loads in contaminated items, although chlorine washes have drawbacks when applied to ready-to-eat fruits and vegetables (such as product bleaching, etc.). However, it is also important to keep in mind that the efficacy of chemical decontamination varies with both virus and produce type.

Education of food handlers and the general public in formal and informal settings is another way to prevent further outbreaks, especially education on proper hygienic practices in relation to at-risk foods. These can be in classroom training sessions in high schools, universities, at work-shops or trade-shows, 4-H camps, etc. Regarding treatment options, human NoV symptoms are known to be reduced by oral administration of bismuth subsalicylate. However, human NoV infections are typically self-limiting (although some recent strains such as GII.4b are known to cause death in the elderly and immunocompromised (Siebenga *et al.*, 2009)). Oral fluid and electrolyte replacement therapy is usually adequate to replenish fluid loss. Parenteral administration of fluids is sometimes needed during cases of severe vomiting and diarrhea, with occasional hospitalization for cases of severe dehydration.

Another issue is that infected food-handlers are known to shed virus for up to 2 weeks after an infection. Hence, the importance of personal hygiene cannot be underemphasized in these cases. Also, appropriate leave from work during illness should be provided to the infected food-handler to decrease the risk of foodborne viral illness to the consumers.

Norovirus is not selective in whom it infects. Public emetic events will continue to occur at sporting events, within institutions and on cruise ships. As the human population continues to rise and more individuals live with compromised immune systems, perhaps more norovirus outbreaks will occur. A continuous battle in combating norovirus is disease transmission by asymptomatic employees. To complicate matters, norovirus genotypes continue to grow and change in their patterns of infectivity. For example, GI viruses were at one time the most commonly infective virus found in humans, but, by 2010, had been replaced by GII.IV viruses. What will be next?

22.6 Troubleshooting approaches and laboratory methods

There are challenges to traceback investigations to determine the origin of contamination, but the greatest challenge here includes the ability to test for and successfully detect norovirus. Some issues surrounding this challenge have been mentioned already in this chapter. Others are listed below. Most processors and small-scale industries (including farmers) cannot afford the cost, labor and equipment needed to perform sophisticated detection assays in a rapid manner for routine testing and surveillance. It would be of great advantage to have field deployable testing devices and assays that are hand-held or portable and do not require skill, labor or expensive equipment. Towards achieving this goal, innovative research is ongoing on development and application of isothermal RNA amplification, where only one temperature is required, making use of only a simple water-bath and eliminating the need for expensive thermocyclers. Novel isothermal methods include reverse-transcriptase loop-mediated isothermal amplification (RT-LAMP) and nucleic acid sequence-based amplification (NASBA). RT-LAMP is a nucleic acid amplification assay that is rapid, specific and sensitive. Incorporation of fluorescence dyes or probes to the assay can convert it into a real-time format for ease of quantification using hand-held fluorometers. This assay has been developed for detection of human noroviruses GI and GII (Yoda *et al.*, 2007, 2009; Fukuda *et al.*, 2006, 2007).

As technology evolves, microarrays have also been designed to detect numerous viral contaminants that could be present simultaneously using a single chip. An oligonucleotide array incorporating 13000 elements representing selected strains of hepatitis A virus (HAV), human coxsackieviruses A and B (CVA and CVB), human norovirus (NV) genogroups I and II, and human rotavirus (RV) gene segments 3,4,10 and 11, has an advantage over PCR in its ability to identify closely related strains (Ayodeji *et al.*, 2009).

22.7 Future trends

Perhaps the greatest impediments to reducing the risk of emetic events caused by norovirus are surveillance and testing standards. Across the European Union and within several states in the USA, the national norovirus outbreak network known as CaliciNet is at work. Virus testing includes determination of the virus genogroup, which can be useful information in surveillance and in monitoring the spread and change of specific viruses. Even within one emetic event there may be different norovirus genotypes, which reinforces the importance of maintaining separation of samples, both environmental and clinical. Even with enhancements in testing performance and ability around the world, norovirus events will not be investigated without surveillance, which is why this remains one of the greatest questions to be answered. How can surveillance be improved? Will persons who are sick with norovirus visit physicians? And as mentioned previously, in order to reduce some of these public emetic events sick food-service employees need to have some security in that they can stay home when ill. How can this be accomplished?

Another important factor to consider is the lack of a susceptible host cell line for human norovirus propagation in the lab and the inability to detect infectious human noroviruses using traditional cell-culture based assays. RT-PCR and molecular methods cannot typically confirm or distinguish the presence of infectious or non-infectious viruses. Several researchers have used various approaches to correlate infectious virus with RNA detection methods. Researchers continue to investigate animal models, like gnotobiotic pigs, and cell culture. Propidium monoazide (which can only enter cells with membrane or structural damage, bind to nucleic acid and block amplification) was added to cell culture combined with reverse transcriptase PCR to distinguish between infectious and non-infectious enteric viruses in water samples (Parshionikar *et al.*, 2010). Therefore, progress using innovative technologies continues to be made towards using rapid detection to control outbreaks. Of course research into testing continues, as described previously.

22.8 Conclusion

Aerosolization of norovirus is a potent characteristic in terms of general public health, in particular combined with low infectious dose, prolonged asymptomatic shedding, environmental stability, substantial strain diversity and lack of lasting immunity. These cases show the importance of good personal hygiene, surveillance, public health, research and education. Each case describes a situation where individuals were put at risk of illness without prior knowledge. Even with heightened surveillance and advanced testing, these risks will not disappear. Public venues including restaurants must be well prepared to prevent spread of norovirus in the case of public emetic events. Places at high risk should consider multiple hurdles and invest in educational tools and resources, especially for challenging areas like restrooms and kitchens.

22.9 Questions for discussion

1. How are 'human' noroviruses transmitted?
2. Describe the routes of human norovirus transmission mentioned in the outbreaks in this chapter; for example during the football game, on the cruise ship and at the resort.
3. What is the infectious dose of human norovirus?

4. How does the low infectious dose impact or affect the spread of norovirus?
5. Why is the incubation period important for the spread of norovirus in the outbreak of norovirus during the football game?
6. What are considered adequate means of hygienic control and sanitation to reduce the spread of norovirus?
7. Describe some of the challenges in controlling emetic events caused by human norovirus.
8. What is meant by secondary attacks of norovirus?
9. Develop your own case study for the transmission of norovirus. What is the role of an emetic event? Could it have been prevented?
10. Research a recent emetic event that resulted in the transmission of norovirus. How many people were affected? Could it have been prevented?

22.10 Acknowledgments

We wish to acknowledge all the researchers and sanitarians who have dedicated time towards the advancement of understanding norovirus and towards a future with reduced risk of illness from public emetic events.

22.11 References

ALFANO-SOBSEY E., D. SWEAT, A. HALL, F. BREEDLOVE, R. RODRIGUEZ, *et al.* 2011. Norovirus outbreak associated with undercooked oysters and secondary household transmission. *Epidemiol. Infect.* 1–7.

AYODEJI, M., M. KULKA, S.A. JACKSON, I. PATEL, M. MAMMEL, *et al.* 2009. A microarray based approach for the identification of common foodborne viruses. *Open Virol. J.* 3:7–20.

BECKER, K.M., C.L. MOE, K.L. SOUTHWICK, and J.N. MACCORMACK. 2000. Transmission of Norwalk Virus during a football game. *NEJM.* 343(17):1223–7.

CENTERS FOR DISEASE CONTROL AND PREVENTION (CDC). 2011. Updated Norovirus Outbreak Management and Disease Prevention Guidelines. *MMWR.* 60(3):1–15.

D'SOUZA, D.H. and X. SU. 2010. Efficacy of chemical treatments against murine norovirus, feline calicivirus, and MS2 bacteriophage. *Foodborne Pathog Dis.* 7:319–26.

D'SOUZA, D.H., A. SAIR, K. WILLIAMS, E. PAPAFRAGKOU, J. JEAN, *et al.* 2006. Persistence of caliciviruses on environmental surfaces and their transfer to food. *Int. J. Food Microbiol.* 108:84–91.

DOMENECH-SANCHEZ A., C. JUAN, J.L. PEREZ, and C.I. BERROCAL. 2011. Unmanageable norovirus outbreak in a single resort located in the Dominican Republic. *Clin. Microbiol. Infec.* 17(6): 952–4.

FUKUDA, S., S. TAKAO, M. KUWAYAMA, Y. SHIMAZU, and K. MIYAZAKI. 2006. Rapid detection of norovirus from fecal specimens by real-time reverse transcription-loop-mediated isothermal amplification assay. *J. Clin. Microbiol.* 44:1376–81.

FUKUDA, S., Y. SASAKI, M. KUWAYAMA, and K. MIYAZAKI 2007. Simultaneous detection and genogroup-screening test for norovirus genogroups I and II from fecal specimens in single tube by reverse transcription-loop-mediated isothermal amplification assay. *Microbiol. Immunol.* 51:547–50.

MACINGA, D.R., S.A. SATTAR, L. JAYKUS, and J.W. ARBOGAST. 2008. Improved inactivation of nonenveloped enteric viruses and their surrogates by a novel alcohol-based hand sanitizer. *Appl. Environ. Microbiol.* 74:5047–52.

MARKS, P.J., I.B. VIPOND, D. CARLISLE, D. DEAKIN, R.E. FEY, and E.O. CAUL. 2000. Evidence for airborne transmission of Norwalk-like virus (NLV) in a hotel restaurant. *Epidemiol. Infect.* 124:481–7.

MATTISON, K., A. SHUKLA, A. COOK, F. POLLARI, R. FRIENDSHIP, *et al.* 2007. Human noroviruses in swine and cattle. *Emerg. Infect. Dis.* 13:1184–8.

PARSHIONIKAR, S., I. LASEKE, and G.S. FOUT. 2010. Use of propidium monoazide in reverse transcriptase PCR to distinguish between infectious and noninfectious enteric viruses in water samples. *Appl. Environ. Microbiol.* 76:4318–26.

SAEED, A.M., and R. NAJI. 2007. Infectious Agents. Chapter in *Epidemiologic Principles and Food Safety*, Tamar Lasky edition. Oxford University, pp. 18–39.

SIEBENGA, J.J., H. VENNEMA, D.P. ZHENG, J. VINJÉ, B.E. LEE, *et al.* 2009. Norovirus illness is a global problem: emergence and spread of norovirus GII.4 variants, 2001–2007. *J. Infect. Dis.* 200:802–12.

TEUNIS, P.F.M., C.L. MOE, P. LIU, S.E. MILLER, L. LINDESMITH, *et al.* 2008. Norwalk virus: How infectious is it? *J. Med. Virol.* 80:1468–76.

WIKSWO, M.E., J. CORTES, A.J. HALL, G. VAUGHAN, C. HOWARD, *et al.* 2011. Disease transmission and passenger behaviors during a high morbidity norovirus outbreak on a cruise ship, January 2009. *Clin. Infect. Dis.* 52:1116–22.

YODA, T., Y. SUZUKI, K. YAMAZAKI, N. SAKON, M. KANKI, *et al.* 2007. Evaluation and application of reverse transcription loop-mediated isothermal amplification for detection of noroviruses. *J. Med. Virol.* 79:326–34. Erratum in: *J. Med. Virol.* 79(6):863.

YODA, T., Y. SUZUKI, K. YAMAZAKI, N. SAKON, M. KANKI, *et al.* 2009. Application of a modified loop-mediated isothermal amplification kit for detecting Norovirus genogroups I and II. *J. Med. Virol.* 81(12):2072–8.

Part V

Case studies in real-life situations

23

Sprout-associated outbreaks and development of preventive controls

M. A. Smith, US Food and Drug Administration, USA

DOI: 10.1533/9781782420279.5.327

Abstract: Fresh produce is an important part of a healthy diet. However, produce is susceptible to contamination. Produce is often consumed raw, or only lightly cooked, which may allow pathogens to survive and cause illness. Sprouts are especially problematic because the conditions that encourage seeds to sprout are also ideal for growth of pathogens, if present. The steps of sprout production do not include processes to eliminate pathogens. This chapter describes an iterative process to better understand hazards associated with sprouted seeds and beans, to identify preventive controls for these hazards and to monitor and promote adoption of effective preventive controls.

Key words: sprouts, sprouted seed, pathogens, outbreaks, preventive controls.

23.1 Introduction

It is currently 2:00 pm Eastern Standard Time, 14 December 2012. Members of the Sprout Safety Alliance, Technical Working Group (SSA TWG) are dialing-in to participate in a call to discuss the second draft of a chapter titled 'Sprout Safety Hazards'.[1] This chapter, to be used as part of a sprout safety training program, provides an overview of different types of food safety hazards and the importance of their control, with emphasis on food safety hazards that have a history of having been associated with sprouts and are considered 'reasonably likely to occur' if not subject to appropriate controls. The SSA workgroup recognizes that specific hazards and their respective controls can differ by product types and processes. They agree that understanding food safety hazards and identifying those hazards that are reasonably likely to occur in sprout operations is the foundation for

[1]Conference call is for illustrative purposes only.

establishing an effective food safety program implementing best practices and sprout-specific preventive controls that will be covered in other training chapters and that have been the subject of earlier calls.

Between 1996 and 2010, there have been at least 35 reported foodborne illness outbreaks in the USA associated with sprouts, resulting in over 2000 laboratory-confirmed illnesses, 123 hospitalizations and one death. Pathogens of concern in sprout-associated outbreaks have been *Salmonella* spp., *Escherichia coli* O157:H7 and *E. coli* O157NM. Some of these outbreaks were quite large and affected people living in many states. Alfalfa sprouts, which are often consumed raw, and are one of the more commonly consumed types of seed or green sprouts in the USA, are also the type of sprout most frequently involved in US sprout outbreaks. However, dozens of types of seeds (and beans) are sprouted in the USA and any type of raw sprout can become contaminated with pathogens and cause illness. Other types of sprouts, including both green sprouts and raw or lightly cooked bean sprouts, such as mung bean sprouts, have also been linked to illness in the USA. Sprouts are one of the most common vehicles identified in produce-associated outbreaks (Fett *et al*., 2006; Painter *et al.*, 2013).

The Sprout Safety Alliance (SSA) is a public–private partnership established in 2012, funded by a cooperative grant through the US Food and Drug Administration (FDA), and coordinated by the Illinois Institute of Technology's Institute for Food Safety and Health (IIT IFSH) (https://www.iit.edu/ifsh/sprout_safety/). The goal of the SSA is to develop a core curriculum, training and outreach programs for stakeholders in the sprout production community to enhance the industry's understanding and implementation of current best practices for promoting sprout safety, as well as to facilitate compliance with pending US sprout safety regulatory requirements, when they are finalized. The goal of the SSA is similar to that of two other Alliances, the Produce Safety Alliance and the Food Safety Preventive Controls Alliance, which, respectively, are developing training courses and materials on preventing contamination of fresh produce on-farm and produce and processed food products at facilities other than farms (see http://producesafetyalliance.cornell.edu/psa.html and http://www.iit.edu/ifsh/alliance/). (For more information on the Food Safety Modernization Act and pending regulations for fresh produce, see Chapter 3 and http://www.fda.gov/Food/FoodSafety/FSMA/default.htm.)

SSA workgroup members include: policy analysts, inspectors and research scientists from state and federal government agencies and representatives from produce associations and the sprout industry, along with academia. As workgroup members join, the committee chair takes roll call. He then asks members to share overall impressions of the current draft chapter. The consensus seems to be that this draft is much more focused on sprout specific hazards compared with the previous draft and the group is appreciative that the editors took their comments into consideration. One member suggested that, in a section of the chapter presenting pathogens that have been involved in sprout-associated outbreaks in the USA, that is *Salmonella* spp., pathogenic *Escherichia coli* and *Listeria monocytogenes*, the TWG follow a format that describes:

1. the characteristics of each pathogen (such as infectious dose and symptoms of illness);
2. risk factors associated with the pathogen (such as sources of seed contamination during growth, harvest and handling, or insanitary conditions and practices in seed or sprout facilities);
3. controls that might be implemented to minimize contamination of seeds and sprouts (such as obtaining seed from reliable sources, examining seed on receipt, treating seed to reduce pathogens, if present, and following appropriate sanitary practices throughout all operations).

Members agreed that it is important that this chapter highlight 'hazards reasonably likely to occur if not subject to appropriate controls' and introduce what those controls might be; that time spent on 'possible but unlikely' hazards could distract attention from known, significant hazards associated with sprout production. As the discussion progressed, someone asked whether the pending produce regulation was likely to impose certain requirements on sprouters, such as hazard assessments at individual operations. It was noted that specific requirements would not be known until the regulation is finalized. Meanwhile, similar to the approach being taken by the Produce Safety Alliance, the TWG should look to existing programs and guidance to develop a curriculum that can provide the best information and tools as quickly as possible to begin to enhance sprout safety.

23.2 Initial sprout safety concerns and recommendations

Whereas the health benefits associated with regular consumption of fresh fruits and vegetables have been clearly demonstrated, it became clear in the mid-1990s that an increasing, although still small, proportion of reported outbreaks of foodborne illness were being attributed to fresh produce. In 1995, FDA asked the National Advisory Committee on Microbiological Criteria for Foods (NACMCF) to investigate and characterize the association between foodborne illness and fresh produce, and to provide recommendations that could be employed to reduce the risk of foodborne illness outbreaks associated with these commodities. In 1999, NACMCF published a report entitled 'Microbiological safety evaluations and recommendations on fresh produce' (NACMCF, 1999a). In that report, NACMCF noted that sprouts shared many common risk factors with other fresh produce commodities, while at the same time identifying sprouts as a special food safety concern among produce commodities because of the potential for pathogen growth associated with the unique way in which sprouts are grown.

Typically, sprouts are produced by first soaking seed in water and then placing seed in a warm, moist, nutrient-rich environment (rotating drums or stationary containers such as large bins or racks of trays) for an average of 3–7 days to promote seed germination and sprout growth, depending on the type of sprout, time of year and process used. The same conditions that encourage seeds to sprout are also nearly ideal for the exponential growth of pathogens if they are present.

It was also noted that there is no inherent step in the production of raw sprouts to reduce or eliminate pathogens (NACMCF, 1999b).

As a follow-up to these findings, NACMCF undertook a review of the current literature on sprout-associated outbreaks to identify the organisms and production practices of greatest public health concern; prioritize research needs; and provide recommendations on intervention and prevention strategies. NACMCF published a second report entitled 'Microbiological safety evaluations and recommendations on sprouted seeds' (NACMCF, 1999b). Based on documented epidemiology and microbial ecology of sprout-associated outbreaks, NACMCF concluded that seed appears to be the most likely source of contamination in sprout-associated outbreaks, and that conditions during sprouting are ideal for amplifying even low numbers of pathogens, if present. However, practices and conditions at the sprouting facility may also impact on the safety of the finished product.

The committee developed a number of specific recommendations, including:

1. enhancing the knowledge of all interested parties pertaining to the microbiological safety of sprouted seeds;
2. implementing Good Agricultural Practices (GAPs) to reduce the potential for microbial contamination of seeds used in sprout production;
3. developing and implementing seed cleaning, storage and handling practices that minimize the potential for microbial contamination;
4. treating seeds with one or more treatments that have been shown to reduce pathogenic bacteria that may be present; coupled with microbiological testing for *Salmonella* spp. and *Escherichia coli* O157;
5. establish Good Manufacturing Practices (GMPs) and food safety systems, including regulatory oversight and improved traceback, that systematically look for means to prevent seeds from serving as the vehicle for foodborne disease;
6. conduct research related to the microbiological safety of sprouted seeds, particularly in the areas of pathogen reduction or elimination, sources of contamination and its prevention, and preventing or retarding pathogen growth during sprouting.

On 27 October 1999, FDA published two guidance documents to inform all parties involved in the production of sprouts – producers, conditioners and distributors of seeds and beans used for sprouting, and sprout producers – that seeds and beans used for sprouting and sprouts have been recognized as an important cause of foodborne illness and to provide recommendations for preventive controls that FDA believed should be taken immediately to reduce the risk of sprouts serving as a vehicle for foodborne illness (FDA, 1999a, b).

The first guidance document, 'Reducing Microbial Food Safety Hazards for Sprouted Seeds' (the Sprout Guide), provides recommendations based largely on recommendations in the NACMCF sprout report (NACMCF, 1999a), that is:

1. seed for sprouting be produced using Good Agricultural Practices (GAPs);
2. seed handling and conditioning be conducted in a manner that minimizes the likelihood of contamination;

3. sprouters employ good sanitation as a standard operating procedure;
4. sprouters treat the seed to reduce pathogens immediately before sprouting;
5. sprouters conduct microbial testing of spent irrigation water (SIW) from every production lot as early as 48 hours into the sprout process to ensure that contaminated product does not enter commerce.

The second companion guidance document entitled, 'Sampling and Microbial Testing of Spent Irrigation Water During Sprout Production' (the Sprout Testing Guide), was intended to assist sprouters in implementing one of the principal recommendations in the broader Sprout Guide, that is producers test SIW for two pathogens, *Salmonella* spp. and *E. coli* O157:H7, before product enters commerce. Although seed disinfection is likely to reduce the level of contamination if pathogens are present in or on seeds, current treatments cannot guarantee total elimination of pathogens. If only a few pathogenic cells survive seed disinfection treatment, they can grow to high levels during sprouting and contaminate the entire batch. Thus, FDA concluded that seed disinfection treatments should be combined with microbial testing to ensure that contaminated sprouts do not enter the food supply. Sampling SIW – or in-process sprouts if SIW is not available – using rapid test kits validated for this use, allows sprouters to obtain results quickly without sacrificing shelf-life of this perishable product.

23.3 Challenges in sprout safety

In its 1999 report, NACMCF found that there was a lack of fundamental food safety knowledge along the continuum from seed production through sprout consumption. Although some sprout industry segments were becoming aware of the potential for this food to be a vehicle for foodborne illness and the need for appropriate controls, such awareness was not universal (NACMCF, 1999b).

When FDA published the Sprout Guides, the agency also announced that it planned to initiate a series of inspections to determine current practices within the industry and to take regulatory action as warranted. (Note: In addition to routine surveillance of food facilities and products, FDA may issue an assignment to its field staff to perform inspections, collect product samples or both. Assignments are targeted to meet specific needs, such as gathering information on current industry practices and conditions to support policy development, education efforts or other regulatory response. Assignments may include specific instructions on selecting firms to visit, what to look for or questions to ask and how samples should be collected and analyzed, if appropriate.) In 2000, FDA issued an assignment to the field to conduct inspections at 137 sprout facilities nationwide. The firms inspected produced anywhere from one to 15 different types of sprouts, with the majority concentrating on one or a few types of sprouts. Mung bean sprouts were the most commonly produced, followed by alfalfa, radish, a variety of sprout mixes, broccoli and clover sprouts. Approximately 37% of firms produced a less common sprout type, such as soybean, onion, pea or lentil.

Inspectors reported significant deficiencies at many of the firms visited, resulting in the issuance of an FDA-483, List of Observations. A warning letter was issued to almost half of the firms for 'failing to have effective controls in place to prevent finished-product sprouts from being rendered injurious to health', in many cases because the microbial testing of spent irrigation or an effective alternative had not been adequately adopted and implemented. Often, firms that failed to implement effective preventive controls were also cited for operating under insanitary conditions.

It should be noted that almost half of the firms inspected in 2000 had been inspected previously in conjunction with an assignment that issued in 1998 and that preceded FDA's Sprout Guides. Of the 63 firms re-inspected in 2000, 41 (65%) were issued an FDA-483 and 27 (43%) were issued warning letters. Although high, this was significantly lower compared with sprout firms whose initial inspection was conducted per the 2000 assignment, of which 77% were issued FDA-483s and 51% received warning letters, indicating that regulatory oversight had value but that more actions were needed.

In addition, many sprout firms tend to be small, with smaller firms potentially having a lower level of food safety knowledge and resources. In an earlier survey conducted by FDA and the California Department of Public Health (then the California Department of Health Services), inspectors visited 45 sprout firms in California. Of these, only 25 firms reported producing over 5000 pounds (2273 kg) of sprouts per day (NACMCF, 1999b).

23.4 Knowledge and research needs

23.4.1 Meeting knowledge needs

FDAs 1999 Sprout Guide provided five broad recommendations for reducing the risk of seeds and sprouts serving as vehicles for foodborne illness. However, with the exception of the companion guidance for microbial testing of spent irrigation water, which, out of necessity, is highly technical, the guidance did not provide significant detail on implementing the recommendations.

In 2000, the California Department of Public Health, Food and Drug Branch (CDPH, FDB) and the US FDA, Center for Food Safety and Applied Nutrition (CFSAN) jointly developed a video 'Safer Processing of Sprouts' in cooperation with the Centers for Disease Control and Prevention, university researchers and industry representatives to assist the industry in producing a safer product (FDA/CDPH, 2000). The video may also be useful for retailers, regulators and anyone working with the industry that wants to better understand the product and current recommendations for best production. Topics covered in the video include: Introduction to Food Safety, Requirements and Guidelines for Sprouters, Growing, Harvesting, Milling and Transportation and Storage of Seed for Sprouting, Sprout Production, Seed Treatment, Sampling and Microbiological Testing, Cleaning and Sanitizing Equipment and Facilities, Personnel Practices and additional resources.

23.4.2 Research needs

FDA also continued to support research by a Sprout Taskforce at the National Center for Food Science and Technology (NCFST) in Chicago (now IFSH) and others, on prevention and intervention strategies to provide a foundation for sprout-specific preventive controls and food safety programs. Some areas represent longstanding challenges to enhancing sprout safety. For example, an analysis of published sprout seed disinfection studies shows that reported treatment efficacy is highly variable. Scientific literature indicates that the 20 000 ppm calcium hypochlorite treatment that has been widely adopted by sprouters who treat seed prior to sprouting produces an average of 2.5 log pathogen reduction with a range of 1.0–6.5 log reduction. This variability seriously complicates comparing efficacy of different treatments across studies. In addition, different types of seeds and beans have differing tolerances for some seed treatments. Seed treatments that significantly reduce germination rate for any given type of seed or bean (e.g. 20 000 ppm calcium hypochlorite and soybean seed) would not be feasible for treating that type of seed. In addition, the microflora in SIW is unique. Microbial testing methods that have not been validated to work on SIW (or sprouts) may not detect pathogens if they are present. Using methods that have not been validated – or using validated methods incorrectly – for every production lot of sprouts may add significantly to costs without adding the intended level of public health protection. Finally, although seed has been identified as the most likely source of contamination in many, but not all, sprout-associated outbreaks, contamination of seed is likely to be low level and sporadic. Testing seed, even when a seed lot has been implicated in an outbreak, often fails to detect pathogens. Establishing measures to prevent contamination of seeds and beans during growing, conditioning and handling is made even harder by the fact that the vast majority of seeds and beans are destined for agricultural use, that is, field planting, rather than use as food (sprouts). Quality considerations for such seed may include varietal purity and similar characteristics, rather than food safety.

23.5 Further developments in sprout safety hazards

FDA has worked closely with the sprout industry to promote adoption and implementation of recommendations in the Sprout Guides. Following release of the Sprout Guides and other early activities (regulatory and educational), the number of outbreaks associated with sprouts, and the number of illnesses associated with outbreaks that did occur, appeared to decline. In fact, there were no US sprout outbreaks reported to FDA in 2005, 2006 or 2007. However, after a number of years of a decline in outbreaks associated with sprouts, outbreaks began to recur and this trend has continued. In late 2008, there was one sprout-associated *Salmonella* outbreak. In 2009, an ongoing *Salmonella* outbreak resulted in more than 200 illnesses. In 2010, sprouts contaminated with *Salmonella* Newport infected 44 people in 11 states. In June 2011, sprouts contaminated with

Salmonella Enteritidis sickened 21 people in five states. In the last few years, sprouts have also been linked to *Listeria* illnesses and positive sample findings.

Recent outbreaks and positive sample findings have led FDA to question: What is the current state of the US sprout industry with respect to consistent and appropriate application of recommended practices in current guidance? Has adherence to these recommendations declined at individual firms? Have new firms joined the sprout industry that are not aware of the recommendations or how to implement them? What is the adequacy of current guidance?

In May, 2009, FDA issued a letter to seed suppliers, distributors and sprouters to make them aware of the agency's concerns regarding continuing outbreaks associated with the consumption of raw and lightly cooked sprouted seed and to urge the industry to review its operations in light of FDA's sprout guidance and other available information (FDA, 2009). FDA strongly encouraged firms in this industry throughout the supply chain – from seed and bean production through sprouting – to review current operations in light of our previously issued guidance documents and to consider other available information for the reduction or elimination of pathogens on fresh produce, and to modify their operations accordingly to ensure they are taking appropriate measures to provide a safe product to consumers.

FDA also met with sprout industry representatives to share information about issues the agency and other partners have observed relative to implementation of the sprout guidance. Our intent was not to imply the extent to which these observations applied to the industry as a whole but, rather, to share things we had seen at one or more locations so that sprouters could use this information to evaluate their own operations and make changes as needed. Areas of concern at one or more locations, based on field assignments and outbreak investigations included:

- lack of seed screening on receipt;
- poor sanitation, wet environment, potential for pathogens to become established and spread;
- aging equipment, cracked drum doors with tape, hard to clean food contact surfaces;
- failure to follow all label directions when mixing and using antimicrobial chemicals approved for reduction of pathogens in seeds or sprouts, including issues with: concentration of seed sanitizing solution, ratio of seed to solution and failure to adequately agitate seed/solution mix during seed treatment and treatment time;
- sample collection, sample size and pooling samples from multiple production lots;
- intermittent (occasional) sampling rather than sampling each production lot and using test methods not validated for spent irrigation water (or sprouts);
- misinterpretation of presumptive positive and or confirmatory test results;
- using the wrong material for confirmatory tests after a presumptive positive is found;

- failure to adequately clean and sanitize facility and equipment after finding a positive.

23.6 Addressing sprout safety hazards

23.6.1 Producer, buyer and regulator collaboration

Similar to other produce, some buyers require audits of sprout facilities as a condition of purchase. Because most sprout operations are conducted in an enclosed facility rather than an open field, such audits tend to focus on GMP rather than GAP issues. More often than not, third party audits of sprout facilities do not include sprout-specific criteria related to recommendations, such as those in FDA's Sprout Guide for seed disinfection treatments and testing sprout spent irrigation water, measures that FDA believes are critical to ensuring the safety of sprouts.

In 2009, the sprout industry convened a Sprout Taskforce to address continuing concerns about sprout safety and to reexamine best practices for sprout production. As a first step, the taskforce began work to draft an audit checklist that builds on the recommendations in the Sprout Guides and other existing sprout safety programs and experiences. This audit checklist could be used by sprout growers to assess their own practices, conditions, and processes and by auditors to conduct audits that include sprout-specific best practices.

As a longer-term action item, the taskforce convened an expert panel to look at a number of key provisions in the Sprout Guide and other programs, to examine the state of the science behind these provisions and to investigate how they might be improved. For example, the taskforce and expert panel are evaluating the comparative efficacy of disinfection treatments applied to seed immediately before sprouting. The taskforce plans to compile a list of treatments that meet certain criteria for inclusion, along with information about conditions of use, such as seed type. They are also reviewing test methods to extend the list of methods cited for testing SIW and sprouts. These lists will be attached as annexes to the Sprout Audit checklist. In addition, the taskforce is looking at the efficacy and feasibility of seed disinfection treatments that may be applied by seed distributors. The expert panel is also tasked with attempting to quantify risk reduction for individual control measures, for example recommendations in FDA's Sprout Guides, along with additional measures, such as seed screening and microbial testing, to develop a model that might be used to assess the cumulative risk reduction of various hurdles or interventions, in various combinations, to create a tool to help tailor interventions or best practices to individual firms or segments of the sprout industry.

23.6.2 Education, outreach and regulation

In March 2012, FDA issued an assignment to conduct inspections of 21 sprout production facilities to: assess the current industry practices using a checklist of sprout-specific practices; collect and analyze samples of SIW and finished product for microbial pathogens (*Salmonella* and *E. coli* O157:H7 in SIW and finished product and *Listeria monocytogenes* in finished product); determine the extent to which sprout producers have adopted FDA guidance; and document conditions and take appropriate regulatory action at firms operating under insanitary conditions. The 2012 assignment builds on information collected by a similar assignment in 2010. Preliminary results from both assignments are being used to assist FDA in policy development (including rulemaking), education and outreach to the sprout community through the Sprout Safety Alliance, and regulatory responses to future inspections at sprout facilities.

On 4 January 2013, FDA released a proposed rule, 'Standards for the Growing, Harvesting, Packing and Holding of Produce for Human Consumption', that would establish science-based minimum standards for the safe growing, harvesting, packing and holding of produce on farms (FDA, 2013). In addition to provisions that would apply to covered produce generally, the proposed rule includes additional sprout-specific requirements. If finalized as proposed, the rule would include requirements to:

- establish measures related to seeds or beans for sprouting to prevent the introduction of hazards, including: if a sprouter has reason to believe that a lot of seeds or beans have been associated with foodborne illness, they must not use it, and a requirement to visually examine seeds and beans and packaging for signs of potential contamination;
- establish measures that must be taken for the growing, harvesting, packing and holding of sprouts, including growing, harvesting, packing and holding sprouts in a fully enclosed building, and cleaning and sanitizing any food contact surfaces that come into contact with seeds or beans;
- treat seeds or beans using a scientifically valid method immediately before sprouting to reduce pathogens;
- test each production batch of SIW or sprouts for *E. coli* O157:H7 and *Salmonella* species and take appropriate follow-up actions if detected;
- test the growing environment for *Listeria* spp. or *L. monocytogenes*;
- establish and implement a written SIW testing and environmental monitoring plan that also includes appropriate follow-up actions if detected.

FDA is providing 300 days for comment on these proposed requirements. The agency is also asking for comment and data to inform a final regulation on specific outstanding questions, such as:

- Whether, and to what extent, it would be feasible to establish additional requirements directed to the production, conditioning and holding of seed before it reaches a sprouter, without unduly imposing a burden on the agricultural sector that produces seed used primarily for growing food or feed crops and not intended for food for human consumption as sprouts?
- Whether, and to what extent, the proposed provisions should be applied to soil grown sprouts? The NACMCF Sprout report and FDA's guidance did not distinguish between hydroponic and soil grown sprouts. However, FDA is not aware of outbreaks associated with soil grown sprouts.

- Whether pathogens in addition to *Salmonella* and *E. coli* O157:H7 should be included in SIW testing requirements, either by specifically listing additional pathogens or by listing criteria for inclusion, such as association in one or more outbreaks linked to sprouts?

23.7 Conclusion

All SSA workgroup members are highly motivated, both industry, and state and federal regulators. As mentioned earlier, there have been at least 35 reported foodborne illness outbreaks in the USA associated with sprouts, resulting in over 2000 laboratory-confirmed illnesses and one death since 1996. In addition, a large outbreak of *E. coli* O104:H4 in Europe in 2011 which resulted in almost 4000 reported cases of illness and 53 deaths, was ultimately linked to sprouts made from fenugreek seed imported from Egypt. This outbreak has had a significant impact on sprouters globally (EFSA, 2011). In response to continued sprout outbreaks, some retailers have announced they are not going to carry sprouts until steps are available to ensure their safety, whereas other groups are instituting new plans to provide verification for science-based operating standards to grow safe sprouts, and invest in scientific research, training and education (Nunes, 2012). The SSA TWG is considering all available, relevant information, whether by US FDA, foreign governments or private entities, in developing sprout safety training materials.

A number of members of the SSA TWG are also members of the Sprout Taskforce. They have first-hand knowledge of the progress being made in the taskforce on activities, such as pilot testing the sprout-specific audit and current industry practices with respect to sprout production, seed treatment and testing SIW. Their knowledge and experience is a critical resource for the SSA TWG in developing training materials for best practices. The sprout taskforce also has first-hand knowledge and experience about the challenges both groups are still grappling with, including identifying seed disinfection treatments capable of reducing pathogens on seed and the conditions, including seed type, in which they will be most effective.

To date, the SSA TWG has developed draft chapters on a number of important preventive controls, including worker health and hygiene, sampling and testing SIW and sprouts, and environmental testing and monitoring. The SSA has also formed another workgroup for education and outreach (SSA EOWG). The SSA EOWG will soon be pilot testing the draft chapters with sprouters, including small sprouters, to get feedback on their clarity and the feasibility of preventive controls as presented (see https://www.iit.edu/ifsh/sprout_safety/). A great deal of work remains ahead but the group members are becoming more comfortable working collaboratively and are hopeful that their efforts now can contribute to significant adoption of best practices for producing sprouts, reduce sprout-associated illnesses and restore confidence in the safety of sprouts.

23.8 Questions for discussion

1. What are the primary sources of contamination for sprouted seeds?
2. Which are the primary pathogens of concern for sprouts?
3. Why is controlling the contamination of seed so problematic?
4. What makes sprouts especially challenging compared with other produce commodities?
5. Discuss the relative impact that factors such as buyer requirements, regulatory oversight and knowledge of hazards and controls, might have on food safety at individual firms, nationally and globally.
6. What is the value of food safety workgroups with diverse perspectives?
7. Why is information from field inspections and investigations so important to policy makers?

23.9 Sources of further information and advice

CANADIAN FOOD INSPECTION AGENCY (CFIA) (2007). Code of practice for the hygienic production of sprouted seeds. Available at: http://www.inspection.gc.ca/english/fssa/frefra/safsal/sprointe.shtml.

CANADIAN FOOD INSPECTION AGENCY (CFIA) (2008). Food safety practices guidance for sprout manufacturers. Available at: http://www.sproutnet.com/Reports/CFIA_FSPG_EN.pdf.

CODEX ALIMENTARIUS COMMITTEE (2003). Code of hygienic practice for fresh fruit and vegetables. Annex II: Annex for sprout production. Available at: www.codexalimentarius.net/download/standards/10200/CXP_053e.pdf.

FDA (1998). Guidance for Industry: Guide to Minimize Microbial Food Safety Hazards for Fresh Fruits and Vegetables. Available at http://www.fda.gov/Food/GuidanceRegulation/GuidanceDocumentsRegulatoryInformation/ProducePlantProducts/ucm064574.htm

FOOD STANDARDS AUSTRALIA NEW ZEALAND (FSANZ). Proposal P1004 – Primary Production and Processing Standard for Seed Sprouts. Available at http://www.foodstandards.gov.au/foodstandards/proposals/proposalp1004primary4361.cfm.

MONTVILLE, R. and SCHAFFNER, D. (2005). Monte Carlo Simulation of Pathogen Behavior during the Sprout Production Process. Applied and Environmental Microbiology 71, no. 2: 746–753. 10.1128/?AEM.71.2.746–753.2005.

STEC WORKSHOP REPORTING GROUP (2012). Experiences from the Shiga toxin-producing Escherichia coli O104:H4 outbreak in Germany and research needs in the field. Berlin, 28–29. *Euro Surveill.* 17(7):pii=20091. Available at: http://www.eurosurveillance.org/ViewArticle.aspx?ArticleId=20091.

23.10 References

EUROPEAN FOOD SAFETY AUTHORITY (EFSA) (2011). Scientific Opinion on the risk posed by Shiga toxin-producing Escherichia coli (STEC) and other pathogenic bacteria in seeds and sprouted seeds. *EFSA Journal* 2011;9(11):2424 [101 pp.]. doi:10.2903/j.efsa.2011.2424.

FDA (1999a). Guidance for Industry: Reducing Microbial Food Safety Hazards for Sprouted Seeds. Available at: http://www.fda.gov/Food/GuidanceRegulation/GuidanceDocumentsRegulatoryInformation/ProducePlantProducts/ucm120244.htm.

FDA (1999b). Guidance for Industry: Sampling and Microbial Testing of Spent Irrigation Water During Sprout Production. Available at: http://www.fda.gov/Food/GuidanceRegulation/GuidanceDocumentsRegulatoryInformation/ProducePlantProducts/ucm120246.htm.

FDA (2009). Letter from FDA to Seed Suppliers, Distributors, and Sprouters. Available at: http://www.fda.gov/NewsEvents/PublicHealthFocus/ucm151758.htm.

FDA (2013). Fact Sheet on the FSMA Proposed Rule for Produce: Standards for the Growing, Harvesting, Packing, and Holding of Produce for Human Consumption. Available at: http://www.fda.gov/Food/GuidanceRegulation/FSMA/ucm334114.htm#summary.

FDA/CDPH (2000). Safer Processing of Sprouts video: Go to http://postharvest.ucdavis.edu/libraries/video/ and search for sprouts, you can find links to all the modules.

FETT, W. F., FU, T. and TORTORELLO, M. L. (2006). Seed Sprouts: the State of Microbiological Safety. In *Microbiology of Fresh Produce*. ASM Press, Washington DC, pages 167–219.

NATIONAL ADVISORY COMMITTEE ON MICROBIOLOGICAL CRITERIA FOR FOODS (NACMCF) (1999a). Microbiological Safety Evaluations and Recommendations on Sprouted Seeds. Available at: http://www.fda.gov/Food/FoodborneIllnessContaminants/BuyStoreServeSafeFood/ucm078789.htm.

NATIONAL ADVISORY COMMITTEE ON MICROBIOLOGICAL CRITERIA FOR FOODS (NACMCF) (1999b). Microbiological Safety Evaluations and Recommendations on Fresh Produce. *Food Control*. 10:117–143.

NUNES, K. (2012). Kroger to no longer sell sprouts. FoodBusinessNews.net. 19 October 2012. Available at: http://www.foodbusinessnews.net/articles/news_home/Food_Safety_News/2012/10/Kroger_to_no_longer_sell_sprou.aspx?ID={E8C37653–995D-46DE-8C33–62D0D0DC7FA7}&cck=1.

PAINTER, J. A. HOEKSTRA, R. M. AYERS, T., TAUXE, R. V., BRADEN, C. R., *et al.* (2013). Attribution of Foodborne Illnesses, Hospitalizations, and Deaths to Food Commodities by using Outbreak Data, United States, 1998–2008. *Emerging Infectious Disease Journal – CDC* 19(3). Available at: http://wwwnc.cdc.gov/eid/article/19/3/11–1866_article.htm.

24

Leafy greens: the case study and real-life lessons from a Shiga-toxin-producing *Escherichia coli* (STEC) O145 outbreak in romaine lettuce

M. A. Baloch, US Centers for Disease Control and Prevention, USA

DOI: 10.1533/9781782420279.5.340

Abstract: In April and May 2010, 33 cases of *Escherichia coli* O145 infections were reported from five US states. An epidemiologic investigation found that the illnesses were associated with the consumption of shredded lettuce. A traceback investigation from the processor led to a farm, Ranch A, that supplied lettuce to the processor. This chapter presents the findings of the field environmental assessment conducted in June 2010 on Ranch A and its surrounding watershed. The environmental assessment findings identify and enumerate environmental antecedents and contributing factors that might have influenced the contamination of the lettuce and the outbreak.

Key words: fresh produce-related outbreaks, *E. coli* O145, leafy greens, foodborne illness outbreak, environmental assessment, systems approach, outbreak investigation, environmental antecedents.

24.1 Introduction

In April 2010, Michigan, Ohio and New York reported several cases of bloody diarrhea in their respective jurisdictions. Laboratory analyses identified matching isolates of Shiga-toxin-producing *Escherichia coli* (STEC) O145 as the infectious agent. An epidemiologic investigation found that the illnesses were associated with the consumption of shredded lettuce. In coordination with Food and Drug Administration (FDA), the states involved conducted traceback investigations and identified a common processor as the source of the shredded romaine lettuce. On 5 May 2010, the New York State Health Department reported isolating *Escherichia coli* (*E. coli*) O145 from the lettuce collected from this processor. On 9 May 2010, the Centers for Disease Control and Prevention's (CDC) PulseNet database confirmed that an isolate from the lettuce was matched by pulsed field gel electrophoresis (PFGE) to the outbreak strain (CDC, 2012a). By 20 May 2010, Michigan, Ohio, New York, Pennsylvania and Tennessee (CDC, 2012b) had reported a total of 33 cases of *E. coli* 0145 infections.

FDA conducted an investigation and did not identify a likely source of contamination at the processor. FDA conducted a traceback investigation from the processor that led to a farm in Yuma County, Arizona, hereafter referred to as Ranch A that supplied lettuce to the processor. This chapter presents the findings of the subsequent field environmental assessment conducted in June 2010 on Ranch A and its surrounding watershed. The environmental assessment findings identify and enumerate environmental antecedents and contributing factors that might have influenced the contamination of the lettuce and the outbreak. Prior to this field environmental assessment, an initial on-farm investigation was also conducted by FDA investigators immediately after identifying Ranch A as the source of suspect lettuce in May 2010. This chapter is an adaptation of the environmental assessment report (Crawford *et al.*, 2010) submitted to FDA and other stakeholders, after the field environmental assessment was conducted.

24.2 Challenges faced by the experts involved

A foodborne outbreak environmental assessment is the systems-based component of a foodborne illness outbreak (FBIO) response that fully describes how the environment contributed to the introduction and/or transmission of agents that are capable of causing human illness. Environment encompasses everything external to the host, including air, food, water, animals, plants, climate, etc., as well as people, social and built environments. All aspects of the environment can be listed as variables that, in relation to transmission are neutral, conducive or protective. From this description, contributing factors and environmental antecedents to an outbreak can be determined (Selman and Guzewich, unpublished).

Therefore, an environmental assessment goes beyond what has traditionally been considered an outbreak investigation. Traditionally, outbreak investigations merely identified a suspected or confirmed agent and/or the vehicle causing illnesses. An environmental assessment digs deeper into the root causes of contamination to identify how and why the contamination occurs by relating contributing factors and environmental antecedents (Higgins and Hartfield, 2004). Identification of environmental antecedents is considered a specifically important outcome of an environmental assessment. Environmental antecedents are elements in the environment that facilitate the occurrence of an outbreak. They may be related to people, equipment, process, food type, economics or other circumstances. They are environmental variables that, under inadequate or in the absence of control, may create conditions known as contributing factors that lead to adverse health outcomes (Gelting *et al.*, 2005). In other words, they are why the contributing factors happened. Environmental antecedents are sometimes referred to as root causes of foodborne outbreaks (Selman and Guzewich, unpublished).

Identifying environmental antecedents is necessary not only to understand the root causes of contributing factors but also to identify mitigation strategies to prevent future outbreaks. Environmental assessment in the aftermath of an outbreak is an emerging concept, and thus is evolving methodologically. Environmental assessments are becoming a standard response to FBIOs associated with contributing factors related to food service establishments and processing plants (Delea *et al.*, 2008). CDC's Environmental Health Specialists Network (EHS-Net) program has demonstrated the effectiveness of such an approach through its involvement by conducting field environmental assessments in response to at least two previous outbreaks (Gelting *et al.*, 2011; CALFERT, 2008).

The environmental assessment conducted in June 2010 is the first to be conducted within the standard outbreak-response mechanism of FDA and, therefore, presented unique challenges. These challenges included incorporating an all-inclusive investigation of contributing factors and environmental antecedents that went beyond a regulatory checklist-based type of inspection.

In essence, an environmental assessment based on a systems approach reconstructs past events that may help identify how a contamination event occurred or was amplified, or both, and therefore is time sensitive. A major challenge in identifying the root causes of this 2010 outbreak was that the field environmental assessment was conducted a month after an initial, on-farm investigation. That initial investigation took a more traditional approach of extensive sampling in and around the growing fields on Ranch A, without taking into account a holistic approach as envisaged in a field environmental assessment. Farm operations and the environmental systems surrounding the farms are very dynamic. A delay such as that occurring after Ranch A's initial investigation, causes loss of key information because of changes in prevalent environmental or weather conditions, or both, during growing and harvesting stages of the crop.

Another major challenge to a systems-based approach at the field level is making an informed decision about defining the boundaries of the system to be assessed. In the past, investigations have often been limited to intensive sampling of produce, irrigation water, and soils and analyzing conditions within the boundaries and the immediate surroundings of the farm fields identified in traceback investigations, which conceptually represents a two-dimensional spatial approach. Such a two-dimensional approach does not provide a comprehensive picture of the different components of an environmental system and their interrelationships and interactions with one another. To understand and identify the potential sources of *E. coli* O145 on Ranch A and its pathways to the produce, all of the relevant components of the system had to be identified and defined in conjunction with their relationships to and influences on each other. These activities necessitate a four-dimensional, spatial-temporal approach that incorporates not only a third spatial dimension of depth but also a fourth dimension of time (Gelting *et al.*, 2011).

This awareness can only be accomplished after a thorough understanding is developed of the environmental circumstances and systems related to the outbreak. In this case, this understanding was achieved by defining the system constituting

Ranch A, the sources of irrigation water, the potential sources of the pathogen in the watershed and the pathways by which the pathogen was transported to the crops.

The systems-based nature of an on-farm field environmental assessment requires the expertise of multiple disciplines such as food safety, microbiology, water quality and hydrology. Such an approach transcends the boundaries of regulatory jurisdiction of a single agency, and may require collaboration with multiple local and state agencies in many areas of focus. Some of these areas of focus could be environmental protection, irrigation-water management, water resources management and watershed management, which would provide the full picture of the influences and environmental causes of the outbreak. The number and nature of the collaborating agencies may vary depending on the type and significance of information that is relevant to an on-farm outbreak investigation; frequently this collaboration constitutes a major challenge. During this particular environmental assessment, the investigation team collaborated with state and local agencies, including the Arizona Department of Agriculture, the Arizona Department of Environmental Quality, the Arizona Department of Health Services, the Wellton-Mohawk Irrigation and Drainage District (WMIDD) and the Yuma County Department of Public Works.

24.3 Significance of the pathogen that caused the case

Shiga-toxin-producing *E. coli*, or STEC bacteria, are associated with human illness and cause symptoms such as bloody diarrhea and hemolytic uremic syndrome. STEC bacteria are grouped by serogroups (e.g. O157 or O145). *E. coli* O157 is the most commonly found STEC serogroup in the USA. Currently, in the USA, non–O157 STEC such as *E. coli* serogroup O145, pose an extensive public health problem and cause an estimated 37000 cases of illness and 30 deaths each year (Fratamico *et al.*, 2009). Increases in the incidence of non-O157 STEC-related cases may be caused by an increased awareness of their role in human illness. The current case study is the first fresh produce outbreak associated with *E. coli* O145 reported in the USA. Between 2004 and 2008, 127 outbreaks associated with fresh produce were reported in the USA. Table 24.1 presents a list of the major fresh produce-related outbreaks in the USA that were reported between 2004 and 2008 (CDC data).

24.4 Geographical and climate factors

According to US Department of Agriculture data (USDA, 2012), lettuce was the leading vegetable crop in 2010 with a total growing area of 267300 acres and a \$2.2 billion production value. Its value constituted 19% of the total value for all fresh vegetable crops in the USA. California (CA) and Arizona (AZ) lead domestic production of lettuce, accounting for about 98% of commercial availability in the

Table 24.1 Major fresh produce-related outbreaks reported in the USA (2004–2008)

Year	State	No. of illnesses	Agent	Food item
2004	New Jersey	6	*E. coli* Shiga-toxin-producing O157:H7	Lettuce, unspecified
2004	CDC	97	*Salmonella enterica* Newport	Iceberg lettuce, unspecified
2004	Arizona	38	Calicivirus norovirus	Lettuce-based salads, unspecified
2005	CDC	34	*E. coli* Shiga-toxin-producing O157:H7	Lettuce, prepackaged
2005	Florida	91	Calicivirus norovirus	Green leaf lettuce
2005	Minnesota	30	Calicivirus norovirus	Salad, unspecified
2006	CDC	238	*E. coli* Shiga-toxin-producing O157:H7	Spinach
2006	CDC	77	*E. coli* Shiga-toxin-producing O157:H7	Lettuce, unspecified
2006	Colorado	50	Calicivirus norovirus	Green salad
2006	Michigan	108	Calicivirus norovirus	Lettuce based salads
2007	Alabama	26	*E. coli* Shiga-toxin-producing O157:H7	Lettuce-based salads, unspecified
2007	CDC	76	*Salmonella enterica* Typhimurium	Spinach
2007	Oregon	25	Calicivirus norovirus	House salad
2007	Washington	128	Calicivirus norovirus	Lettuce-based salads
2008	CDC	74	*E. coli* Shiga-toxin-producing O157:H7	Iceberg lettuce, unspecified
2008	California	51	Calicivirus norovirus	House salad
2008	Connecticut	30	Calicivirus norovirus	Lettuce-based salads, unspecified
2008	Florida	29	Calicivirus norovirus	Lettuce-based salads, unspecified

USA. Typically, between November and March, most domestic lettuce is grown in Yuma (Arizona) and in California's Imperial Valley. Most growers and harvesters move to California to grow a new lettuce crop from November to October in the Salinas Valley. The romaine lettuce implicated in the 2010 outbreak was grown in November 2009 in four fields of Ranch A in the Wellton-Mohawk Valley in Yuma County, Arizona, and was harvested in April or May 2010.

According to the US Geological Survey (USGS) classification of hydrologic units, the valley is located in the Lower Gila watershed. The Lower Gila watershed

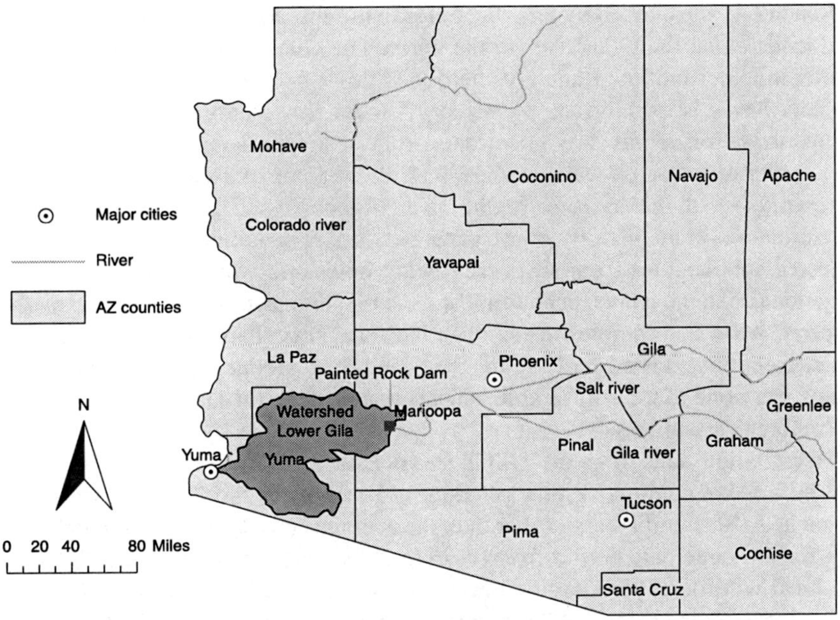

Fig. 24.1 Map of Arizona state and location of Lower Gila watershed.

comprises the drainage area of the Gila River below Painted Rock Dam, an earthfill structure southwest of Phoenix built in 1959 as shown in Fig. 24.1. Immediately downstream of Painted Rock Dam, the Gila River changes course from northwesterly to southwesterly. The river then runs east to west through the center of the watershed, flowing into the Colorado River at Yuma, Arizona.

The Wellton Canal supplied irrigation water for the implicated lettuce crop. The Wellton Canal is a subcanal of the Wellton-Mohawk main canal, which is operated by WMIDD. Wellton-Mohawk's water comes from the Colorado River; water flow is diverted at the Arizona abutment of the Imperial Dam into the Gila Canal, a joint-use facility shared by five Yuma-area irrigation districts. At Gila Canal Mile 15, WMIDD's share of the flow diverts into the Wellton-Mohawk Canal, flowing from west to east in the Lower Gila watershed. At three pumping stations, pumps deliver water to the canal headwaters in stages. Approximately 378 miles of main canals, lateral canals and return flow channels make up the WMIDD. The canals and the laterals are concrete lined to prevent seepage, except for 8 miles of the main canal upstream (west) of the first main pumping plant—upstream of the Wellton-Mohawk area where Ranch A lettuce farms are located. WMIDD also operates a network of 90 evenly spaced drainage wells. Drainage wells allow excess water to empty into collector channels that flow into a main conveyance channel discharging into the Colorado River near Yuma. This process prevents waterlogging, keeps the depth of the water table below root level and disposes of the salts that accumulate resulting from evapotranspiration from crops (WMIDD, 2010).

Ranch A is surrounded by growing fields to the east and west, and undeveloped wild lands along the Gila River to the north. The Gila River is approximately a quarter mile north of the Ranch A's fields and flows from east to west. The bed of the Gila River is usually completely dry—water flows only after precipitation events, irrigation-return flow or releases from upstream dams; the Gila River is not a source of irrigation water (AZDWR, 2009). Figure 24.2 shows the schematic representation of the location of the river, Ranch A and potential sources of contamination in the vicinity of the canal network. The initial farm investigation collected soil samples from suspect growing fields and water samples from the irrigation canal, the conveyance canal and the two drainage wells identified on the property. All water samples tested negative for STEC. Three soil samples from two suspect fields tested positive for Stx1- and Stx2-producing non-O157 STEC strains, but none of the soil sample isolates matched the O145 Stx2-producing *E. coli* outbreak strain.

Precipitation data from the USGS's Dateland Station upstream of Ranch A showed a mean cumulative precipitation of 3 inches for the January–April data period in 2010, significantly higher than the average cumulative precipitation of 1 inch for the same data period from 1996 to 2009. Information provided by state and local wildlife experts verified minimal movement of wildlife into Ranch A during this growing season; this circumstance was likely to be because of abundant water and vegetation found elsewhere because of increased rainfall during the growing season as noted earlier. Furthermore, no STEC was detected in any of the coyote, bobcat, rabbit, or rodent feces samples collected on and around Ranch A during the initial on-farm investigation. Hence, the initial on-farm investigation did not identify any significant potential sources of *E. coli* on Ranch A. The

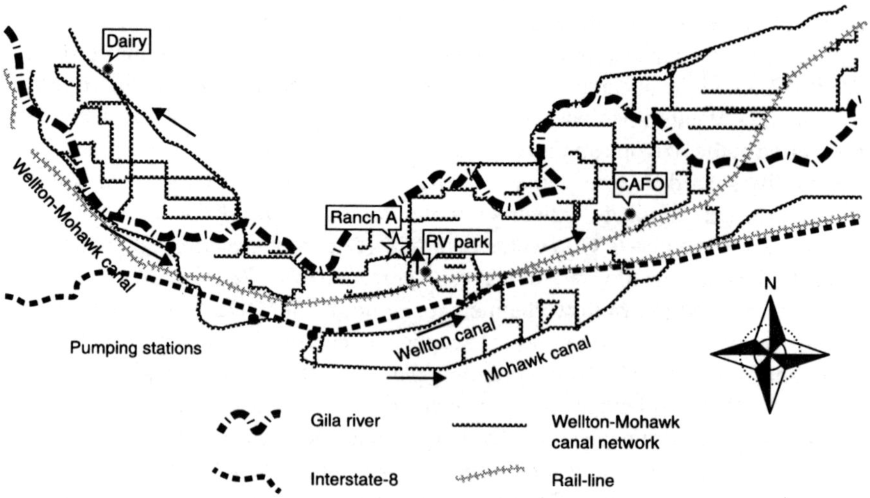

Fig. 24.2 Schematic representation of the canal network, Ranch A and potential sources of contamination (arrows show direction of flow in the canal network).

evidence led to the key assumption that water used for irrigating crops and dilution of pesticides and fertilizers was the most likely vehicle of pathogen transmission onto and on the farm and, thus, to the produce.

During the investigation the growers reported that the suspect romaine lettuce at Ranch A was irrigated solely with surface water from the WMIDD canal system and this same water was used to apply pesticides and fertilizers and in harvesting. Additionally, at least two drainage wells, used to extract excessive plant-root–zone water to avoid salinity and waterlogging, exist on Ranch A, but neither were used for irrigation.

Thus, the field environmental assessment investigation included the following activities to identify the environmental antecedents encompassing the environmental assessment approach described above:

- identifying the watershed, including major water bodies, weather stations and flow-measuring stations in the watershed pertinent to this outbreak;
- characterizing and understanding the source or sources of irrigation water for the farms of concern on Ranch A;
- identifying and collecting background information about the hydrology of the watershed;
- characterizing the hydrologic conditions in the watershed prevailing at and around the time of outbreak;
- identifying potential sources of contamination in the watershed;
- helping identify critical areas for prioritizing sample location;
- identifying potential hydrologic routes of contamination to the water system;
- recommending precautionary measures to prevent future similar outbreaks.

24.5 Regulatory and economic aspects

24.5.1 Regulatory aspects

FDA is the authority to protect and promote public health by regulating and supervising the safety of food that moves in interstate commerce, excluding meat, poultry and processed egg products. FDA, state and local regulatory agencies conduct regulatory inspections to identify violations based on the appropriate regulations and collect supporting documentation of those violations. Environmental assessments and regulatory inspections incorporate similar techniques, such as observing, sampling, reviewing records and interviewing managers and staff, at the time of the visit; however, these techniques are applied differently in the two different situations.

A regulatory inspection identifies risks of contamination based on the specific requirements and standards set forth by laws, regulations and guidance. It is an organized and formal process focused on monitoring the facility's exact performance problems in complying with laws and regulations during one point in time. The outcomes of a regulatory inspection may be used to support regulatory actions if observations and documented practices provide evidence that the firm is not meeting current regulatory standards for food safety.

A foodborne outbreak environmental assessment, on the other hand, is a description of what happened in the past. This description captures all aspects of the location under review including those factors that may or may not have influenced potential contamination, proliferation and survival pathways, or may have exerted a neutral influence on events. Environmental assessments determine contributing factors and environmental antecedents, and can also provide information on regulatory violations that may or may not have been relevant to the outbreak. As described earlier, an environmental assessment relies on consolidation and analyses of observation and assessment findings, laboratory results, traceback and epidemiological investigations.

Thus, an environmental assessment involves an iterative process that integrates laboratory and epidemiologic data as they evolve over the process of the outbreak investigation. Additionally, because environmental assessments identify the underlying factors causing illness, they can be used to develop science-based interventions. Finally, to be most effective, environmental assessments must be an integral component of a FBIO investigation to help holistically understand the problem at hand and develop interventions that can prevent the situation from recurring.

24.5.2 Economic aspects

Although specific economic impacts of the 2010 *E. coli* O145 outbreak have not been researched, estimates suggest the fresh produce industry suffered significant losses through recalls, lost sales and legal expenses. We can infer the loss by using the economic impact of the 2006 *E. coli* O157 outbreak in spinach, which resulted in an estimated farm-level loss in spinach sales of about $12 million and more than a $63 million loss in retail sales (Paggi, 2009). In addition to these direct costs to the industry, indirect healthcare costs associated with foodborne illness also must be recognized. The estimated total cost of foodborne illness to the USA is almost $152 billion a year including lost quality of life, medical costs and lost life expectancy (Scharff, 2010). From 1996 to 2007, approximately 72 reported outbreaks of foodborne illness were associated with approximately 20 fresh produce commodities, both domestic and imported. Of this total, 13 outbreaks were associated with tomatoes, 11 with melons and 24 with leafy greens such as lettuce and spinach. More illness is likely to result from each outbreak related to produce (an average of 48 cases) than from the average outbreak associated with poultry (30 cases), beef (27 cases) or seafood (10 cases).

24.6 Industry and market aspects

In the wake of the 2006 spinach-related outbreak, the leafy greens industry in California initiated California Leafy Greens Handler Marketing Agreement (CA LGMA) in 2007, which was followed by initiation of the Arizona Leafy Greens Shipper Marketing Agreement (AZ LGMA) later in 2007. These Leafy Greens

Marketing Agreements (LGMAs) consist of member companies that voluntarily comply with food safety practices that have been adopted from academic and industry research and are accepted by their respective boards. Compliance is verified by mandatory audits, and non-compliance with food safety practices is dealt with by issuing penalties. These industry-wide initiatives claim to represent diverse geographic and farming practices, including different sizes of operations and diverse methods of traditional or organic growing (LGMA, 2012).

Inexplicably, the lettuce grown on Ranch A complied with AZ LGMA's audit requirements. The investigation team shared its findings with the representatives of the industry in a conference call. It is expected that the findings of the environmental assessment will help the leafy greens industry adopt the principles of an environmental assessment in its field audit practices.

24.7 Addressing the outbreak

A multidisciplinary team consisting of a microbiologist, two food safety specialists and a hydrologist conducted an environmental assessment to characterize the Lower Gila watershed to identify the environmental antecedents and potential contributing factors that caused the *E. coli* O145 contamination of lettuce on Ranch A. The environmental assessment focused on key elements related to the environmental system (watershed, irrigation canal system, sources of contaminants), people (farm workers, users of irrigation canal network), equipment (growing, harvesting and irrigation) and processes (hydrologic, growing, harvesting, irrigation). Characterizing the watershed involved identifying its boundaries and analyzing hydrologic data, including flow data and rainfall data at USGS gauging stations on the Gila River upstream and downstream of Ranch A. The key findings from the hydrologic analysis of watershed conditions when the suspect crops were grown and harvested included the following:

- Surface water originally from the Colorado River from the WMIDD was the only source of water for the irrigation of the suspect crop.
- Cumulative precipitation was higher than average in the watershed during the late growing season. Consequently, this high volume of precipitation and the rainfall intensity in January may have created higher run-off potential and increased loading of microbial contamination to the canal network.
- No evidence of influence from the Gila River during high flow conditions was found on and around the farms. Similarly, no hydrologic connection was found that might have influenced the microbiologic quality of the water in the canal network from the Gila River.
- A working hypothesis was developed using all the data and information gathered from the site visit and all other pieces of the investigation (epidemiologic, regulatory visits, etc.). The hypothesis posits that potential run-off because of higher-than-average cumulative precipitation in the watershed might have created a pathway for land-based microbial contaminants

to contaminate the lettuce fields at Ranch A directly. Additionally, the evidence suggested that contamination also could have occurred indirectly by potential loading of microbes into the water in the irrigation canal.

Thus, a reconnaissance survey was conducted as part of the environmental assessment in the Lower Gila watershed to identify and evaluate potential sources of STEC. The focus of this survey was to identify areas with some evidence of a hydrologic connection to the fields at Ranch A. The survey revealed that the main Wellton-Mohawk irrigation canal coming from the Colorado River splits into two canals near the town of Wellton. This is known as the Wellton-Mohawk split; the Wellton Canal delivers water to the Wellton Valley where Ranch A is located, and the Mohawk Canal delivers water to the Mohawk Valley. Considering the hydrologic connection between the Wellton Canal and its lateral canals with Ranch A, the irrigation canal system and transportation pathways near the irrigation canal system were thoroughly inspected to identify any potential significant sources of STEC. The potential sources of contamination were determined based on evidence of run-off and flow direction. Soil and water samples were collected from the identified potential sources.

As a result of the reconnaissance survey and inspection of Wellton Canal and its lateral canals, the following potential permanent sources of STEC were identified and evaluated:

1. a Concentrated Animal Feeding Operation (CAFO) located east of Ranch A in the immediate vicinity of the Gila River;
2. a small dairy co-located with a small beef cattle feedlot sharing a storm water retention pond located west of Ranch A;
3. a subdivision located on a mesa on both sides of the main Wellton canal;
4. an RV park located on a mesa above the valley floor adjacent to and sloping towards a lateral of the Wellton irrigation canal.

The locations of potential sources of contamination are shown in Fig. 24.2. The CAFO and the small dairy co-located with a small beef cattle feedlot were inspected, and the most recent state inspection reports for the facilities were reviewed. No evidence was found of uncontrolled drainage from the CAFO, the dairy or the beef cattle feedlots, or of any other potential mode of escape for the STEC from the confines of the facilities, other than via birds and wind. The state inspection reports verified that the CAFO and the dairy were in compliance with waste-handling regulations (the number of cattle held at the beef cattle operation does not require a permit and it is not inspected by the state, thus no state inspection reports were issued).

Soil samples collected from the Gila River bed, both upstream and downstream of the natural potential drainage route from the CAFO, tested positive for Stx1- and Stx2-producing STEC, but were not a match for the outbreak strain. The source of the STEC contamination was not determined. Similarly, water samples collected from the storm-water–retention pond shared by the dairy facility and beef cattle facility also tested positive for Stx1- and Stx2-producing

STEC, but were not a match for the outbreak strain. The STEC in the storm-water–retention pond was likely shed by the cattle from one or both operations; however, this was not confirmed. Because a pathogen transportation mechanism could not be identified, these facilities were reasonably deemed unlikely sources of the outbreak contamination.

The subdivision located on a mesa on both sides of the main Wellton Canal was upstream from the lateral canal split that supplied irrigation water to Ranch A. The elevation of the irrigation canal was lower than that of the subdivision. The subdivision was using on-site, underground, activated sludge-treatment–package plants to treat its wastewater. Effluent from the treatment systems was used to irrigate the golf course and maintain two artificial ponds in the subdivision for aesthetic landscaping. The wastewater-treatment plant and ponds were located approximately 300 yards south of the main Wellton Canal, and no evidence of drainage from the subdivision into the Wellton Canal was observed. Water samples collected from one of the ponds tested negative for STEC. Also, an unsecured pump that appeared to draw water from a submerged inlet pipe in the Wellton Canal was located between the south side of the main Wellton Canal and the subdivision. Because of its location above the main Wellton Canal immediately upstream of the lateral canal split, and the submerged inlet plumbing in the Wellton Canal, this subdivision could not be completely ruled out as a potential source of the outbreak contamination.

The RV park was adjacent to the Wellton Canal upstream of the lateral canal that serviced the fields on Ranch A. The park also depended on on-site wastewater-treatment systems to treat its sewage using multiple septic systems with multiple drainage fields. Areas that sloped toward the irrigation canal along the side of the property adjacent to the Wellton Canal were conspicuous, and moist soil indicated drainage toward the canal. No surface source for the soil moisture was observed, indicating that the moisture was from a subterranean source and, thus, may be overflow from the RV park septic systems. Soil samples from these moist areas tested negative for STEC. Drag swabs and mud collected from the lateral irrigation canal around the RV park tested positive for Stx1- and Stx2-producing STEC, but were not a match for the outbreak strain.

The RV park was likely to be the source of the STEC detected in the drag swabs and mud collected from the adjacent lateral canal, but this could not be confirmed. Soil-survey findings for the RV park site demonstrated that the soil type was not suitable for septic absorption fields because of poor drainage capacity. This evidence suggested that the RV park potentially could have been the source of the outbreak contamination; however, no definitive confirming evidence was found.

In addition to the permanent sources of contamination described earlier, the investigation team heard anecdotal reports of sheep grazing on seasonal harvested alfalfa and hay fields in the vicinity of Ranch A in the fall and winter. However, on further investigation, these reports were found to be speculative and no definitive evidence or report of sheep grazing in the vicinity of Ranch A or near the canal providing water to Ranch A was found. Nevertheless, because of the lack of comprehensive information, seasonal sheep grazing cannot be completely ruled out as a potential source of the outbreak pathogen. In addition, an unsecured pump and hose on the main Wellton Canal at the lateral canal split supplied water to Ranch A and was another potential contamination point not associated with a permanent source of STEC. Erosion by run-off drainage into the main Wellton Canal was evident at this location; soil samples collected from the run-off pathway tested positive for Stx2-producing STEC, but did not match the outbreak strain. Because the pump and hose were unsecured, an RV owner could easily have dumped and rinsed out their vehicle's waste system into the main Wellton Canal, although this activity was neither reported nor observed during the investigation. Similarly, water discharging from a float chamber pipe that controls the gates that feed water at the Wellton-Mohawk split tested positive for Stx2-producing STEC, but did not match the outbreak strain.

24.8 Troubleshooting approaches and laboratory methods

The environmental assessment underscored the significance of a systems-based approach for investigating an illness outbreak to identify and enumerate environmental antecedents and contributing factors. These prominent potential environmental factors typically would not be explored by conventional investigation methods. Discovering the environmental antecedents and potential contributing factors to an outbreak may help regulators and the produce industry develop mitigation strategies to prevent outbreaks in the future. Critical environmental antecedents, including unusual weather events and human recreational activity, that were uncovered during the course of this environmental assessment greatly affected the microbiologic quality of waters used in agricultural production operations.

Potential sources of contamination found in the hydrologically connected environs of Ranch A constituted environmental antecedents that could have interacted and been influenced by other environmental antecedents. These environmental antecedents include excessive rainfall and soil conditions in the absence of controls such as monitoring septic systems drainage fields for overflow conditions. That interaction may have created conditions known as contributing factors (e.g. contaminated irrigation water) that might have caused adverse health outcomes related to this *E. coli* O145 outbreak.

The following potential troubleshooting strategies were recommended to reduce the potential for future contamination and outbreaks similar to the specific circumstances associated with this environmental assessment. These recommendations were shared with the relevant federal, state and local government agencies and the growers to help them initiate control and mitigation steps.

1. A thorough assessment of the RV park's septic system is needed to determine its integrity and the potential for drainage to the Wellton Canal during both normal and high rainfall circumstances. The assessment should identify

system-failure points and recommend adequate repair and maintenance to prevent failure in the future.
2. Grazing livestock should be restricted to fields that are not hydrologically connected to fresh produce fields or irrigation canals. This may necessitate installing barriers along canals located in grazing fields to prevent livestock from contaminating irrigation canals.
3. The canal water should be protected from contamination events such as misuse at locations with unsecured pump and hose facilities. To prevent contamination, pumps and hose facilities along the main Wellton Canal should be secured and inaccessible to unauthorized users.
4. Strategies and protocols for microbiologic monitoring of irrigation-water quality should be developed. A starting point may be increased water-quality monitoring by growers and harvesters during, and immediately after, an extreme precipitation event upstream of the water source.

24.9 Future trends

This outbreak investigation highlights the role of a comprehensive environmental assessment in identifying environmental antecedents and contributing factors. However, more research needs to be done to refine and develop a standard methodology for a field environmental assessment applicable to diverse situations and settings. Furthermore, contamination of fresh produce in the field can come from many sources, including water, soil and soil amendments, workers, wildlife and adjacent land. Little research has focused on assessing the spatial and temporal variability of pathogen occurrence in irrigation water as a function of watershed land uses and weather events. Therefore, research should be conducted to focus on: a thorough characterization of how rainfall and other weather events impact the microbial loads of a watershed; the typical background levels of microbial contaminants in the canal system; and all potential sources of contamination to the watershed. Such research may help identify conditions that may warrant extra protective or preventive measures, or both, by growers and harvesters who use the canal as their water source.

24.10 Questions for discussion

1. What is the difference between a traditional outbreak investigation and an environmental assessment investigation?
2. What are some of the challenges faced by investigators conducting a post-outbreak field environmental assessment?
3. How do you define an environmental antecedent?
4. How do you define an environmental contributing factor?
5. What is the role of environmental assessment as a food safety investigation and prevention tool?

6. Why is it necessary to incorporate a systems-based approach in disease outbreak investigations?
7. What is the difference between a regulatory inspection and an environmental assessment?
8. Why is it necessary to work with a team consisting of experts from a number of disciplines while conducting an environmental assessment?
9. How do watershed conditions affect the quality of irrigation water, and how may that quality affect produce safety?
10. How does this chapter help you understand and define the relationship between extreme weather events and food safety?

24.11 Acknowledgments

This chapter presents the findings of the field environmental assessment initiated by the FDA in collaboration with CDC in response to the *E. coli* O145 outbreak in 2010. The author was the CDC participant on the field environmental assessment team consisting of FDA officials Willette Crawford, PhD, and Kevin Gerrity. This book chapter is adapted from the field environmental assessment report available from the FDA's website (Crawford *et al.*, 2010).

I would also like to thank Richard Gelting, PhD, PE; Kristin Delea, MPH; Max Zarate-Bermudez, PhD; and Carol Selman, MPH, for their review and in-depth comments on the first draft of this book chapter. The following organizations provided assistance in conducting the environmental assessment described in this book chapter:

- California and Arizona Leafy Green Marketing Agreements;
- Arizona Department of Agriculture;
- Arizona Department of Environmental Quality;
- Arizona Department of Water Resources;
- Arizona Department of Health Services;
- Wellton-Mohawk Irrigation and Drainage District.

24.12 Sources of further information and advice

E-LEARNING ON HOW TO CONDUCT A FOODBORNE ILLNESS OUTBREAK ENVIRONMENTAL ASSESSMENT (US CDC) http://www.cdc.gov/nceh/ehs/eLearn/EA_FIO/index.htm.

ENVIRONMENTAL HEALTH SPECIALISTS NETWORK (US CDC) http://www.cdc.gov/nceh/ehs/EHSNet/.

ENVIRONMENTAL ASSESSMENTS (US FDA) http://www.fda.gov/Food/RecallsOutbreaks Emergencies/Outbreaks/ucm235425.

GELTING R. and M.A. BALOCH. 2012. The Food-Water Nexus: Irrigation Water Quality, Risks to Food Safety, and the Need for a Systems-Based Preventive Approach. *Journal of Environmental Health* 75 (3). http://www.cdc.gov/nceh/ehs/Docs/JEH/2012/Oct_Food_Water_Nexus.pdf (last accessed on 12 October 2012).

TODD E.C.D., C.A. BARTLESON, J.J. GUZEWICH, A. TAN, M. LEE, and M. NAZAROWEC-WHITE, 2011. Procedures to Investigate Foodborne Illness, Sixth Edition 2011. Des Moines, International Association for Food Protection.

24.13 References

AZDWR ARIZONA DEPARTMENT OF WATER RESOURCES. 2009. Arizona Water Atlas, Volume 7, Lower Colorado River Planning Area.

CALFERT – CALIFORNIA FOOD EMERGENCY RESPONSE TEAM, 2008. Investigation of the Taco John's *Escherichia coli* Associated with Iceberg Lettuce – Final Report. California Department of Public Health, Food and Drug Branch Sacramento, CA 95899–7435.

CDC. 2012a. Timeline of Infections: Multistate Outbreak of *E. coli* O145 Infections Associated with Romaine Lettuce—United States, 2010. http://www.cdc.gov/ecoli/2010/ecoli_O145/EcoliO145Timeline.pdf (last accessed on 28 August 2012).

CDC. 2012b. Investigation Update: Multistate Outbreak of Human *E. coli* O145 Infections Linked to Shredded Romaine Lettuce from a Single Processing Facility Updated May 21, 2010 (FINAL Update). http://www.cdc.gov/ecoli/2010/ecoli_o145/index.html (last accessed on 28 August 2012).

CRAWFORD W., M.A. BALOCH, and K. GERRITY. 2010. Environmental Assessment: Non-O157 Shiga Toxin-Producing *E. coli* (STEC). http://www.fda.gov/Food/RecallsOutbreaks Emergencies/Outbreaks/ucm235477.htm (last accessed on 30 August 2012).

DELEA K.C., C. SELMAN, and EHS-NET. 2008. Understanding Foodborne Disease Outbreaks Using Environmental Assessments. Poster presentation at the International Conference on Emerging Infectious Disease Annual Meeting, Atlanta, GA, 16–19 March 2008.

FRATAMICO P.M., C. DEBROY, T. MIYAMOTO, and Y. LIU. 2009. PCR detection of enterohemorrhagic *Escherichia coli* O145 in food by targeting genes in the *E. coli* O145 O-Antigen gene cluster and the Shiga toxin 1 and Shiga toxin 2 genes. *Foodborne Pathogens and Disease*. 6(5):605–11.

GELTING R.J., M.A. BALOCH, M.A. ZARATE-BERMUDEZ, and C. SELMAN. 2011. Irrigation water issues potentially related to the 2006 multistate E. coli O157:H7 outbreak associated with spinach. *Agricultural Water Management*. 98:1395–402.

GELTING R., J. SARISKY, C. SELMAN, C. OTTO, C. HIGGINS, *et al.* 2005. Use of a systems-based approach to an environmental health assessment for a waterborne disease outbreak investigation at a snowmobile lodge in Wyoming. *Int. J. Hyg. Environ. Health*. 208(1–2):67–73.

HIGGINS C.L. and B.S. HARTFIELD. 2004. A systems-based food safety evaluation: an experimental approach. *J. Environ. Health*. 67(4):9–14.

LGMA – THE CALIFORNIA LEAFY GREENS MARKETING AGREEMENT. 2012. http://www.caleafygreens.ca.gov/about-us/faqs (last accessed on 5 September 2012).

PAGGI S.M. 2009. Testament to the United States Department of Agriculture Hearings on the Proposed National Marketing Agreement No. 970 Leafy Green Vegetables Handled in the United States, 23 September 2009, Monterey, California.

SCHARFF R.L. 2010. Health-Related Costs from Foodborne Illness in the United States. Washington: The Produce Safety Project at Georgetown University, an Initiative of the Pew Charitable Trusts.

SELMAN C. and J. GUZEWICH, Unpublished. Foodborne Illness Outbreak Environmental Assessment.

USDA – US DEPARTMENT OF AGRICULTURE. 2012. Vegetables 2011 Summary. http://usda.mannlib.cornell.edu/MannUsda/viewDocumentInfo.do?documentID=1183 (Last accessed on 30 August 2012).

WMIDD – WELLTON-MOHAWK IRRIGATION AND DRAINAGE DISTRICT. 2010. Available at: http://wmidd.org (accessed 30 July 2010).

25

The case of the European *Escherichia coli* outbreak from sprouts*

P. Luber, Federal Office of Consumer Protection and Food Safety (BVL), Germany

DOI: 10.1533/9781782420279.5.356

Abstract: This chapter describes an investigation into the European *Escherichia coli* outbreak in May–June 2011. An outbreak caused by the pathogenic Shiga-toxin-producing *E. coli* O104:H4 occurred in Germany, causing the highest reported cases of haemolytic-uremic syndrome in humans associated with a single outbreak. From the start of the outbreak, fresh produce had been suspected as the cause; however, special efforts and tools were needed to narrow the list of suspicious vegetables to sprouts. A second small outbreak caused by the same pathogen in France initiated a European investigation, which resulted in evidence that both outbreaks were caused by contaminated fenugreek seeds.

Key words: STEC, outbreak, crisis management, tracing, sprouts, sprouting seeds.

25.1 Introduction: investigation of foodborne disease outbreaks in Germany

Germany is a Federal Republic made up of 16 federal states (Bundesländer) which are individually responsible for the implementation of European regulations and directives in the area of food safety, including the investigation and control of foodborne outbreaks. In outbreak situations where more than one Bundesland is affected, federal food safety and public health authorities also participate in the outbreak investigation. A crisis situation centre at the Federal Office of Consumer Protection and Food Safety (BVL) is activated, which takes care of crisis

---

* **Disclaimer**: The findings and conclusions in this publication are those of the author alone and do not necessarily represent the views, decisions or policies of the Federal Office of Consumer Protection and Food Safety.

management on the food safety side of the outbreak investigation. The information gathered by the federal states during an outbreak investigation, such as results obtained from the microbiological analysis of foods, protocols from risk-based inspections of food business operators and food supply chain information, will be sent to BVL. Crisis managers at BVL then collate and analyze the data to produce an overview of the outbreak situation and report to the Federal Ministry of Nutrition, Agriculture, and Consumer Protection (BMELV) and the European Commission.

Different tools are used to identify contaminated food(s). On the public health side of the outbreak investigation, the Robert Koch Institute (RKI) performs interviews with patients and, depending on the situation, conducts epidemiological case-control or cohort studies to identify the causative food(s). On the food safety side, BVL coordinates the supra-regional outbreak investigation. The responsible authorities of the federal states undertake risk-based inspections of food business operators, sample and analyze foods and analyze food supply chains in order to identify the contaminated food that caused the outbreak.

After a contaminated food has been found, the second step is to identify the source of contamination. To find out where and when the causative food was contaminated with a pathogen, the responsible authorities of the federal states perform risk-based inspections including sampling and microbiological analyses along the food chain. Together with food business operators, the food safety authorities take measures to address the source of contamination. Moreover, results from the investigation of foodborne outbreaks are used to implement preventive measures and improve food safety to avoid future outbreaks.

25.2 The German *E. coli* outbreak from detection to highly likely clarification

On Friday 20 May 2011, the German human health authorities detected several human EHEC cases caused by an infection with the Shiga-toxin-producing *E. coli* of the serotype O104:H4. Cases occurred in several federal states in northern Germany and a foodborne outbreak was suspected. The Robert Koch Institute (RKI) informed the national food safety management authority BVL, who activated the crisis situation centre and informed the responsible food safety authorities of the 16 German federal states. The outbreak investigation and the search for the contaminated food began immediately. That weekend, a team of RKI epidemiologists travelled to Hamburg, where a large number of cases had been hospitalized, to interview patients and to start a case-control study. The food safety authorities performed risk-based inspections at outbreak clusters (e.g. restaurants, hotels and canteens). Information on suppliers of foods that had been used at clusters was collected, and samples from foods and the environment were taken for microbiological analysis. Moreover, food samples for microbiological analysis were taken from wholesalers and retail outlets in order to identify the contaminated food that had caused the outbreak.

On Wednesday 25 May 2011, the results of the RKI case-control study became available. The epidemiological study revealed that significant numbers of the EHEC cases had eaten more cucumbers, tomatoes and lettuce than healthy control persons. As a first measure to stop the outbreak, that same day a consumption advice was published asking consumers to avoid consumption of these three kinds of vegetables in northern Germany. Food supply chain analysis on the food safety side of the outbreak investigation showed that many of the cucumbers, tomatoes and lettuces that had been consumed at outbreak clusters had been traded at the wholesale market in Hamburg. The food safety authority of Hamburg took samples of fruits and vegetables at the market and identified Shiga-toxin-producing *E. coli* (STEC) on some cucumber samples. On Monday 30 May 2011, a large group of food safety inspectors did a sweeping inspection of the wholesale market and took large numbers of environmental samples. None of these samples tested positive for the outbreak strain so it was possible to rule out that an environmental contamination of vegetables on the market had occurred. Furthermore, molecular typing of the STEC strains isolated from cucumbers later revealed that these pathogens had a different serotype from the O104:H4 STEC outbreak strain, and thus the cucumbers were not the cause of the outbreak.

25.2.1 The problem of identifying the food causing the outbreak

Meanwhile the number of identified human EHEC cases rose steadily, and the first infected patients had died. Crisis managers, food inspectors and microbiological laboratories worked around the clock but could not find STEC O104:H4 in any of the many hundreds (later these became several thousands) of food samples they analyzed. Mysteriously, the outbreak strain was not found at any of the outbreak clusters, and risk-based inspections along the food chain were without positive findings. Ongoing questioning of EHEC patients did not reveal new insights about common food consumption patterns. Ten days into the outbreak investigation, the need to identify the food causing the outbreak had become increasingly urgent.

Thus, to develop new approaches for the food safety side of the outbreak investigation, the German Federal Ministry of Nutrition, Agriculture, and Consumer Protection (BMELV) called into action an EHEC taskforce. This taskforce brought together experts from the Federal States (Lower Saxony, Schleswig-Holstein, Mecklenburg-West Pomerania, Hamburg, Bavaria), from BVL and from the Federal Institute of Risk Assessment (BfR). The work of the taskforce was coordinated by BVL. Additionally, experts from the RKI, scientists from the European Food Safety Authority (EFSA) and risk managers from the European Commission participated in the taskforce.

The EHEC taskforce developed two new strategic approaches to identify the contaminated food. Both strategies were executed in parallel. The first approach was a detailed traceback analysis for five outbreak clusters where several customers developed the EHEC disease. All of the salad ingredients and toppings that had been used at each cluster during the exposure timeframe were traced back along the food supply chains. The goal of this approach was to identify common food sources and supply chains for all five outbreak clusters.

The second approach was a traceforward analysis of the supply chains of a suspicious sprout producer located in Lower Saxony. During the early stages of the outbreak, investigation by the food safety authorities of Lower Saxony had found that this producer had delivered sprouts to several outbreak clusters. Although sprout samples that were available at the time when the inspections took place tested negative for the outbreak strain, the taskforce decided to further analyze the supply chains of the sprout producer in order to find out if he had delivered sprouts to any other outbreak clusters (increasing numbers of clusters were identified over time by the public health side of the outbreak investigation).

25.2.2 Seed sprouts

After 4 days of investigative work, the EHEC taskforce had identified 26 outbreak clusters where sprout mixtures from the suspicious producer had been delivered. At the same time, results from a cohort study performed by RKI for one of the restaurants associated with several EHEC cases became available (Robert Koch Institut, 2011). This epidemiological study revealed that only the consumption of sprouts at this restaurant was significantly associated with the development of EHEC disease. Based on these new findings, the presidents of BVL, BfR and RKI called a press conference on Friday 10 June 2011 to inform the public about the results of the outbreak investigation and to give new consumption advice. Consumers were advised to avoid eating raw sprouts, while cucumbers, tomatoes, and lettuce were declared free of suspicion.

On 14 June 2011, the traceforward analysis was complete. It revealed that 41 outbreak clusters were linked with sprouts from the producer in Lower Saxony. The detailed traceback analysis for five outbreak clusters revealed that during the respective exposure timeframe, all clusters had used either the sprout mixture 'Milde Mischung' or the mixture 'Würz Mischung', which had been delivered by the suspicious sprout producer.

In conclusion, both new investigative tracing approaches taken by the EHEC taskforce revealed that contaminated sprouts from a producer in Lower Saxony most likely had caused the outbreak.

25.2.3 The cause of the sprout contamination

Although evidence from tracing the food supply chains indicated that sprouts were the cause of the outbreak, it was not clear at this point how the sprouts had been contaminated. The experts of the EHEC taskforce analyzed the available data, and concluded that seeds or beans used for sprouting, the water used in the sprouting process, the environment at the sprout processor or the personnel could have been the source of the contamination. The responsible food safety authorities of Lower Saxony took numerous samples from the food business operator for microbiological analysis, and the public health authorities questioned and tested the personnel. However, not all suspicious seeds could be tested for *E. coli* O104:H4. Some of the batches of seeds that had been used during the presumptive production time that was associated with the outbreak had been used up and were not available for testing. As a precaution, a traceback analysis was performed for all batches of seeds and beans that had been used to produce sprouts for the two sprout mixtures that had been delivered to the outbreak clusters. All German sprout-producing businesses were contacted by the respective responsible food safety authority to find out if any had the same batches of seeds or beans in stock as those in use at the processor that was highly likely to have caused the outbreak. From this, should a positive result be found in the ongoing testing of seeds and beans from the sprout producer in Lower Saxony, quick management action would be possible. The results from this traceback analysis of seeds proved to be very valuable later, when a second outbreak occurred and the European outbreak investigation started.

A detailed report on the work of the EHEC taskforce with a network graphic of the sprout supply chains and several figures has been published in the *Journal of Consumer Protection and Food Safety* (Die Task Force EHEC, 2011).

25.3 Challenges encountered during the clarification of this outbreak

The epidemiological curve of the STEC O104:H4 outbreak and further epidemiological analysis of the case have shown that this was a point source outbreak where the contaminated food was on the market for only a short period of time (Robert Koch Institut, 2011; Die Task Force EHEC, 2011). The outbreak investigation was hindered by the fact that the causative agent was characterized by an unusually long incubation period. On average, there was a period of 8 days between ingestion of the contaminated food and the development of bloody diarrhoea. Thus, the outbreak itself and clusters where several people were infected were detected relatively late. The most likely source of the outbreak was contaminated fenugreek sprouts.

In general, plant foods have a very short shelf-life. By the time the food safety authorities had learned of the outbreak and inspected outbreak clusters such as restaurants, the contaminated food was no longer there. The contaminated sprouts had either been used up or perished and had been discarded. During the outbreak investigation, thousands of food samples had been analyzed but none were positive for the outbreak strain of STEC O104:H4. A possible explanation might be that there were no more contaminated sprouts on the market or available as retain samples in restaurants on which to do microbiological testing.

Furthermore, challenges were encountered on the public health side of the outbreak investigation. Sprouts are usually not the main ingredient in salads and were often being used as decorative elements only, for example on sandwiches. When cases were questioned about the food they had consumed in the time before the onset of disease, they often did not recall that they had eaten sprouts. The large proportion of women among the cases, trawling questionnaires and the case-control study performed during the first days of the outbreak investigation indicated that a salad ingredient or topping was the cause of the outbreak. However, because of the specific situation, neither of the classical tools for outbreak investigations (microbiological testing and epidemiological studies with questionnaires) worked to identify the causative food.

25.4 Significance of the pathogen that caused the case

The STEC O104:H4 outbreak strain had novel epidemiologic and clinical features. A higher proportion of infected persons developed haemolytic-uremic syndrome (HUS) than described in other STEC outbreaks. Despite HUS being considered a disease that affects primarily children, the majority of cases in the outbreak were adults (Robert Koch Institut, 2011). Molecular analysis of the outbreak strain revealed that it was not a classical enterohaemorrhagic *Escherichia coli* (EHEC), but an enteroaggregative haemorrhagic *E. coli* (EAHEC). The combination of a strong attachment to the intestine with the production of Shiga toxins made this outbreak strain particularly fatal.

25.5 Burden of disease and geographical and regional significance

This outbreak of EHEC infections was the largest recorded to date in Germany and, based on the high number of HUS cases, this is so far the largest reported outbreak of its kind worldwide. A total of 855 cases of HUS and 2987 cases of EHEC gastroenteritis (without development of HUS) were reported; hence a total of 3842 cases were linked to the outbreak in Germany. Among the HUS cases, 68% were female and the median age was 42 years (range from 0 to 91 years). Among HUS patients, 35 (4.1%) died and among the patients with EHEC gastroenteritis, 18 (0.6%) died. Within the outbreak time period, cases were reported from all 16 German federal states; however, the five most northern federal states (Hamburg, Schleswig-Holstein, Bremen, Lower Saxony and Mecklenburg-Western Pomerania) were most severely affected (HUS disease incidences from 1.8 to 10.0 cases per 100 000 persons – all other federal states had incidence rates <1 cases per 100 000 persons).

France reported a local EHEC O104 outbreak in Bordeaux that, interestingly, was not associated with any travel to Germany. This outbreak comprised 15 cases. Additional cases that have been linked directly or indirectly to travel into Germany also occurred in Denmark, Greece, the UK, Luxembourg, the Netherlands, Norway, Austria, Poland, Sweden, Spain, the Czech Republic, Canada, Switzerland and the USA (Robert Koch Institut, 2011).

25.6 Regulatory aspects

The European legal framework to prevent, monitor and manage foodborne outbreaks is multidisciplinary. It contains requirements for public health and food safety managers, risk assessors and risk communicators. The need for coordination between these professionals and between different levels (local, national, EU and international) is laid down in EU law. An overview is provided in Annex 3 of the DG SANCO Commission staff working document SANCO/13004/2011 (2012). The general principles, including the risk analysis concept, the precautionary principle and the protection of consumer interest, are laid down in the general food law, Regulation (EC) No 178/2002. Moreover, the general food law provides rules on the management of food crises, the adoption of emergency measures and the operation of the Rapid Alert System for Food and Feed (RASFF). The general rules for prevention of outbreaks through hygiene requirements are outlined in Regulation (EC) No 852/2004 – Hygiene of foodstuffs, which is part of the hygiene package of regulations. The food hygiene regulations are supported by the EU Regulation on microbiological criteria, Regulation (EC) no 2073/2005 (as amended). In this Regulation on microbiological criteria, a *Salmonella* criterion for sprouted seeds, an *E. coli* criterion for ready-to eat pre-cut fruit and vegetables and criteria for *Listeria monocytogenes* in ready-to-eat foods have been established.

The Directive 2003/99/EC on the monitoring of zoonoses and zoonotic agents ensures that monitoring of foodborne pathogens is carried out at the most relevant stage of the production chain, and that results from the member states are reported and submitted to EFSA for trend analysis. The directive also sets out the legal base for investigating foodborne outbreaks and requires member states to ensure communication at the national level with all involved sectors (public health, food safety, feed, etc.) and to keep the commission updated on the respective national contact points.

The human health community network under Decision No 2119/98/EC of the European Parliament and of the Council of 24 September 1998 has set up a network for epidemiological surveillance and control of communicable diseases in the community. This network oversees epidemiological surveillance and early alert and response coordination to serious events caused by communicable diseases. The Early Warning and Response System application (EWRS) allows for the sharing of information which is essential in the notification of outbreaks, and coordination of measures between the member states and the commission to respond to serious cross-border events caused by communicable diseases.

25.7 Market and economic issues

The industrial and market aspects of the case have been evaluated by the European Commission (DG SANCO, 2012). A detailed economic analysis of the outbreak is presented in Annex 2 of this document. During the first 2 weeks of the outbreak, the losses for European farmers in the fruit and vegetable sector caused by lost sales, low prices and overcapacity were estimated to be at least 812 million €. In addition to this, a temporary export ban of vegetables from the European Community to Russia occurred, constituting an annual value of 600 million €. This global estimation refers to the first 2 weeks of the crisis only. Thus, the total economic loss was expectedly higher. Arguably, the biggest damage was to the image of fresh produce. Consumers usually associate fruit and vegetables with healthy nutrition, not with food poisoning. The European Commission supported the fresh vegetables producers with exceptional measures on market intervention for a total value of 227 million €. Such measures concerned the products most directly affected by the crisis, that is tomatoes, cucumbers, lettuces and certain endives, courgettes and sweet peppers. Significant numbers of consumers had refrained from buying these vegetables from the beginning of the outbreak because of the uncertainty of the source and because of recommendations not to consume these products raw. In the aftermath of the outbreak, several promotion activities were also launched in order to win back consumer trust in fruits and vegetables.

25.8 The second *E. coli* O104:H4 outbreak in France

All of the hundreds of samples that were available at the sprout producer in Lower Saxony when the outbreak investigation started, tested negative for the outbreak strain. Thus, only the strong evidence resulting from the traceback and traceforward analysis was at hand to prove that sprouts had caused the *E. coli* O104:H4 outbreak in Germany. Then a second disease outbreak with the same pathogen happened in France.

On the evening of Friday, 24 June 2011, a phone call from the commission reached the crisis situation centre at BVL. A sprout-related *E. coli* O104:H4 outbreak had been reported in France. Germany was requested to contact the French authorities. By this time, the outbreak in Germany had been under investigation for 4 weeks. German authorities telephoned their French colleagues and, within a couple of minutes, fenugreek seeds had been identified as the main suspect for being the vehicle of the outbreak strain.

The spread of the outbreak to France had occurred at a social event, in which three kinds of sprouts had been on the menu. The sprouts had been produced in a private setting and the seeds used for sprouting had been bought at a retail outlet. Initial investigations in France showed that the sprouting seeds had been imported from the UK. No travel or any other connection existed to the outbreak in Germany. During the telephone call it became apparent that there was only one overlap between the lists of suspicious sprouts that had been consumed in Germany and France: fenugreek sprouts. Thus, the fenugreek seeds that had been used in both outbreak settings most likely were the common source of *E. coli* O104:H4.

The commission immediately started a European outbreak investigation and gave a mandate to EFSA to coordinate a Europe-wide traceback of the seeds that had been used for sprouting in France and Germany. First, priority was given to the traceback of fenugreek seeds. Now it proved helpful that the German authorities had the traceback information already at hand. On 28 June, it became clear that fenugreek seeds at both outbreak sites had been delivered by the same supplier in Germany, who had imported three batches of fenugreek seeds from Egypt (European Food Safety Authority, 2011). The next morning EFSA and ECDC published a joint rapid risk assessment which concluded that fenugreek seeds exported from Egypt had caused both *E. coli* O104:H4 outbreaks. After the traceback of fenugreek seeds was finished and a report was issued, European risk management measures were taken. On 6 July 2011 a Commission Decision on emergency measures applicable to fenugreek seeds and certain seeds and beans imported from Egypt was published. The implicated batches of fenugreek seeds were recalled and removed from the market across Europe.

Although the outbreak was now clarified and risk management measures were taken, the investigation did not stop. Ongoing analyses and research into what had happened revealed, for example, that the outbreak strain was not an EHEC strain and that it was characterized by an unusually long incubation period. The outbreak investigators in France conducted a retrospective cohort study among all adults, including food handlers, attending the outbreak event on 8 June 2011 near Bordeaux. The result of this cohort study was that fenugreek was the only sprout type with an independent association to illness (RR 5.1) in multivariable analysis (King *et al.*, 2012). Although it was not possible to generate a positive microbiological test result for STEC O104:H4 in sprouts or fenugreek seeds during the investigation of both outbreaks, the sum of epidemiological evidence showing that fenugreek sprouts produced from contaminated seeds had caused the outbreaks is overwhelming.

For both outbreak settings, cohort studies prove that only the consumption of (fenugreek) sprouts at the clusters was significantly associated with the disease. The backward and forward tracing of sprouts completed during the German outbreak investigation clearly links outbreak clusters with deliveries of sprouts from the producer in Lower Saxony. Finally, this sprout producer had used fenugreek seeds for sprouting that originated from the same batch as the seeds that had been used for sprouting in the French outbreak setting.

25.9 Conclusion

Every foodborne outbreak is different, and several tools are available for their investigation and clarification. Sometimes epidemiological investigations, such as questioning of patients, deliver hints for the identification of contaminated foods, whereas in other cases positive test results from a microbiological laboratory help to identify the cause of an outbreak. Tracing of foods and the analysis of food supply chains is an important element of any outbreak investigation. However, during the acute search for the contaminated food and cause of the EHEC outbreak in Germany, the newly developed outbreak investigation strategy of the EHEC taskforce, with the use of epidemiological tracing tools to analyze food supply chains, was the only strategy that worked to identify the contaminated food.

This newly developed tool of a detailed traceback analysis for food batches that have been used at outbreak clusters during the exposure timeframe in order to identify common foods and/or supply chains (delivery knots) is deemed very useful for the clarification of outbreaks. Traceforward and backward analyses can be specifically used in the retrospective clarification of point source outbreaks which had been caused by plant foods and other perishable foods with a short shelf-life that are no longer available for microbiological testing once an outbreak investigation starts.

The close collaboration between European member states, European authorities and between food safety authorities, health authorities and scientists has enabled the identification of sprouts as the cause of this outbreak. In conclusion, the approach taken in 2011 for the investigation of the European *E. coli* outbreak was a recipe for success and can be a model for future foodborne outbreak investigations.

25.10 Questions for discussion

1. Why do epidemiologists analyze the gender and age distribution of cases during an outbreak investigation?
2. What information can be derived from an epidemiological curve?
3. What are the available tools for outbreak investigations to identify the causative food?
4. Why is typing of microorganisms important during outbreak investigations?
5. Which microbial (human) pathogens are often associated with foods of plant origin? Why is this?
6. What are the risk management options for controlling microbiological risks during sprout production?
7. What is the precautionary principle, and what role did it play during the 2011 European *E. coli* outbreak?
8. Although hundreds of samples from the processing environment, process water and seeds used for sprouting taken at the producer in Lower Saxony have been tested, the outbreak strain was not detected. Does this rule out the possibility that sprouts produced in these premises had caused the outbreak?
9. Which authorities or institutions are involved in foodborne outbreak investigations in your country? Describe their roles.

25.11 Sources of further information and advice

More information about the European aspects of the outbreak investigation including, for example, detailed publications on the traceback of fenugreek seeds and background information on Shiga-toxin-producing *E. coli* can be found on the homepage of EFSA, the European Food Safety Authority: http://www.efsa.europa.eu/en/topics/topic/ecolioutbreak2011.htm.

An international standard addressing the hygienic production of sprouts can be found in the Codex Alimentarius: Codex Alimentarius (2003). Code of hygienic practice for fresh fruits and vegetables. Annex II: Annex for sprout production. CAC/RCP 53-2003. http://www.codexalimentarius.org/standards/list-of-standards/en/?provide=standards&orderField=fullReference&sort=asc&num1=CAC/RCP.

The World Health Organization has published guidelines showing best practice approaches for the investigation and control of foodborne outbreaks: WHO (2008). Foodborne disease outbreaks: Guidelines for investigation and control. http://whqlibdoc.who.int/publications/2008/9789241547222_eng.pdf.

25.12 References

DG SANCO (2012). Commission staff working document SANCO/13004/2011: Lessons learned from the 2011 outbreak of Shiga toxin-producing Escherichia coli (STEC) O104:H4 in sprouted seeds. http://ec.europa.eu/food/food/biosafety/salmonella/docs/cswd_lessons_learned_en.pdf.

DIE TASK FORCE EHEC: B. BERNARD, M. BISPING, B. BROSCHEWITZ, M. BUCHER, A. FETSCH, *et al.* (2011). Ergebnisbericht der Task Force EHEC zur Aufklärung des EHEC O104:H4 Krankheitsausbruchs in Deutschland. *Journal für Verbraucherschutz und Lebensmittelsicherheit* 6:483–95.

EUROPEAN FOOD SAFETY AUTHORITY (2011). Shiga toxin-producing *E. coli* (STEC) O104:H4 2011 outbreaks in Europe: Taking Stock. *EFSA Journal* 9(10):2390–422 doi:10.2903/j.efsa.2011.2390. www.efsa.europa.eu/efsajournal.

KING, L.A., F. NOGAREDA, F.-X. WEILL, P. MARIANI-KURKDJIAN, E. LOUKIADIS, *et al.* (2012). Outbreak of shiga toxin-producing *Escherichia coli* O104:H4 associated with organic fenugreek sprouts, France, June 2011. *Clinical Infectious Diseases* 54(11):1588–94.

ROBERT KOCH INSTITUT (2011). Final presentation and evaluation of epidemiological findings in the EHEC O104:H4 Outbreak Germany 2011. http://www.rki.de/EN/Home/EHEC_final_report.html.

26

Case study on the safety and sustainability of fresh bottled coconut water

E. H. M. Walter, Embrapa Food Technology, Brazil, A. Y. Kuaye, State University of Campinas, Brazil and J. Hoorfar, Technical University of Denmark, Denmark

DOI: 10.1533/9781782420279.5.367

Abstract: This chapter presents a case study on the safety and sustainability of fresh coconut water in Brazil, now widely available in convenient plastic bottles. The authors intend to show that fresh coconut water safety and liability insurance is an opportunity and responsibility for the Brazilian government and society, as well as for other tropical countries. The safety of fresh coconut water is not only a subject of public health, but also an image presentation of a healthy and pleasant product that generates revenue for many street workers and micro-entrepreneurs.

Key words: coconut water, bottled fresh juice, microbiology, safety, sustainability.

26.1 Introduction

Coconut water is one of the main beverages enjoyed in tropical countries, especially in the coastal regions, where palm trees are part of the landscape. The beverage is mildly sweet with a unique aerated feel when freshly extracted from young green coconuts. In addition, it is highly appreciated as a natural isotonic drink by sports practitioners in Brazil. Table 26.1 details coconut water's comparative composition with commercially available sport drinks.

The World Health Organization (WHO, 2010) recommends ingestion of coconut water as a source of potassium in rehydration treatment in cases of cholera with acute diarrhea. As a result, these market opportunities have created an increasing consumer demand, particularly for freshly made coconut water. The beverage therefore has major economic, societal and environmental impacts,

Table 26.1 Carbohydrates and minerals in coconut water compared with a sports drink (mg/100 ml)

Parameter	Coconut water*	Sports drink†
Carbohydrates	4710	5800
Calcium	27	1
Phosphorus	5	9
Sodium	2	46
Potassium	204	8
Magnesium	6	3

* Beverage obtained from young green fruits (Yong *et al.*, 2009). Like any fruit, soil and climatic conditions of production may affect its composition. Coconut water is considered a fruit juice which is directly extracted from the coconut without expressing the coconut meat (endosperm), and is therefore different from coconut milk.

† Taken from FAO (2007). Composition may change according to the producer.

which have led to research into the safety of appropriate packaging and preservation of fresh coconut (Walter *et al.*, 2009), which is the focus of this chapter.

In Brazil, coconut palm trees (*Cocos nucifera* L.) are found in two main varieties: giant and dwarf. Each variety is subdivided into green, yellow and red varieties; however, the coconut water from the green variety is the most consumed as it is sweeter than that from the yellow and red varieties. The young (tender) coconut (7 months) is preferred over the mature fruit, which is ripened for 12 months. The young fruit also has a larger volume of liquid.

Although thermo-processing, ultra-filtration and other technologies can render coconut water a microbiologically safe product with an extended shelf-life, the freshly made beverage obtained straight from the fruit in nature is preferred by consumers. The most popular form is when the beverage is consumed through a straw from the green coconut; however, this presents difficulties as the green coconut pericarp (husk) is large and heavy, weighing approximately 60% of the fruit mass. 'The best we can get is the water naturally packaged in the green coconut; however, it is not the best to take.' As a result, an alternative approach has been developed, where fresh coconut water is conveniently filled in plastic bottles and these have been made widely available throughout Brazil.

Although fresh coconut water was initially restricted to consumption in beach areas, this development has allowed for the consumption of the fresh beverage all over the country, and Figs 26.1–26.3 show examples of places where fresh coconut water is sold in bottles. Increasingly, the world market is starting to get a taste for the industrialized coconut water; the fresh beverage is becoming part of the back-to-nature eating trend. Another major advantage of this premium beverage in Brazil is the price, which is similar to, and in some cases less expensive than, the industrialized one.

Fig. 26.1 Fresh bottled coconut water on the street.

Fig. 26.2 Fresh bottled coconut water in a supermarket.

Fig. 26.3 Fresh bottled coconut water in a shopping mall.

The FAO has sponsored research projects in other coconut-producing countries and revealed similar scenarios, one of which culminated in a training guide entitled *Good practice for the small-scale production of bottled coconut water* (Rolle, 2007). In Brazil, Embrapa has generated several agricultural and processing technologies for fresh and industrialized coconut water. Recommended documents are presented in Aragão, 2002; Cabral *et al.*, 2005; Fontes *et al.*, 2003.

26.1.1 Characterization of coconut water

Coconut water is considered to be sterile while in the hermetic cavity of fruits; however, the beverage is susceptible to microbial contamination during extraction and refrigeration. Furthermore, it is particularly prone to the survival and growth of microorganisms, especially bacteria, because its main constituents are sugars and minerals.

Coconut water from the young green coconuts of dwarf palms is approximately 5% sugar (fructose, glucose, sucrose and sugar alcohols) and 1% ash (mainly potassium, calcium, magnesium, phosphorus and sodium) when the fruit reaches 7 months of maturation, the ideal point of harvest. Coconut water is necessary for the maturation and germination of the fruit, and its composition changes during these processes. Vitamins, lipids, amino acids, nitrogenous compounds, organic acids, enzymes and phytohormones are also present but in low concentrations (Aragão, 2002; Yong *et al*., 2009).

The sugar contents and other soluble solids in coconut water provide carbohydrates for microbial growth and, together with the minor constituents, play different roles in microbial as well as plant and human systems, because of their distinct biological properties. Finally, it is worth noting the pH (4.6–5.6) and the water activity (≈0.99) of coconut water, which classify it as a low-acid/high-water activity product that supports rapid bacterial growth under optimal conditions.

26.2 Safety control challenges and foodborne outbreaks

26.2.1 Safety control challenges

The main challenges for the microbial control of bottled fresh coconut water have been in defining and applying a safety strategy. The country of Brazil is as large as a continent, and has a complex production chain. The fresh beverage is prepared and sold in a wide variety of facilities and infrastructures, including snack bars, kiosks, stalls and mobile carts by street vendors. Coconut water is also processed in industries with different production scales and technologies.

As previously mentioned, although coconut water is considered to be sterile inside the fruit, it can become contaminated during the extraction, refrigeration or packaging processes. Figure 26.4 shows a typical process flow diagram for bottled fresh coconut water. There are many possible routes of contamination of coconut water from air and different surfaces of contact as coconut water can become contaminated by any microorganism, including foodborne pathogens, and there are usually no inactivation steps to eliminate the pathogens. This is, in fact, the standard challenge for all fresh produce.

It is also worth noting that the initial sterility of coconut water can result in a very low microbial background flora during the production of the bottled beverage.

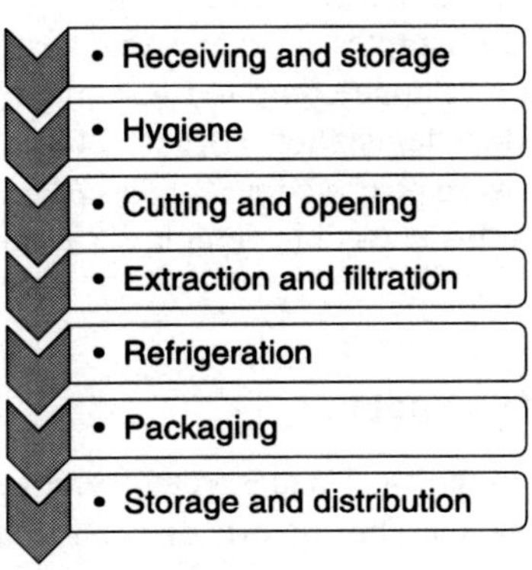

Fig. 26.4 Process flow diagram for bottled fresh coconut water.

This reduces microbial competition and leaves substantial space for pathogens in an environment without antimicrobial substances that are produced by spoilage microorganisms. In theory, this is a scenario of an unspoiled product that can, in extreme cases, become unsafe for consumption.

26.2.2 Foodborne outbreaks associated with coconut

Different foodborne outbreaks associated with the consumption of some forms of coconut products have been reported internationally (Strawn *et al.*, 2011). A *Salmonella* outbreak in South Wales was traced back to dried coconut and spanned several months (Wilson and Mackenzie, 1955). In Liverpool, UK, an outbreak of salmonellosis was related to coconut (Semple, 1962). An outbreak of 167 cases of paratyphoid fever A occurred all over Singapore and was linked to consumption of de-shelled coconuts (Teoh *et al.*, 1997). Here, the attack rate was highest among the Indian ethnic group between the ages of 25 and 34 years. A shigellosis outbreak associated with a coconut milk dessert prepared at an elementary school occurred in Suan Phung, Thailand, which was caused by a nalidixic acid-resistant *Shigella dysenteriae* type 1 (Hoge *et al.*,1995). In Maryland, USA, there was an outbreak of *Vibrio cholera* traced back to frozen, but freshly made, coconut milk (Taylor *et al.*, 1993). An epidemiological investigation showed infection in four out of six persons who had consumed the product imported from Thailand.

Up to now, there have been no outbreak reports of microbial foodborne diseases resulting from the consumption of coconut water. Here it should be mentioned that outbreaks are only a minor part of food safety issues; most foodborne diseases are happening quietly and continuously without giving rise to spectacular outbreaks. For coconut water, the lack of outbreak reports could be attributed to the lack of a systematic surveillance system in coconut-producing countries, but also limited detection capabilities.

Interestingly, *Escherichia coli* and *Salmonella* spp. have been detected in fresh coconut water stored under refrigeration (Melo *et al.*, 2003), and populations of up to 1.6×10^5 cfu/ml of *Bacillus cereus* (Leite *et al.*, 2000) and 8.0×10^4 cfu/ml of *Staphylococcus aureus* (Hoffmann *et al.*, 2002) have also been found. Fortes and others (2006) analyzed 20 samples of refrigerated coconut water packaged in plastic bottles and noted that none complied with the microbiological standards established by the Brazilian legislation (Brasil, 2001). The samples had a pH between 5.0 and 5.2, and were positive for *Escherichia coli*. These data imply that fresh coconut water might be a possible vehicle for foodborne diseases.

26.3 Sustainability aspects

The coconut palm is one of the most widespread fruit trees in the world, occurring naturally in tropical continents. This historic dispersion and adaptability has led to its prominence in many countries not only in economic aspects, but also social and environmental. Every part of the coconut palm has a use, which makes it an important plant resource for humanity. It is dubbed 'the Tree of Life' because it gives humans everything they need, from food to shelter.

The commercial use of coconuts is restricted to approximately 90 countries, as the palm tree has optimal growth conditions in sandy soils, intense solar radiation and high precipitation. Coconut palms are grown in most tropical countries mainly for the high oil content of the endosperm (copra), which is widely used in both food and non-food industries (e.g. margarine and soap). In Brazil, coconuts are mainly used to produce coconut water, dried coconut and coconut milk.

World production of coconuts has recorded a steady increment over the past 50 years, particularly accelerating in the last three decades. The overall production rate has increased by more than 35% without substantial alteration of the harvest area; an area increase of around 13% (FAO, 2012). Indonesia, the Philippines and India are the major producers, and in 2010 these three countries were responsible for almost 75% of the world production of coconuts, approximately 60 million tonnes. The total land mass covered under coconut cultivation is around 11 million hectares, the size of a country like Colombia.

The current advancement of Brazilian agriculture is remarkable: in the 1990s the country held 10th position in the world ranking, and Brazil is currently the fourth-largest producer of coconuts, producing almost 3 million tonnes a year. Coconut production doubled in the last two decades, and production is concentrated at the coastline of the northeast and parts of northern Brazil, but all regions have shown an increase (IBGE, 2012). In terms of productivity, Rio de Janeiro and Espírito Santo are the states with higher incomes, but they are closer to the major consumer market.

In small- or large-scale production of coconut water, the main residue is the husk; in the green coconut it corresponds to 80% of the gross weight. However, unlike the mature coconut husk, which is traditionally used for the production of fiber and powder, green coconut husk is discarded. The material is disposed in landfills and dumps, which causes a huge problem for municipal garbage collection because of the large volume.

Embrapa, together with the metallurgical FORTALMAG, have developed a mechanical process for obtaining fiber and powder from green coconut husk (Rosa, 2012). The fiber can be used in the manufacture of various agricultural products and crafts, as well as in the industry and construction field, replacing other natural and synthetic fibers. The coir dust is a culture medium used for 100% natural seed germination, plant propagation in nurseries and in the growing of flowers and vegetables.

26.4 Economic issues

26.4.1 Market drivers

The motivation for the production scenario in Brazil lies mainly in the improved agro-industry and the varieties of coconut explored and its final use. The production of green coconuts (for coconut water) has been encouraged, and cultivations of dwarf and hybrids of dwarf and giant palm trees have been introduced, instead of giant palm trees, which are less productive and cultured in semi-extractive systems (Fontes and Wanderley, 2006). Giant palm tree plantations are mostly abandoned, as producers are discouraged by the low price of dried coconuts (Fontes, 2010).

It is noteworthy that about 90% of the world's coconut production comes from small farmers with areas up to 5 acres. Most of this production is internally consumed in the producing countries. In Brazil, roughly 70% of properties have up to 10 ha. Unlike many other crops, coconuts can be produced throughout the year and can provide the producer with a continuous income during the productive life of the tree (Martins and Jesus Júnior, 2011).

The largest exporters of fresh coconut in the world are Vietnam, Indonesia, Sri Lanka, Thailand, India and the Dominican Republic. On the import side, China is the largest importer, with approximately 35% of the market share, followed by the USA, Netherlands, the UK and Germany. Together, these countries accounted for 50% of global imports in 2009. This market has changed compared with the previous decade, as in 1999 the major importers were the USA, Hong Kong, United Arab Emirates and China (FAO, 2012).

The size and importance of the coconut water trade in the Brazilian market can be estimated according to fruit production and, specifically, the growth in recent years of dwarf green coconut palm tree cultivation. There are still very few market statistics for coconut water, but it is estimated that fresh coconut water, sold as the fruit and then freshly bottled, accounts for 80% of the amount consumed in the country (Cunha, 2011). In 2008, the market for industrialized coconut water was about 40 million liters, double what it was in 2004, and it is estimated that between 2011 and 2016 more than US$175 million will be invested in the coconut production chain, with the main focus on coconut water (Casado, 2011). The beverage industry is focused on the domestic and international market, and new companies are coming into the country.

In order to gain an idea of the possibilities of getting fresh coconut water beyond the beaches of tropical countries, it is important to look at what is happening in the industrialized coconut water beverage market (see the report 'Coconut Water 2012: Innovation and Natural Health Benefits Drive a New Category' at www.researchandmarkets.com).

1. The market for coconut water beverages grew 100% in 2011, taking the combined retail sales value of the US and European markets to over $265 million (~196 million). This achievement is all the more impressive because it is a premium-priced category, and the market is expected to grow another 50% in 2012. The report sets out which brands are successful and why, and also defines the key drivers for the coconut water phenomenon.
2. Coconut water has a massive advantage as an ingredient as it connects to consumers' desire for 'naturality', the biggest worldwide trend in the food and beverage business.
3. Coconut water meets consumers' expectations for products that are 'free-from' artificial colors, preservatives or additives, and allergens.

4. Coconut water is also 'naturally functional': it naturally has 15 times the amount of potassium found in the equivalent volume of most sports drinks. While being isotonic and naturally sweet, coconut water is also low in calories, with only 19–24 calories per 100 ml, compared with 43 calories per 100 ml for orange juice.

26.4.2 From waste to money

Interestingly, before the 1990s coconut water was actually a waste product for the industry. All manufacturers were in the industry to produce dried coconut and coconut milk, and coconut water was used to water coconut plantations or was treated, at a high cost, before being dumped into rivers. The story began to change when a traditional coconut company began to apply a more suitable aseptic processing (ultra-high temperature) and packaging to allow for an aseptic product to be available all over the country, without any need for refrigeration or any type of additive.

26.4.3 Societal significance

An interesting story reported in a daily journal demonstrates the societal significance of this product: 45-year-old Danilo Albuquerque is a micro-entrepreneur and a native from the northeastern state of Paraíba, based in Rio de Janeiro (Rio, 2011). Danilo left his job as a security guard to run his own kiosk selling fresh coconut water and snacks in Rio de Janeiro, and his main customer base is the neighborhood hospital. He began the business with a loan of US$250 and he managed to buy his apartment last year. At that time, Danilo was a guarantor for a friend, who owns a kiosk similar to his, in the modality of individual microcredit. 'I helped Joel, who is a young worker and he also got the credit.'

This is why the World Health Organization (WHO, 1996) has outlined the following advantages for food from street vendors:

- a source of inexpensive, convenient and often nutritious food for urban and rural poor;
- a source of attractive and varied food for tourists and the economically advantaged;
- a major source of income for a vast number of persons, particularly women;
- a chance for self-employment and the opportunity to develop business skills with low capital investment.

26.5 Regulatory aspects in Brazil

In opposition to those potential benefits, it has also been recognized that street-vendors are often poor, uneducated and lack appreciation for safe food handling. As a result, street foods are perceived to be a major public health risk. If a community is to have the full benefits of street-vended foods with minimal risk of foodborne diseases, government intervention is required to ensure that the standard of safety for such foods is best attainable in the context of the prevailing local situation.

Street-vended foods may pose significant public health problems because of:

- the lack of basic infrastructure and services, such as potable water supplies;
- difficulty in controlling the large number of street food vending operations because of their diversity, mobility and temporary nature;
- insufficient resources for inspection and laboratory analysis;
- a general lack of factual knowledge about the microbiological status or the precise epidemiological significance of many street-vended foods;
- street vendors' poor knowledge of basic food safety measures;
- inadequate public awareness of hazards posed by certain street foods.

Interest in promoting this important sector of the food supply system has increased in many countries, even those which had previously ignored or attempted to suppress it. As a consequence, over-regulation and the imposing of too many overhead costs are viewed as counter-productive and may substantially reduce the sector's capacity to provide a variety of nutritious foods at affordable prices.

Brazil's specific legislation for industrialized coconut water is under the control of the Ministry of Agriculture, Livestock and Food Supply (Brasil, 2009), and fresh coconut water prepared and/or sold in snack bars, kiosks, stalls and by street vendors, as well as similar places and producers, is under the control of the Ministry of Health (Brasil, 2005). In addition, legislation applicable to fresh coconut water also applies to food and beverages prepared with any type of fresh produce. It includes substances or mixtures of substances obtained from the pulp or other parts of vegetables, with or without the addition of other ingredients intended for consumption, such as coconut water, sugarcane juice, pulp and salad fruits, fruit and vegetable juices, mixed drinks of fruit with alcohol beverages (i.e. Caipirinha) and similar products.

The sanitary-hygienic procedures of the legislation for fresh products address three main points: purchase, reception, and storage of raw materials, ingredients, packaging, and supplies; food handlers; and preparation and exposure to sale of the food and beverages (Brasil, 2005). This legislation is based on other Brazilian legal standards, as well as the Codex Alimentarius 2001 'Code of Hygienic Practice for the Preparation and Sale of Street Foods' and the World Health Organization's *Essential Safety Requirements for Street-Vended Foods* (WHO, 1996).

The legislation for industrialized coconut water defined it as an undiluted (and not fermented) beverage obtained from the liquid portion of the coconut palm fruit by means of appropriate technological processes (Brasil, 2009). The beverage is classified as cooled, pasteurized, frozen, sterilized, concentrated, dried and reconstituted. Cooled coconut water is the result of product being packed immediately after extraction and, without discontinuity, is subjected to an appropriate refrigeration process. Cooled coconut water, similar to pasteurized coconut water, should be packed, maintained and commercialized under

Table 26.2 Legal standards for cooled, pasteurized or frozen coconut water in Brazil

Parameter	Minimum	Maximum
pH	4.30	4.50
Soluble solids (Brix)	n.d.	6.70
Potassium (%w/w)	140.00	230.00
Sodium (%w/w)	2.00	30.00
Yeast and molds (cfu/ml)	n.d.	20
E. coli or coliforms at 45 °C (cfu/ml)	n.d.	1
Salmonella spp.	Absent in 25 ml	n.d.

Source: Brasil (2009).
n.d., no standard defined.

refrigeration conditions at a maximum temperature of 5 °C. Frozen coconut water can be previously pasteurized or not.

Cooled coconut water is considered an industrialized beverage as it has to comply with legal standards and may need pH adjustments. Table 26.2 shows the physic-chemical and microbial parameters of cooled, pasteurized and frozen coconut water. To adjust the physic-chemical parameters of industrialized coconut water, food additives approved for fruit juice may be added. The product must be processed, packaged, stored and transported according to the requirements of the Good Manufacturing Practice established in specific legislation.

26.6 Technical control strategies

Control of foodborne pathogens in fresh coconut water requires a combination of surface contact hygiene and time/temperature control. This strategy follows the principles of hurdle technology, where two or more preservation technologies are used to prevent growth of microorganisms in foods. Good hygiene practices for surfaces that come into contact with the fruit are well described in the FAO's training guide (Rolle, 2007).

In 1993, WHO applied a survey to its member states in order to assess the current situation of street-vended foods and to understand the position of the relevant authorities with regards to the hazards posed by street-vended foods and their contributing factors, as well as hazard management approaches. Over 100 countries participated in this survey, making it the most extensive report on street-vended food available to date. Among key findings, the majority of countries reported cross-contamination of foods (from raw food, infected handlers and inadequately cleaned equipment) as well as time and temperature abuse to be the major factors that contribute to foodborne disease.

When time and temperature is used as a control, the shelf-life of the product should be equal to or less than the lag phase of the pathogen(s) of concern. An extended lag-time is the major focus of the non-lethal preservative methods applied to foods, and in order to establish the time/temperature control for fresh coconut water and its shelf-life, a challenge test was performed (Walter *et al.*, 2009).

The selected microorganism for the challenge test was *Listeria monocytogenes*, a psychotropic and ubiquitous foodborne pathogen. Coconut surfaces can be contaminated by *L. monocytogenes* resulting from agricultural practices, such as when the fruits are placed directly on the soil (the natural habitat of the bacteria). Considering that the bottled fresh beverage is normally sold and consumed cold, *L. monocytogenes* could find adequate conditions to grow relatively fast under refrigeration and reach infective levels. The ability of *L. monocytogenes* to grow in low-acid fruits/fruit pulps, such as melon, watermelon and papaya (Penteado and Leitão, 2004) also has been reported.

Behavior of *L. monocytogenes* in fresh beverages stored at 4 °C, 10 °C and 35 °C was studied, and coconut water was aseptically extracted from young green coconuts and samples were inoculated in triplicate with a mixture of five strains of *L. monocytogenes* with a mean population of approx. 3 $\log_{10}$ cfu/ml. The kinetic parameters of the bacteria were estimated from the Baranyi model and compared with predictions of the Pathogen Modelling Program (PMP, version 7.0) from the United States Department of Agriculture (USDA).

Table 26.3 shows the *L. monocytogenes* kinetic parameters in fresh coconut water. The shortest lag-time of the bacteria at 4 °C was 12.7 days, at 10 °C 4.2 days and at 35 °C 3 hours. The reduction of the storage temperature from 35 °C to 4 °C resulted in approximately a 100 times increase in lag-time, showing that storage temperature is an effective way of controlling *L. monocytogenes* in fresh coconut water. Reducing temperature from 35 °C to 10 °C and then to 4 °C, led to a reduction in growth rate around 15 and 70 times, respectively, and these data can be used to estimate the shelf-life of fresh coconut water. It should always be considered that refrigeration prolongs the lag-time but does not inhibit growth of the bacteria, and temperature abuse can considerably increase the level of potential hazards.

In the Pathogen Modelling Program, pH and water activity (aw), values determined for fresh, young and green coconut water (pH 4.88 and aw 0.995), were defined for *L. monocytogenes* growth in Brain Heart Infusion (BHI) broth. The bacteria presented around half of the lag-time in fresh coconut water at 10 °C and 35 °C compared with BHI broth and approximately the same lag-time at 4 °C. Therefore, fresh coconut water can be considered a culture medium.

Although the implementation of good agriculture and manufacturing practices is essential in reducing coconut water contamination, it cannot entirely eliminate the possibility of the presence of pathogens. In addition to hygiene and time/temperature controls, consumer education is essential in order to assure fresh coconut water safety. Pregnant women, the elderly and people with compromised immune systems should only consume coconut water that has been obtained directly from the fruit; however, sterilized coconut water would also be an alternative, despite its sensory alterations.

Under street conditions or at any type of marketing place, time/temperature control is essential to safety and sensory preservation of fresh coconut water;

Table 26.3 Growth kinetic parameters of *L. monocytogenes* in fresh coconut water stored at different temperatures and comparison with prediction given by the Pathogen Modelling Program (PMP)

Storage temperature (°C)	Growth rate [(log_{10}cfu/ml)/h]		Lag time (h)	
	Fitting to the Baranyi model	*PMP prediction	Fitting to the Baranyi model	PMP prediction
4	0.0043 ± 0.00044	0.0059 (0.0052–0.0068)	357.16 ± 184.62a	353.3 (291.4–428.5)
10	0.0198 ± 0.0017	0.0191 (0.0172–0.0210)	76.70 ± 23.83b	128.8 (112.2–147.8)
35	0.3044 ± 0.0390	0.3010 (0.2508–0.3762)	3.50 ± 0.60c	8.5 (6.8–10.6)

Source: adapted from Walter *et al.* (2009).

* PMP prediction for pH 4.88 and aw 0.995.

however, other strategies must also be in place as microbiological, chemical and physical contamination of the beverage are possible. WHO (1996) reports state that efforts must be made (a) to educate the personnel involved; (b) to improve the environmental conditions in which the trade is practiced; (c) to provide the essential services to assist street food vendors in assuring the safety of their wares; (d) to implement simple techniques for regulation of street foods; and, finally, (e) to adopt a HACCP-based strategy to prioritize food control measures specific to the safety of street-vended foods.

26.7 Conclusion and future trends

The sustainability of the coconut production chain, combined with the market demand for a premium beverage such as fresh coconut water, is an opportunity and responsibility for the Brazilian government and society and other tropical countries. The safety of fresh coconut water is not only a subject of public health, but also an image presentation of a healthy product that generates revenue for many street workers and micro-entrepreneurs, besides being an opportunity to provide a pleasant beverage for the society and our guests. The Brazilian legislation is well established, and this leaves us with a classic case in need of technical support, application of best practices and sanitary surveillance.

26.8 Questions for discussion

1. What makes bottled fresh coconut water a beverage that needs to be treated carefully by the authorities in Brazil?
2. What is the relevance of fresh coconut water for non-producing countries?
3. Why fresh coconut water is not just a safe issue?
4. What is making people replace soft drinks, sport drinks and conventional juices and nectars for coconut water in Brazil and the importing countries?
5. Which strategy would you take for the safe control of fresh coconut water?

26.9 Acknowledgments

The authors are grateful to Dr Virginia Martins da Matta, from Embrapa Food Technology (Brazil), for valuable reviews and ideas in the elaboration of this chapter.

26.10 References

ARAGÃO, W. M. (2002) *Coco*: Pós-colheita. Brasília: Embrapa Informação Tecnológica. 76 p. (Frutas do Brasil; 29).

BRASIL. (2001) Ministério da Saúde. Resolução RDC n° 12, de 02 de janeiro de 2001. *Aprova o Regulamento Técnico sobre padrões microbiológicos para alimentos*. Available at: http://bvsms.saude.gov.br/bvs/saudelegis/anvisa/2001/res0012_02_01_2001.html. Accessed 26 September 2013.

BRASIL. (2005) Ministério da Saúde. Resolução RDC n° 218, de 29 de julho de 2005. *Dispõe sobre o Regulamento Técnico de procedimentos higiênico-sanitários para manipulação de alimentos e bebidas preparados com Vegetais*. Available at: http://portal.anvisa.gov.br/wps/wcm/connect/7b8719004745974 19fabdf3fbc4c6735/RDC_218.pdf?MOD=AJPERES. Accessed 26 September 2013.

BRASIL. (2009) Ministério da Agricultura, Pecuária e Abastecimento. Instrução Normativa n° 27, de 22 de julho de 2009. *Estabelecer os procedimentos minimos de controle higiênico-sanitário, padröes de identidade e características minimas de qualidade gerais para a água de coco*. Available at: http://sistemasweb.agricultura.gov.br/sislegis/action/detalhaAto.do?method=consultarLegislacaoFederal. Accessed 26 September 2013.

CABRAL, L. M. C., PENHA, E. M., MATTA, V. M. (2005) Água de coco verde refrigerada. Brasília: Embrapa Informação Tecnológica. 34 p. (Agroindústria Familiar).

CASADO, L. (2011) Indústria investe na produção de água de coco. *Valor Econômico*. São Paulo, 11 January 2011. Available at: http://www.valor.com.br/. Accessed 5 September 2012.

CODEX ALIMENTARIUS. (2001) CAC/RCP 43-1997, Rev. 1-2001. *Revised Code of Hygienic Practice for the Preparation and Sale of Street Foods*. (Regional Code – Latin America and the Caribbean). Available at: http://siweb.dss.go.th/standard/Fulltext/codex/CXP_043E.pdf. Accessed 26 September 2013.

CUNHA, L. (2011) Pepsico tenta aumentar o consumo de água de coco. *Valor Econômico*. São Paulo, 11 January 2011. Available at: http://www.valor.com.br/. Accessed 5 September 2012.

(FAO) FOOD AND AGRICULTURE ORGANIZATION OF THE UNITED NATIONS. (2007) *How to bottle coconut water*. Available at: http://www.fao.org/ag/magazine/0701sp1.htm. Accessed 5 September 2012.

(FAO) FOOD AND AGRICULTURE ORGANIZATION OF THE UNITED NATIONS. (2012) *FAOSTAT*. Available at: http://faostat.fao.org/. Accessed 5 September 2012.

FONTES, H. R. (2010) Caracterização do quadro atual e principais ameaças à produção de coco seco no nordeste do Brasil. Available at: http://www.infoteca.cnptia.embrapa.br/bitstream/doc/880194/1/Caracterizacao0001.pdf. Accessed 5 September 2012.

FONTES, H. R., RIBEIRO, F. E., FERNANDES, M. F. (2003) *Coco*: Produção. Brasília: Embrapa Informação Tecnológica. 106 p. (Frutas do Brasil; 27).

FONTES, H. R., WANDERLEY, M. (2006) Situação atual e perspectivas para a cultura do coqueiro no Brasil. Aracaju: Embrapa Tabuleiros Costeiros. 16 p.

FORTES, E. P., LIMA, A., CRONEMBERG, M. G. O., CRISPIM, L. S. (2006) Qualidade físico-química e microbiológica das águas-de-coco envasadas e comercializadas em Teresina, Piauí. *Hig. Aliment*., 20: 87–90.

HOFFMANN, F. L., COELHO, A. R., MANSOR, A. P., TAKAHASHI, C. M., VINTURIM, T. M. (2002) Qualidade microbiológica de amostras de água de coco vendidas por ambulantes na cidade de São João do Rio Preto – SP. *Hig. Aliment*., 16: 87–92.

HOGE, C. W., BODHIDATTA, L., TUNGTAEM, C., ECHEVERRIA, P. (1995) Emergence of nalidixic acid resistant *Shigella dysenteriae* type 1 in Thailand: An outbreak associated with consumption of coconut milk dessert. *Intl. J. Epidemiol*., 24: 1228–32.

(IBGE) INSTITUTO BRASILEIRO DE GEOGRAFIA E ESTATÍSTICA. (2012) Produção Agrícola Municipal. Available at: http://www.ibge.gov.br/home/estatistica/pesquisas/pesquisa_resultados.php?id_pesquisa=44. Accessed 5 September 2012.

LEITE, C. C., ASSIS, P. N., SILVA, M. D., SANT'ANA, M. E. B., SANTANA, L. R. R. (2000) Avaliação microbiológica da água de coco produzida e comercializada na cidade de Salvador – BA. *Hig. Aliment*., 14: 64–6.

MARTINS, C. R., JESUS JÚNIOR, L. A. (2011) *Evolução da produção de coco no Brasil e o comércio internacional*: panorama 2010. Aracaju: Embrapa Tabuleiros Costeiros. 28 p.

MELO, N. P. M., CARDONHA, A. M. S., OLIVEIRA, A. C. F. (2003) Qualidade microbiológica das águas de coco envasadas e comercializadas na cidade de Natal – RN. *Hig. Aliment.*, 17: 113–14.

PAULO, D. S. (2011) Antes dos anos 90, água de coco era lixo. *Valor Econômico*. São Paulo, Nov. 17th 2011. Available at: http://www.valor.com.br/. Accessed 5 September 2012.

PENTEADO, A. L., LEITÃO, M. F. F. (2004) Growth of Listeria monocytogenes in melon, watermelon and papaya pulps. *Intl J. Food Microbiol.*, 92: 89–94.

RIO, D. (2011) Ex-segurança investe em quiosque de água de coco e compra casa própria. *Valor Econômico*. São Paulo, October 10th 2011. Available at: http://www.valor.com.br/. Accessed 5 September 2012.

ROLLE, R. (2007) *Good practice for the small-scale production of bottled coconut water*. Rome: Food and Agriculture Organization of the United Nations. 36 p.

ROSA, M. F. (2012) *Beneficiamento da casca de coco verde*. Available at: http://www.cnpat.embrapa.br/home/portfolio/tecnologia.php?id=10. Accessed 15 October 2012.

SEMPLE, A. B. (1962) Annual report of the Medical Officer of Health. *Brit. Food J.*, 64: 173–6.

STRAWN, L. K., SCHNEIDER, K. R., DANYLUK, M. D. (2011) Microbial safety of tropical fruits. *Crit. Rev. Food Sci. Nutr.*, 51: 132–45.

TAYLOR, J. L., TUTTLE, J., PRAMUKUL, T. (1993) An outbreak of cholera in Maryland associated with imported commercial frozen fresh coconut milk. *J. Infect. Dis.*, 167: 1330–5.

TEOH, Y. L., GOH, K. T., NEO, K. S., YEO, M. (1997) A nationwide outbreak of coconut-associated paratyphoid: A fever in Singapore. *Annl. Acad. Med. Singapore.*, 26: 544–8.

WALTER, E. H. M., KABUKI, D. Y., ESPER, L. M. R., SANT'ANA, A. S., KUAYE, A. Y. (2009) Modelling the growth of *Listeria monocytogenes* in fresh green coconut (*Cocos nucifera* L.) water. *Food Microbiol.*, 26: 653–7.

(WHO) WORLD HEALTH ORGANIZATION. (1996) Essential Safety Requirements for Street-Vended Foods. Rev. Edition. WHO/FNU/FOS/96.7.

(WHO) WORLD HEALTH ORGANIZATION. (2010) First steps for managing an outbreak of acute. Available at: http://www.who.int/topics/cholera/publications/en/first_steps.pdf. Access at: 10 October 2012.

WILSON, M. M., MACKENZIE, E. F. (1955) Typhoid fever and salmonellosis due to the consumption of infected desiccated coconut. *Symp. Food Microbiol Public Health*. pp. 510–21.

YONG, J. W. H., GE, L., NG, Y. F., TAN, S. N. (2009) The chemical composition and biological properties of coconut (*Cocos nucifera* L.) water. *Molecules*, 14: 5144–64.

27

Control of fresh produce safety in Denmark

J. Hoorfar, Technical University of Denmark, Denmark

DOI: 10.1533/9781782420279.5.383

Abstract: Substantial effort has been made in Denmark to establish food safety systems by both producer organizations and public bodies. A centralized, one-agency administration has been set up to carry out all food and veterinary inspections. The majority of outbreaks are discovered through a centralized patient notification system. The aim of this chapter is to provide the reader with an understanding of the Danish food control model and how it can be used to cope with future challenges in local and global production and trade. A case study is included to illustrate how source tracing works in real-life situations.

Key words: food control, food safety system, contamination, sampling, primary production.

27.1 Introduction

Food safety in Denmark is achieved through implementation of and adherence to systems, rather than a reliance on testing. Such systems are adapted to local food production, food habits and culture. As Denmark is one of the world's largest food exporters, food production is a major part of the national economy, and so food safety has always been a political priority. This has made the country a pioneer in the development of new systems to ensure the production of safe and healthy food. Therefore, it is interesting to study how the Danish model was built up, and how it functions in practice.

27.1.1 The system in Denmark

Denmark adopted the Single Control Agency System following a major reform. The reform began in 1997 and resulted in centralization. The new agency, the

Danish Food Administration (DFA), consolidated all food safety functions, including inspection of the control of food originating from animals. Previously, inspection functions were distributed between the Ministry of Health and the Ministry of Agriculture and the Municipalities (see Fig. 27.1).

The main purpose of the reform was to improve the system, rendering it more efficient through reduction in overlapping responsibilities. Prior to reform, municipal inspectors were part of a small system. The DFA now carries out a wide range of controls regarding food safety and food hygiene. Control activities include control of animal health and welfare, as well as control of food manufacturers, food businesses and restaurants (Anon., 2010).

One of the strengths of the Danish model is the close collaboration between public health authorities (Statens Serum Institut, SSI), the DFA and the National Food Institute at the Technical University of Denmark (DTU Food). This collaboration is formalized in the Danish Zoonoses Center (www.food.dtu.dk), located at DTU Food. The center coordinates and reports on national efforts to combat foodborne pathogens. The case study in section 27.7 illustrates a real-life outbreak situation, and the work that ensued.

27.1.2 Ensuring inspection quality

Inspection quality is controlled and secured in various ways. The DFA has quality strategies for inspections: one for food inspection, and one for veterinary inspection. The key elements are the development of inspector competences, inspection values and supervision quality. Inspection principles are also of importance. All inspections have the same effect, no matter which inspector is responsible. The regions follow the inspectors' work and development. Reports made by the inspector may be checked, and quality supervisions performed.

Quality supervision is where the nearest superior follows an inspector during inspections. This is done to check whether the inspection corresponds to the quality requirements for the inspection in practice, and to enable dialogue on a specific task performance. After the inspection, and away from the site, the supervisor and the inspector discuss and evaluate, producing a report reflecting agreements concerning skills, enhancement and other factors. Each year approximately 50% of inspectors are quality supervised in this way.

27.1.3 Relationships between the head office, regional offices and the EU

Throughout the year, the DFA head office follows the success with which regions reach the targets mentioned in a 'results contract'. The head office assesses the regions annually or biannually. Assessment considers whether the regions are following the appropriate guidelines, and whether their management performance is good in their fields of responsibility.

The DFA publishes a report of main findings and results on their homepage, http://www.foedevarestyrelsen.dk/Sider/forside.aspx. The regions are responsible for following up on the recommendations or deficiencies identified. The supervision team follows up on recommendations and requests. The DFA head office also checks inspection report samples in order to ensure clarity, consistency and legal accuracy.

An External Audit Unit, working under the Ministry of Food, Agriculture and Fisheries, provides an additional level. According to the Danish Food Act, this unit examines whether food control is carried out in accordance with legislation, control plans and strategies, and whether it is carried out in an economically efficient manner.

The audit unit reports to the Commission for Food Control, consisting of six experts in food control matters. The Minister appoints members on the basis of nominations made by the Danish Consumer Council and the main Danish food industry organizations. The Commission on Food Control must approve audit plans. There is one audit program, containing the annual audit, based on a 5-year rolling plan and specific, separate audits. The reports are published on the Ministry homepage (www.fvm.dk).

The Food and Veterinary Office of the EU Commission carries out missions in all EU member states. The EU Commission checks whether rules and inspection of food, animal health and welfare, are in accordance with EU rules. Missions may be focused on certain foodstuffs, or on more horizontal issues.

27.2 Inspection principles

Both in Denmark and elsewhere, consumers, establishments, primary producers and inspection authorities must be able to rely on the Danish control system. For this reason, the DFA and its regional offices are subject to the following fundamental principles.

27.2.1 Primary producers and establishments bear the responsibility

Responsibility for complying with the rules lies with the primary producers and establishments. This includes compliance with requirements determined by EU Executive Orders and any requirements emanating from third countries. Establishments and primary producers are required to set up own-check programs based on the HACCP principles. This means that only in a very few areas, such as the inspection of animals for slaughter, do the authorities themselves approve products, or perform internal control for establishments or primary producers.

27.2.2 The control must be preventive

Providing information to establishments and primary producers is an important aspect of the control process. Such provision is to ensure that the reasoning behind the regulations is understood, and that establishments and primary producers are motivated to follow these regulations. This principle does not change the fact that establishments and primary producers are responsible for familiarizing themselves with the relevant regulations.

27.2.3 Establishments must be authorized

Establishments that process or sell foodstuffs or non-food animal products must be authorized, approved or registered before they may sell or produce their products. Upon authorization or approval, an assessment of the company is carried out and a number of conditions are laid down, including the obligation for the company to inspect its own products (own-check program).

27.2.4 Inspection is the cornerstone

With respect to establishments and primary producers, inspection is the cornerstone of the control process. The starting point for inspection is the establishments' handling and sale of foodstuffs. Analytical control should act as a supplement to such inspection.

27.2.5 Analytical control

Analytical control should be aimed at verifying that the establishment or primary producer can manage the handling of foodstuffs or primary production. Analytical control should be determined by inspection authorities. In special areas, plans for specimen selection are worked out centrally; for example, with inspection for veterinary medicine and pesticide residues.

27.2.6 Need-oriented inspections

Inspection should be both regular and address need. It should also be dynamic, and most intensively applied where need is greatest. Once a problem has been solved, inspection should be moved elsewhere. The results of the inspection, and the new knowledge acquired (e.g. from surveillance and charting studies) should be used to keep the control process goal-oriented, and to further improve regulations.

27.2.7 Seek the source of the problem

Inspection should seek out the source of the problem, and should be concentrated as close as possible to the links from which problems are arising in the chain from 'stable to table'. If problems are discovered in links further down the chain, the responsible party should be contacted with a view to carrying out an inspection of the establishment or primary producer concerned, and this is where the problem should be solved.

27.2.8 Reaction

The control process should avail itself of any necessary sanctions to ensure that regulations are observed. On the one hand, the reaction should be no more radical than necessary. On the other, it must have sufficient impact to ensure respect for the regulations. If the establishment or primary producer concerned fails to comply with the control authority's sanctions, then sanctions should be escalated.

27.2.9 Uniform effects

The effects of inspection must be uniform, both geographically, and within the various branches of the food industry. This means, for example, that imported foods and local production must be subject to the same requirements, and that regulations must be uniformly enforced nationwide.

27.2.10 Document its own reliability

The inspection process must be able to document its own reliability and effectiveness, and these records must be made available to the public. Inspection results should be clearly visible in places where consumers purchase foodstuffs, and on the Internet, in order to make it easier for consumers to evaluate both the establishments involved and the official control process.

27.3 How foodservice and food establishments are controlled

All food establishments in Denmark must be either registered or approved by the DFA. The DFA uses various control tools when inspecting food businesses, including:

1. ordinary control, according to fixed frequencies;
2. extraordinary control, subsequent to the controls that have resulted in sanctions;
3. other inspections, such as are related to certain subjects with fixed frequencies;
4. control in relation to approval or authorization;
5. prioritized basic control, supplementary ordinary controls focused on specific areas, such as compliance level activities;
6. control campaigns.

With a few exceptions, controls are carried out unannounced. Sampling for laboratory analyses may supplement control activity. The frequency of ordinary control activities in the various food business sectors is determined by a general risk assessment and an evaluation of needs in relation to the individual food operator (Tables 27.1 and 27.2). In this way, the control is focused on sectors and businesses with the highest level of risk and consequent need for control.

Table 27.1 Wholesalers with and without production. The table shows the relations between risk groups and Standard and Elite frequencies for the number of ordinary controls per year for wholesale and retail businesses

Risk group	Standard frequency	Elite frequency	Elite 2 frequency	Examples of businesses
Very high	5	3	2	Slaughterhouse
High	3	1	–	Brewery
Medium	1	0.5	–	Warehouse with food of vegetable origin
Low	0.5	0.5	–	Wholesale trade of wine and liquor
Very low	As and when required	–	–	Agency trade of food

Table 27.2 Control frequency of retailers

Risk group	Standard frequency	Elite frequency	Examples of businesses
Very high	–	–	–
High	2	1	Butcher
Medium	1	0.5	Baker
Low	0.5	0.5	Grocer's shop
Very low	As and when required	–	Wine shop

All foodservice sectors such as slaughterhouses, warehouses, butchers and bakers, are placed in one of five risk groups, based on seven microbiological and chemical risk factors, and a general factor dependent on sector activity. Each risk group has a standard control frequency. The standard control frequencies vary from five ordinary controls per year, for example in establishments producing meat products and minced meat, to controls decided by need, for example in wine shops.

Evaluation of need is carried out at the establishment level. Differences between establishments regarding their ability and intent to respect food legislation will cause: less control for business designated 'Elite', with less frequencies compared with the ordinary control of businesses; more control for establishments designated as 'Extra', with (follow-up) inspections for businesses that have incurred sanctions in the past.

In Denmark, meat control takes place in all slaughterhouses. No animal can be slaughtered for human consumption without prior inspection by a veterinarian. Veterinarians and auxiliaries perform ante- and post-mortem controls. They also audit in-house, self-check programs, and carry out other procedures and inspections such as hygiene and animal welfare.

27.3.1 Sanctions

When an inspector finds an establishment in violation of food law, he or she can choose from the following actions:

1. no sanction. If the violation is so small that sanctioning would be disproportionate, the inspector can then issue guidance;
2. warning;
3. injunction;
4. prohibition order;
5. administrative fine;
6. reporting to the police for prosecution;
7. withdrawal of authorization.

Inspectors choose the level of sanction deemed necessary to ensure that non-compliance is corrected. Continued non-compliance with food law entails escalating sanctions.

27.3.2 The smiley-scheme

The smiley-scheme is popular among Danish consumers and food enterprises, and smileys have been effective in improving food safety. The Danes have seen smileys on inspection reports in all their retail food enterprises since 2001. The four different smileys signal how well the business complies with food regulation.

In 2008 the 'elite-smiley' was introduced. Businesses with the best inspection records now use this in their marketing materials. The smileys appear at the top of the food inspection report (Fig. 27.2). The reports must be displayed for consumers

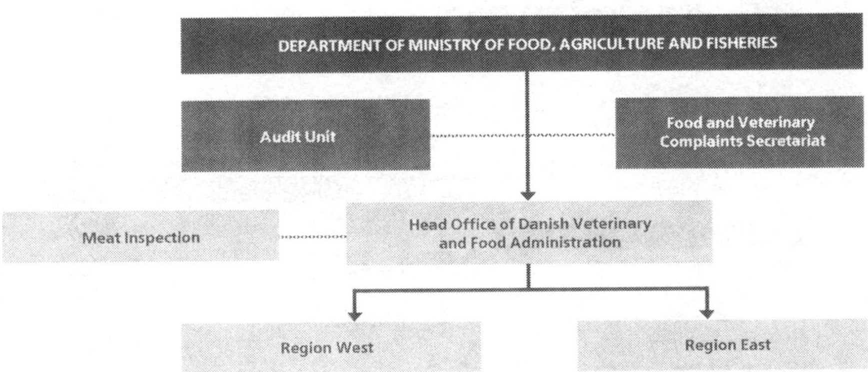

Fig. 27.1 Organization of the Danish Food Administration (DFA) with approximately 1100 employees (apart for those employed at meat inspection). Source: Ministry of Food, Agriculture and Fisheries of Denmark, Danish Veterinary and Food Administration, http://www.foedevarestyrelsen.dk/english/Aboutus/Organization/Pages/default.aspx.

to read, before deciding whether to enter a shop or restaurant. The reports must also be easily found on homepages. Usually, all reports for the last four inspections are available for the public to view at: www.findsmiley.dk.

Publication philosophy

The basic idea behind the Danish smiley-scheme is to:

- Make inspection results public and the content easily understood.
- Enable consumers to use the results to make a more informed choice on where to shop.
- Give establishments another powerful incentive to secure excellent food safety.

Shops and restaurants responsible

In Denmark, shops and restaurants are responsible for complying with food regulations. During unannounced inspections, public food inspectors check for compliance.

All shops, restaurants and other enterprises selling food and beverages to the public are regularly inspected, typically one to three times a year. Frequency is based on a risk evaluation of all lines in the business.

Smiley-reports (Fig. 27.2) should be posted in all supermarkets, groceries, bakeries, butchers, kiosks, restaurants, pizzerias, canteens, hospital kitchens, kindergartens, nurseries and elderly people's homes. Even the hot-dog stand in the street must display them.

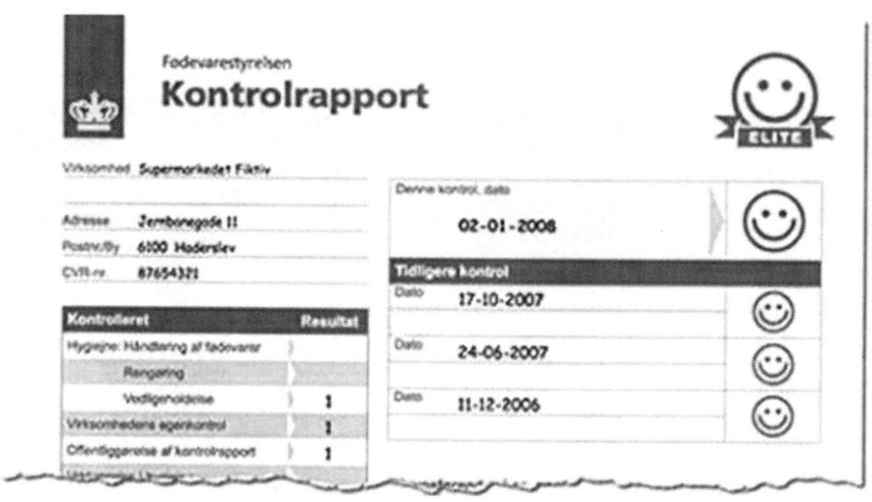

Fig. 27.2 A typical smiley report, visible at Danish food establishments. The report reflects the hygienic conditions and compliance with the mandatory own-check program. The cost of inspection and the report is met by the food establishment.

What do the smileys symbolize?
There are four different smileys. They symbolize that the inspector:

☺ had no remarks.
☺ has emphasized that certain rules must be obeyed.
😐 has issued an injunction order or a prohibition.
☹ has issued an administrative fine, reported the enterprise to the police or withdrawn approval.

Enterprises evidencing hazardous health conditions are closed down until problems are corrected.

Elite-smiley for Elite establishments
Businesses in sectors with a standard frequency of 0.5 (control every other year) or more, achieve Elite status if they have received no sanctions on the last four consecutive control reports, and have further received no sanctions within the past 12 months. Businesses in sectors with a standard frequency of 1 or more will have fewer controls if they have achieved Elite status. Businesses that have achieved Elite status can use an Elite-smiley in their marketing. You can read more about smileys and elite businesses on www.findsmiley.dk.

27.3.3 Campaigns and laboratory projects

Every year the DFA performs a number of campaigns and laboratory projects. Campaigns are directed towards areas where there is a suspicion or knowledge of problems or high risks. Most laboratory projects are centrally coordinated and aimed at monitoring and control, but screening projects also occur where necessary.

Inspectors might take samples for analysis if they have a suspicion of non-compliance, or as a follow-up. These projects are regionally coordinated. Reports from campaigns and laboratory projects are published on the DFA homepage: www.fvst.dk – Fødevareundersøgelser. The reports are used to evaluate the need for amended inspection procedures, guidelines or other information.

27.3.4 Internal checks in food establishments

Food establishments must perform their own checks in order to comply with general food safety rules and standards. EU legislation on this is based on the principle that food establishments are responsible for compliance. This applies to all parts of the food chain, from 'stable to table'.

Food establishments are required to perform their own checks. All must use good manufacturing practice, and most establishments must also have a written self-check program. The written program is required for areas where a direct risk to food safety or special requirements in the legislation exist, for example in the case of export to third countries or preparation of organic food.

Those parts of the self-check program aimed at food safety must be based on the principles of HACCP (Hazard Analysis Critical Control Points). This includes, for example, risk analysis, documented monitoring and corrective action. Analyses might be part of the program, to demonstrate its effect. Many establishments' self-check programs do not require analytical control, apart from basic measurements of temperature and weight.

The food industry has compiled so-called sector codes for good hygienic practice, or national guidelines. Many food industry sectors have produced guidelines describing how companies can be organized and run in a hygienically responsible way and implement efficient self-check programs. These guidelines help companies tailor their own programs.

For those establishments following good manufacturing practice without a written self-check program, inspectors check compliance through inspection. For establishments with a written self-check program, inspectors perform an audit. The inspector checks if the self-check program is sufficient and well-implemented. In case of shortcomings, the inspector requires improvements to the implementation of the program, or a revision.

27.3.5 Animal feed

Animal feed is controlled twice a year in production establishments and once a year on farms that have implemented HACCP. The focus is on:

- HACCP/GAP implementation;
- traceability;
- end products;
- storage, separation and transport;
- cleaning.

Imported feed is controlled only if arriving from a source that has proven unsafe. Otherwise a check of documents is performed.

27.3.6 Sampling

During inspection visits, inspectors may take samples of raw material, additives and semi-finished material, and send these to laboratories affiliated to DFVA in order to check compliance. Sampling can be a part of the national plan, created at the central level with suggestions from the regional level. A multi-year control plan is usually prepared and subsequently approved by the EC.

The sampling plan for each year considers the projected annual production of foodstuffs to be sampled. It also takes into consideration both new pathogens and threats from known pathogens.

In 2007, the administration took 56252 samples of various contaminants and components, with 22000 samples taken at the regional level. Some 370 samples were taken from genetically modified organisms. A further 600–700 samples of feed are taken annually for feed and seed testing. Eight laboratories carry out the national plan for control of residues: three in Denmark and five in the EU. The total number of samples in 2011 was 37000. There is a planned reduction to

27 000 in the coming years. For more information about analytical methods, please see Chapter 20 of this book.

27.4 How Denmark controls fruit and vegetable safety

An increasing number of Danes become ill, mainly from imported fruit and vegetables, because products are contaminated and/or they are not cleaned properly before consumption. This is a general trend both in Denmark and throughout the Western world. The supervisory role in control of fruit and vegetables belongs to the DFVA.

The general policy holds that food businesses, and thus importers and distributors of fruit and vegetables, are responsible for the safety of their products. Companies must ensure that ready-to-eat fruit and vegetables are free of pathogenic bacteria, and should do so by making demands on suppliers, including the use of clean water for irrigation, appropriate toilet facilities for employees, and the requirement that no pets be allowed in fields where vegetables are grown.

Companies must self-monitor, and both companies and importers must use laboratory tests to check the compliance of their suppliers and to ensure the safety of products for distribution. The DFVA monitors and implements control campaigns and takes samples for analysis from manufacturers, importers and fruit and vegetable distributors to verify such operator compliance.

The DFA takes regular inspection samples of fruit and vegetables to test for harmful bacteria. Sample selection in individual companies is under the supervision of the company itself where deemed necessary and appropriate. Companies are free to publish their test results but are not required to do so. Most controls are carried out as part of a specific control project. Over the years, there have been several control projects on foodborne pathogens in fruit and vegetables, including frozen berries, salads and herbs.

Examples of control projects include a 2011 campaign focused on hygiene among producers of strawberries, cucumbers, peas and the like, across the country. The main focus was on toilets, clean equipment, keeping animals and pests out of production, the use of clean water for irrigation or washing and the training of staff in health risks. The campaign showed that three in 19 were not compliant with all the rules. Particularly, toilet facilities were inadequate, there was a lack of prevention of animal/pest access to container storage and a lack of water analysis.

In 2011, the DFA carried out two control campaigns on importers and traders. In 2010, a campaign was directed against illegal importers of food across Danish borders. All three campaigns focused on control of non-animal foods. Inspection campaigns in 2010 found illegal imports of fruit and vegetables, including herbs. As a result, such products are now the main focus of investigation.

In May 2009, the DFVA initiated ongoing studies of *Salmonella*, *Campylobacter* and *E. coli* in Danish and imported ready-to-eat vegetables and herbs. In the period 2009–2011, it studied about 3400 samples. In 2012, the DFVA investigated a further 2000 samples. Products under investigation include greens, herbs, sprouts, legumes and baby corn. The results are published at www.fvst.dk. Overall, studies have shown that *Salmonella* and *Campylobacter* are present in low levels in lettuce, sugar snap peas, baby corn and herbs.

In 2011, the DFVA prepared a report on initiatives in control with fresh produce. The report lists opportunities to improve Danish control with fresh produce, and makes a number of recommendations, mostly for primary producers and importers. Recommendations include demands on boiling frozen raspberries, demands on bean sprout production and greater supervision of producers and importers of fresh produce.

From 1 October 2012 it is now mandatory for food companies, foodservices and catering industries to cook frozen raspberries that are used in products such as smoothies, cakes and unbaked desserts. The reason for this is that many Danish consumers have become ill through eating frozen raspberries. Consumer education campaigns have also been launched.

27.5 Lessons from the Danish control model

Experience from Denmark, as well as from other countries, shows that in the wake of major foodborne outbreaks, it is often the case that political pressure mounts for a greater degree of control based on the sample analyses of finished foods. It should be remembered, however, that in most cases we achieve the most effective prevention through changes in agricultural practice at the level of primary production. Previous unilateral use of control analyses of end products for the prevention of foodborne diseases have, in practice, proven inefficient.

This realization was the basis for a paradigm shift towards a 'farm-to-fork' philosophy in the mid-nineties. The concept is now internationally accepted, but in fact has only been introduced in a small number of countries, including Denmark. This has meant that a number of the relevant food pathogens are now fought in primary production, when it comes to meat-derived foods, for example; with considerable success in the Danish control of *Salmonella* (Wegener, 2010).

Initially relating to intestinal pathogens in meat, and not the occurrence of similar pathogens in fruit and vegetables, nevertheless the problem of fruit and vegetables often results from product contamination in primary production as well, in this case in the fields, caused by agricultural practices (such as inappropriate use of manure or sewage sludge on crops, or fecal contamination of irrigation water). The most effective solutions to such problems are often found in a primary production system that, because of globalization, may well lie outside Denmark—and even the EU. The importance of proper education and training for both the management and workforce involved in primary production should not be underestimated, and several specific cases of fruit- and vegetable-related outbreaks have led to an effective follow-up, including training and education in exporter countries.

27.6 References

ANON. (2010). Food Safety Inspections: Lessons learned from other countries. Bundes Ministry of Finance, Berlin, Germany.

WEGENER HC. (2010). Danish initiatives to improve the safety of meat products. *Meat Sci.* 84:276–83.

27.7 Appendix: the largest outbreak

The following is the story of an outbreak published in a Danish newspaper.

The child wanted to get out. Contractions were strong and came at short intervals. She leaned against the elevator, while a wave of pain spread across her abdomen. It should not be now. It was her first child, and it was premature. When she had called to the hospital with severe stomach cramps, she had been told to take a hot bath and stay home. No pangs, no access to the maternity ward.

She had cried for fifteen minutes in the shower before she and her boyfriend decided to drive to Horsens Hospital. The contractions started in the car. When they reached the elevator the contractions were so powerful that Lotte Rosendal was uncertain whether the birth was in progress; '*It never happens so fast. It will go away*,' said the midwife who received her at the hospital. But the contractions came back immediately. The midwife put a sensor on Lotte's stomach to measure the contraction interval, but the sensor's needle vibrated in one large oscillation.

Then more staff came, and Lotte Rosendal began her struggle.

Eventually, the sound of a child screaming . . . They put the girl on her stomach.

Forty-five minutes after entering the hospital, Lotte Rosendal became a mother. It was difficult for her to understand. Why had it happened so fast?

The girl was named Naia. She was beautiful, shapely, and unaffected by the great speed with which she had come into the world. Lotte Rosendal and Naia were on the maternity ward for two days to gather strength. But the night before they went home Lotte began to shiver, and even an extra blanket was no help. A doctor thought it was flu, but took a blood test all the same before Lotte and Naia were allowed to leave.

Lotte was called in again the next morning. Her infection counts were unusually high. Doctors scanned her because they suspected debris from the placenta left in the uterus. If so, curettage. But there was nothing. Afterward Lotte was sent home. She was exhausted, and during the long night her fever increased to over 40°C. Lotte Rosendal could no longer distinguish one day from another. But she remembers that they took many blood tests; and the doctors finally found the answer to why she was so sick: She had Salmonella. Not only had it hit her stomach, but had colonized her entire body for weeks and caused a blood infection: bacteremia (septicemia; blood contamination), while she was pregnant. Doctors believed that salmonella had triggered the explosive birth: Lotte Rosendal's body had pushed little Naia out before time, to protect the child.

27.7.1 Detective work

In the heart of Copenhagen, there is an imposing white plastered house, which for more than a century has housed the Statens Serum Institut (SSI), the Danish public body charged with securing national preparedness towards infectious diseases. It was from here that Dr. Thorvald Madsen, in early 1900, extracted serum from horses and used it to produce diphtheria vaccines. Today, horses and syringes are replaced by electron microscopes and petri dishes. Here doctors and hospitals from across the country send samples if they suspect that a patient has been seriously ill and that the infection comes from a major disease outbreak.

On 22 September 2011 the epidemiology department received an e-mail:

> '*We have just serotyped Salmonella Strathcona from three patients in three different hospitals: Hillerød, Hvidovre and Herning. This serotype is extremely rare, and the geographic distribution of the cases is worrying.*'

To the untrained eye the email is dry and incomprehensible. But to Steen Ethelberg, who is a Senior Scientist at the department, it was tantamount to a bomb going off. It told him that three Danish patients had been so seriously ill that a doctor had tested their blood or stool. Salmonella Strathcona was found in all three samples.

This serotype is not a frequent 'visitor' in Europe, since Salmonella Typhimurium and Salmonella Enteritidis are the main serotypes that each year send thousands of people in Denmark onto the toilet and into bed with diarrhea, abdominal cramps, and fever. Salmonella Strathcona is among the rarest of the approximately 2500 known Salmonella serotypes. In many years, only two people have had it in Europe. The fact that the three patients in Hillerød, Hvidovre, and Herning were infected with Salmonella Strathcona at the same time was not just 'worrying'. Steen Ethelberg knew that when something as rare as this serotype hits three Danes in three regions of Denmark at the same time, an unusual outbreak has begun.

27.7.2 The stowaways

Food has gone global. The seasons no longer exist in the supermarket, where we expect to find the same fresh vegetables and fruits all year round. We can buy blueberries from Chile, asparagus from Peru or apples from New Zealand. Demand has multiplied imports of foreign fruits and vegetables. In recent years, Denmark has imported 45 mill kg of tomatoes and 73 mill kg of apples annually.

This development has serious consequences for food safety, because *cargo ships can contain stowaways*: Small microbes and pathogens can come in with imported fruit and vegetables. Things like salmonella, campylobacter, VTEC, and listeria can cause serious illness. And that has indeed happened. As Steen Ethelberg from Statens Serum Institute puts it:

> '*Earlier you went down with traveling diarrhea while on vacation. Now, you get it at home. There are more fresh vegetables from exotic countries in supermarkets, and there are far greater risks of products you get in from Kenya, than those produced here in Denmark.*'

The Danish Food Administration (DFVA) stresses that the basics are healthy to eat: fruits and vegetables. However, from 2005 to 2010 the Administration has recorded 24 outbreaks where fruits and vegetables were involved in making Danes ill. Officially 2275 were registered sick in these outbreaks, but the actual figure is estimated to be 10–20 times higher, as many as 45 000 Danes may have been affected. This figure is only for the known outbreaks. Many outbreaks go undetected.

In all outbreaks of fruit and vegetables in Denmark since 2005, the source of infection has been fruit or vegetables from abroad. This is because we import more from countries where farmers use water contaminated with bacteria from animals or sewers to irrigate. We also eat more green. More than the daily 600 grams, which is the official recommendation, there is a focus on health and green food.

'The Danes have been traditionally meat lovers, and therefore meat has until recently mostly filled in the disease picture. But now people eat much greener, we are eating more imported green and the raw-food trend is starting to catch up. In the United States, where health trends started, infections from fruit and vegetables consumption is the main food safety issue. We therefore expect that the problem will grow in Denmark,' says Prof. Hoorfar from National Food Institute, Technical University of Denmark.

2011 was an *annus horribilis* for food safety. In the first half of 2011 alone, there were 10 outbreaks and 297 very sick patients in Denmark; a number that can be multiplied by a factor of 20. But the worst was in Germany, where 50 people died, 900 went down with kidney failure, and more than 3000, mostly German patients, were seriously ill when they ate VTEC-infected bean sprouts from a German manufacturer. After that, Danish cucumbers, tomatoes, and lettuce were initially suspected, but the infected vegetable was shown to be fenugreek seed imported from Egypt (see Chapter 25).

27.7.3 Following the patient's food consumption

Lotte Rosendal was patient number 14 with Salmonella Strathcona, which Statens Serum Institut registered. This was the 6 October 2011, when five new cases came in. The outbreak was now nationwide, and at the same time nine patients were reported as infected in Germany with the same salmonella.

For two weeks, SSI, the DFA and the National Food Institute at Technical University of Denmark (DTU Food) attempted to trace the source of the infection. But no luck. So the food, which was probably the cause of the outbreak – whether it was fowl, fish or fruit – could still be bought and eaten by Danes across the country. The authorities had interviewed all patients about what they ate, drank and cooked three weeks before they were sick: Have you been traveling? Have you eaten at a restaurant, had take-away in a cafeteria or at a hot dog stand? Did you eat chicken, eggs, beef? And so on.

The interviews gave no clear answer. But they showed that patients were severely affected with high fever, diarrhea with 20–30 toilet visits a day, and admissions to hospital. The most common sources of infection – charter trips, sporting events and canteens – were excluded. And the most common meats such as chicken and beef were not the villain. There were slightly more women than men among the sick. And those infected did not eat as much meat and cold cuts as the average Dane. In fact, they ate more fruits and vegetables, and suspicion was therefore aroused.

But whether the infection came from grapes, salad, or something else, Steen Ethelberg could not tell. The only evident common denominator in many interviews was that those infected had shopped in the supermarket chain Rema1000. If you could track what they had bought there, the source of infection could be traced. All the patients were therefore given a phone call.

Lotte Rosendal, after 4 weeks in the hospital and on antibiotics, was about to begin eating properly. She was asked to log on to her online banking account and see when she had shopped at Rema1000 supermarket. From the information, Rema1000 could use a central database to find receipts for items purchased, and send these on to SSI.

This was time consuming, and while authorities searched for the source of the infection the outbreak came to an end. The last Danish patient – No. 43 – was recorded in the middle of October. The control system had worked so slowly that the product which had made at least 43 Danes seriously ill had managed to remain on sale for several weeks, and was probably eaten by several hundred Danes. Neither the supermarkets nor the Danish Food Administration had detected that the product contained Salmonella Strathcona.

27.7.4 Not the first time

This is far from a unique problem. *Berlingske* (the newspaper) has reviewed the cases of fruits and vegetables since 2006, when a supermarket, importer, or the Food and Drug Administration, has withdrawn a product because it was found to contain things such as salmonella, campylobacter, listeria or pesticides. In every third case, the withdrawal has happened just before or after the last date of sale. So, unfortunately, the consumers have had ample time to purchase and eat the contaminated product.

On 25 October, the DFA and SSI were convinced that the infection was caused by a product from Rema1000. Eight out of 10 affected families had purchased the same product in the green department: Italian Datterino cherry tomatoes at Danish supermarkets that received 19 800 kilograms of these tomatoes from a tomato producer in Sicily, through a wholesaler in Verona. The tomatoes had been for sale at Rema1000 all over Denmark in the period the Danes were sick. The manufacturer had in the same period delivered the same type of tomato to supermarkets in Germany and Austria, which also had patients with Salmonella Strathcona.

However, the authorities were left with a substantial problem: The villain was found, but the evidence was gone. Although the DFA and SSI were confident that the Sicilian tomatoes were the source of infection for 2011's largest salmonella outbreak, they could not prove it. It had taken so long to identify the source of infection – more than a month – that the tomatoes had long ago been either eaten or spoiled, and it was therefore impossible to test and showcase a proof – a Datterino-tomato with Salmonella Strathcona on it.

For the same reason the Danish authorities did not reveal the name of the Italian supplier and tomato producer when the public was told in November 2011 that tomatoes from Italy had infected 43 Danes. The Danish Food Administration has not disclosed the name of the manufacturer.

27.7.5 Back to the farm

Remagrønt A/S Vejle; it says on the sticker on the floor of a warehouse in the small Italian town of Trevenzuolo, not far from Verona. *Berlingske* journalists have travelled to Verona to follow the tomatoes all the way from farm to fork, and trace the leads to Univeg Italia, which is one of Italy's largest food exporters. From Univeg's cool storage Italian fruit and vegetables are sent all over Europe, and it was from here the salmonella-suspected tomatoes were sent Denmark.

Univeg Italia had received the 20 tonnes of tomatoes, which were delivered to the Danish supermarket, from a single supplier: A tomato grower from the town of *Vittoria* in southern Sicily, which Univeg has used for many years, and against which no previous allegations have been made, according to the company.

In the middle of November 2011 Univeg received a message that the DFA was confident that the company had sent salmonella contaminated tomatoes to Denmark. The Italian food authorities in Verona and Sicily became involved. Univeg's warehouse in Verona was examined and the local Sicilian police, the Carabinieri, were sent out to the tomato producer.

All this happened on 25th November 2011, more than three months after the tomatoes from the site were likely infected with salmonella, and long after the salmonella outbreak had gone away. When *Berlingske* contacted the police in Sicily, it was explained that samples of soil, tomatoes and water were taken from the manufacturer. The journalists were also told that the tomatoes came from cooperative Piano Stella, who has a post address in Vittoria. *Berlingske* tried for two weeks to get a comment from Piano Stella about the tomato case, but had no luck.

27.7.6 Who to blame?

However, Univeg says that there was no trace of Salmonella, either at the producer's site in Sicily or in the storage warehouse in Verona. That in itself is not so surprising, since the samples were taken two months after the salmonella outbreak. Despite this fact, Univeg contended that the controls acquit the company and the producer; the Salmonella contamination cannot have stemmed from Italy, but rather from Denmark: 'Our technical assessment is that the salmonella outbreak may be due to local conditions in Denmark,' points out Mr. Benvenuti, who is responsible for food safety at Univeg. He takes aim at the Danish authorities: 'Often food authorities prefer to identify foreign sources of infection first, rather than looking at their own local sources.'

In Denmark the authorities are convinced that the outbreak originated from Italy. SSI and DFVA have in their final outbreak report listed 10 reasons why the infection may have come from the Italian tomatoes. For example, nearly all the sick Danes had bought the same type of Italian tomato at the same supermarket chain. In addition, the same type of tomato was delivered to Germany and Austria, where people became sick of the same rare salmonella serotype. However, since the outbreak took so long to trace, the evidence was gone, either eaten or spoiled. And without a tomato with salmonella, it is difficult to convince the billion Euro Italian vegetable wholesalers that the company must accept blame.

27.7.7 What now?

Thus ends the saga, without consequences for those responsible; the DFA can issue fines for companies that impair food security, but only in Denmark. This summer saw a catering business in the center of Copenhagen (Halmtorvet) pay DKK 40 000 in fines after 44 Danes contracted salmonella from sliced roast beef. However, the Italian wholesaler and manufacturer, which seems to have infected at least 43 Danes with salmonella last year, gets away with neither penalty nor consequence.

So what is the outcome? After several months of effort to hunt down the source of infection for the Salmonella Strathcona outbreak: not much. Emergency preparedness officer Kim Sigsgaard from the DFA:

> *'We had to go through many stages in the investigation, and the process could not go much faster. When we found the source, it was such a long time gone that the tomatoes were not on the market anymore. We may then find it difficult to determine liability. However, what we learned from the tomato case was that we became aware of a new type of salmonella. So the next time we see one, we can hopefully respond faster.'*

Jeffrey Hoorfar is doubtful: *'This is not enough.'*

He, in addition to being a professor at the National Food Institute, is leader of an international group of scientists working to improve food safety in fruits and vegetables, and he believes that control systems react too slowly:

> *'The control systems have not kept pace with the increasing number of outbreaks caused by imported fruits and vegetables. This is not prioritized correctly. It is time to go out proactively and take samples. I do not think that the authorities today take enough samples and keep a close enough eye with this. Authorities react only when people become sick,'*

says Jeffrey Hoorfar, adding:

> *'Control of animal products such as meat and eggs are much tougher, and for animal products we have Source Attribution Accounts. This means that we are proactive and monitor the animal production closely. That is missing for fruits and vegetables. One must simply get on with it now.'*

Source: *Berlingske* (the largest Danish newspaper), Sunday 30 September 2012.

28

Mushroom production in China: the illegal use of fluorescent whitening agents (FWAs) and related outbreaks

Z. Xing and X. Zhao, Shanghai Academy of Agricultural Sciences, China

DOI: 10.1533/9781782420279.5.401

Abstract: Mushrooms are highly nutritious food sources, which can be cultivated on cheap and usually readily available raw materials. China is the world's major producer and consumer of mushrooms, and output in 2011 (22.6 million tons) accounted for more than 75% of the global total. A major obstacle to the marketability of white mushrooms (e.g. *Agaricus bisporus*) is browning of the fruit body that occurs after harvesting. Illegal use of fluorescent whitening agents (FWAs) to offset the adverse effects of browning has now been reduced to very few sporadic cases as a result of measures introduced by the Chinese Ministry of Agriculture.

Key words: mushroom, *Agaricus bisporus*, fluorescent whitening agents, shelf-life.

28.1 Introduction

Mushrooms are a highly nutritious food source, a valuable cash crop, a source of high-value metabolites with medicinal properties and can be cultivated on raw materials that are cheap and usually readily available. In China, these qualities have been recognized for many centuries and descriptions of wild mushrooms are recorded in literary works dating back before the Qin and Han Dynasties (~ 210 BC). What is thought to be the first record describing the artificial cultivation of *Flammulina velutipes* (Jin Zhen Gu, 金针菇 in Chinese, the 'golden needle' or 'winter mushroom') appears in the ancient Chinese agricultural text 'the collection of main agricultures in four seasons' (Si Shi Zuan Yao, 四时纂要). The book, published during the Tang Dynasty around 900 AD (Lv, 2006), describes in detail the preparation of the growth substrate, spawning (inoculation with fungal spores), watering and eventual harvesting of the mushroom fruit bodies.

China is estimated to have between 1500 and 2000 indigenous edible mushroom species, of which 981 species have been identified (Mao, 2004). Following the introduction of Reform and Open Policies in 1978, the Chinese mushroom industry expanded along with other industries and the country is now the major producer and consumer of mushrooms. According to the latest data available, over 90 species have been domesticated and some 60 species have been cultivated on an industrial scale. Since 1978, total mushroom production in China has increased from 60 000 tonnes (5.7% of total mushroom production worldwide) to 21.5 million tons (>75% of total global output) in 2010 (Fig. 28.1). Mushroom production is now the sixth-largest agricultural-based industry in China, and highly important for reconstruction in rural areas, particularly in relatively remote and poverty-affected regions with comparatively abundant natural resources.

Although the bulk of production is consumed domestically, China's contribution to the international mushroom trade has also increased dramatically during the last three decades. In 2011, overseas exports to Japan, the USA, Korea, the European Union, the East Asian Union, Russia and elsewhere amounted to 500 000 tonnes. Major exported mushroom products include dried or fresh *Lentinula edodes* (Xianggu, 香菇), dried *Auricularia auricula* (Mu'er, 木耳, wood ear), *Agaricus bisporus* (canned or preserved in brine), dried *Tremella fuciformis* (Yin'er, 银耳, white jelly ear), medicinal mushroom species such as *Ganoderma* and *Cordyceps*, and mushrooms collected from the wild (e.g. fresh *Tricholoma matsutake* (Songrong, 松茸); dried *Boletus edulis*). In recent years, dried mushroom exports have increased, whereas sales of canned and brine-preserved products have decreased.

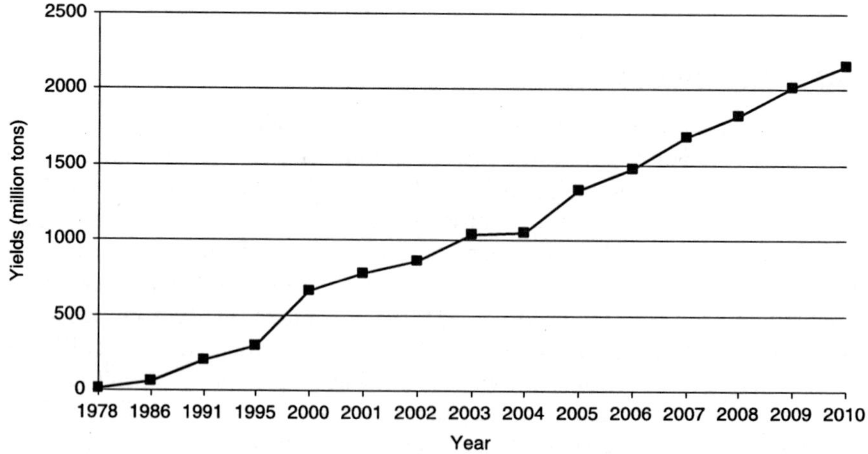

Fig. 28.1 Annual mushroom yields in China (1978–2010).

28.2 Contamination of mushrooms with fluorescent whitening agents (FWAs)

Agaricus bisporus (Lange) Imbach (the white button mushroom, champignon) is the most commonly cultivated and consumed mushroom in Western Europe and North America. It contains high levels of proteins, carbohydrates, minerals (e.g. potassium, iron and phosphorus) and vitamins (e.g. vitamin C, niacin and thiamine), and is low in fat and calories (Clarke, 2007; Wu *et al.*, 2006). The earliest record of button mushroom cultivation dates back to France in the early eighteenth century, and it was introduced into China in the 1930s. Commercial-scale cultivation has undergone a major expansion since the 1980s and, in terms of output, *A. bisporus* now ranks third (after *Pleurotus* spp. and *Lentinula edodes*, and above *Flammulina velutipes* and *Auricularia*) spp. among the mushroom species produced in the country. However, although there are now many highly regulated industrial-scale button mushroom cultivation facilities in China, much of the country's output is still produced by small-scale farmer/family-type operations where the normal levels of official supervision have yet to be fully implemented. This raises the possibility of potential food safety risks, such as contamination with pesticide residues and quality deterioration after harvesting.

Mushroom appearance is the most important criteria influencing consumer purchase, and supermarkets and fresh markets in China demand fresh, white, firm, closed-cap *A. bisporus* of good flavour. However, mushrooms have a relatively short shelf-life compared with most ready-to-use vegetables because of their higher respiration rate and the absence of barriers to protect them from water loss or microbial attack. A key feature affecting consumer acceptability is the browning reaction affecting various fresh-cut produce including mushrooms. Browning occurs when endogenous polyphenoloxidase enzymes (e.g. tyrosinase) oxidize monophenols (also present in the mushroom tissue) into *o*-diphenols and *o*-diphenols into quinones, which undergo further non-enzymatic polymerization leading to the formation of brown melanin-like pigments. The process is accelerated when mushrooms are bruised, cut or subjected to microbial attack during cultivation, picking, transportation and/or storage. Browning is a major obstacle to the business survival of many small-scale mushroom farmers in China, who lack the resources required to implement the necessary preventative measures available to the large-scale producers.

Rumours surrounding the application of fluorescent whitening agents (FWAs) to button mushrooms to offset the adverse effects of browning apparently began circulating almost 10 years ago, although they received little attention at the time. It is believed that FWAs were initially used in 2002 when an advertisement for a chemical named 'fresh-keeping and whitening agent of mushroom' appeared in edible fungi magazines. This illegal activity first came to the attention of the Testing Center for Edible Fungi Quality and Safety of the Agricultural Ministry of China (Ministry of Agriculture, MOA) in Shanghai in 2005.

Following the detection of FWAs on button mushrooms, the MOA immediately initiated an investigation and, that same year, a comprehensive survey of mushroom quality and safety was conducted throughout China. Mushroom samples (not just *A. bisporus*) were collected from fresh food markets and supermarkets located in 11 cities, and contaminated samples were found in each case. Affected mushrooms included not only samples of *A. bisporus*, but also other white mushrooms such as *Hericium erinaceus* (Lion's Mane Mushroom) and *Pleurotus nebrodensis* (Funcia di basilicu). In cases where the original test results were confirmed, the MOA instructed local ministry departments to investigate further and ensure that the illegal application of FWAs to mushrooms was strictly prohibited. In the 3 years following the original tests (2006–2008), MOA-directed examinations for detecting FWA-contaminated mushrooms were introduced to 38 cities and, from 2009 extended further to include fresh markets and supermarkets in 32 provinces' cities throughout the country. The policy has been highly successful as, in 2012, only a few sporadic cases of FWA-contaminated mushrooms were recorded throughout the whole of China.

28.2.1 Challenges

The case of FWA-contaminated mushrooms has parallels with the 2008 Chinese milk scandal involving the adulteration of baby milk formula with melamine. Prior to 2008, national and industry milk standards did not require testing for melamine. Similarly, specific regulations relating to the application of FWAs to mushrooms were not in place as the use of such agents was an abuse and completely illegal. Consequently, there was no methodology available for the qualitative and quantitative determination of FWAs in mushrooms. Based on methods adopted by pulp and paper, we drafted the Chinese Agricultural Industry Standards of Qualitative Analysis Method for FWA on mushrooms in 2006. However, there are several difficulties associated with designing a fully comprehensive quantitative test applicable to FWAs in mushrooms. There are literally hundreds of FWAs available, and the chemical structure of each FWA found on mushrooms would first need to be determined. Quantitative analysis would also require procedures for extracting FWAs from contaminated mushroom tissue, data on extraction efficiencies and FWA standards of appropriate purity. It would also be useful to conduct a food safety risk assessment exercise linking FWA contamination levels and mushroom consumption data. However, the focus of short-term concerns is to prevent contaminated mushrooms reaching the consumer and, as FWAs are not readily removed over time, evidence of adulteration can easily be detected by exposure of the fruit bodies to UV light at wavelengths of 254 nm and 365 nm.

28.3 Significant factors affecting the outbreak

28.3.1 Significance of the chemical agent responsible

FWAs, also called optical brighteners, are used in detergents as textile whiteners, for paper brightening, and as colour-correcting or brightening additives in cosmetics such as shampoos, conditioners and eye makeup. They absorb energy in the UV part of the spectrum (340–370 nm) and re-emit it in the blue portion of the visible spectrum (420–470 nm), making any treated surface appear whiter, brighter and fresher (Smulders *et al.*, 2002). The Color Index includes approximately 400 types, of which less than a quarter are produced commercially.

FWAs can effectively offset the adverse effects of browning and produce the whitened appearance of fresh button mushrooms at very low cost, which is why the illegal application of FWAs began in many cities in China.

28.3.2 Geographical and regional significance, climate and general consumer perceptions

China, with its large land mass, complex topographical features and diverse climate, provides ideal conditions for cultivating many different mushroom species. Its manifold agricultural activities also generate huge quantities of lignocellulosic wastes that can be used as substrates for mushroom cultivation. In the 1980s, the button mushroom was cultivated mainly indoors in eastern and southeastern China and, as most of the fresh product was processed (canned) and exported to overseas markets, FWAs were not an issue. Growing seasons were then confined to spring and autumn but, after 1995, button mushroom cultivation spread to northern China where the cooler summer climate permitted cultivation throughout the summer months. Now, by controlling the temperature, humidity and illumination in the mushroom cultivation house, button mushrooms are produced all year round throughout China. However, because cultivation is usually carried out in rural areas, and most fresh mushrooms are consumed in urban areas, it can take at least 3 days to transport the mushrooms from the farms to the supermarkets. Even supermarkets, in order to reduce overheads, seldom store fresh mushrooms under refrigerated conditions and, combined with a limited capacity cold-chain transportation system, quality deterioration of fresh mushrooms became a serious problem. Usage of FWA was perceived as a solution by unscrupulous traders, even though the practice is illegal. Treated mushrooms had a fresh, white appearance even when the inner core of the fruit body had begun to deteriorate. Contamination was more prevalent in major cities such as Shanghai and Beijing, and in supermarkets compared with fresh wholesale markets.

Consumers are now often able to tell if button mushrooms have been improperly treated based on appearance. Even in 2005, they noted that 'dry' mushrooms were more expensive than 'wet' mushrooms (FWA-treated), but were unaware of the reason. In order to reduce costs and store the mushrooms for longer, restaurant owners also preferred to buy 'wet' mushrooms. Realizing the negative impact FWA-treatment would have on their livelihood, some mushroom dealers attempted to demonstrate the authenticity of their mushrooms by providing them together with mycelium and casing material (a mixture of peat and limestone added as a surface layer to the compost to promote mushroom formation). Unfortunately, this represented a potential source of other forms of contamination and food safety problems.

28.4 Regulatory and economic aspects

28.4.1 Regulatory aspects

The Chinese mushroom industry is presently under the supervision of several government departments. The MOA is responsible mainly for mushroom cultivation and production, and plays a pivotal role in the promotion and development of the mushroom industry. Since 2007, the MOA has organized and supported a group of senior mushroom scientists and experimental stations, known as the National Mushroom Industry Technological System Project. This project provides continuous and stable financial support for mushroom research and cultivation for 10 years, and represents a major stimulus to the mushroom industry.

According to the Food Safety Law of China implemented in June, 2009, the Ministry of Health (MOH) was given overall responsibility for coordinating the management of food safety matters in the country. However, the MOA is charged with maintaining mushroom quality and safety before processing and cooking. The General Administration of Quality Supervision, Inspection and Quarantine of China (AQSIQ) oversees the operations of the mushroom processing companies, the State Administration for Industry and Commerce of China (SAIC) is responsible for mushroom products in the markets, and the State Food and Drug Administration of China (SFDA) supervises the quality of mushroom dishes served in restaurants and staff canteens. In February 2010, the State Council of China established a new central government unit, the National Food Safety Office, with responsibility for all food safety issues in China.

Only food additives listed in the National Food Safety Standards for Food Additives Application (GB 2760) can be legally applied to certain foodstuffs, and the amount of residue allowable in each case is subject to strict limits. FWAs have never been included in GB 2760, and it is illegal to apply FWAs to mushrooms under the Food Safety Law of the PRC (2009) and the Agri-products Quality and Safety Law of the PRC (2006).

28.4.2 Economic aspects

Although mushroom cultivation in China has a long history, the mushroom industry itself has flourished only during the past 35 years. Its positive impact has been felt particularly in relatively remote and poorer regions of the country where it has made a major contribution to the development of the Chinese rural economy. Over 30 million workers are now directly engaged in the mushroom industry, which not only has increased per capita incomes but has improved the general well-being of the populace through the introduction of new nutritional and medicinal resources. Many varieties of cultivated and wild species are exported. For example, some 2.8 million tonnes of fresh button mushrooms were produced in China in 2010, of which 270 000 tonnes (canned or fresh) were exported. As a result of heavy investment, Chinese companies can now produce mushrooms in controlled environmental facilities (temperature, relative humidity, illumination).

In Shanghai alone, several companies produce more than 10 tonnes per day, with the largest output reaching 30 tonnes. In 2011, some 450 industrial-scale facilities were operational, with a total daily output of 3183 tonnes and an annual production of 990.4 thousand tonnes, representing over 4.8% of the total Chinese mushroom production (20 427 thousand tonnes) in that year. It is estimated that the output of these companies will reach 30% of total production within the next decade. Hence, any safety issue that causes the Chinese authorities to impose restrictions on production and sales while an investigation is conducted would have a disastrous effect on the industry. Furthermore, as Chinese consumers tend to regard 'mushrooms' collectively in the context of food safety (i.e. they do not differentiate between individual species), any safety issue affecting one species will adversely affect mushroom consumption across the board and have an immense negative impact on the market for this agricultural product.

28.5 Production and market issues and further developments

The quality and safety of fresh mushrooms is a main focus of concern of the Chinese regulatory authorities and the mushroom industry as a whole. Immediately the central government became aware of the illegal use of FWAs, local food and agricultural departments were instructed to organize the checking of fresh white button mushrooms in markets and mushroom cultivation facilities. All fresh wholesale markets and supermarkets were required to examine each batch of fresh mushrooms.

As the cultivation of a particular mushroom species is often confined to a few specific geographical locations within China, a major problem for mushroom growers is to maintain the postharvest quality of their products when, in many cases, these have to be transported over large distances before reaching the marketplace. Although refrigeration is most often used to reduce postharvest deterioration, the cold chain system for fresh mushrooms is relatively underdeveloped in China at the present time. Both the government and the industry recognize the dangers posed by the use of illegal additives and, even though the development of an effective cold chain system will impose a considerable financial burden, implementation is seen as a key future requirement.

28.6 Troubleshooting approaches and laboratory methods

During the investigation, FWAs were found to bind tightly to the mushroom surface, and soaking contaminated fruit bodies in distilled water for 30 minutes failed to release any of the chemical. FWA levels can be measured quantitatively using fluorescence spectrophotometry or high performance liquid chromatography. However, both these methods require some degree of sample preparation as well as reference standards to determine the exact content of FWA. Contamination can readily be detected qualitatively simply by exposing mushrooms to a UV lamp operating at wavelengths of 254 nm and 365 nm.

28.7 Future trends

The eminent Chinese mushroom biologist, Professor Shu-Ting Chang, has emphasized that both the mushrooms themselves and derived products must be of good quality and free from potentially harmful substances. Three levels of control are envisaged affecting: the quality and safety of the raw materials used in production; the cultivation and manufacturing processes; and the quality and safety of the final products. Good quality control is essential for maintaining consumer confidence and thereby creating a stable and expanding market. This involves closer scrutiny of mushroom-growing operations that ensure high-quality output on a regular and reliable basis. Cultivating mushrooms under more environmentally friendly conditions characterized, for example, by clean air and water supplies, and the absence of insect pests and harmful chemicals, is vital. Important areas for future research include mushroom breeding, cultivation technology, pest and disease control, and longer-term maintenance of 'freshness' by improved packaging, storage and transportation systems, all of which will lead to improvements in quality and safety.

28.8 Acknowledgements

We thank Professor Yu Li, Dr Qi Tan and Dr Yehan Cui for valuable discussions. This work was supported by grants from the Ministry of Agriculture of China and the Ministry of Science and Technology of China (No. 2008BADA1B08-3).

28.9 Sources of further information and advice

VAN GRIENSVEN, L.J.L.D. 1988. *The cultivation of mushrooms*. Darlington Mushroom Laboratories, Rustington, UK.

KRAMER, J.B. 1992. Fluorescent whitening agents. In N.T. DE OUDE (ed.), *The Handbook of Environmental Chemistry, Vol 3, Part F, Anthropogenic compounds. Detergents*. Springer-Verlag, Berlin.

28.10 References

CLARKE, J. 2007. *Mushrooms: the new super food*. Report for the Mushroom Bureau, UK. pp. 1–12.

LV, Z.Z. 2006. *Mushroom cultivation*. Higher Education Press, Beijing, China.

MAO, X.L. 2004. Chinese mushroom species names and their major characters. *International Agriculture Trade*, 90:48–51.

SMULDERS, E., W. RYBINSKI., E. SUNG., W. RÄHSE., J. STEBER., *et al.* 2002. *Laundry Detergents. Ullmann's Encyclopedia of Industrial Chemistry*, Wiley-VCH, Weinheim.

WU, S.L., X.M. SUN., B. WANG., F. YANG, and G.P. GU. 2006. Analysis of main nutrition components in *Agaricus brunnescens* fruitbodies. *Chinese Wild Plant Resources*, 2:47–52.

29

The case of lemons in caves: a sustainable storage system for Turkish lemons

A. I. Nicolau, Dunarea de Jos University of Galati, Romania

DOI: 10.1533/9781782420279.5.409

Abstract: This chapter presents details of the storage of Turkish lemons in warehouses carved out of tuffs (soft rock) in Cappadocia, Turkey. The challenges encountered when using this storage method are also discussed, including the importance of applying Good Storage Practices and preliminary programmes such as storage facility hygiene, personnel hygiene and waste management to prevent losses caused by *Penicillium*. Comments on the interest of fruit and vegetable companies to carve modern warehouses in the volcanic rocks of Cappadocia are also made. The chapter also reviews the advantages that Turkey has for producing, storing and exporting lemons, and regulatory and economic aspects of storing lemons in tuff warehouses are briefly reviewed.

Key words: lemons, caves, *Penicillium*, Cappadocia, warehouse, storage.

29.1 Introduction

A visit by the author to Cappadocia in 2008 revealed a natural system for storing fruits and vegetables on an industrial scale. Cappadocia is situated in the centre of Turkey on the Eastern part of the Anatolian plateau, which is considered to be Turkey's most visually striking region because of the caves, clefts, chimneys and folds that erosion has formed in the volcanic rock. On one such rock, the words 'Lemon house' were painted. A thick layer of rotten lemons was found opposite to the entrance of a cave, inside which, in 2 m high rooms, about 10 workers were sorting, grading and packaging the lemons under moderately intense electric lights. Two more workers were piling little wooden crates with lemons individually packed in wax paper.

The cave was a store house, carved out of the rock to function as a naturally cool storage depot. Such natural storage facilities are one of the most important sources of wealth in the region.

29.2 Microbiological problems during storage

The most common and serious problems occurring during storage of citrus fruit in tuff warehouses like the one described above are green and blue moulds, represented by *Penicillium digitatum* and *Penicillium italicum* respectively. Over 75% of citrus rot is attributed to these moulds, which can cause significant economic losses to lemon distributors. This particular lemon house had a significant problem with mould and had lost substantial quantities of lemons to penicilli development, as a result of the lack of fungicide use.

29.3 Troubleshooting approaches

Lemon losses during the storage period are to be expected in any storage system, but are difficult to predict. The magnitude of the loss is dependent on several factors, including the variety and source of the fruit, management of the storage facilities and duration of storage. Even if all these factors are constant, it is possible to sustain very different levels of loss from year to year. The increase in total sales revenue must be greater than the cost of storage loss plus storage costs in order for the storage to be considered profitable. Minimizing loss is therefore of paramount importance to a fruit business.

Because the spores produced by *P. digitatum* and *P. italicum* are easily propagated in the air or by contact among fruits, contamination and environmental conditions that favour mould growth must be avoided. Inside the lemon house in question, the air was circulated by convection. However, the optimum storage temperature and humidity was not attained in all areas, because crates containing lemons were laid directly on the floor without space between the crates and the walls (Fig. 29.1). In order to ensure that all areas of the warehouse are properly ventilated, pallets should be placed underneath the crates, which should be piled at least 10 cm from the walls. A waste management system should also be created and adhered to. The decomposing lemons placed close to the door outside the lemon house in Cappadocia served to spread *Penicillium* spores around the facility (Fig. 29.2).

Recommendations for improvement in this particular instance could include:

- collection of spoiled lemons in closed containers;
- regular and frequent removal of waste material from inside the warehouse;
- relocation of the area designated to produce waste to a safe distance from the storage facility;
- use of the storage area for temporary storage only of produce waste.

In addition, workers should not touch a healthy lemon after touching one spoiled by penicilli. Workers should be made aware that species such as *P. italicum* are able to penetrate even the uninjured peel of lemons, and can spread much more easily from one lemon to another than *P. digitatum*.

The clothing worn by the workers (Fig. 29.1) also demonstrated that a safety and quality system was not being implemented correctly. The author recommended

Fig. 29.1 Interior of a tuff warehouse in Cappadocia. The crates with lemons are laid down directly on the floor and placed against the walls, blocking the proper circulation of the air and generating areas where condensation occurs. Workers are not wearing any special equipment.

Fig. 29.2 Site designated for waste near the warehouse. Moulded lemons are kept in the open air, facilitating the spread of *Penicillium* spores.

to the lemon house manager that he start by implementing a preliminary safety and quality programme and train the workers in Good Storage Practices, hygiene in the storage facility, personal hygiene and waste management. No feedback has been received from the manager of the lemon house at the time of writing.

29.4 Challenges for lemon storage

Maintaining the quality and safety of lemons necessitates proper temperature and relative humidity management as well as care during postharvest handling operations. These actions help to slow down fruit respiratory processes, and to minimize microbial contamination and physical injury.

Lemons should be kept at temperatures between 7 °C and 12 °C, and air relative humidity between 85% and 95%, in order to successfully store them for off-season availability (University of California, Davis, n.d.). Storage in such conditions assures a maximum shelf-life of 6 months, a period long enough to last over the winter before harvest of a new crop. Adequate ventilation must be maintained to promote the removal of ethylene from storage rooms and reduce senescence and incidence of fungal decay. It is recognized that proper storage of lemons also improves juice content, flavour and colour (Gross and Smilanick, 2004).

From this point of view, the caves in Cappadocia offer the ideal conditions for the storage of lemons. These man-made rock structures are carved in soft tuffs that have good thermal isolation properties. Temperature and humidity measurements in different areas of such warehouses have indicated that variations in climatic conditions within the caves are very slight when compared with those outside on the surface (Aydan and Ulusay, 2003). Temperatures in the caves are fairly stable at 10–15 °C throughout the year (eNotes, n.d.) with an average temperature of 12.8 °C (Janick, 2005).

The different layers of tuffs provide varying internal atmospheric conditions that are used in a variety of ways according to need. Because climatic conditions differ from one storage area to another, different caves are suited to different food products. For example, certain caves are indicated for the storage of citrus fruits, whereas others are more suited to storing potatoes, apples, grapes, dried vegetables, cheese or flat bread (*yufka*) (Emge, 2011).

The use of caves for storing fruit may have initially inspired food technologists to develop storage techniques using modified atmospheres (MAS) and controlled atmospheres (CAS).

29.5 Geographical and regional significance, climate and general consumer perceptions

It is considered that the Turkish citrus fruit supply chain is 'sitting on a gold mine' because of the country's geographic position. Turkey has several regions favourable for growing citrus fruits, including lemons. Secondly, as explained above, seasonal fruits and vegetables can be stored in the Cappadocia region at very low cost. Road links are also favourable to the fruit trade, with fruit exporters able to supply Romania in 2 days, Ukraine in 4 days and Russia in 5 or 6 days. Another advantage is represented by the organization of export companies, which often maintain a family structure and run sales offices in the recipient countries. In addition, the range of produce available matches the present expectations of Romanian, Ukrainian and Russian markets, where price is a very important factor (*FRuiTRoP*, 2008).

Romania is a traditional market for Turkish lemons, but these fruits are also present to a lesser extent on the market in other EU countries such as Germany, the Netherlands and France. The absence of product variety demanded in the Western EU markets and the existence of competing countries (Spain, Greece and Cyprus) do not make Turkey the preferred exporter for lemons in this area. However, Turkish lemons are often accepted in the Western EU when other competitors are not able to meet the market demands (Zenginoglu and van Dijk, 2006).

29.6 Regulatory aspects

Storage is an important link in the food chain for ensuring the quality and safety of citrus fruits. In all member states of the European Union, as well as in candidate countries such as Turkey, the storage of food and food products should comply with a consolidated version (2009) of the Regulation (EC) No 852/2004 of the European Parliament and of the Council on the hygiene of foodstuffs. Because the tuff warehouses of Cappadocia are not conventional storage facilities, they cannot entirely fulfil the requirements of this regulation. For example, the tuff warehouses cannot have 'impervious walls made of painted plaster, epoxy resin or similar coatings, ceramic tiles, PVC', but they can have surfaces that are 'smooth, durable, light in colour and capable of being cleaned without deterioration'. Therefore, those who store lemons in tuff warehouses must prove that hygienic standards are maintained inside and that this type of storage does not put consumer safety at risk.

Lemons exported to EU countries should comply with the relevant EU marketing standards. Thus, although Turkey is not yet a member of the EU, Turkish companies have to prove that their lemons conform to marketing standards for the EU market to accept their product. Conformity can be demonstrated with a Certificate of Conformity (CoC) and accompanying documents (invoices or other documents indicating the country of origin, the quality class, the variety or a Certificate of Industrial Use, if lemons are intended for processing). EU Regulation 543/2011 contains the 10 specific marketing standards applying to citrus (including lemons) and other fruits. These 10 specific standards are in line with the internationally agreed UNECE (United Nations Economic Commission for Europe) standards.

29.7 Economic aspects

Turkey grows 787 000 tonnes of lemons annually and exports around 311 000 tonnes. Every year, tonnes of citrus fruits are brought to the Cappadocia tuff stores from Mediterranean orchards, to be stored before being exported (Fruit Logistica, 2011). Storage in such warehouses brings benefits of millions of lira to the region every year.

Underground or semi-underground warehouses such as those found in Cappadocia are considered more advantageous than above-ground facilities. The only energy costs associated with storage in such facilities are air conditioning and lightning. In some caves, fans are used for air conditioning, whereas others use natural shafts to help the air circulate through convection.

The warehouses provide 30 million Turkish liras per year for the regional economy, and the cost of carving out a 5000-ton capacity warehouse is around 1 million liras. In contrast, building a similar warehouse above ground would cost around 2 million liras (Cingi, 2010). This storage system is equally adequate for fruits and vegetables and is an example of a sustainable system that rivals conventional modern ones from the point of view of both construction and exploitation.

29.8 Production and market issues

Lemon variety is one of the factors that influences loss during storage. Varieties that have better storage and transport properties should therefore be preferred.

Interdonato lemons are commonly grown in Turkey and this early variety represents the most of the citrus fruit exported from the country. The other important lemon variety is Lamas, which is grown only in a limited area of the Mediterranean region. Lamas, which is also an early variety, is considered high quality because it is medium sized with few seeds (three to five) and much juice (34%), has outstanding flavour and is convenient for storage and transport. As a result of its better storage and transport properties, the interest in this variety has begun to rise. For example, the Turkish company Aksun (fruit exporter) have begun to choose it in preference to Interdonato (FreshPlaza, 2007a). Lamas lemons are exported to all countries, but mainly to Eastern European countries such as Russia, Ukraine, Romania, Macedonia and Montenegro and to the Middle East (FreshPlaza, 2007b).

29.9 Further developments

According to an article published in the Turkish newspaper *Radikal* (Cingi, 2010), areas of soft rock in Cappadocia are now being hollowed out to provide companies with underground warehouses.

Based on a decision by the Turkish Council of Monuments, it would be possible to pass from the primitive warehouses with capacities between 50 and 100 tons, which have been carved in the past by people in the region, to modern depots, with warehousing capacities reaching 20 000 tons (Cingi, 2010). The decision of the Turkish Council of Monuments is necessary to make a distinction between protected and unprotected areas. Tuff warehouses are advantageous because they have a high degree of isolation because of the thickness of the walls (2–3 metres), they offer a more stable inner temperature (throughout the year temperature span is between 4 °C and 10 °C) and are versatile (in the off-season, lemons or other citrus fruits could be replaced by other fruits or vegetables).

These big depots have to have reliable air conditioning systems in order to face the challenges of storing fruits and vegetables. According to Cingi (2010), nearly 50 companies have been looking for warehouse space in the region.

29.10 Research and training needs

TÜBİTAK, the Scientific and Technological Research Council of Turkey, is coordinating and supporting extensive research on different means of prolonging storage and reducing storage disorders of fruits, including lemons. Major emphasis is given to pre-cooling technologies in the South-eastern Anatolia region (GAP[1] area) under very hot climatic conditions, but cold storage and controlled atmosphere are also being studied, mainly in the cases of citrus and deciduous fruits (apples and pears). Another research direction focuses on protecting the crop in the field and postharvest using integrated pest and disease management, with emphasis on the development of biological control and the use of safe and environmentally friendly pesticides.

Besides research, training for people working in the fruit and vegetable industry is a real necessity in Turkey. A report of academics from Cukurova University (Adana, Turkey) stated that training is informal in Turkey and depends on demonstrations carried out by state extension agencies, whereas Turkish exporters should be responsible for providing producers with information and training (Yesdloglu *et al.*, 2007). The same report revealed that universities, provincial directorates of Ministry of Agriculture and Rural Affairs, Exporters Unions and Chamber of Trade, occasionally arrange seminars and meetings and several research institutions publish journals and pamphlets (ALATARIM of Alata Research Institute, DERĐM of BATEM, and MEYED of Fruit Juice Association) in order to keep the representatives of the fresh fruit chain updated.

Since 2007, when this report was made, and 2008, when the author visited the Lemon House, the situation has improved. However, there is still room for further improvement. Better results could be obtained by the entire Turkish fruit and vegetable industry if training were performed on a regular basis by the representatives of each link in the fresh fruit chain.

[1] The Southeastern Anatolia Project (Turkish: *Güneydoğu Anadolu Projesi, GAP*) is a multi-sector integrated regional development project based on the concept of sustainable development for the 9 million people (2005) living in the Southeastern Anatolia region of Turkey.

29.11 Questions for discussion

1. What does sustainability mean in this context?
2. Which preharvest factors and pre-storage treatments could improve the storage and post-storage quality of citrus fruits?
3. Why do lemons and other citrus fruits spoil, and how can spoilage be prevented?
4. What are the best storage conditions for citrus fruits?
5. How can we differentiate the types of moulds? How can we distinguish the different species of *Penicillium*?
6. What are the human health consequences of the consumption of fungi-infected citrus, and how can we fight them?
7. How can we control the microbiological quality of air?
8. What are modified atmosphere and controlled atmosphere storage (MAS and CAS)?
9. What is ethylene and how can it influence the quality of fruit?
10. What is a marketing standard?
11. What does the 'consolidated version' of a regulation refer to?

29.12 Sources of further information and advice

LADANIYA, M. S. 2007. *Citrus Fruit: Biology, Technology and Evaluation*, 1st Edition, Academic Press (ISBN: 9780123741301).

MCKAY, A. H., FÖRSTER, H., ADASKAVEG, J. E. 2012. Efficacy and application strategies for propiconazole as a new postharvest fungicide for managing sour rot and green mold of citrus fruit. *Plant Dis.* 96: 235–42.

PAULL, R. E. 1999. Effect of temperature and relative humidity on fresh commodity quality, *Postharvest Biology and Technology* 15: 263–77.

THOMPSON, A. K. 2010. *Controlled Atmosphere Storage of Fruits and Vegetables*, CABI, Second edition, ISBN-10: 1845936469, ISBN-13: 978–1845936464.

OECD. 2010. *International Standards for Fruit and Vegetables: Citrus Fruits*, OECD, ISBN: 9789264083745 (pdf), 9789264083738 (print).

29.13 References

AYDAN, O., ULUSAY, R. 2003. Geotechnical and geo-environmental characteristics of man-made underground structures in Cappadocia, Turkey, *Engineering Geology*, 69 (3–4): 245–72.

CINGI, S. 2010. Investments in Cappadocia moving into 'underground' economy, *Radikal* (Istanbul), 31 October 2010. http://www.hurriyetdailynews.com/default.aspx?pageid=438&n=cappadocia-earns-money-from-the-underground-2010–10-31.

EMGE, A. 2011. At home in 'Non-being'. Understanding Cappadocia's cave dwellings, Fairy Chimneys and Underground Cities in Central Anatolia, *Cappadocia Academy Working Paper Series*, 1: 1–18.

ENOTES. n.d. *Storage of Food*. http://www.enotes.com/storage-food-reference/%20storage-food (accessed August 2012).

FRESHPLAZA. 2007a. *Turkey: lemon exporter Aksun switching to Lamas*. http://www.freshplaza.com/news_detail.asp?id=11126#SlideFrame_1 (accessed August 2012).

FRESHPLAZA. 2007b. *Production of Turkish Interdonato lemons -70%.* http://www.freshplaza.com/news_detail.asp?id=11877#SlideFrame_1 (accessed August 2012).

FRUIT LOGISTICA. 2011. Turkey: Partner country plans best-ever trade fair presentation. http://www.fruitlogistica.de/en/PressService/PressReleases/index.jsp?lang=en&id=28672 (accessed August 2012).

FRUITROP. 2008. June issue, no 157. http://passionfruit.cirad.fr/index.php/download/(id)/3798/(langue)/eng/(type)/complet (accessed August 2012).

GROSS, K. C., SMILANICK, J. L. 2004. Lemon, In: Gross, K.C., Wang, C.Y., Saltveit, M. 2004. *The Commercial Storage of Fruits, Vegetables, and Florist and Nursery Stocks*. Draft revised. USDA, ARS. Agricultural Handbook, Number 66 (http://www.ba.ars.usda.gov/hb66/082lemon.pdf).

JANICK, J. 2005. The origin of fruits, fruit growing and fruits breeding. *Plant breeding Rev.* 25:12.

UNIVERSITY OF CALIFORNIA, DAVIS. n.d. *Lemon: Recommendations for Maintaining Postharvest Quality*. http://postharvest.ucdavis.edu/PFfruits/Lemon/ (accessed August 2012).

YESDLOGLU, T., EMEKSDZ, F., TUZCU, O., ALEMDAR, T. 2007. National citrus sector analysis: Turkey, Report for EuroMedCitrusnet Project, Deliverable 9. The European Union 6th Framework Programme, Food Quality and Safety. 33 pages.

ZENGINOGLU, A., VAN DIJK, G. 2006. The structure of Turkey's citrus fruit export from the standpoint of Turkey's membership in the EU, European Association of Agricultural Economists, 98th Seminar, 29 June–2 July 2006, Chania, Crete, Greece http://ageconsearch.umn.edu/bitstream/10107/1/sp06ze03.pdf.

Index

Lightning Source UK Ltd.
Milton Keynes UK
UKOW04n1856250214

227159UK00006B/132/P

9 781782 420187